ELEVENTH EDITION

Laboratory Manual in
PHYSICAL
GEOLOGY

ELEVENTH EDITION

Laboratory Manual in
PHYSICAL
GEOLOGY

PRODUCED UNDER THE AUSPICES OF THE

American Geosciences Institute
www.americangeosciences.org

AND THE

National Association
of Geoscience Teachers
www.nagt.org

Vincent S. Cronin, Editor
Baylor University

ILLUSTRATED BY

Dennis Tasa
Tasa Graphic Arts, Inc.

 **Pearson**

330 Hudson Street, NY NY 10013

Executive Editor, Geosciences Courseware: *Christian Botting*
Courseware Director, Content Development: *Ginnie Simione Jutson*
Courseware Specialist, Content Development: *Jonathan Cheney*
Courseware Director, Portfolio Management: *Beth Wilbur*
Portfolio Management Assistant: *Emily Bornhop*
Managing Producer, Science: *Mike Early*
Content Producer: *Becca Groves*
Rich Media Content Producer: *Ziki Dekel*
Production Management & Composition: *Christian Arsenault, SPi Global*
Copyeditor: *JaNoel Lowe*
Illustrations: *Dennis Tasa*
Interior and Cover Designer: *Cenveo Publishing Services*
Rights & Permissions Project Manager Management: *Matt Perry*
Photo Researcher: *Danny Meldung, Photo Affairs, Inc.*
Senior Procurement Specialist: *Stacy Sweinberger*
Manufacturing Buyer: *Maura Zaldivar-Garcia*
Executive Marketing Manager: *Neena Bali*
Senior Marketing Manager, Field: *Mary Salzman*
Marketing Assistant: *Ami Sampat*

Cover Photo Credit: *Buena Vista Images/Getty Images*

Library of Congress Cataloging-in-Publication Data
Laboratory Manual in Physical Geology / produced under the auspices of the American Geosciences Institute, and the National Association of Geoscience Teachers; Vincent S. Cronin, editor, Baylor University; illustrated by Dennis Tasa, Tasa Graphic Arts, Inc. – Eleventh edition.
     pages cm.
1. Physical geology–Laboratory manuals.   I. Cronin, Vincent S.   II. Tasa, Dennis   III. American Geological Institute   IV. National Association of Geology Teachers.
     QE44.L33 2018
     551.078

                                         2016051967

2   17

ISBN 10: 0-13-444660-7
ISBN 13: 978-0-13-444660-8
(Student edition)

 **Pearson**   www.pearsonhighered.com

Contributing Authors

Thomas H. Anderson
University of Pittsburgh

Harold E. Andrews
Wellesley College

James R. Besancon
Wellesley College

Jane L. Boger
SUNY–College at Geneseo

Phillip D. Boger
SUNY–College at Geneseo

Claude Bolze
Tulsa Community College

Richard Busch
West Chester University of
Pennsylvania

Jonathan Bushee
Northern Kentucky University

Roseann J. Carlson
Tidewater Community College

Cynthia Fisher
West Chester University of
Pennsylvania

Charles I. Frye
Northwest Missouri State
University

Pamela J.W. Gore
Georgia Perimeter College

Anne M. Hall
Emory University

Edward A. Hay
De Anza College

Charles G. Higgins
University of California, Davis

Michael F. Hochella, Jr.
Virginia Polytechnic Institute
and State University

Michael J. Hozik
Richard Stockton College
of New Jersey

Sharon Laska
Acadia University

David Lumsden
University of Memphis

Richard W. Macomber
Long Island University, Brooklyn

Garry D. Mckenzie
Ohio State University

Cherukupalli E. Nehru
Brooklyn College (CUNY)

John K. Osmond
Florida State University

Charles G. Oviatt
Kansas State University

William R. Parrott, Jr.
Richard Stockton College
of New Jersey

Raman J. Singh
Northern Kentucky University

Kenton E. Strickland
Wright State University

Richard N. Strom
University of South Florida,
Tampa

James Swinehart
University of Nebraska

Raymond W. Talkington
Richard Stockton College
of New Jersey

Margaret D. Thompson
Wellesley College

James Titus*
U.S. Environmental Protection
Agency

Nancy A. Van Wagoner
Acadia University

John R. Wagner
Clemson University

Donald W. Watson
Slippery Rock University

James R. Wilson
Weber State University

Monte D. Wilson
Boise State University

C. Gil Wiswall
West Chester University
of Pennsylvania

*The opinions contributed by this
person do not officially represent
opinions of the U.S. Environmental
Protection Agency.

Contents

About Our Sustainability Initiatives

Pearson recognizes the environmental challenges facing this planet, and acknowledges our responsibility in making a difference. This book has been carefully crafted to minimize environmental impact. The binding, cover, and paper come from facilities that minimize waste, energy consumption, and the use of harmful chemicals. Pearson closes the loop by recycling every out-of-date text returned to our warehouse.

Along with developing and exploring digital solutions to our market's needs, Pearson has a strong commitment to achieving carbon neutrality. As of 2009, Pearson became the first carbon- and climate-neutral publishing company. Since then, Pearson remains strongly committed to measuring, reducing, and offsetting our carbon footprint.

The future holds great promise for reducing our impact on Earth's environment, and Pearson is proud to be leading the way. We strive to publish the best books with the most up-to-date and accurate content, and to do so in ways that minimize our impact on Earth. To learn more about our initiatives, please visit **https://www.pearson.com/sustainability.html**

Image & Text Credits*

CHAPTER 1
p.2: Fig 1.1 NASA Johnson Space Center; p.6: Fig 1.3 Jim Peaco/National Park Service; p.8: Fig 1.6 U. S. Geological Survey; p.12: Fig 1.9b Ralph A. Clavenger/Getty Images; p.17: Fig 1.11c NOAA; p.33: Fig A1.6.3 NOAA.

CHAPTER 2
p.37: NASA Johnson Space Center/Image Science & Analysis Laboratory; p.39: Fig 2.1 Jet Propulsion Laboratory/NASA; p.48: Fig 2.12 UNAVCO; p.55: Fig A2.1.1 Marine Geoscience Data System; p.65: Fig A2.6.1 GPlates (www.gplates.org) Seton, M., Muller, R.D., Zahirovic, S., Gaina, C., Torsvik, T., Shephard, G., Talsma, A., Gurnis, M., Turner, M, Maus, S, and Chandler, M., 2012, Global Continental and Ocean Basin Reconstructions Since 200 MA (Earth-Science Reviews, v. 113, p. 212-270).

CHAPTER 3
p.94: Fig 3.23 USGS Mineral Commodity Summaries/U.S. Department of the Interior.

CHAPTER 4
p.107: SuperStock/Alamy Stock Photo.

CHAPTER 5
p.123: Ammit/Alamy Stock Photo; p.124: Fig 5.1a Joel Sorrell/iStock/Getty Images; p.124: Fig 5.1b M.L. Coombs/Alaska Volcano Observatory/U.S. Geological Survey; p.129: Fig 5.9 Based on Le Bas and Streckeisen, 1991, The IUGS Systematics of Igneous Rocks (Journal of the Geological Society, Vol. 148, pp. 825-833); p.135: Fig 5.16 Based on Klein and Hurlbut, 1993 (Manual of Mineralogy, p. 561).

CHAPTER 6
p.149: Pulsar Images/Alamy Stock Photo; p.167: Fig 6.26b NASA; p.184: Fig A6.9.1 U.S. Geological Survey.

CHAPTER 7
p.185: Orsolya Haarberg/National Geographic/Getty Images; p.193: Fig 7.11 Dennis Tasa; p.198: Fig A7.2.1.5 Thomas J. Mortimer; p.205: Fig A7.5.4 Based on the IUGS; p.206: Fig A7.5.5 Dennis Tasa.

CHAPTER 9
p.231: OpenTopography.org; p.257: Fig A9.2.1 US Topo and Historical Topographic Map Collection/U.S. Department

of the Interior; p.259: Fig A9.4.1 US Topo and Historical Topographic Map Collection/U.S. Department of the Interior; p.262: Fig A9.5.1 US Topo and Historical Topographic Map Collection/U.S. Department of the Interior.

CHAPTER 10
p.265: Robert Simmon & Jesse Allen/NASA Earth Observatory; p.274: Fig 10.9 Jesse Allen/U.S. Geological Survey/NASA; p.284: Fig A10.2.1.2 N.J. Silberling/USGS; p.284: Fig A10.2.1.3 W.B. Hamilton/USGS.

CHAPTER 11
p.293: EcoPhotography.com/Alamy; p.295: Fig 11.1a Jacques Descloitres/MODIS Rapid Response Team/NASA/GSFC; p.296: Fig 11.2 U.S. Geological Survey; p.300: Fig 11.9 NASA; p.306: Fig 11.15 NASA; p.310: Fig A11.2.2 US Topo and Historical Topographic Map Collection/U.S. Department of the Interior; p.311: Fig A11.2.3 US Topo and Historical Topographic Map Collection/U.S. Department of the Interior; p.315: Fig A11.4.1 US Topo and Historical Topographic Map Collection/U.S. Department of the Interior.

CHAPTER 12
p.325: Arco Images GmbH/Alamy Stock Photo; p.332: Fig 12.7 U.S. Geological Survey.

CHAPTER 13
p.347: Jesse Allen/U.S. Geological Survey/NASA; p.352: Fig 13.3 NASA; p.364: Fig 13.18a U.S. Geological Survey.

CHAPTER 14
p.375: NASA/JPL/UAriz; p.379: Fig 14.3 NASA; p.381: Fig 14.5 U.S. Geological Survey; p.382: Fig 14.6 Jeff Schmaltz/MODIS Rapid Response Team/NASA/GSFC; p.385: Fig 14.9 NASA Earth Observatory; p.386: Fig 14.1 NASA Earth Observatory; p.389: Fig A14.2.1 Robert Simmon/U.S. Geological Survey/NASA; p.390: Fig A14.2.2a U.S. Geological Survey.

CHAPTER 15
p.397: Fig 15.1a-d NOAA; p.397: Fig 15.1g Albert E. Theberge/ NOAA; p.400: Fig 15.4 NASA.

CHAPTER 16
p.409: David Ramos/Getty Images; p.411: Fig 16.2 U.S. Geological Survey.

Laboratory Manual in Physical Geology is produced under the auspices of the American Geosciences Institute (AGI) and the National Association of Geoscience Teachers (NAGT). For decades it has been the most widely adopted manual available for teaching laboratories in introductory geology and geoscience. This 11th edition is more user-friendly than ever, with an effective pedagogical format and many more teaching and learning options. It is supported by MasteringGeology—the most effective and widely used online homework, tutorial, and assessment platform in the Geosciences, including an eText version of the lab manual—as well as by GeoTools (ruler, protractor, UTM grids, sediment grain size scale, etc.), an *Instructor Resource Manual*, and many other online resources.

The idea for this jointly sponsored laboratory manual originated with Robert W. Ridky (past president of NAGT and member of the AGI Education Advisory Committee), who envisioned a manual made up of the "best laboratory investigations written by geology teachers." To that end, this edition represents the cumulative ideas of more than 225 contributing authors, 31 years of evolution in geoscience and geoscience education, the comments of faculty peer reviewers and geoscience professionals, and important input from students and instructors who have used past editions.

About the 11th Edition

Why did AGI and NAGT develop an 11th edition of the *Laboratory Manual in Physical Geology*? We are in a time of dramatic change within the geosciences, in society, and in the environment of our planetary home—Earth. New technologies, new data, and new hypotheses are flooding into the geosciences, as geoscientists strive to apply new knowledge to provide expertise to a society trying to cope with challenges related to water, energy, mineral resources, natural and human-induced environmental hazards, and global change. An ever-changing *Laboratory Manual in Physical Geology* is essential for AGI and NAGT to fulfill their missions in helping to educate and inform the public and to facilitate the development of the next generation of geoscientists.

Through a nationwide search conducted collaboratively by AGI and NAGT, Professor Vincent S. Cronin was selected as editor of the 11th edition. Dr. Cronin brings a mix of practical instructional knowledge and expert geological background. He has been a geoscientist and university-level geoscience educator for about three decades, and has experience teaching with the *Laboratory Manual in Physical Geology*. His research includes topics in engineering geology, structural geology, and tectonics. He is also Co-Chair of the U.S. Section of the International Association for Promoting Geoethics.

The team that developed the new edition of this classic laboratory manual sought to preserve features that are familiar and valued by the geoscience education community, while making revisions that were requested by users or deemed necessary to better reflect current developments in specific areas of the geosciences. We have also developed new online resources for students, such as a glossary to facilitate the learning process of students using this textbook, as well as enhancing access to images and videos. In this edition, we have embarked on a process that will ultimately make all of our illustrations more accessible to people with color blindness or other vision-related issues. We are committed to a geoscience community that is diverse, inclusive, and welcoming to all. And we recognize that there is an essential ethical dimension to our work in geoscience, which we explore in many ways throughout the text.

Supporting Text

The text that appears before the Activity section of each chapter serves two goals. One is the practical goal to provide essential information to students working on the Activities. The second is to provide students with a coherent body of information that will remain after the Activities are completed, and after the Activity pages are torn from the book.

Most of the supporting text in the 11th edition is new or has been revised. These changes have been made on the basis of reviews by faculty and students, and based on current trends in the geosciences. Great care has been taken to compose supporting text that is scientifically correct, uses the appropriate geoscience terms correctly, is comprehensible by undergraduate college students, and is well supported with illustrations. Whereas the target audience for this Laboratory Manual is a diverse population of students who will ultimately pursue knowledge and careers in other fields, we also want this edition to provide a sound foundation for those who pursue additional study in geoscience.

Vocabulary and Geoscience Terminology

We have continued the tradition of using vocabulary appropriate to undergraduate students in the 11th edition, and have sought to keep geoscience jargon to a necessary minimum. Rock and mineral terms are used in a way that is consistent with the published standards of the International Mineralogical Association and the International Union of Geological Sciences, as well as with the latest edition of the American Geoscience Institute (AGI) *Glossary of Geology*. The complete AGI *Glossary of Geology* is available in print, as an E-book for Kindle and Nook, as an app for mobile devices (available at the Apple Store and at Google Play), and online for universities and companies (but not for individuals; www.americangeosciences.org/pubs/glossary). An abbreviated glossary developed for the *Laboratory Manual in Physical Geology* is also available online.

Art

Dennis Tasa's brilliant artwork reinforces the visual aspect of geology and enhances student learning. Many figures have been revised for the 11th edition, and all of the figures in both the text and in the activities are numbered to help students navigate to the resources they need efficiently. For example, labels and overlays have been added to photographs and other images to facilitate students' understanding.

- **Photographs.** There are almost two hundred new photographs or satellite images in the 11th edition. Some of the new photographs of rock and mineral specimens are the result of very high resolution macro photography enhanced by focus-stacking technology.
- **Maps.** There are about three dozen new maps in this edition. New topographic maps are based on the most current U.S. topographic map product published digitally by the USGS. Many of the maps have been simplified to reduce irrelevant elements and improve clarity.
- **Illustrations.** Many of the graphics retained from previous editions are revised, and there are dozens of entirely new illustrations and tables.

Activities

Of the 96 activities spread across 16 chapters, 10 are new and many of the rest are revised to improve content and clarity. Having access to such a large number of activities allows an instructor to select and adapt activities according to course content and level of difficulty. And because many activities do not require sophisticated equipment, they can also be assigned for students to complete as pre-laboratory assignments, lecture supplements, homework, or recitation topics.

Math

Geoscience in the 21st Century relies largely on quantitative information, although many definitions and descriptions of geological materials involve qualitative information, too. Nearly four hundred years ago, Galileo wrote that the book of Nature is written in the language of mathematics. While it is true that essential insights are gained through descriptive or qualitative observations, we cannot progress far in geoscience without quantitative measurements, mathematical expressions that link quantitative observations together, and statistical analysis. We assume that students have an average understanding of basic high-school mathematics, and develop useful math skills as they are needed to understand the material or to complete an activity.

Outstanding Features

This edition contains the tried-and-tested strengths of ten past editions of this lab manual that have been used by faculty and teachers over more than three decades, with updates to maintain its position as a reflection of current geoscience thinking. The outstanding features listed below remain a core part of this manual.

Pedagogy for Diverse Styles/Preferences of Learning

Hands-on multisensory-oriented activities with samples, cardboard models, and GeoTools appeal to *concrete/kinesthetic learners*. High quality images, maps, charts, diagrams, PowerPoints™, cardboard models, and visualizations appeal to *visual/spatial learners*. Activity sheets, charts, lists, supporting text, and opportunities for discourse appeal to *linguistic/verbal/read-write learners*. Presentation graphics (PowerPoint) and video clips appeal to *auditory/aural learners*. Numerical data, mathematics, models, graphs, systems, and opportunities for discourse appeal to *logical/abstract learners*.

Pre-Lab Videos

Pre-Lab videos are found on the chapter-opening spreads of each lab, and are accessed via a Quick Response (QR) code or direct web-link. These videos allow students to come to lab better prepared and ready to immediately benefit from their engagement with lab exercise. No longer do instructors have to spend the first portion of hands-on lab time lecturing. The videos can be viewed during the students' own preparatory time, and review key concepts relevant to the lab activities. The videos, created by Callan Bentley (Northern Virginia Community College), are personable and friendly, and assure students that they will be able to successfully complete the lab activities by following a clear series of steps. Students can download free QR reader apps from the Apple App Store or Google Play.

Format and Pedagogical Framework

- **Big Ideas and Engaging Chapter Openers.** Every laboratory opens with an engaging image and a statement of *Big Ideas*, which establish the overall conceptual themes upon which the laboratory is based. *Big Ideas* are concise statements that help students understand and focus on the lab topic.
- **Think About It—Key Questions.** Every activity is based on a key question that is linked to the *Big Ideas*. *Think About It* questions function as the conceptual "lenses" that frame student inquiry and promote critical thinking and discourse.
- **Guided and Structured Inquiry Activities.** Many of the laboratories begin with a guided inquiry activity. These inquiries are designed to be engaging and to help students activate cognitive schemata that relate to the upcoming investigations. These can be used for individualized or cooperative learning. The guided inquiry activity is followed by activities that are more structured, so students can develop their understanding of specific geoscience concepts and principles.
- **Reflect & Discuss Questions.** Every activity concludes with a *Reflect & Discuss* question designed to foster greater accommodation of knowledge by having students apply what they learned to a new situation or to state broader conceptual understanding.

- **Continuous Assessment Options.** The pedagogical framework and organization provides many options for continuous assessment such as *Think About It* questions and guided inquiry activities that provide options for pre-assessment, activity worksheets, and the *Reflect & Discuss* questions. When students tear out and submit an activity for grading, their manual will still contain the significant text and reference figures that they need for future study. Grading of students' work is easier because all students submit their own work in a similar format. Instructors save time, resources, and money because they no longer need to photocopy and distribute worksheets to supplement the manual.

Enhanced Learning Options

- **Transferable Skill Development and Real-World Connections.** Many activities have been designed or revised for students to develop transferrable skills and make connections that are relevant to their lives and the world in which they live. For example, they learn how to obtain and use data and maps that will enable them to make wiser choices about where they live and work. They evaluate their use of Earth resources in relation to questions about resource management and sustainability. They learn to use resources provided by the U.S. Geological Survey, JPL-NASA, NOAA, Google Earth™, and other online sources of reliable data and analysis about Earth's resources, hazards, changes, and management.

- **The Math You Need (TMYN) Options.** Throughout the laboratories, students are referred to online options for them to review or learn mathematical skills using *The Math You Need, When You Need It* (TMYN). TMYN consists of modular math tutorials that have been designed for students in any introductory geoscience course by Jennifer Wenner (University of Washington–Oshkosh) and Eric Baer (Highline Community College).

- **Mobile-Enabled Media and Web Resources.** Quick Response (QR) codes give students with smartphones or other mobile devices instant access to supporting online media content and websites.

- **Enhanced Instructor Support.** Instructor materials are available online in the Instructor Resource Center (IRC) at www.pearsonhighered.com/irc. Resources include the enhanced *Instructor Resource Manual* (answer key and teaching tips), files of all figures in the manual, PowerPoint presentations for each laboratory manual in JPEG and PowerPoint formats, the Pearson Geoscience Animation Library (over 120 animations illuminating the most difficult-to-visualize geological concepts and phenomena), and MasteringGeology options.

MasteringGeology

The MasteringGeology platform delivers engaging, dynamic learning opportunities—focused on course objectives and responsive to each student's progress—that are proven to help make course material accessible and to help them develop their understanding of difficult concepts. Robust diagnostics and unrivalled gradebook reporting allow instructors to pinpoint the weaknesses and misconceptions of a student or class to provide timely intervention.

- **Pre-lab video quizzes** help students come to lab better prepared and ready to immediately get started with the lab exercise.
- **Post-lab quizzes** assess students' understanding and analysis of the lab content.

Learn more at www.masteringgeology.com.

Learning Catalytics

Learning Catalytics™ is a "bring your own device" student engagement, assessment, and classroom intelligence system. With Learning Catalytics you can:

- assess students in real time, using open-ended tasks to probe student understanding.
- understand immediately areas in which adjustments to instruction will be helpful to students.
- improve your students' critical-thinking skills.
- access rich analytics to understand student performance.
- add your own questions to make Learning Catalytics fit your course exactly.
- manage student interactions with intelligent grouping and timing.

Learning Catalytics is a technology that has grown out of twenty years of cutting edge research, innovation, and implementation of interactive teaching and peer instruction. Available integrated with MasteringGeology. To learn more, go to www.learningcatalytics.com.

Materials

Laboratories are based on samples and equipment normally housed in existing geoscience teaching laboratories (page xxi).

GeoTools, GPS, and UTM

Rulers, protractors, a sediment grain size scale, UTM grids, and other laboratory tools are available to cut from transparent sheets at the back of the manual. No other manual provides such abundant supporting tools! Students are introduced to GPS and UTM and their application in mapping. UTM grids are provided for most scales of U.S. and Canadian maps.

Support for Geoscience!

Royalties from sales of this product support programs of the American Geosciences Institute and the National Association of Geoscience Teachers.

New & Updated in the 11th Edition

1.1: A View of Earth from Above (new)
New—coordinate systems and Google Earth used to locate places and geologic features.
A1.1.1, A1.1.2, A1.1.3, A1.1.4: new tables for data and student answers

1.2: Latitude and Longitude or UTM Coordinates of a Point (new)
New—map scales, measure on map, and calculate representative fraction.
A1.2.1: new diagram; **A1.2.2:** new map with UTM coordinates

1.3: Plotting a Point on a Map Using UTM Coordinates (new)
New—practice in plotting points precisely using UTM coordinates and Google Earth.
A1.3.1: map with UTM coordinates

1.4: Floating Blocks and Icebergs (new)
New—measure volume and mass, calculate density, apply Archimedes principle.
A1.4.1: new art relating volume, mass, and density; **A1.4.2:** new art showing buoyancy

1.5: Summarizing Data and Imagining Crustbergs Floating on the Mantle (10e: 1.6)
Revised to clarify concept of isostasy; calculate standard deviations of rock-sample densities
A1.5.1, A1.5.2: revised data tables for densities mode and standard deviation

1.6: Unit Conversions, Notation, Rates, and Interpretations (10e: 1.4)
Revised to focus on rates of change (erosion, geothermal gradient, and changing atmospheric CO_2 levels).
A1.6.1, A1.6.2, A1.6.3: revised data and graphs with updated atmospheric CO_2 data

1.7: Scaling, Density, and Earth's Deep Interior (new)
New—draw Earth's interior structure to scale based on graph of density changes beneath surface.
A1.7.1: new table; **A1.7.2:** new diagram and graph of Earth's interior structure

2.1: Reference Frames and Motion Vectors (new)
New—uses concepts of reference frames and vectors to explore plate motions.
A2.1.1: new map of Juan de Fuca plate; **A2.1.2:** new graphic (template) for drawing vectors

2.2: Measuring Plate Motion Using GPS Time Series (10e: 2.1)
Revised to walk students step-by-step through activity; **Part D:** New *Reflect & Discuss* question; use of UNAVCO online calculator.
A2.2.2: revised map

2.3: Hotspots and Plate Motions (10e: 2.8)
Part A: Question #4 added; Question #5 revised; **Part B:** added questions and steps; use of Pythagorean theorem to determine speed of plate motion.

2.4: How Earth Materials Deform (10e: 2.3)
Added steps and questions help students grasp how activities model Earth processes.

2.5: Paleomagnetic Stripes and Seafloor Spreading (10e: 2.4)
A2.5.1: revised map of Juan de Fuca plate

2.6: Atlantic Seafloor Spreading (10e: 2.5)
Heavily revised: Concept of isochrons on map of Atlantic seafloor added; new questions lead students through an in-depth exploration.
A2.6.1: new basemap of seafloor spreading

2.7: Using Earthquakes to Identify Plate Boundaries (10e: 2.6)
Questions and procedures edited to clarify and add background info.

3.1: Mineral and Rock Inquiry (10e: 3.1)
Questions edited for clarity; **Part C:** New *Reflect & Discuss* question based on students' observations.

3.2: Mineral Properties (10e: 3.2)
Part I: New *Reflect & Discuss* questions on distinguishing crystal systems, forms, and habits.

3.5: Mineral Dependency Crisis (10e: 3.5)
Part C: New *Reflect & Discuss* question on foreign vs. domestic sources of essential minerals.

4.1: Rock Inquiry (10e: 4.1)
Part C: New *Reflect & Discuss* question on how rocks form.
A4.1.3: new photo of gneiss to go with new question

4.2: What Are Rocks Made Of? (10e: 4.2)
A4.2.1: new improved photo #5 of gabbro

4.3: Rock-Forming Minerals (10e: 4.3)
A4.3.1: 8 new photos of rock-forming minerals

4.4: What Is Rock Texture? (10e: 4.4)
A4.4.1: new photos of labradorite gabbro

4.5: Rock and the Rock Cycle (10e: 4.5)
A4.5.2: table revised (because obsidian is not considered a rock)

5.5: Estimate Percentage of Mafic Minerals (new)
New—describe igneous rock samples and estimate mafic-mineral content.
A5.5.1: new scale of percent mafic minerals; **A5.5.2:** new photos of igneous rock samples for analysis

5.6: Estimate Composition of a Phaneritic Rock by Point Counting (new)
New—point-counting to estimate mineral composition; calculate averages and standard deviations for point-count data.
A5.6.1: new photo of granite sample and art of its mineral composition; **A5.6.2:** new grids used in point counting exercise

6.2: Sediment from Source to Sink (new)
New example walks students step-by-step through analysis of how sediment changes.
A6.2.1: new sequence of photos by author showing changes in sediment along river

6.3: Clastic Sediment (10e: 6.3)
Questions edited to clarify and add background information.

6.4: Bioclastic Sediment and Coal (10e: 6.4)
Revised to use preferred terms *bioclastic* and *siliciclastic*. Questions edited to clarify; **Part C:** New *Reflect & Discuss* question on classifying types of coal.

6.7: Grand Canyon Outcrop Analysis and Interpretation (10e: 6.7)
Questions edited to clarify.
A6.7.1: labels and lines added to photo of sedimentary strata

7.2: Minerals in Metamorphic Rock (new)
New exercise in mineral identification.
A7.2.1: set of 6 minerals that students identify based on descriptions in the text

7.3: Metamorphic Rock Analysis and Interpretation (10e: 7.2)
Questions edited to clarify and add background information.

7.5: Metamorphic Grades and Facies (10e: 7.4)
Questions edited to clarify and add background information; **Part C:** New *Reflect & Discuss* question on interpreting A7.5.4.
A7.5.4: new PT diagram of Barrow's metamorphic zones; **A7.5.5:** revised block diagram with improved legibility and detail

8.1: Geologic Inquiry for Relative Dating (10e: 8.1)
A8.1.2, A8.1.3: adds grayscale duplicates to photos so that students can trace the contacts

8.3: Use of Index Fossils to Date Rocks and Events (10e: 8.3)
Terms added to enable more precise relative dating of index fossils.

8.4: Numerical Dating of Rocks and Fossils (10e: 8.4)
Part C Revised to clarify and add background information.

8.5: Infer Geologic History from a New Mexico Outcrop (10e: 8.5)
A8.5.1: adds labels to photo to clarify the geology of the strata

8.6: Investigating a Natural Cross Section in the Grand Canyon (10e: 8.6)
Part A Revised to clarify *Reflect & Discuss*.
A8.6.1, A8.6.2: adds grayscale duplicates to photos so that students can trace the contacts

9.2: Map Locations, Distances, Directions, and Symbols (10e: 9.2)
Parts D, F, G: Revised to clarify and add background information; new questions about magnetic declination, grid north and true north.
A9.2.1: adjust map scale for accuracy, add UTM coordinates and improve legibility

9.3: Topographic Map Construction (10e: 9.3)
Background information added to questions about contour lines and hachures.

9.4: Topographic Map and Orthoimage Interpretation (10e: 9.4)
A9.4.1: improve legibility of streams, contour lines, and labels; **A9.4.2:** topo map with improved legibility

9.5: Relief and Gradient (Slope) Analysis (new—although based on 10e, 9.5)
Uses real-world example; new questions on using map scale and measurements to calculate gradient and interpret slopes.
A9.5.1: uses part of a real USGS topo map of Yosemite N.P.

9.6: Topographic Profile Construction (10e: 9.6)
A9.6.1: replaces graph paper with more student friendly template for drawing topographic profile

10.1: Map Contacts and Formations (10e: 10.3)
Revised activity now focuses on Grand Canyon example, and provides more guidance for students in tracing the contacts.
A10.1.2: replaces the Grand Canyon topo map with adapted version of more legible map by the geologist Billingsley

10.2: Geologic Structures Inquiry (10e: 10.1)
Part C: New *Reflect & Discuss* question involves determining strike and dip of inclined strata.
A10.2.1: new, clearer photo of Grand Canyon (upper left); **A10.2.2:** new photo of inclined strata; includes a grayscale duplicate so that students can draw dip vector and label photo

10.3: Fault Analysis Using Orthoimages (10e: 10.7)
Part C: New *Reflect & Discuss* question about vertical as well as horizontal components of fault slip.
A10.3.1, A1.3.2: scale and north arrow added to orthoimages

10.4: Appalachian Mountains Geologic Map (10e: 10.8)
A10.4.1: geologic cross section revised to clarify.

10.5: Cardboard Model Analysis and Interpretation (10e: 10.5)
Lightly edited to clarify questions and use correct terminology.

10.6: Block Diagram Analysis and Interpretation (10e: 10.6)
A10.6.1: Edits to captions on the block diagrams to clarify the questions.

11.1: Streamer Inquiry (10e: 11.1)
Lightly edited to make instructions more precise; **Part C:** New questions on a stream's discharge, channel width and volume.

11.2: Drainage Basins, Patterns, Gradients, and Sinuosity (10e: 11.2)
Revised to clarify and shorten; 10e **Parts D** and **E** used as the basis for Activity 11.3.
A11.2.1: map contour lines and labels revised to improve legibility; **A11.2.2:** improved legibility of basemap; **A11.2.3:** improved legibility of topo map

11.3: Mountain Stream (10e: 11.2)
Made up of **Parts D** and **E** of 11.2; focuses on analysis of specific streams in Ennis, MT quadrangle.
A11.3.1: improved labeling of topo map and profile box

11.4: Escarpments and Stream Terraces (10e: 11.3)
"Starter" points added to profile box to aid students in constructing profile; location coordinates added.
A11.4.1: improved legibility of basemap of Voltaire, ND

11.6: Retreat of Niagara Falls (10e: 11.5)
Background information on Niagara gorge added;
Part A: Measure Niagara gorge on map; location coordinates added.

A11.6.1: cross section of Niagara Falls moved to point-of-use in Activity 11.6

11.7: Flood Hazard Mapping, Assessment (10e: 11.6)
Concepts of gage datum and gage height added; questions and procedures revised to clarify.
A11.7.1: improved legibility of labels; north arrow and scale added; **A11.7.2:** flood data table revised; **A11.7.4:** labels, north arrow, and scale added

12.1: Groundwater Inquiry (10e: 12.1)
Part A: New Experiment 2 compares rates of groundwater flow; **Part D:** New *Reflect & Discuss* question on Experiment 3 (former 2)
A12.1.2: art showing setup of tubes in Experiment 2; **A12.1.3:** revised data in table

12.2: Where Is the Nasty Stuff Going? (new)
New—construct a contour map of water table and trace pollutants' flow lines through the aquifer.
A12.2.1: art showing water table elevations used in constructing map of groundwater flow

12.3: Using Data to Map the Flow of Groundwater (new)
New—use piezometer data to determine total head values, map them, and infer flow lines.
A12.3.1: cross section showing total head values at different points within an aquifer

12.6: Land Subsidence from Groundwater Withdrawal (10e: 12.4)
Part A: New questions 8b–f on changes in water level based on hydrograph data (Fig. 12.6.X); **Part B:** New *Reflect & Discuss* question on effects of land use change on subsidence.
A12.6.2: graphs of land subsidence in Santa Clara Valley in relation to water table and groundwater use.

13.1: The Cryosphere and Sea Ice (10e: 13.1, 13.6)
Part A: Revised to clarify questions and shorten; **Parts B** and **C:** Based on 10e, Activity 13.6, with new *Reflect & Discuss* in which students compare climate change in Arctic and Antarctic.
A13.1.1: data table and graphs (where student's plot the data) on Arctic sea ice extent

13.2: Mountain Glaciers and Glacial Landforms (10e: 13.2)
Revised to focus on glaciation in Yosemite, with added background information.
A13.2.1, A13.2.2: new profile boxes for constructing profiles; **A13.2.3:** new topo map with transect line and profile box

13.3: Nisqually Glacier Response to Climate Change (10e: 13.5)
Parts A and **B:** Revised to clarify questions, with step-by-step directions on plotting data.
A13.3.1: revised data table, bar graph, and blank graph of changes in glacier

13.4: Glacier National Park Investigation (10e: 13.4)
Part D: Revised to explain Continental Divide; new question 3 on correlating glaciers with topography and slope orientation.

13.5: Some Effects of Continental Glaciation (10e: 13.3)
Revised to shorten: Focuses on Whitewater, WI topo map; new question 7 on identifying glacial features on map.

14.1: Dry Land Inquiry (10e: 14.1)
Parts B and **D:** Edited for clarity; **Part C:** Mostly new—use GoogleEarth to analyze a fault-bounded valley.

14.2: Sand Seas of Nebraska and the Arabian Peninsula (10e: 14.3)
Parts A and **B:** Edited to clarify questions and add background info; location coordinates added.
A14.2.2: new aerial photo and topo map of White Lake, NB quadrangle

14.3: Dry-Land Lakes of Utah (10e: 14.4)
Part F: Edited to clarify and remove references to 10e's use of a stereogram image.
A14.3.1: locator map and scale information added to map of Wah Wah Valley, UT

14.4: Death Valley, California (10e: 14.2)
Part C: New—identify and trace normal faults on map.: **Part D:** New—consider plate motions in relation to pull-apart basins.
A14.4.1: labels edited and added to map of Death Valley

15.1: Coastline Inquiry (10e: 15.1)
Part A: Question 1 revised to focus on geology of the coastal areas; **Part B:** *Reflect & Discuss* questions made more specific.

15.2: Introduction to Coastlines (10e: 15.2)
Edited to simplify terminology.

15.3: Coastline Modification at Ocean City, Maryland (10e: 15.3)
Edits to clarify historical changes; **Part F:** *Reflect & Discuss* on the future of Ocean City given its elevation and rate of local sea-level rise.

15.4: The Threat of Rising Seas (10e: 15.4)
Added background information on Hurricane Sandy's storm surge on Staten Island, NY.

16.1: Earthquake Hazards Inquiry (10e: 16.1)
Part A: Revised to clarify directions and rephrase *Reflect & Discuss*; **Parts B, C,** and **D:** Revised to clarify questions and add background information.

16.2: How Seismic Waves Travel Through Earth (10e: 16.2)
Edited to clarify terms and concepts relating to seismic waves.

16.3: Locate the Epicenter of an Earthquake (10e: 16.3)
Part C: Revised question 3 refers students to a USGS website with an interactive map of active faults.

16.4: San Andreas Fault Analysis at Wallace Creek (10e: 16.4)
Part A: Revised to facilitate tracing and interpreting the fault; **Part B:** New background info and steps for calculating displacement rate.
A16.4.1: improved shaded relief map of topography in Wallace Creek area; **A16.4.2:** new table for students answers

16.5: New Madrid Seismic Zone (10e: 16.5)
New background information on faults in the New Madrid Seismic Zone (NMSZ).
A16.5.1: revised map to show the primary fault in the NMSZ and identify the epicenter

Acknowledgments

Development and production of this revised 11th edition of *Laboratory Manual in Physical Geology* required the expertise, dedication, and cooperation of many people and organizations, to which we want to express our sincere appreciation. As editor of several prior editions, Richard Busch provided a strong foundation for this edition through his thoughtful work and knowledge of geology. Revisions in the 11th edition are based on generous suggestions from faculty and students using the manual, market research by Pearson, and more than 100 expert reviews contributed by geoscience professionals from over two dozen of the member organizations of the American Geosciences Institute (AGI), who are named in the following pages. Katherine Ryker (Eastern Michigan University) and Jenn Wenner (University of Wisconsin, Oshkosh) served as an editorial panel on behalf of the National Association of Geoscience Teachers (NAGT), each providing in-depth reviews of labs as they were being revised.

The very talented publishing team at Pearson Education led the effort. Executive Editor Christian Botting's knowledge of market trends, quest to meet the needs of faculty and students, and dedication to excellence guided the 11th edition. Jonathan Cheney's pre-revision memos and developmental editing framed the revision goals for each topic and ensured that all writing was practical and purposeful. Emily Bornhop managed accuracy reviews of revision drafts. Deepti Agarwal and Becca Groves set revision schedules, tracked revision progress, managed the production process, and efficiently coordinated the needs and collaborative efforts of team members. Their expertise and dedication to excellence enabled them to locate, manage, and merge disparate elements of lab manual production. The team at SPi, lead by Christian Arsenault, was responsible for page design, proofing, and compositing pages for publication. We thank Christian for addressing every challenge and achieving our product goals.

We thank the following individuals for their constructive criticisms and suggestions that led to improvements for this edition of the manual:

Kathleen Browne *Rider University*
Lee Ann Burrough *Prairie State College*
Lorraine Carey *Houston Community College–Central Campus*
Ruth Coffey - *Adelphie University*
Bernie Housen - *Western Washington University*
David King - *Auburn University*
Elena Miranda - *California State University, Northridge*
Sherry Oaks - *Front Range Community College*
Jessica Olney - *Hillsborough Community College*
Mark Ouimette - *Hardin-Simmons University*
Suki Smaglik - *Central Wyoming College*
Jesse Thornburg - *Temple University*
Henry Turner - *University of Arkansas*
Robin Wham - *California State University, Sacramento*
Wendi Williams - *NorthWest Arkansas Community College*

We thank Carrick Eggleston (University of Wyoming), Tomas McGuire, and Tom Mortimer for the use of their personal photographs. Andrew Fountain (Portland State University) provided information about Nisqually Glacier, and Tom Holzer (USGS) helped us with information related to the failure of the Cypress Structure during the Loma Prieta earthquake. Photographs and data related to St. Catherines Island, Georgia, were made possible by research grants to the editor from the St. Catherines Island Research Program, administered by the American Museum of Natural History and supported by the Edward J. Noble Foundation. Christopher Crosby of OpenTopography.org and the leaders and staff of UNAVCO (Meghan Miller, Donna Charlevoix, Shelley Olds, Beth Pratt-Sitaula) have been extremely generous with their time, expertise, graphics, and online resources. Cindy Cronin reviewed thousands of pages of draft text and page proofs, tolerated the conversion of her house into a photo studio for rocks and minerals, and kept the editor's body and soul together throughout the attenuated revision cycle for this edition.

Maps, map data, photographs, and satellite imagery have been used courtesy of the U.S. Geological Survey; Canadian Department of Energy, Mines, and Resources; Surveys and Resource Mapping Branch; U.S. National Park Service, U.S. Bureau of Land Management; JPL-NASA; NOAA; and the U.S./Japan ASTER Science Team.

We again thank Jennifer Wenner (University of Washington, Oshkosh) as well as Eric Baer (Highline Community College) for making possible the online options for students to review or learn mathematical skills using *The Math You Need, When You Need It* (TMYN) modules.

The continued success of this laboratory manual depends on the constructive criticisms, suggestions, and new contributions from everyone who uses it. We sincerely thank those who contributed to this project in various ways, and welcome all comments related to this edition. With your input, the *Laboratory Manual in Physical Geology* can continue to evolve and improve for the benefit of the geoscience community served by AGI and NAGT. Please continue to submit your suggestions and constructive criticisms directly to the editor:

Vince Cronin (Vince_LM_Editor@CroninProjects.org).

Allyson K. Anderson Book, *Executive Director, AGI*
Cathryn Manduca, *Executive Director, NAGT*
Edward Robeck, *Director of Education and Outreach, AGI*
Vincent S. Cronin, *Editor*

The following members of AGI member societies provided expert reviews of one or more sections:

Hossam Abdel-hameed (Tanat University), Society for Sedimentary Geology

Jay Austin (Duke University), The Clay Minerals Society

Nina L. Baghai-Riding (Delta State University), Society for Sedimentary Geology

Meghan E. Black (University of Alberta), Society for Sedimentary Geology

John Brady (Smith College), Mineralogical Society of America

Colin J.R. Braithwaite (University of Glasgow), Society for Sedimentary Geology

Rolando Bravo (Southern Illinois University), American Institute of Hydrology

William R. Brice (Professor Emeritus, University of Pittsburg at Johnstown), Petroleum History Institute

Paul Burger (National Park Service, Alaska Regional Office), National Speleological Society

Peter Burgess (School of Environmental Sciences, University of Liverpool), Society for Sedimentary Geology

Richard W. Carlson (Carnegie Institution for Science), The Geochemical Society

Christopher Coughenour (University of Pittsburg at Johnstown), Petroleum History Institute

Tom Crawford (University of West Georgia), National Association of State Boards of Geology

David Decker (University of New Mexico), National Speleological Society

Rebecca L. Dodge (Midwestern State University), American Association of Petroleum Geologists

Chelsie R. Dugan (Bureau of Land Management, New Mexico), National Speleological Society

Jim Flis (Houston Geological Society), Society for Sedimentary Geology

Devin Galloway (U.S. Geological Survey), International Association of Hydrogeologists–U.S. National Chapter

Vicki Harder (New Mexico State University), Mineralogical Society of America

Caitlin Hartig (University of North Dakota), Association of Earth Science Editors

Mrinal Kanti Roy (University of Rajshahi), Society for Sedimentary Geology

Leonard Konikow (U.S. Geological Survey, retired), International Association of Hydrogeologists–U.S. National Chapter

David C. Kopaska-Merkel (Geological Survey of Alabama), Society for Sedimentary Geology

Sandip Kumar Roy (Consultant, Oil and Gas Exploration, India), Society for Sedimentary Geology

Eve Kuniansky (U.S. Geological Survey), International Association of Hydrogeologists–U.S. National Chapter

Lewis Land (New Mexico Institute of Mining and Technology), National Cave and Karst Research Institute

Dennis Markovchick (U.S. Geological Survey), Association of Earth Science Editors

Ben McKenzie, Society for Sedimentary Geology

John Mylroie (Mississippi State University), National Speleological Society

Carl Olson (Towson University), Geoscience Information Society

Bogdan P. Onac (University of Southern Florida), National Speleological Society

Thomas Oommen (Michigan Technological University), Association of Environmental & Engineering Geologists

Paul H. Pausé (Geoscience Education Organization), Society for Sedimentary Geology

Douglas Rambo, National Association of State Boards of Geology

Stacy Lynn Reeder (Schlumberger-Doll Research), Society for Sedimentary Geology

Fredrick Rich (Georgia Southern University), The Palynological Society

Ana Maria Rojas (Colombian Observatory of Medical and Forensic Geology), International Medical Geology Association

Robert W. Scott (The University of Tulsa), Society for Sedimentary Geology

Rebecca Smith, Society for Mining, Metallurgy, and Exploration

Daniel Sturmer (The University of Cincinnati), Society for Sedimentary Geology

Al Taylor (Nomad Geosciences LLC), Society of Independent Professional Earth Scientists

Carolina Tenjo (Universidad Nacional de Colombia), International Medical Geology Association

Ethan Theuerkauf (University of North Carolina), Society for Sedimentary Geology

Thelma Thompson (University of New Hampshire), Geoscience Information Society

Joshua Villalobos (El Paso Community College), American Geophysical Union

Alejandro Villalobos Aragón (Universidad Autónoma de Chihuahua), International Medical Geology Association

Carole Ziegler (Southwestern College), Association of Earth Science Editors

Measurement Units

People in different parts of the world have historically used different systems of measurement. For example, people in the United States have historically used the English system of measurement based on units such as inches, feet, miles, acres, pounds, gallons, and degrees Fahrenheit. However, for more than a century, most other nations of the world have used the metric system of measurement. In 1975, the U.S. Congress recognized that global communication, science, technology, and commerce were aided by use of a common system of measurement, and they made the metric system the official measurement system of the United States. This conversion is not yet complete, so most Americans currently use both English and metric systems of measurement.

The International System (SI)

The International System of Units (SI) is the decimal system of measurement adopted by most nations of the world, including the United States (http://www.bipm.org/en/publications/si-brochure/). Each SI unit can be divided or multiplied by 10 and its powers to form the smaller or larger units. Therefore, SI is a "base-10" or "decimal" system.

SYMBOL	NUMBER	NUMERAL	POWER OF 10	PREFIX
T	one trillion	1,000,000,000,000	10^{12}	tera-
G	one billion	1,000,000,000	10^9	giga-
M	one million	1,000,000	10^6	mega-
k	one thousand	1000	10^3	kilo-
h	one hundred	100	10^2	hecto-
da	ten	10	10^1	deka-
	one	1	10^0	
d	one-tenth	0.1	10^{-1}	deci-
c	one-hundredth	0.01	10^{-2}	centi-
m	one-thousandth	0.001	10^{-3}	milli-
μ	one-millionth	0.000001	10^{-6}	micro-
n	one-billionth	0.000000001	10^{-9}	nano-
p	one-trillionth	0.000000000001	10^{-12}	pico-

Examples

1 meter (1 m) = 0.001 kilometers (0.001 km), 10 decimeters (10 dm), 100 centimeters (100 cm), or 1000 millimeters (1000 mm)

1 kilometer (1 km) = 1000 meters (1000 m)

1 micrometer (1 μm) = 0.000,001 meter (.000001 m) or 0.001 millimeters (0.001 mm)

1 kilogram (kg) = 1000 grams (1000 g)

1 gram (1 g) = 0.001 kilograms (0.001 kg)

1 metric ton (1 t) = 1000 kilograms (1000 kg)

1 liter (1 L) = 1000 milliliters (1000 mL)

1 milliliter (1 mL or 1 ml) = 0.001 liter (0.001 L)

Abbreviations for Measures of Time

A number of abbreviations are used in the geological literature to refer to time. In this edition, we use the abbreviation "yr" for years, preceded as necessary with the letters "k" (kyr) for thousand years, "M" (Myr) for million years, or G (Gyr) for billion years.

Mathematical Conversions

To convert:	To:	Multiply by:	
kilometers (km)	meters (m)	1000 m/km	LENGTHS AND DISTANCES
	centimeters (cm)	100,000 cm/km	
	miles (mi)	0.6214 mi/km	
	feet (ft)	3280.83 ft/km	
meters (m)	centimeters (cm)	100 cm/m	
	millimeters (mm)	1000 mm/m	
	feet (ft)	3.2808 ft/m	
	yards (yd)	1.0936 yd/m	
	inches (in)	39.37 in/m	
	kilometers (km)	0.001 km/m	
	miles (mi)	0.0006214 mi/m	
centimeters (cm)	meters (m)	0.01 m/cm	
	millimeters (mm)	10 mm/cm	
	feet (ft)	0.0328 ft/cm	
	inches (in)	0.3937 in/cm	
	micrometers (μm)	10,000 μm/cm	
millimeters (mm)	meters (m)	0.001 m/mm	
	centimeters (cm)	0.1 cm/mm	
	inches (in)	0.03937 in/mm	
	micrometers (μm)	1000 μm/mm	
	nanometers (nm)	1,000,000 nm/mm	
micrometers* (μm)	millimeters (mm)	0.001 mm/μm	
nanometers (nm)	millimeters (mm)	0.000001 mm/nm	
miles (mi)	kilometers (km)	1.609 km/mi	
	feet (ft)	5280 ft/mi	
	meters (m)	1609.34 m/mi	
feet (ft)	centimeters (cm)	30.48 cm/ft	
	meters (m)	0.3048 m/ft	
	inches (in)	12 in/ft	
	miles (mi)	0.000189 mi/ft	
inches (in)	centimeters (cm)	2.54 cm/in	
	millimeters (mm)	25.4 mm/in	
	micrometers* (μm)	25,400 μm/in	
square miles (mi^2)	acres (a)	640 acres/mi^2	AREAS
	square km (km^2)	2.589988 km^2/mi^2	
square km (km^2)	square miles (mi^2)	0.3861 mi^2/km^2	
acres	square miles (mi^2)	0.001563 mi^2/acr	
	square km (km^2)	0.00405 km^2/acr	
gallons (gal)	liters (L)	3.78 L/gal	VOLUMES
fluid ounces (oz)	milliliters (mL)	30 mL/fluid oz	
milliliters (mL)	liters (L)	0.001 L/mL	
	cubic centimeters (cm^3)	1 cm^3/mL	
liters (L)	milliliters (mL)	1000 mL/L	
	cubic centimeters (cm^3)	1000 cm^3/L	
	gallons (gal)	0.2646 gal/L	
	quarts (qt)	1.0582 qt/L	
	pints (pt)	2.1164 pt/L	
grams (g)	kilograms (kg)	0.001 kg/g	WEIGHTS AND MASSES
	pounds avdp (lb)	0.002205 lb/g	
ounces avdp (oz)	grams (g)	28.35 g/oz	
ounces troy (ozt)	grams (g)	31.10 g/ozt	
pounds avdp (lb)	kilograms (kg)	0.4536 kg/lb	
kilograms (kg)	pounds avdp (lb)	2.2046 lb/kg	

To convert from degrees Fahrenheit (°F) to degrees Celsius (°C), subtract 32 degrees and then divide by 1.8. To convert from degrees Celsius (°C) to degrees Fahrenheit (°F), multiply by 1.8 and then add 32 degrees.

*Sometimes called microns (μ).

LABORATORY EQUIPMENT

Also refer to the GeoTools provided at the back of this laboratory manual.

Acid bottle

Hand lens

Crucible tongs

Streak plate

Geologist's chisel tip pick

Pocket knife with steel blade

Dropper

Wash bottle

Internet access

Ruler

Protractor

Drafting compass

Beaker

Graduated cylinder

Cleavage goniometer

Hot plate

Digital electronic balance

Safety goggles

Triple beam (platform) balance

Beaker tongs

Pan balance

ELEVATION IN METERS

1000 t
2000–4000
500–1,999
200–499
0–199
Sea Level
Below sea level

NORTH AMERICA
Political & Physical Map

⊛ ● Metropolitan areas more than 20 million
⊛ ● Metropolitan areas 10–20 million
⊛ ● Metropolitan areas 5–9.9 million
⊛ • Metropolitan areas 1–4.9 million
⊛ ○ Selected smaller metropolitan areas
⌐┘ Plate boundaries

Filling Your Geoscience Toolbox

Pre-Lab Video 1

https://goo.gl/c3zAzr

Contributing Authors

Cynthia Fisher • *West Chester University of Pennsylvania*   C. Gil Wiswall • *West Chester University of Pennsylvania*

▲ Geoscience student documenting a small fault in the Raton Formation west of Trinidad, Colorado (37.12641°N, 104.76647°W).

BIG IDEAS

Geology is the science of Earth. Society needs reliable information about Earth as it confronts challenges related to resources, natural hazards, environmental health, and sustainable development. Geoscientists observe Earth using many technologies, from sophisticated airborne or orbital sensors to laboratory instruments and basic fieldwork. We map Earth's surface and describe locations using several coordinate systems. Mapping Earth helps us document change over time and identify where useful resources occur. Mathematics is an important language we use to communicate ideas in geoscience.

FOCUS YOUR INQUIRY

Think About It   What do we see when we look at different parts of Earth?

Think About It   How is the elevation of a solid block floating in a more dense fluid related to the relative densities of the two materials?

Think About It   How can we convert many observations into useful summary data?

Think About It   How can we represent data to help us interpret trends and implications?

Think About It   How does a variation in density help us understand the broad structure of Earth?

1

Introduction

On December 7, 1972, three astronauts in Apollo 17 looked back on Earth from a distance of ~29,000 km as they glided toward their rendezvous with the Moon. Either Jack Schmitt—an astronaut who is also a geologist—or astronaut Ron Evans took a photograph of our home in full sunlight (Fig. 1.1). Earth is a breathtakingly beautiful sight, cloaked in a swirling atmosphere that reveals the South Atlantic and Indian oceans, Africa, Arabia, Antarctica, and the island of Madagascar near the center of the image. It is a portrait of a dynamic planet. Below those slowly moving clouds are oceans of liquid water circulating much more slowly than the atmosphere. There are plates of solid lithosphere in which the continents are embedded and that extend below the ocean basins. The plates are also in motion, gliding over the top of a solid mantle that flows even more slowly. Below the rocky mantle and almost 3000 km below Earth's surface is the liquid iron core whose circulation is responsible for the magnetic field that protects Earth from many of the most dangerous effects of solar radiation. Even the solid inner core is interpreted to be in motion, rotating slightly faster than the outer part of the planet.

This dynamic Earth is the stage on which all of the acts of life are played. It is the source of all the resources we need to survive. As vast as Earth is from our perspective as individuals living out our lives on its surface, Earth is an oasis as viewed from space. It is worth the time for us to gain a basic understanding of our home.

In developing this laboratory manual, we provide you with the opportunity

- To see good examples of the materials that make up the surface of earth
- To work with maps and imagery and other forms of data used in the geosciences
- To experience using some simple mathematics to help us understand relationships within the physical world
- To learn about this planet that we share.

The labs in this course are for actively *doing* things. We want you to experience science and, in particular, to engage in geoscience.

You live during a remarkably productive time in the geosciences. Today, it is increasingly common for data to be shared soon after it is collected and for all types of raw data about Earth and its systems to be routinely collected in great quantities. This has been encouraged if not mandated by major funding agencies and has yielded tremendous benefits to science and society. In this set of lab chapters, we will tap into online sources of research-quality data collected by field-based geoscientists as well as by automated networks of seismographs and GPS receivers, Earth-observing satellites, and stream gages. Moderate to high-resolution imagery of Earth are now available online for free as are new digital topographic maps and archival maps for most of the United States. You won't just hear the stale factoid that Los Angeles and San Francisco are moving toward each other because they are located on different plates, but you will learn how to access the GPS velocity data and will acquire the knowledge needed to explore motion along that remarkable plate boundary yourself.

This lab book is a gateway to your exploration of Earth.

How Is Each Laboratory Chapter Organized?

There are important features in this lab manual you might miss if they were not pointed out to you. Square black-and-white QR codes are printed throughout each chapter that can be scanned with a smartphone that has an app for reading QR codes. There are many such apps available on the web for little or no cost. Scanning the QR code links you to a web resource that will likely be helpful.

Each chapter begins with some general information, followed by an *Activity Box* that introduces you to the first of the lab activities. Each Activity Box alerts you to an upcoming activity, describes the objective of the activity, and makes you aware of any special data, tools, or resources you will need in order to complete the activity successfully. **The text that follows an Activity Box is closely related to the activity, so you should read the**

Figure 1.1 The blue marble. Photograph of Earth taken by the crew of Apollo 17 on their way to the Moon in December 1972. Earth's restless, dynamic nature is well expressed in this image of our home.

relevant section of text before you begin the activity. As you are working on an activity at the back of the chapter and need to find relevant information in the text, the corresponding Activity Box functions like a bookmark. Just find the corresponding Activity Box and start reading down from there.

Terms that are particularly important for you to know are printed in a **bold** type face. Those terms are usually defined in the text, but additional information about these special words is available in an online glossary prepared for use with this lab manual.

Science as a Process for Learning Reliable Information

An essential goal of any introductory geoscience course is to help you deepen your understanding of science and of its importance to our lives. Of course, many books and essays have been written in an attempt to describe the nature of science from various viewpoints and worldviews. The geoscientists who have collaborated in developing this laboratory manual would like you to understand at least some of the essential characteristics of science:

- Science is a way of learning reliable information about the world.
- Reliable information is derived from reproducible observations.
- All scientific observations involve some degree of uncertainty, and the proper assessment and reporting of that uncertainty is a fundamental responsibility of scientists.
- We will refer to reproducible observations with their associated uncertainties as **scientific facts**. Scientific facts are always subject to refinement.
- Scientific explanations of the relationships between scientific facts—**hypotheses**—must be testable.
- Many or most preliminary hypotheses are eventually found to be incomplete or simply wrong. Hypotheses in science generally have a short life span because they are replaced by better, more complete hypotheses. Science is a winnowing process that helps us identify the false leads and dead ends so that we can focus on potentially fruitful areas of inquiry.
- Scientific reports are critically reviewed by appropriate scientific experts prior to publication in the peer-reviewed journals that form the communication backbone of science.
- Science involves the work of individual scientists and teams of scientists in the context of a worldwide community of scientists involved in the process of reproducing data, assessing uncertainty, testing hypotheses, reviewing scientific reports made by other scientists, and adding reliable information to our models of how the world works.

- Mathematics is a fundamental language in science because of the clarity and efficiency with which it describes the relationships among scientific facts.

Geoscience and Geoethics

Geoscience is the branch of science that is primarily concerned with the natural history, materials, and processes of Earth and, by extension, of other planetary and subplanetary bodies within our observational reach. Geoscientists study Earth through many different types of inquiry. We go out into the natural world, collect specimens, observe processes, take measurements, and record descriptions. We conduct chemical and physical tests on geological materials in a laboratory setting, making use of all the analytical tools of a chemist or physicist. Geologists engage in theoretical modeling to define the mathematical relationships that govern the behavior of geological materials and processes. In each of these areas of inquiry, the geoscientist's work must be reproducible or else it is not scientifically valid or useful.

Some envision science as a search for truth, although that begs the classic philosophical question, "What is truth?" Most scientists are content to leave notions of truth to philosophers to debate while we work daily to enhance our store of reliable information acquired through the methods of science. Scientific knowledge can be very reliable, but it never fully escapes its provisional nature. We are always improving our knowledge by testing our hypotheses with new data.

Albert Einstein wrote, "Truth is what stands the test of experience" in an essay on ethics. Ethics plays a fundamental role in science in several ways, including the practice of science, the application of science, and the motivation behind the work of scientists. Virtually all of the major professional organizations in science and in geoscience have developed codes of ethics (**http://www.americangeosciences.org/community/agi-guidelines-ethical-professional-conduct**). **Geoethics** is an expanding field of inquiry and thought.

Geoscientists have important responsibilities toward society because of our unique knowledge of Earth (**Fig. 1.2**). Geoscientists must act ethically to provide society with the reliable information needed to make good choices related to energy, mineral resources, water management, environmental health, natural hazards, climate change, and many other issues of public policy. Only reliable information is useful as we confront our many challenges. In addition to our responsibilities toward society, geoscientists are also responsible for acting as caretakers of the only habitable planet that we have useful access to—our home, Earth.

The fact that you are privileged to have access to a college course in physical geology imparts ethical responsibilities to you. Take full advantage of your opportunity

Figure 1.2 Society needs the input of geoscientists. A landslide destroyed this important access road in southern California that cost many millions of dollars to repair. Geoscientists provide society with the expertise needed to effectively address challenges related to energy, supply of industrial minerals, fresh water, climate change, and natural hazards ranging from landslides and floods to earthquakes and volcanic eruptions.

to think, to question, and to learn about Earth. To paraphrase James Blaisdell of Pomona College, you bear your added riches in trust for all of humanity.

Learning to Think Like a Geoscientist

As you complete exercises in this laboratory manual, think and act like a geoscientist. Focus on questions about Earth materials and history, natural resources, processes and rates of environmental change, where and how people live in relation to the environment, and how geology contributes to sustaining the human population. Conduct investigations and use your senses and tools to make observations. As you make observations, record the data you develop. Engage in critical thinking—apply, analyze, interpret, and evaluate the evidence to form tentative ideas or conclusions. Engage in discourse or collaborative inquiry with others (exchange, organization, evaluation, and debate of data and ideas). Communicate inferences—write down or otherwise share your conclusions and justify them with your data and critical thinking process.

These components of geoscience work are often not a linear "scientific method" to be followed in steps. You may find yourself doing them all simultaneously or in odd order. For example, when you observe an object or event, you may form an initial interpretation about it. Those initial impressions need to be expanded and formalized into testable hypotheses, and that process of developing hypotheses often inspires the collection of new data. Your tentative ideas are likely to change as you acquire additional information.

When making observations, you should observe and record **qualitative data** by describing how things look, feel, smell, sound, taste, or behave. You should also collect and record **quantitative data** by counting, measuring, or otherwise expressing in numbers what you observe. Carefully and precisely record your data in a way that others could understand and use it.

Your instructor will not accept simple yes or no answers to questions. He or she will expect your answers to be complete statements justified with data, sometimes accompanied by an explanation of your critical thinking. Show your work whenever you use mathematics to solve a problem so your method of thinking is obvious.

ACTIVITY 1.1

A View of Earth from Above, (p. 19)

Think About It   What do we see when we look at different parts of Earth?

Objective   Learn how to use Google Earth to view Earth from above, and then use that skill to investigate our planet.

Before You Begin   Read the following section: Getting to Know Your Planet.

Plan Ahead   This activity requires that you use Google Earth to find several features on Earth's surface given their map coordinates. You will need to have access to the web and to Google Earth during the lab period, or this activity will have to be completed outside of lab time. Google Earth is a free application, and current information about how to access and use Google Earth is available online at **earth.google.com**.

Getting to Know Your Planet

We gained the ability to rise a significant distance above Earth's surface when the Montgolfier brothers in France developed hot-air balloons large enough to carry us aloft in 1783. Alfred and Kurt Wegener were avid balloonists in Germany and set the world endurance record in 1906, less than three years after the Wright brothers' first flight in a powered airplane at Kitty Hawk, North Carolina. (Alfred Wegener would later propose the idea of continental drift that evolved into our current understanding of plate tectonics.) As anyone who has had a window seat on an aircraft on a clear day will tell you, Earth's surface is endlessly fascinating (Fig. 1.3). We are privileged to live at a time when there are many technologies available to help us study Earth's surface from above.

Throughout this course, you will encounter opportunities to look at Earth's surface using Google Earth, which is a free web application that you can access through earth.google.com. The first task you will need to be able to perform using Google Earth is to find a particular place on Earth's surface once given the coordinates of that point. That involves entering the map coordinates of the point into the Search box in Google Earth and clicking the "Search" button. You will also need to learn how to activate different features (e.g., Places, Borders, Photos), to zoom in and out of the image, to determine the height above Earth that is depicted in the image, and generally how to move from place to place across the surface. Different versions of Google Earth operate differently. The Help resource on your particular version of Google Earth will guide you in learning how to use its various functions to study the surface of Earth.

First, we need to learn about two ways we have developed to specify the location of a given point on Earth, using map coordinates. The two general methods are the geographic coordinates of latitude and longitude, and the UTM coordinates.

Geographic Coordinates (Latitude and Longitude)

The location of a point on Earth can be specified using its latitude and longitude (Fig. 1.4). **Latitude** is measured relative to the **equator**, which is the circle around Earth located in the tropics exactly halfway between the north and south poles. The latitude is 0° at the equator, 90° at the north pole, and −90° at the south pole. The poles are the places where Earth's spin axis intersects the ground surface to the north (**north pole**) and south (**south pole**). Semicircles that wrap around Earth from the north pole to the south pole are called **meridians**. We measure the latitude of a given point along the meridian that passes through that point. The latitude is the angle from the point where that particular meridian crosses the equator to the center of Earth and back out to the given point.

Longitude is measured relative to a particular meridian that passes through a specific point on the grounds of the Royal Observatory at Greenwich, England. A group called the *Earth Rotation and Reference Systems Service* keeps track of the practical definition of the international reference meridian (the **prime meridian**) for us so we don't have to worry about it. The angle between the prime meridian and the meridian that passes through a given point is that point's longitude. Longitudes measured to the east of the prime meridian (east longitudes) are considered *positive* longitudes, and longitudes measured to the west of the prime meridian (west longitudes) are considered *negative* longitudes. So longitudes range from 180° (180°E) to 0° along the prime meridian to −180° (180°W). With the exception of several of the western Aleutian Islands in Alaska, all of the states in the United States have west, or negative, longitudes.

If you are moving either due north or due south, a change of one degree in latitude (at the same longitude) is a difference of ~111.2 km across Earth's surface.

Figure 1.3 Grand Prismatic Spring. This beautiful pool at Yellowstone National Park is one of the largest hot springs on Earth and is home to bacteria capable of living under extremely harsh conditions. The solid Earth, liquid water, water vapor, a hint of Earth's hot interior, and evidence of life are all visible in this image, illustrating the interconnectedness of Earth's systems. (Photo by Jim Peaco of the National Park Service.)

(The symbol "~" is used here to indicate an approximate number.) One degree of latitude is 1/360th of the distance around Earth whose average radius is ~6,371 km and whose circumference is ~40,030 km. The meridians that mark different longitudes all converge at the north and south poles and are spaced their maximum distance of ~111.2 km per degree of longitude

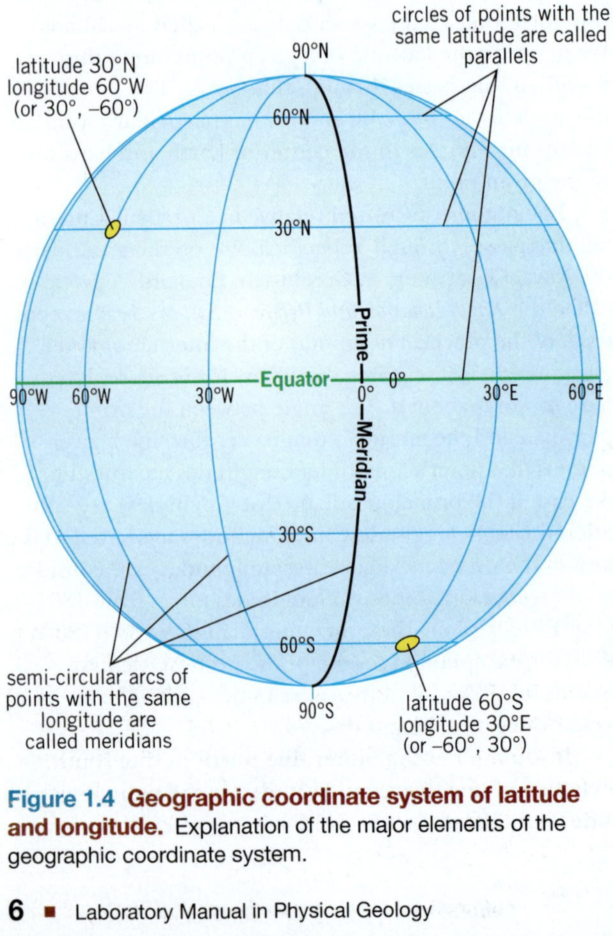

where they cross the equator. So it would take a commercial jetliner over 7 minutes to fly across 1 degree of longitude at the equator (~111.2 km), but it would take a motivated sugar ant about a quarter of a second to walk 1 degree of longitude (~1.7 centimeters [cm]) if the ant was just a meter away from the north or south pole.

Geoscientists frequently use **decimal degrees** to indicate position when we use the geographic coordinate system of latitudes and longitudes. For example, the great obelisk of the Washington Monument in Washington, D.C., is located at latitude 38.889469°N, longitude 77.035258°W. When it is made clear that the number pair refers to latitude and longitude, we can simply write 38.889469, −77.035258 without the degree symbol or the letters N and W because positive-signed latitude is understood to be north latitude and negative-signed longitude is understood to be west longitude. You can find the Washington Monument in Google Earth by entering 38.889469, −77.035258 in the Search box. You can also find it by entering 38.889469°, −77.035258° or 38.889469°N, 77.035258°W, but then you would need to remember how to insert the degree (°) symbol using your web-enabled device.

An alternative to decimal degrees that has persisted for a very long time is the **degrees-minutes-seconds** system that subdivides each degree into 60 minutes of arc and then each minute of arc into 60 seconds of arc. Our use of degrees, minutes, and seconds for angles and our use of 24 hours of 60 minutes and 60 seconds for daily time have ancient roots. The "60" originated approximately 5000 years ago in the ancient Sumerian system of counting, which was a base-60 system. Not being ancient Sumerians, we generally use a base-10 system today, but old habits are difficult to break. Maps produced by the U.S. Geological Survey (USGS) are based on the

Figure 1.4 Geographic coordinate system of latitude and longitude. Explanation of the major elements of the geographic coordinate system.

degree-minute-second way of expressing latitude and longitude.

It is useful to know how to convert from the degrees-minutes-seconds system to decimal degrees. Let's say we have an angle of a degrees, b minutes, and c seconds, which is commonly written as $a°\ b'\ c''$. The decimal-degree equivalent is $[a + (b/60) + (c/3600)]$ degrees, noting that $60 \times 60 = 3600$. Here's an example.

$14°\ 38'\ 52''$ is the same as
$[14 + (38/60) + (52/3600)]° \cong 14.647778°$

Google Earth can locate points expressed in the degree-minute-second, degree-decimal minute, and decimal degree systems of expressing longitude and latitude.

Universal Transverse Mercator (UTM) Coordinates

Most handheld GPS receivers can provide us with locations in either a latitude–longitude format or a UTM format. Latitude–longitude is a bit easier to explain, but UTM has some practical advantages. UTM coordinates are based on a projection of Earth's surface onto a plane that has a coordinate grid in meters that is aligned (approximately) north–south, east–west. It is more useful to people trying to navigate from point A to point B to know how many meters they need to go in a given direction than to know how many degrees of longitude or latitude.

Although variations exist, the UTM coordinates of a given point usually have four components listed in the following order: **zone number**, **latitude band**, **easting**, and **northing** (**Fig. 1.5**). An example that can be directly interpreted by Google Earth's Search function is 12T 370730 4608526. Try it out and see where it leads you.

UTM Zones. Earth is divided into 60 zones starting at longitude 180°—the **international dateline** halfway around Earth from the prime meridian—and proceeding to the east. Each zone is 6° of longitude wide and extends from 80°S to 84°N latitudes. Zone 1 is from 180°W to 174°W followed by zone 2 from 174°W to 168°W. The continental United States extends from UTM zone 10 in the west to 19 in the east. Each zone has a **central meridian**, which is halfway between the two zone boundaries (**Fig. 1.5**). For example, the central meridian of zone 2 is longitude 171°W.

UTM Bands. Each zone is divided into 20 latitude bands that are 8° of latitude tall. The bands are lettered from C (between 80° and 72°S latitude) to X (between 72°N and 84°N latitude). Zone X is the only zone that is larger than 8° high; it was extended to 12° high to cover all of the major continental areas above sea level. The bands from N to X are in the northern hemisphere (**Fig. 1.5**). The continental United States is in bands R, S, T, and U.

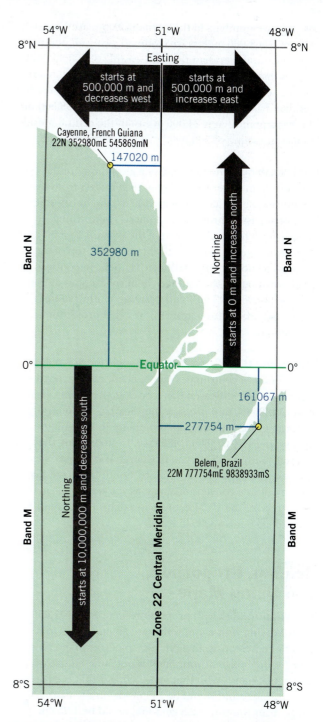

Figure 1.5 Universal Transverse Mercator (UTM) coordinate system. UTM zone 22, latitude bands M and N along the equator near the mouth of the Amazon River, northeastern South America. UTM coordinates of two cities are shown as examples.

UTM Easting. The easting within a given zone is measured in meters perpendicular to the zone's central meridian whose easting is defined as 500,000 meters, or 500,00 mE (**Fig. 1.5**). (The abbreviation *mE* is expressed in words as "meters east" or "easting.") It seems odd to express position in this way because we normally define position relative to some point whose coordinate is defined as 0, but the originators of UTM coordinates did not want to have any negative numbers in their system. Zones are a maximum of ~668,000 meters (m) wide, so giving all points along a central meridian the same easting of 500,000 mE ensures that no east–west coordinate in the zone has a negative value. A point that is 1500 m to the *west* of the central meridian has an easting of 500,000 − 1500 = 498,500 mE, whereas a point that is 24,000 m east of the central meridian has an easting of 500,000 + 24,000 = 524,000 mE.

UTM Northing. The northing of a given point in the northern hemisphere is its distances in meters from the equator measured along the meridian—the north–south line—that passes through the point. UTM northings in the northern hemisphere start at the equator (0 mN) and increase northward to ~9,300,000 mN at the top of the UTM projection at 84°N latitude (**Fig. 1.5**). In contrast, UTM northings in the southern hemisphere start at the equator (10,000,000 mS) and decrease southward to ~1,100,000 mS at the bottom of the UTM projection at 80°S latitude.

Expressing UTM Coordinates. Google Earth is able to interpret UTM coordinates (zone, band, easting, northing) typed into its Search box. For example, typing 18S 323482 4306480 and pressing the "Search" button will send you to the Washington Monument in Washington, D.C. Typing 13T 623805.6 4859546.9 and pressing the "Search" button will get you to George Washington's sculpted nose at Mount Rushmore, South Dakota. Fully written out in a conventional way, as you would in a geoscientific report, the UTM coordinates of his nose would be 13T 623805.6 mE 4859546.9 mN

Scaling, Proportion, and Using Maps

The ability to work with proportions is an essential skill when working with maps, which are an important tool used by geoscientists. Imagine a map of your hand at ½ scale, so the image of your hand is just 50% the size of your actual hand. With that scale, a finger that is 2 cm wide on your hand would be 1 cm wide on the map image. Your thumb, at 2.5 cm wide, would be 1.25 cm wide on the map. Your little finger is 7 cm long but just 3.5 cm long on the map. Your middle finger is 9 cm long, but

the image of that finger is 4.5 cm long on the map. The ratio of the length of the feature on the map to the length of the corresponding feature on your hand is always 1 to 2,

$$\frac{1}{2} = \frac{1.25}{2.5} = \frac{3.5}{7} = \frac{4.5}{9}$$

Representative Fraction Scale. Printed on some maps is a **representative fraction scale** (or simply a **fractional scale**) which is fundamentally a ratio. Examples of the two most common styles of fractional scales are 1/24,000 and 1:24,000. These scales are identical to each other—they represent two styles of presenting the same information. The meaning of "1:24,000" is that one unit of some sort (centimeters, for example) on the map is equal to 24,000 of that same unit (cm) on the ground in the mapped area. The problem with representative fractional scales is that they are of no use if the map is enlarged or reduced in size during reproduction. The "1:24,000" will still be printed on the map even if the map is reproduced to the size of a postage stamp or billboard.

Bar Scale. Many maps contain some form of a thin rectangle called a **bar scale** that represents a specific length on the map. An example of the style of bar scale used by the USGS is shown in **Fig. 1.6**. Bar scales are used extensively in this book and throughout the geosciences in part because the printed bar expands and contracts along with the map, so the map scale can always be interpreted. Let's work an example. Imagine that you have a map with a bar scale that represents a length of 2000 m (2 kilometers [km]) on the ground in the mapped area. You measure the bar with a rule, and find that 2 km on the ground is the same as 4 cm on the map because the distance from 0 to 2 km on the map's bar scale is 4 cm long. You want to take a 5 km stroll, 2.5 km down the road and 2.5 km back. What is the length of a line on the map that represents 2.5 km along the road?

We can frame this problem as a proportion or ratio problem: 4 cm is to 2 km as c (the unknown map distance) is to 2.5 km. This is a type of problem we would like to be able to solve every time we encounter it regardless of the specific numbers involved. So rather than use numbers, let's use a unique letter to represent each of the numbers in our problem, rather than just the unknown map distance (c), and restate the problem: a

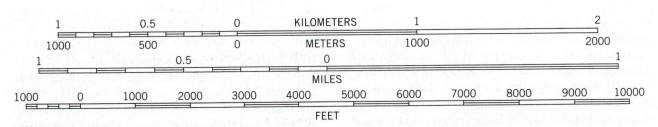

Figure 1.6 Typical bar scale used by the USGS. This bar scale was copied from a USGS topographic map published in 2015. It shows a metric bar scale in kilometers and two types of scales that use the English units miles and feet.

is to *b* as *c* is to *d*. In other words, the ratio of *a* to *b* is the same as the ratio of *c* to *d*, or

$$\left(\frac{a}{b}\right) = \left(\frac{c}{d}\right)$$

This equation can be rearranged as $(a \times d) = (b \times c)$. In our original problem, the value of variable *c* was unknown. We can rearrange the equation again so that variable *c* is isolated by itself on one side of the equation.

$$c = \frac{(a \times d)}{b}$$

Now, let's insert the numbers from our problem.

$$c = \frac{(a \times d)}{b} = \frac{(4 \times 2.5)}{2}, \text{ so } c = 5$$

You would look on your map, mark a point 5 cm down the road on your map, find a recognizable landmark like a street corner near your mark, and start walking there and back for your stroll.

What if you saw something interesting on the same map (like an all-night ice cream store) and measured a map distance of 7 cm from your current location to that point. How far would you have to walk to get there? This is just another form of the same problem except that this time variable *c* is known but *d* is unknown. Recall that $(a \times d) = (b \times c)$, so we can rearrange this equation to isolate *d* on one side of the equation and then replace the variables with the actual numbers from our problem.

$$d = \frac{(b \times c)}{a} = \frac{(2 \times 7)}{4}, \text{ so } d = 3.5$$

You would need to walk 3.5 km there and another 3.5 km back. The ability to manipulate these simple equations and solve proportion problems is a very useful skill.

About Mathematics in Geoscience. Geoscience is a branch of science that integrates certain aspects of physics, chemistry, biology, computer science, and related technologies in developing our understanding of Earth. Science is a quantitative enterprise. To paraphrase Galileo, mathematics is the language of science. You might think that mathematics seems like a large unfriendly animal with teeth and claws that you encounter along your path to learning about science. But math doesn't bite and is more like a set of tools. Some tools, like a hammer, are easier to use than others. Regardless of the work you will pursue in your life, it will take some time and effort to learn how to use the math tools you will need for your work.

It would be a disservice to college students to introduce you to the geosciences in a way that completely avoids quantitative ideas. The geoscientists who developed this laboratory manual thought it reasonable to use simple math at an average high school level in some of the text and activities. Much of the time, we will just use arithmetic. If you need additional help beyond that provided in this laboratory manual, ask your teacher. Help for some topics is available online through a

resource called The Math You Need, When You Need It (**http://serc.carleton.edu/ mathyouneed/index.html**), from the Khan Academy (**https://www.khanacademy.org**), and from other similar resources.

https://goo.gl/r1Kwfe

Rearranging Equations—The Math You Need

You can learn more about how to rearrange equations to solve for a given variable (including practice problems) at **http://serc.carleton.edu/mathyouneed/ equations/index.html** featuring The Math You Need, When You Need It tutorials for students in introductory geoscience courses.

https://goo.gl/eaVJq7

ACTIVITY 1.4

Floating Blocks and Icebergs, (p. 26)

> **Think About It**   How is the elevation of a solid block floating in a more dense fluid related to the relative densities of the two materials?

Objective   Practice working with lengths, volumes, and density while learning more about buoyancy and Archimedes' Principle.

Before You Begin   Read the following sections: Measuring Earth Materials and Archimedes' Principle.

Plan Ahead   It is best if you have an opportunity to actually observe and measure a rectangular wooden block floating in water to work out its density. Ask your teacher about setting up that experiment if it is not already available to you.

Measuring Earth Materials

Observation and measurement are fundamental parts of the scientific process. Here, we review a few basics of measurement that will be needed throughout this laboratory course.

SI Units of Length Measurement

The International System of Units (SI) is based on seven basic units of measure from which other units are derived. The SI system is used throughout science with few exceptions. The standard of length measurement is the **meter** (m), and useful units derived from that base unit include the **kilometer** (km; 1 km = 1000 m), **centimeter** (cm; 100 cm = 1 m), **millimeter** (mm; 1000 mm = 1 m), **micrometer** or **micron** (μm; 1,000,000 μm = 1 m). Since 1983, the official SI definition of a meter "is the length of the path travelled by light in vacuum during a time interval of 1/299,792,458 of a second." Practically speaking, we use a manufactured secondary standard such as a ruler or a tape measure to determine length most of the time, although very accurate measurements of length are now made using lasers, radar, and other technologies.

People who grew up in the United States are probably more familiar with units of measurement that we inherited from England, and a vestige of that remains in the topographic maps of the USGS whose topographic contours are defined in feet (ft.) and whose bar scales sometimes include miles (mi.). Because of this, you will need to know that 1 mile is 5280 ft., each foot has 12 inches (in.), and the international definition of 1 inch is a length equal to 2.54 cm *exactly* (**Fig. 1.7A**). From that one conversion factor

can be obtained all other conversions in lengths from metric to the old English system.

Review the measurement examples in **Fig. 1.7A** to be sure that you understand how to make *reliable* or *reproducible* metric measurements. Note that the length of an object may not coincide with a specific centimeter or millimeter mark on the ruler, so you may have to estimate the fraction of a unit *as exactly as you can*. The length of the red rectangle in **Fig. 1.7A** is between graduation marks for 106 and 107 mm, so the most reasonable measurement of this length is 106.5 mm. Also be sure that you measure lengths starting from the zero point on the ruler, *not necessarily from the end of the ruler*.

A. LINEAR MEASUREMENT USING A RULER

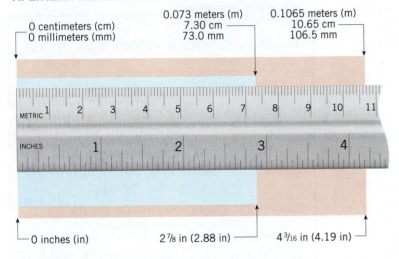

B. VOLUME OF A BLOCK MEASURED IN CUBIC CENTIMETERS: cm³

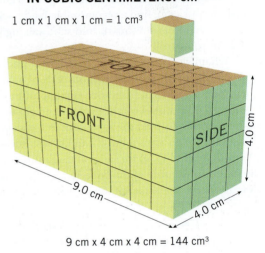

1 cm x 1 cm x 1 cm = 1 cm³

9 cm x 4 cm x 4 cm = 144 cm³

C. FLUID VOLUME MEASURED WITH GRADUATED CYLINDERS IN MILLILITERS: mL or ml

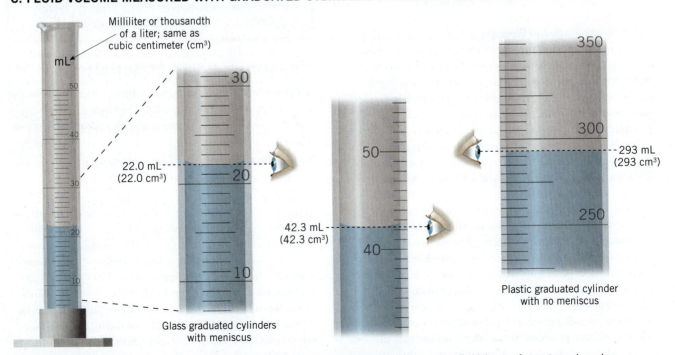

Figure 1.7 **Tools and scales of measurement. A.** Measurement of length using a ruler. **B.** Volume of a rectangular prism measured in cubic centimeters. **C.** Liquid volume measured with a graduated cylinder.

There will be times when you will need to convert a measurement from one unit of measure to another. This can be done with the aid of the mathematical conversion charts like the one printed in the front of many versions of this laboratory manual. For example, to convert mm to m, divide the measurement in mm by 1000 (because there are 1000 mm per m):

$$\frac{106.5 \text{ mm}}{1000 \text{ mm/m}} = 0.1065 \text{ m}$$

Thus, 106.5 mm is the same as 0.1065 m.

Unit Conversion—The Math You Need

You can learn more about unit conversion (including practice problems) at **http://serc.carleton.edu/mathyouneed/units/index.html** featuring The Math You Need, When You Need It tutorials for students in introductory geoscience courses.

https://goo.gl/Tiy3ft

Area and Volume

The SI base unit used in the measurement of area and volume is the meter. Area is described using square meters (m^2, km^2, cm^2, mm^2, etc.), and volume is described in cubic meters (m^3) or **liters** (L) where 1 liter is equal to 1000 cm^3 and 0.001 m^3 (**Figs. 1.7B** and **1.7C**). The abbreviation for liter can be either the lowercase or capital L, and we will generally use the capital option in order to avoid confusion with the number 1. It is convenient to remember that 1 mL (1 mL = 1/1000 L) is the same volume as 1 cm^3. The graduated cylinders used in many geoscience labs are marked in mL.

Most natural materials have an irregular shape, so their volumes cannot be calculated accurately from measurements made with rulers. However, the volumes of these odd-shaped materials can be determined by measuring the volume of water they displace. This is often done in the laboratory with a *graduated cylinder* (**Fig. 1.7C**), a vessel used to measure liquid volume. Most graduated cylinders are marked in mL. When you pour water into a graduated cylinder made of glass, the surface of the liquid is usually a curved *meniscus*, and the volume is read at the bottom of the curve (**Fig. 1.7C**, middle and left-hand examples). In some plastic graduated cylinders, however, there is no meniscus. The water level is flat (**Fig. 1.7C**, right-hand example).

If you place a rock in a graduated cylinder full of water, the rock takes up space previously occupied by water at the bottom of the graduated cylinder. This displaced water has nowhere to go except higher into the graduated cylinder. Therefore, the volume of an object such as a rock is exactly the same as the volume of water that it displaces.

The water displacement procedure for determining the volume of a rock is illustrated in **Fig. 1.8**. Start by choosing a graduated cylinder into which the rock will fit easily, and add enough water to be able to totally

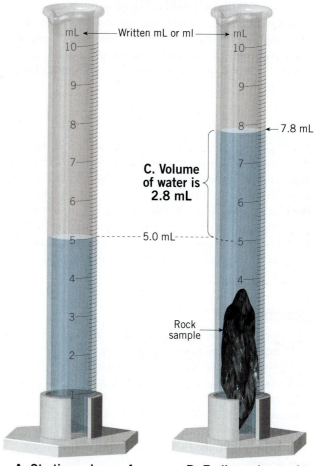

WATER DISPLACEMENT METHOD FOR DETERMINING VOLUME OF A ROCK SAMPLE

Written mL or ml

C. Volume of water is 2.8 mL

7.8 mL

5.0 mL

Rock sample

A. Starting volume of water

B. Ending volume of water

PROCEDURES

A. Place water in the bottom of a graduated cylinder. Add enough water to be able to totally immerse the rock sample. It is also helpful to use a dropper bottle or wash bottle and bring the volume of water (before adding the rock sample) up to an exact graduation mark like the 5.0 mL mark above. Record this starting volume of water.

B. Carefully slide the rock sample down into the same graduated cylinder, and record the ending volume of the water (7.8 mL in the above example).

C. Subtract the starting volume of water from the ending volume of water to obtain the displaced volume of water. In the above example: 7.8 mL – 5.0 mL = 2.8 mL (2.8 mL is the same as 2.8 cm^3). This volume of displaced water is the volume of the rock sample.

Figure 1.8 Procedure for determining volume of a rock sample by water displacement.

immerse the rock. It is also helpful to use a dropper or wash bottle to raise the volume of water (before adding the rock) up to an exact graduation mark (5.0 mL mark in **Fig. 1.8A**). Record this starting volume of water. Then carefully slide the rock sample down into the same graduated cylinder and record this ending level of the water (7.8 mL mark in **Fig. 1.8B**). You might need to guide the specimen down into the water with a thin wire to avoid a splash that would cause a loss of water, spoiling your measurement. Subtract the starting volume of water from the ending volume of water to obtain the displaced volume of water (2.8 mL, which is the same as 2.8 cm³). This volume of displaced water is also the volume of the rock sample.

Mass

The SI base unit for mass is the **kilogram** (kg), which is the same as 1000 **grams** (g). One kg is the mass of 1 L of water at 4°C at which temperature liquid water has a density of exactly 1 kg/L = 1 g/mL = 1 g/cm³. So the common 1-L bottle of water or soda has a mass of about 1kg. Mass is measured using a laboratory balance. Research balances are calibrated so that they accurately measure the mass of a standard object of known mass called a *reference mass*.

Density

The amount of mass in a known volume of material is the **density**. In the SI system, density is expressed typically in either kg/m³ or g/cm³, depending on the context. It is easy enough to demonstrate that 1 kg/m³ = 1000 g ÷ 1,000,000 cm³ = 0.001 g/cm³. For example, the density of a typical crystal of the mineral *quartz* has a density of 2650 kg/m³ or 2.65 g/cm³. Scientists and engineers use the Greek character rho (ρ) to represent density.

Calculating Density—The Math You Need

You can learn more about calculating density (including practice problems) at **http://serc.carleton.edu/mathyouneed/ density/index.html** featuring The Math You Need, When You Need It tutorials for students in introductory geoscience courses.

Archimedes' Principle

Scientists have wondered for centuries about how the distribution of Earth materials is related to their density and gravity. Curious about buoyancy, the Greek scientist and mathematician Archimedes experimented with floating objects around 225 BC. When he placed a block of wood in a bucket of water, he noticed that the block floated and the water level rose (**Fig. 1.9A**). When he pushed down on the wood block, the water level rose even more. And when he removed his fingers from the wood block, the water pushed it back up to its original level of floating. Archimedes eventually realized that every floating object is pulled down toward Earth's center by gravity, so the object displaces fluid and causes the fluid level to rise. However, Archimedes also realized that every floating object is also pushed upward by a buoyant force that is equal to the weight of the displaced fluid. This is now called **Archimedes' Principle**.

Archimedes wrote that an object immersed in a fluid is subjected to an upward-directed force—a **buoyant force**—that is equal to the weight of the fluid displaced by the object. In 1687, Isaac Newton defined **force** as the product of **mass** times **acceleration**, and the acceleration that affected the blocks and fluids of interest to

A. FLOATING WOOD BLOCK

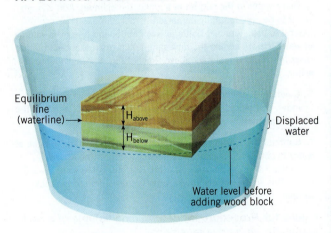

B. ICEBERG

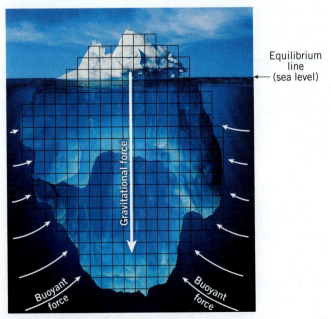

Figure 1.9 Buoyancy. Both a floating block of wood (**A**) and an iceberg (**B**) float in a denser liquid in a manner described by Archimedes' Principle. Refer to text for discussion. (Iceberg image © Ralph A. Clavenger/Getty Images. All rights reserved.)

Archimedes is the acceleration of gravity. Because both the block and the fluid are subject to the same acceleration, we can state an important result of Archimedes' Principle: the mass of a block that is floating in a fluid is equal to the mass of the fluid displaced by the block.

An object will sink if it is denser than the fluid it displaces. An object will float if it is less dense than the fluid it displaces. When the object achieves a motionless floating condition, it is balanced between the downward pull of gravity and the upward push of the buoyant force.

Isostasy

In the 1880s, geologists began to document the abundant evidence that levels of shoreline along lakes and oceans had changed often throughout geologic time in all parts of the world. Geologist Edward Suess proposed the hypothesis that changes in sea level can occur *if the volume of ocean water changes* in response to climate. Global atmospheric warming leads to sea level rise caused by melting of glaciers, and global atmospheric cooling leads to a drop in sea level as more of Earth's water gets stored in thicker glaciers. However, the American geologist Clarence Dutton suggested that shorelines can also change *if the level of the land changes* and the volume of water remains the same.

Dutton reasoned that if blocks of Earth's crust are supported by a viscous mantle beneath them that is capable of slowly flowing even though it is in the solid state (as Silly Putty™ can flow even though it is a solid), then the crustal blocks must float on the mantle according to Archimedes' Principle. Building on ideas first expressed in the 1850s by John Pratt and George Airy, Dutton proposed that Earth's crust consists of buoyant blocks of rock that float in gravitational balance in the top of the mantle. Dutton called this floating condition **isostasy**—a term derived from Greek for "equal standing." Loading a crustal block by adding lava flows, sediments, glaciers, or liquid water will decrease its buoyancy, and the block will sink (like pushing down on a floating wood block). Unloading materials from a crustal block will increase its buoyancy, and the block will rise.

You can think of isostasy as the equilibrium condition between any floating object and the more dense fluid in which it is floating. Gravity pulls the iceberg in **Fig. 1.9** down toward Earth's center, so the submerged root of the iceberg displaces water. The water is also acted on by gravity, generating a fluid pressure that increases with increasing depth. The base of the iceberg is pushed up by the fluid pressure, creating Archimedes' buoyant force. **Isostatic equilibrium** occurs when the buoyant force balances the downward force exerted by the iceberg. An **equilibrium line** separates the iceberg's submerged root from its exposed top.

Equations—The Math You Need

You can learn more about equations (including practice isostasy problems) at **http://serc.carleton.edu/mathyouneed/ equations/ManEqSP.html** featuring The Math You Need, When You Need It tutorials for students in introductory geoscience courses.

https://goo.gl/bt7D0Q

ACTIVITY 1.5

Summarizing Data and Imagining Crustbergs Floating on the Mantle, (p. 29)

Think About It How can we convert many observations into useful summary data?

Objective Contribute data to a dataset and gain some practice in obtaining descriptive statistics.

Before You Begin Read the following sections: Basic Number Management, Measuring Earth Materials, and Archimedes' Principle.

Plan Ahead This activity requires a laboratory balance for measuring mass, a graduated cylinder for measuring volume, and small samples of granite and basalt that can fit into the graduated cylinder.

ACTIVITY 1.6

Unit Conversions, Notation, Rates, and Interpretations of Data, (p. 31)

Think About It How can we represent data to help us interpret trends and implications?

Objective Practice analyzing, graphing, and interpreting data.

Before You Begin Read the following sections: Basic Number Management and Measuring Earth Materials.

Basic Number Management

Many scientists spend much of their investigative time working with numbers, so there are some fundamental skills that are worth learning about. If our goal and responsibility as scientists is to acquire reliable information about our world, we must always be concerned with the question of how reliable our information is likely to be. What is its uncertainty? And we must be careful about not representing quantitative information in a way that misleads others into thinking that we know something much better than we actually know it.

Significant Figures

There are three aspects of the broad topic of defining and interpreting significant figures that deserve particular attention as we begin our work in the physical geology labs: measurement, calculation, and interpretation of results.

Our ability to make accurate measurements is always limited by the maximum resolution of the method we use to measure, and perhaps by other factors related to measurement procedure. If all you have is a stick 1 m long and are asked to measure the height of a classmate, you won't be able to say much more than her or his height is between 1 and 2 m, about 2 m, or a bit greater than 2 m. If you received an upgrade to a tape measure with mm gradations marked on it, the resolution of your basic measuring device is improved. Now you need to consider whether you measure your classmate while standing or lying down (people are longer when lying down than when standing), in the morning or later in the day (people are longer in the morning after sleeping for hours), with the top of their head shaved or with hair (hair affects the measurement), and so on. The answer to "How tall are you" is not a single number because our height is variable within a small range every day.

The significant figures of the results of a calculation or series of calculations are limited by the number of significant figures in the numbers used in those calculations. Imagine that a person's length is measured independently by five careful observers using a mm scale, and their results are 1778 mm, 1779 mm, 1778 mm, 1777.5 mm, and 1778.5 mm. While most observers reported results to the nearest mm, two reported to the 10ths of a mm. The observer who reported a length of 1777.5 probably meant "somewhere between 1777 and 1778" rather than "between 1777.4 and 1777.6." If we use a calculator to compute the average of these lengths, the result it returns is 1778.2. If we report only significant figures, what number should be reported as the average? The resolution of the result is only as good as the input data with the least resolution, so the average should be reported as 1778. All of the observers would agree on the first three figures—177—but there is an uncertainty of about ± 1 in the last digit.

There is uncertainty in all measurements. By convention—an agreed-upon practice—scientists generally report measurements so that the last digit has an uncertainty of about ± 1 unit in that decimal place. For example, a measurement of 5.23 that is reported to the proper number of significant figures to the right of the decimal point should be interpreted to mean 5.23 ± 0.01. We infer that the most accurate value would lie somewhere between 5.22 and 5.24.

Some numbers are known exactly with no uncertainty. A number can be known exactly because of the way it is defined. For example, 1 in. is equal to 2.54 cm exactly, and so we could add an infinite number of zeros to the right of the "4" and all would be significant digits. Simple counting can yield exact numbers as well. For example, at a given time and date three live baby birds were observed in the nest. There were exactly 3, not 3 ± 1.

Rounding Numbers

Scientists generally follow a convention for rounding numbers to an appropriate number of significant figures. In the age of calculators that mindlessly present many digits more than are significant, knowing how to round numbers "to the nearest tenth" or "to the nearest million" is essential. We round the results of calculations prior to reporting them but do not round values obtained within a series of calculations that lead to the results.

Imagine that we have a set of numbers that we need to round to the nearest integer—the nearest whole number without any digits to the right of the decimal point and, in fact, without the decimal point.

1. If the first digit to the right of the decimal point is *smaller than 5*, then drop all digits right of the decimal point and leave the first digit to the left of the decimal point unchanged. For example, each number in the following set is rounded to 137: {137.0, 137.1, 137.2, 137.3, 137.4}.

2. If the first digit to the right of the decimal point is *larger than 5*, drop all digits to the right of the decimal point and add 1 to the first digit to the left of the decimal point (i.e., round up). For example, each number in the following set is rounded to 138: {137.6, 137.7, 137.8, 137.9}.

3. If the first digit to the right of the decimal point *is 5* and there are nonzero numbers to the right of the 5, then drop all digits right of the decimal point and add 1 to the first digit to the left of the decimal point (i.e., round up). For example, 134.51 rounds to 135.

4. If the first digit to the right of the decimal point *is 5*, there are no nonzero numbers to the right of the 5, and the digit to its left is an *even* number, then drop all digits right of the decimal point and leave the first digit to the left of the decimal point unchanged. Some examples follow.

130.5 rounds to 130	132.5 rounds to 132
134.5 rounds to 134	136.5 rounds to 136
138.5 rounds to 138	

5. If the first digit to the right of the decimal point *is 5*, there are no nonzero numbers to the right of the 5, and the digit to its left is an *odd* number, then drop all digits right of the decimal point and add 1 to the first digit to the left of the decimal point (i.e., round up). Some examples follow.

131.5 rounds to 132	133.5 rounds to 134
135.5 rounds to 136	137.5 rounds to 138
139.5 rounds to 140	

This convention minimizes the effect of rounding errors.

Exponential Notation (Base^{Exponent})

One consequence of living in the modern world is the need to be able to work with numbers that are very big and very small. The annual budget of the United States of America is reported to be on the order of 3.5 trillion dollars ($3,500,000,000,000). The size of a disease-causing bacterium is measured in microns (micrometers, one millionth of a meter), and our upcoming discussions of atoms and molecules will consider lengths on the picometer (one trillionth of a meter) scale.

Scientists use exponential notation to express very large and very small numbers. The number 10^5 is said to be in exponential form because it uses an exponent (5) to indicate that 5 copies of the base number (10) are multiplied together: $10 \times 10 \times 10 \times 10 \times 10$. Hence, the exponential number 10^5 is equal to 100,000.

Scientific Notation (Number × 10^Exponent)

A number represented in scientific notation has one nonzero number to the left of the decimal point, an appropriate number of significant figures to the right of the decimal point, and is multiplied by the appropriate base-10 exponential number. There is a distinct advantage to using the number "10" as a base. When we use 10 as our base, the exponent tells us the number of places we move the decimal point to the right or left of its original position in order to recover the explicit full-blown representation of the number. For example,

$1 \times 10^2 = 100$ move the decimal 2 places to the right of the 1

$1 \times 10^1 = 10$ move the decimal 1 place to the right of the 1

$1 \times 10^0 = 1$ don't move the decimal at all—leave it to the right of 1

$1 \times 10^{-1} = 0.1$ move the decimal 1 place to the left of the 1

$1 \times 10^{-2} = 0.01$ move the decimal 2 places to the left of the 1

But what if we need to use a number that is not expressed as a power of 10, such as our current best estimate of the age of the universe: 13,799,000,000 years $\pm$ 21,000,000 years? The age of the universe expressed in scientific notation is 1.3799×10^{10} years, $\pm 2.1 \times 10^6$ years, or expressed differently as $(1.3799 \pm 0.0021) \times 10^{10}$ years. To recover the age in explicit terms, find the exponent (10) and move the decimal point that number of places (10 places) to the right of its current position between the 1 and the 3. The result is 13,799,000,000 years.

We can also "go small" with scientific notation. The atomic radius of an oxygen atom has been measured as 0.000000000060 m, or 60 picometers. In scientific notation, the radius of an oxygen atom is 6.0×10^{-11} m. To recover the explicit form of the radius expressed in meters, we start by finding the exponent (-11) in the scientific notation, note that it is a negative number (which indicates that we need to move the decimal point to the left), and then move the decimal 11 places to the left from its current position between the 6 and 0. The result is 0.000000000060 m.

Simple Descriptive Statistics

We are unconsciously comfortable with the idea of summarizing sets of numbers into one or a few numbers. Meteorologists talk about the average temperature for this calendar day, students worry about their grade point average, baseball fans pay attention to batting average and on-base percentage, and so on. These are called **descriptive statistics**.

The **average** or **mean** is probably the most commonly used descriptive statistic. Imagine that we collect a large set of measurements of, say, the length of fossil specimens of a given extinct species. The average is simply the sum of all the lengths divided by the number of values in the set. The average has its limitations as a descriptive statistic. The average does not help us understand whether there was a wide range of lengths or whether they were all pretty much the same. What is the distribution of measurements on either side of the average? There are many questions one could ask about the data set that are not well answered by the average.

For example, the average per capita income in North Carolina is currently estimated to be between ~$25,000 and $28,000 per year. Michael Jordan's average salary during the 14 years after he graduated from the University of North Carolina with a BA degree in cultural geography is reported to have been over $6.5 million per year. If you were to determine the average salary of BA graduates in cultural geography from UNC during their first decade out of school, that average would be massively skewed by MJ's income, rendering the average effectively meaningless.

Among the most common ways we describe a set of data are the number of values in the set (N or n), the *extrema*—maximum value and minimum value—the most common value (the **mode**), the middle value of a set that is sorted by size (the **median**), the average value (the **mean**), and the **standard deviation** of the set of values. Most of these are quite simple to specify, especially if the set of numbers has been sorted by size. The standard deviation requires a bit more effort to calculate and is a measure of how widely the numbers in the set diverge from the average. If all values in a set of numbers were the same, the standard deviation would be 0, but the more they differ from the average, the greater the standard deviation.

The **sample standard deviation** (s) is the version of the standard deviation that is used when you are not working with an extremely large set of numbers and is given by the equation

$$s = \sqrt{\left(\frac{1}{N-1}\right)\left(\sum_{i=1}^{N}(x_i - \overline{x})^2\right)}$$

For many, that equation is about as clear as mud, so let's simply write out a series of steps to compute the sample standard deviation.

1. Breathe normally and smile. Hum a jaunty tune quietly if you wish. Know that you are not alone in this world, and that you are loved, so relax.

2. Count the number of values in the set. We will represent that number by the variable N.

3. Add all of the values together and divide the sum by *N*. That is, find the average of the set of numbers. We will represent the average by the symbol $\bar{x}$.

4. Subtract the average $\bar{x}$ from the first value in the set and then square the result (i.e., multiply the result by itself). Write the resulting number down. Do the same thing (subtracting $\bar{x}$, squaring the result, and writing down the resulting number) with each of the other numbers in the set. You have now written-down a new, second set of numbers.

5. Add the values in the second set of numbers (the ones you computed in step 4) together, resulting in one number—the sum of the second set of numbers. In the equation, this step is indicated by the use of the Greek capital letter sigma: Σ.

6. Divide the sum computed in step 5 by the quantity $(N-1)$.

7. Compute the square root of the answer from step 6. Nobody computes square roots by hand anymore, so find a calculator and press the appropriate buttons.

Let's practice finding these basic descriptive statistics using a small set of numbers that are already sorted in order of increasing size: {198, 234, 239, 242, 243, 243, 254, 270}. The results are presented as you might see them on a calculator screen.

- Number of values in the set (*N*): 8
- *Extrema*: minimum is 198, maximum is 270
- Average or mean: 240.375, rounded to 240
- Median (middle value): halfway between the middle pair of values, 242 and 243, or 242.5. (Had this set contained an odd number of values, the median would have been the middle value of the sorted set.)

ACTIVITY 1.7

Scaling, Density, and Earth's Deep Interior, (p. 35)

Think About It   How does a variation in density help us understand the broad structure of Earth?

Objective   Practice scaling from thousands of kilometers to millimeters on a page, and discover a few fundamental characteristics of Earth's interior structure.

Before You Begin   Read the following sections: Isostasy and Earth's Global Topography, Measuring Earth Materials, and Basic Number Management.

Plan Ahead   You will need to work with a drafting compass, metric ruler, and a calculator.

- Mode (the number that appears the most times in the set): 243
- Sample standard deviation (*s*): 20.4166 or 20, rounded to the nearest integer.

You will have a chance to practice using descriptive statistics in Activity 1.5 in which you will describe the density of specimens of granite and basalt (**Fig. 1.10**). Granitic rock (granite, granodiorite, diorite, tonalite) is commonly used to represent the average composition of Earth's solid outermost layer, the crust, in the continental areas that are above sea level. Basaltic rock (basalt, gabbro, diabase) is

basaltic rock

granitic rock

0   5 cm

Figure 1.10 Average composition of oceanic and continental crust. A. Below a covering of sediment and sedimentary rock, the oceanic crust is composed of rock with a composition like this dark basalt. **B.** Continental crust has an overall composition similar to this light-toned granitic rock, although the crust is composed by many different rock types.

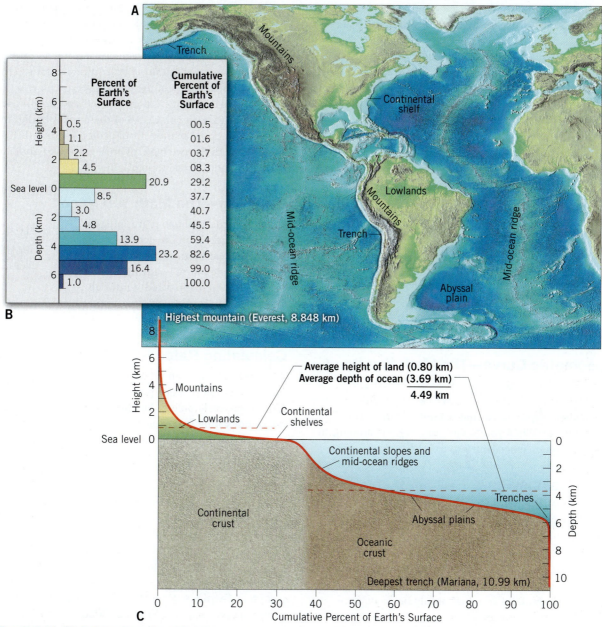

Figure 1.11 Global topography of Earth. A. Relief map of the solid upper surface of Earth with different elevations marked by different colors. **B.** Histogram of global topography. **C.** Hypsometric curve (or hypsographic curve) of Earth's global topography.

characteristic of the oceanic crust that is beneath most of the ocean basins and forms more than 60% of Earth's outer surface. We will learn more about these rock types in future labs.

Isostasy and Earth's Global Topography

Clarence Dutton applied his isostasy hypothesis in 1889 to explain how the shorelines of lakes or oceans could be elevated by vertical motions of Earth's crust. At that time, little was known about Earth's mantle or topography of the seafloor. Modern data show that isostasy has broader application for understanding global topography.

Global Topography: The Hypsometric Curve

A variety of technologies carried aboard aircraft and orbital satellites now measure the shape of Earth's outer surface very exactly. We can even use satellite data to provide general bathymetric maps of the seafloor, although accurate seafloor maps require shipborne sensors. These improved digital elevation models of Earth allow us to produce relief maps like the image in **Fig. 1.11A** of the solid outer surface of Earth.

https://goo.gl/9QMxsc

Color is used to indicate elevation on **Fig. 1.11A**, and the explanation for the different colors is contained within the bar chart (histogram) in **Fig. 1.11B**. For example, all of the points on Earth's solid surface between sea level and 1 km above sea level are shown in green, composing about 20.9% of Earth's total surface. Notice that the histogram has two bars that are longer than the rest, so we say that the distribution of elevations is *bimodal*. One of the elevation modes corresponds to elevations between sea level and 1 km above sea level and corresponds to most areas on continents and islands. The other elevation mode occurs between 4 and 5 km below sea level and corresponds to most of the ocean floor.

Fig. 1.11C is called a **hypsometric curve** (or *hypsographic curve*) and shows the cumulative percentage of Earth's surface that occurs at specific elevations or depths in relation to sea level. (The word "hypsometric" is based on the Greek *húpsos* or *hypsos* for height.) This curve is not the profile of a continent because it represents Earth's entire surface. Notice that the cumulative percentage of land is only 29.2% of Earth's surface, and most of the land is lowlands. The remaining 70.8 cumulative percent of Earth's surface is covered by ocean, and most of the seafloor is more than 3 km deep.

Hypsometric Curve—The Math You Need

You can learn more about the hypsometric curve and how to read and use it at **http://serc.carleton.edu/mathyouneed/hypsometric/index.html** featuring The Math You Need, When You Need It tutorials for students in introductory geoscience courses.

Global Isostasy

The average elevation of the continents is about 0.8 km above sea level, but the average elevation of the ocean basins is 3.69 km below sea level. Therefore, the difference between the average continental and ocean basin elevations is 4.49 km! If the continents did not sit so much higher than the floor of the ocean basins, then Earth would have no dry land and there would be no humans. What could account for this elevation difference? One clue may be the difference between crustal granite and basalt in relation to mantle peridotite.

The outermost of Earth's solid layers, the crust, occurs in two different types, each with a distinctive composition. *Basaltic rock* forms the crust of the oceans beneath a thin veneer of sediment. The average composition of continental crust is thought to be similar to a granitic rock, although the crust is actually composed of many different rock types along with a thin veneer of sediment. (We will learn much more about these broad types of rock in future labs.) You might think of the continents (green and brown in **Fig. 1.11A**) as granitic islands surrounded by a low sea of basaltic ocean crust (blue). All of these rocky bodies rest on the solid upper mantle. Could compositional differences between the two types of crust help us to understand Earth's bimodal global topography?

Graphing—The Math You Need

You can learn more about graphing and how to use graphs in the geosciences at **http://serc.carleton.edu/mathyouneed/graphing/index.html** featuring The Math You Need, When You Need It tutorials for students in introductory geoscience courses.

https://goo.gl/6CENFY

Calculating Rates—The Math You Need

You can learn more about calculating rates (including practice problems) at **http://serc.carleton.edu/mathyouneed/rates/index.html** featuring The Math You Need, When You Need It tutorials for students in introductory geoscience courses.

https://goo.gl/5NHJfl

MasteringGeology™

Looking for additional review and test prep materials? Visit the Study Area in MasteringGeology to enhance your understanding of this chapter's content by accessing a variety of resources, including Pre-Lab Videos, Self-Study Quizzes, Geoscience Animations, Mobile Field Trips, *Project Condor* Quadcopter videos, *In the News* articles, glossary flashcards, web links, and an optional Pearson eText.

Activity 1.1

Name: _____ Course/Section: _____ Date: _____

This activity requires use of Google Earth on a device that has an active connection to the Internet. You will need to be able to search, zoom in and out of the image, and find the apparent altitude at which the Google Earth image is being viewed. You will also need to be able to input a degree symbol (°) for which you can get help from your teacher. It is useful to choose to view the labels for various places and to look at photos that have been posted on Google Earth because of the additional information that sometimes accompanies the photos.

A Four types of map coordinates are provided in **Fig. A1.1.1**: three versions of latitude and longitude and one UTM. For each, input the coordinates into Google Earth using its Search function, press the "Search" button to go to the spot, identify the feature at that location, note where it is (country, continent, or island), and write a brief description in the space provided in the last column. You will probably want to zoom out to view the surroundings after Google Earth finds each location for you.

Coordinate Style	Coordinates		Google Earth Seach Input	Description—Be sure to zoom out and look around before answering
	Latitude	Longitude		
decimal degrees	27.9881°N	89.9253°E	27.9881, 86.9253	Continent/Island: Country: What did you find there?
degrees–decimal minutes	77°50.680′S	166°40.603′E	–77°50.680′, 166°40.603′	Continent/Island: Country: What did you find there?
degrees–minutes–seconds	63°04′10.2″N	151°00′26.64″W	63°04′10.2″, –151°00′26.64″	Continent/Island: Country: What did you find there?

Coordinate Style	Zone	Latitude Band	Easting	Northing	Google Earth Seach Input	Description—Be sure to zoom out and look around before answering
Universal Transverse Mercator (UTM)			Easting	Northing	31U 448252 5411935	Continent/Island: Country: What did you find there?
	31	U	448252mE	5411935mN		

Figure A1.1.1

B Find an interesting place on Earth, and let us know what you found by providing the information requested in **Fig. A1.1.2**. Share your discoveries with other students and your teacher.

Longitude & Latitude	Best-View Altitude (km)	Continent/Island	Description

Figure A1.1.2

C The latitude–longitude coordinates for several interesting sites are provided in **Fig. A1.1.3** in decimal degrees along with the best altitude to view each of the features. Choose a few sites to investigate using Google Earth, and record your results in **Fig. A1.1.4**.

Site #	Latitude, Longitude	Best-View Altitude	Site #	Latitude, Longitude	Best-View Altitude	Site #	Latitude, Longitude	Best-View Altitude
1	40.5230, –112.1510	5–15 km	9	32.2186, 35.5661	10 km	17	–3.0647, 37.3584	10–75 km
2	–21.1492, 14.5775	50 km	10	35.2715, –119.8275	2 km	18	35.2505, –75.5288	1–2000 km
3	19.4726, –155.5918	15–150 km	11	4.1744, 73.5097	2–600 km	19	0.2220, –50.3950	300 km
4	56.3333, –79.5000	150 km	12	–30.5450, 138.7280	25–100 km	20	45.9764, 7.6583	15–25 km
5	21.8462, 54.1514	1–200 km	13	40.8216, 14.4260	2–10 km	21	37.5940, –122.4240	10 km
6	37.7460, –119.5336	5–25 km	14	35.0275, –111.0228	5 km	22	57.2688, –4.4921	50 km
7	36.0999, –112.0994	10–75 km	15	59.0850, –136.0615	5–20 km	23	21.1240, –11.4020	100 km
8	24.9652, –76.4201	75 km	16	29.1200, 25.4300	1–150 km	24	62.7924, –164.1667	95–500 km

Figure A1.1.3

Site #	Continent/ Island	Country (± State)	Description — Be sure to zoom out to the "best-view altitude" and look around before answering.

Figure A1.1.4

D **REFLECT & DISCUSS** Turn off the function in Google Earth that displays national or state borders and place names. Navigate to 54.1291, −7.3064 and examine the area from a eye altitude of ~4 km.

1. Where do you think the national border is in this image?

2. Have Google Earth display the border, and zoom out to a higher elevation. What countries are separated by this border? _____ and _____. What is your impression of this national border?

3. Now turn off the borders again, navigate to −9.8396, −66.3362, and zoom out to an eye altitude of ~100 km. How does land use differ across the border?

4. Have Google Earth display the border, and determine which countries are separated by the border: _____ and _____. What natural feature marks the border on the landscape?

Name: _____ **Course/Section:** _____ **Date:** _____

A Air samples have been collected at the Mauna Loa Observatory (MLO) on the main island of Hawaii since the late 1950s and analyzed to determine the concentration of carbon dioxide (CO_2) in the atmosphere. Located at 3397 m above sea level on the north slope of the Mauna Loa Volcano, this remote sampling site is high enough so that contamination is minimized. The location of MLO is shown by the yellow-filled circle in **Fig. A1.2.1**, which is a map that uses the geographic coordinate system of decimal latitude and longitude. One of the orange lines through the site extends north–south along a meridian—an arc of equal longitude—and the other extends east–west along a parallel—a circle of equal latitude (**Fig. 1.4**).

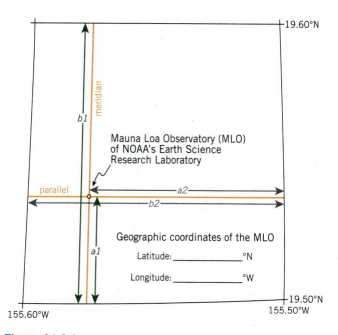

Figure A1.2.1

1. Latitude 19.50°N is the minimum latitude shown on the map. Carefully measure the distance along the orange line from the 19.50°N parallel to the MLO site. We will call that map distance $a1$. $a1 =$ _____ cm

2. Measure the total map distance along that same orange line from the 19.50°N parallel through the MLO site to the 19.60° parallel and call that map distance $b1$. $b1 =$ _____ cm

3. We know the difference in latitude from the bottom of the map at 19.50°N to the top of the map at 19.60°N and call that distance $d1$. $d1 =$ _____ °

4. We would like to know the difference in latitude between 19.50°N and the MLO site and will call this unknown quantity $c1$. The ratio of $a1$ to $b1$ is the same as the ratio of $c1$ to $d1$. Use what you learned about proportions by reading the text earlier in this chapter to find $c1$. $c1 =$ _____ °

5. Determine the latitude of the MLO site (19.50°N + $c1$) _____ °N

6. Just in case you come across this situation again in the (near) future, write a general equation to solve this type of equation directly using the variable names a, b, c, and d: $c =$ _____

   Follow the same general procedure to find the longitude of the MLO site.

7. Measure the distance from longitude 155.50°W to MLO ($a2$), recognizing that longitude 155.50°W is the minimum longitude shown on the map: $a2 =$ _____ cm.

8. Map the distance from 155.50°W through MLO to 155.60°W ($b2$): $b2 =$ _____ cm.

9. The difference in longitude between 155.50°W and 155.60°W ($d2$): $d2 =$ _____ °

10. Use your general equation from part **A6** to find the value of $c2$ (that is, the difference in longitude between 155.50°W and the MLO site). $c2 =$ _____ °

11. Determine the longitude of the MLO site (155.50°W + $c2$). _____ °W

B Part of the USGS 7.5-minute topographic quadrangle map of the Lower Geyser Basin, Wyoming (2015), is reproduced in **Fig. A1.2.2**. On the topographic map, put a dot in the center of the blue oval representing the *Grand Prismatic Spring* (**Fig. 1.3**). We are going to build on what you learned in part A to find the UTM coordinates of the center of the Grand Prismatic Spring.

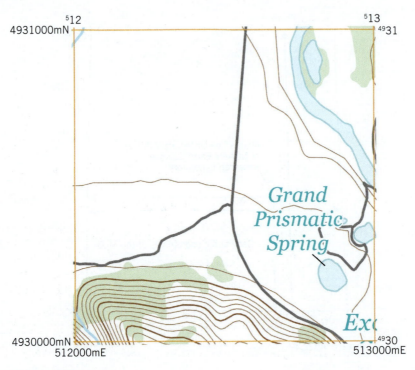

Figure A1.2.2

1. Regarding a first look at the map in **Fig. A1.2.2**

   (a) What is the minimum easting on the map? _____ mE. What edge of the map is bounded by the minimum easting (top, bottom, left, right)? _____

   (b) What is the minimum northing on the map? _____ mN. What edge of the map is bounded by the minimum northing (top, bottom, left, right)? _____

   (c) Is the central meridian for this zone to the east (right) or west (left) of this map area? (Refer to **Fig. 1.5** for assistance.) _____

2. The yellow square that surrounds most of the map in **Fig. A1.2.2** represents a horizontal area on the ground at Yellowstone National Park that is 1000 m by 1000 m (1 km × 1 km) in dimension. We can use a side of that square like a bar scale.

   (a) As accurately as you can, measure the length of a side of the yellow square on the map in cm; we'll use the variable name $a3$ for this map length. $a3 =$ _____ cm

   (b) We will use the variable name $b3$ for the ground distance along one of the sides of the yellow square—1 km—but expressed *in the same units* as we used for $a3$. That is, $b3$ is the number of cm in 1 km. There are 100 cm in 1 m, and 1000 m in 1 km. $b3 =$ _____ cm

   (c) The fractional scale of this map can be found by dividing $b3$ by $a3$—we'll call that result $e3$. The fractional scale is $1/e3$ or $1:e3$. $1/$_____ or $1:$_____.
   That means that 1 length unit of any sort (cm, for example) on the map represents $e3$ length units of the same sort (cm) on the ground.

3. Find the UTM coordinates of the middle of Grand Prismatic Spring.

  (a) Measure the map distance from the side that represents the minimum easting of the yellow square ($c3$).

   $c3 =$ _____ cm

  (b) Use the fractional scale of the mathematical expression you developed in part **B2(c)** to find the ground distance between the minimum easting and the middle of Grand Prismatic Spring ($d3 = c3 \times e3$). $d3 =$ _____ cm

  (c) Convert $d3$ into meters, and call the result $d3'$. $d3' =$ _____ m

  (d) What is the easting of Grand Prismatic Spring (minimum easting $+d3'$)? _____ mE

   Follow the same general procedure to find the northing of Grand Prismatic Spring.

  (e) Measure the map distance from the side that represents the minimum northing of the yellow square ($c4$).

   $c4 =$ _____ cm

  (f) Use the fractional scale of the mathematical expression you developed in part **B2(c)** to find the ground distance between the minimum northing and the middle of Grand Prismatic Spring ($d4 = c4 \times e3$).

   $d4 =$ _____ cm

  (g) Convert $d4$ into meters, and call the result $d4'$. $d4' =$ _____ m

  (h) What is the northing of Grand Prismatic Spring (minimum northing $+d4'$)? _____ mN

  (i) The UTM coordinates of the middle of Grand Prismatic Spring are

   12T _____ mE _____ mN

A Part of the USGS 7.5-minute topographic quadrangle map of Old Faithful, Wyoming (2015), is reproduced in **Fig. A1.3.1**. We are going to plot the location of the active vent of the Old Faithful geyser. As of mid-August 2015, the active vent was located approximately at UTM coordinates 12T 513671 mE 4923032 mN.

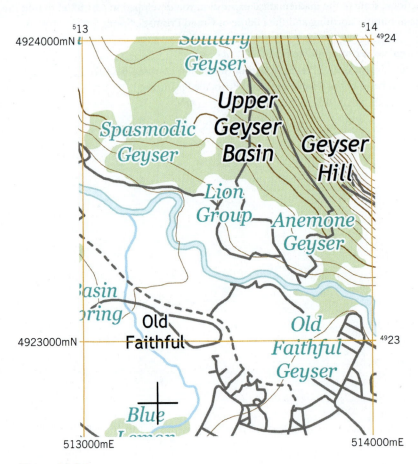

Figure A1.3.1

1. Regarding a first look at the map in **Fig. A1.3.1**

   (a) What is the minimum easting on the map? _____ mE. What edge of the map is bounded by the minimum easting (top, bottom, left, right)? _____

   (b) What is the minimum northing listed on the map? _____ mN. What edge of the map is bounded by the minimum northing (top, bottom, left, right)? _____

   (c) Is the central meridian for this zone to the east (right) or west (left) of this map area? (Refer to **Fig. 1.5** for assistance.) _____

2. The yellow square that surrounds most of the map in **Fig. A1.3.1** represents a horizontal area on the ground that is 1 km by 1 km in dimension, so we can use a side of that square as a bar scale.

   (a) As accurately as you can, measure the length of a side of this square in cm; we'll use the variable name $a5$ for this length. $a5 =$ _____ cm

   (b) We will use the variable name $b5$ for the number of cm in 1 km. $b5 =$ _____ cm

   (c) The fractional scale of this map can be found by dividing $b5$ by $a5$—we'll call that result $e5$. The fractional scale is $1/e5$, or 1:$e5$. 1/_____ or 1:_____. That means that 1 length unit of any sort (cm, for example) on the map represents $e5$ length units of the same sort (cm) on the ground.

3. Plotting the vent location: 12T 513671 mE 4923032 mN.

   (a) Based on the UTM coordinates of the vent, how far north of the 4923000 mN line is the vent *on the ground* at Yellowstone National Park? _____ m

   (b) Using the fractional scale, how far up from the yellow 4923000 mN line is the vent *on the map*? _____ cm

   (c) Draw a line on the map that is that distance above the 4923000 mN line.

   (d) Based on the UTM coordinates of the vent, how far east of the 513000 mE line is the vent? _____ m

   (e) Using the fractional scale, how far to the right of the yellow 513000 mE line is the vent on the map? _____ cm

   (f) Draw a line on the map that is that distance to the right of the 513671 mE line.

   (g) The active vent is located where the two lines you drew intersect. Put a small circle around that point, and add an appropriate label to the map near that point to identify it as the active vent.

B REFLECT & DISCUSS Use Google Earth to navigate to the active vent at 12T 513671 4923032 (or latitude 44.46046, longitude −110.82815). Think like a geoscientist and ignore all of the human-built modifications to the landscape. Now, observing just the natural landscape, what do you see around the Old Faithful geyser? What is the color or tone of the material around the geyser? What do the patterns of small streams around the geyser look like? Use your description to find other geysers in the area, and record the location (UTM or lat–long) of at least one other geyser you found using your description of Old Faithful.

Name: _____ Course/Section: _____ Date: _____

A Imagine that you have a solid that is square on the top and bottom and either a square or rectangle on all the other sides (i.e., it is a cube or rectangular prism) like the one shown in column A of **Fig. A1.4.1**, whose sides have lengths a, b, and c, and the b and c are of equal length.

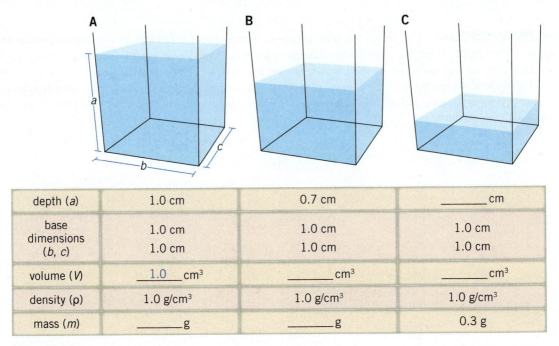

	A	B	C
depth (a)	1.0 cm	0.7 cm	_____ cm
base dimensions (b, c)	1.0 cm 1.0 cm	1.0 cm 1.0 cm	1.0 cm 1.0 cm
volume (V)	__1.0__ cm³	_____ cm³	_____ cm³
density (ρ)	1.0 g/cm³	1.0 g/cm³	1.0 g/cm³
mass (m)	_____ g	_____ g	0.3 g

Figure A1.4.1

1. Now imagine that you know the volume of the solid (V, measured in cm³) and the length of two of the three sides, b and c (measured in cm). Explain in words or with an equation how you can calculate the length of the remaining side, a.

2. The relationship between mass, density, and volume is that mass (m, measured in grams, g) is the product of density (ρ, measured in g/cm³) times volume (cm³): $m = \rho \times V$. Recall that the Greek letter rho (ρ) is used to represent density.

   (a) If you know mass and density, you can calculate the volume. How?

   (b) If you know volume and density, you can calculate mass. How?

3. Use the relationships you just described to compute the values needed to fill in the blanks in the table in **Fig. A1.4.1**.

4. Imagine a rectangular prism of pure water whose density (ρ) is 1.0 g/cm³. The base of this imaginary prism is 1 cm on a side, and the remaining side of the prism is the water depth (a) measured in cm. The mass (m) of that prism of water, expressed in grams, is related to the water depth (a) in the following way: the value of a is _____ the value of m.

B One way of restating Archimedes' Principle is that the mass of a block that is floating in a fluid is equal to the mass of the fluid displaced by the block. Imagine that we have a solid block in which all of the sides are 1 cm long—a cube with a volume of 1 cm³ (**Fig. A1.4.2A**). We set it in a beaker of pure water and find that, when the block and water come to rest, 70% of the block is submerged under water, so a = 0.7 cm and b = 1.0 cm (**Fig. A1.4.2B**).

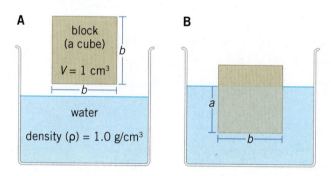

Figure A1.4.2

1. What is the volume of water displaced by the block? _____ cm³

2. The density of the water (ρ_{water}) is 1.0 g/cm³, so what is the *mass* of water displaced by the block? Multiply the volume of water by the density of the water. The mass of water displaced by the block is _____ g.

3. Using the previous answer, apply Archimedes' Principle to find the *mass* of the block. The mass of the entire block (m) is _____ g.

4. The block is a cube with a volume of 1 cm³. Now that you know the mass and the volume of the block, what is the *density* of the block (ρ_{block})? ρ_{block} = _____ g/cm³

5. The submerged part of the block is 70% of the total volume of the block. What is the ratio of the density of the block to the density of the water; that is, what is ρ_{block} divided by ρ_{water}? Express your answer as a percentage (e.g., 0.7 = 70%). $\rho_{block}/\rho_{water}$ = _____ %

6. Using words rather than numbers, compare the ratio of the volume of the submerged part of the block ($V_{submerged}$) to the total volume of the block (V_{total}) with the ratio of the density of the block (ρ_{block}) to the density of the water (ρ_{water}). The ratio $V_{submerged}/V_{total}$ is _____ the ratio $\rho_{block}/\rho_{water}$.

C If your teacher has provided a block of wood and a beaker or bowl of water, try the floating-block experiment yourself.

1. Carefully measure the three dimensions of the wood block and record the lengths.

_____ cm _____ cm _____ cm

2. Determine the volume of the block. V_{block} = _____ cm^3

3. If one is available, use a pan balance to measure the mass (m) of the block. m_{block} = _____ g. Now determine the density of the block (ρ_{block}), which is equal to the volume divided by the mass (V_{block}/m_{block}). ρ_{block} = _____ g/cm^3

4. Gently place the block in water and wait until it comes to rest. Mark the waterline on the block, and measure the distance the block extended below the waterline.

_____ cm

5. Determine the volume of the block that was submerged. _____ cm^3

6. Calculate the ratio of the volume of the submerged part of the block to the total volume of the block. _____

7. Use Archimedes' Principle to find the density of the block. _____ g/cm^3

8. Compare your answers for parts **C3** and **C7**, and comment on which seems like it might be a more reliable or simpler method to measure the density of the block.

D The density of water ice in icebergs is 0.917 g/cm^3. The average density of ocean water varies with temperature and salinity (saltiness), but we will assume a density of 1.025 g/cm^3.

1. Use Archimedes' Principle to calculate how much of an iceberg is submerged below sea level. Show your work.

2. Use Archimedes' Principle to calculate how much of an iceberg is exposed above sea level. Show your work.

3. Notice the graph paper grid overlay on the picture of an iceberg in **Fig. 1.9B**. Use this grid to determine and record the cross-sectional area of this iceberg that is below sea level and the cross-sectional area that is above sea level by adding together all of the whole boxes and fractions of boxes that overlay the root of the iceberg or the exposed top of the iceberg. Use these data to calculate the percentage of the iceberg that is below sea level and the percentage that is above sea level. How do your results compare to your calculations in steps **D1** and **D2**?

4. What do you think might happen as the top of the iceberg melts?

E **REFLECT & DISCUSS** How much does the melting of an iceberg floating in the ocean contribute to sea level rise. (*Hint:* Does the liquid level change when an ice cube floating in a glass of liquid water melts?)

Summarizing Data and Imagining Crustbergs Floating on the Mantle

Activity 1.5

Name: ______________________ **Course/Section:** ______________ **Date:** __________

A *As exactly as you can,* determine the mass and volume of a small sample of basalt. Use a laboratory balance to measure the sample's mass, and use the water-displacement method with a graduated cylinder to determine the volume (**Fig. 1.8**). Add your data to the basalt density chart (**Fig. A1.5.1**). Calculate the density of your sample of basalt to 10ths of a g/cm^3. Then determine the descriptive statistics for all 10 lines of sample data in the basalt density chart.

BASALT DENSITY CHART

Basalt Sample Number	Sample Mass (g)	Sample Volume (cm³)	Sample Density (g/cm³)
1	40.5	13	3.1
2	29.5	10	3.0
3	46.6	15	3.0
4	31.5	10	3.2
5	37.6	12	3.1
6	34.3	11	3.1
7	78.3	25	3.1
8	28.2	9	3.1
9	55.6	18	3.1
10			

Density of basalt

Average or Mean = ______ g/cm³ Median = ______ g/cm³ Mode = ______ g/cm³ Standard Deviation = ______ g/cm³

Figure A1.5.1

B *As exactly as you can,* determine the mass and volume of a small sample of granite. Follow the same methodology for measuring the granite that you used in the previous section. Add your data to the granite density chart (**Fig. A1.5.2**). Calculate the density of your sample of granite to 10ths of a g/cm^3. Then determine the descriptive statistics for all 10 lines of sample data in the granite density chart.

GRANITE DENSITY CHART

Granite Sample Number	Sample Mass (g)	Sample Volume (cm³)	Sample Density (g/cm³)
1	32.1	12	2.7
2	27.8	10	2.8
3	27.6	10	2.8
4	31.1	11	2.8
5	58.6	20	2.9
6	62.1	22	2.8
7	28.8	10	2.9
8	82.8	30	2.8
9	52.2	20	2.6
10			

Density of granite

Average or Mean = ______ g/cm³ Median = ______ g/cm³ Mode = ______ g/cm³ Standard Deviation = ______ g/cm³

Figure A1.5.2

C Geoscientists who study the propagation of earthquake energy through Earth's interior—seismologists—indicate that the average density of the uppermost upper mantle is around 3.3 g/cm³. Where it is hot enough, the mantle is able to flow.

1. Seismologists indicate that the average thickness of basaltic ocean crust is about 7 km. As a thought experiment, imagine that a particular bit of solid oceanic crust with the same bulk density that you calculated for basalt is floating in a sea of upper mantle material that has a bulk density of 3.3 g/cm³ and that is hot enough to flow. (In fact, most of the oceanic crust is on top of a relatively cool, solid part of the upper mantle that does not flow, but let's not spoil the fun with details.) Use what you have learned about Archimedes' Principle, buoyancy, and isostasy to estimate (i.e., to calculate) how high (in km) basalt would float in a viscous, flowing upper mantle. Show your work.

   The upper surface of oceanic crust in this imaginary world would be about _____ km above the level of the upper mantle.

2. Seismologists indicate that the average thickness of granitic continental crust is about 35 kilometers. Make the same kind of calculation you just completed in section **C1**, but this time, use the average density you determined for granite to estimate how high granitic continental crust would float in a viscous, flowing upper mantle. Show your work.

   The upper surface of continental crust in this imaginary world would be about _____ km above the level of the upper mantle.

3. What is the difference (in km) between your answers in **C1** and **C2**?

4. How does this difference between **C1** and **C2** compare to the actual difference between the average height of continents and average depth of oceans on the hypsographic curve (**Fig. 1.11C**)?

D **REFLECT & DISCUSS**   Reflect on all of your work in this laboratory so far. Explain why Earth has a bimodal global topography.

E **REFLECT & DISCUSS**   How is a mountain like the iceberg in **Fig. 1.9B**?

Name: _____ **Course/Section:** _____ **Date:** _____

A Make the following unit conversions using the Mathematical Conversions chart on page xx.

1. 10 mi. = _____ km

2. 1 ft. = _____ m

3. 16 km = _____ m

4. 25 m = _____ cm

5. 25.4 mL = _____ cm³

6. 1.3 L = _____ cm³

B Write these numbers using scientific notation

1. 6,555,000,000 = _____

2. 0.000001234 = _____

C RATES

1. Our current best estimate is that the western Grand Canyon in Arizona began to be excavated by river erosion about 6 million years ago. The greatest depth of the Grand Canyon is about 1.6 km.

   (a) What is the mean (average) rate that the Grand Canyon has been eroded into the Colorado Plateau during the past 6 million years, expressed in millimeters per year? Show your work below.

   (b) If erosion of the Grand Canyon proceeded at that rate during the most recent century, how much deeper is it today than it was on the day you were born? Show your work.

2. During Earth's very early history, our planet was whacked by meteorites large and small, and eventually grew to its current size. Earth probably went through a period in which it was molten from near the surface to its center. Throughout its history, Earth has been hotter at its center than at its outer surface.

   (a) In caves just below Earth's ground surface in continental crust, the temperature tends to be a constant ~15°C (~59°F). (That's why people use caves for wine storage.) The deepest mine on Earth is currently the Mponeg gold mine just southwest of Johannesburg, South Africa, which reaches just over 4 km below the surface. At the bottom of that mine, the rock temperature reaches 66°C (~151°F). Using the cave temperature to represent the surface temperature of the crust at ~0 km depth, what is the rate at which temperature changes in Earth between 0 and 4 km—the near-surface *geothermal gradient*? _____ °C/km

(b) The temperature at the bottom of the lithosphere in Earth is often inferred to be around 1300°C. If we assume a depth to the base of the lithosphere of 100 km, what is a reasonable estimate for the geothermal gradient between 0 km and ~100 km? _____ °C/km

(c) The center of Earth at an average depth of 6371 km below the surface has a temperature that has been estimated to be approximately 6000°C. What is the average geothermal gradient from Earth's surface to its center? _____ °C/km

(d) Write a brief statement of a hypothesis you think might best explain the variation in the geothermal gradient within Earth.

D **SINGLE-LINE GRAPH** The amount of CO_2 in the atmosphere has been monitored at Mauna Loa Observatory, Hawaii, since the late 1950s, initiated by Dave Keeling of the Scripps Institution of Oceanography and continued in cooperation with the U.S. National Oceanic and Atmospheric Administration (NOAA). A selection of data from this ongoing study is presented in **Fig. A1.6.1**, showing how the concentration of CO_2 in ppmv (parts per million volume) has changed per decade since 1959. The source of the data is **http://www.esrl.noaa.gov/gmd/ccgg/trends/**.

1. Round each of the CO_2 concentration values to the nearest integer value, and write the rounded numbers on the lines provided in the table of **Fig. A1.6.1**. To get you started, the first rounded value is 316.

2. Plot the data onto the graph as carefully as you can.

3. Use a ruler to draw a line that follows the trend of the points. To do this, you will need to visually approximate by drawing the line so that it passes through the middle of the points with about as many points above as below the line.

4. What does the slope of the line you drew through the data tell you about the change in the atmospheric concentration of CO_2 as observed at the Mauna Loa Observatory?

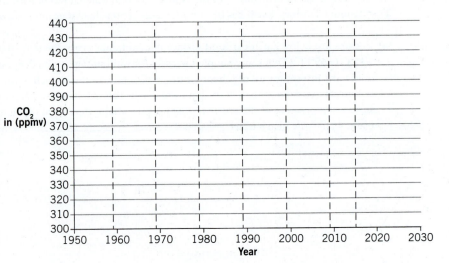

Annual Average Concentration of Atmospheric Carbon Dioxide (CO_2) at Mauna Loa Observatory, Hawaii		
Year	CO_2 (ppmv)	Rounded to Integer
1959	315.97±0.12	_____
1969	324.62±0.12	_____
1979	336.78±0.12	_____
1989	353.07±0.12	_____
1999	368.33±0.12	_____
2009	387.37±0.12	_____
2015	400.83±0.12	_____

Figure A1.6.1

E **BAR GRAPH** We will use a different type of graph to derive additional information from the CO_2 data (**Fig. A1.6.2**).

Average Yearly Rate of Increase in the Concentration of Atmospheric Carbon Dioxide (CO_2) at Mauna Loa Observatory, Hawaii	
Time interval	Rate of increase per year
1959–1969	0.9
1969–1979	
1979–1989	
1989–1999	
1999–2009	

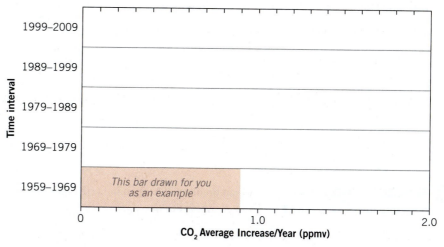

Figure A1.6.2

1. Calculate the average rate of increase in CO_2 concentration per year for the time intervals 1959–1969, 1969–1979, 1979–1989, 1989–1999, and 1999–2009, and write the results in the spaces provided in **Fig. A1.6.2**. The value for 1959–1969 is provided for you as an example.

2. Plot the results as a bar graph. The 1959–1969 bar is plotted for you.

3. Briefly describe how you interpret the information you plotted in the bar graph.

F **TWO-LINE GRAPH** Two different datasets are plotted as a function of time in **Fig. A1.6.3**, both obtained by analysis of an ice core from Vostok Station, Antarctica (**https://www.ncdc.noaa.gov/data-access/paleoclimatology-data/datasets/ice-core**). The blue line tracks the change in temperature at Vostok Station relative to the present, and the relative

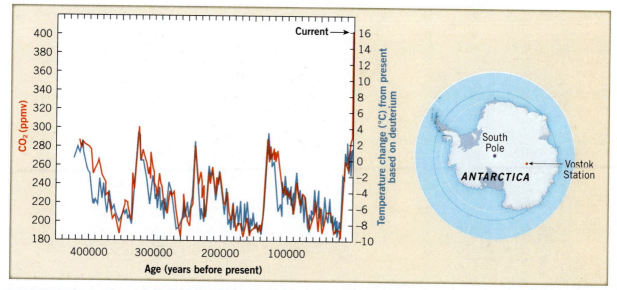

Figure A1.6.3

temperature scale is printed on the right vertical axis of the graph. Positive values indicate warmer than present, and negative values are cooler than present. The red line tracks the concentration of carbon dioxide at Vostok Station, and the corresponding scale is on the left vertical axis. Both plots share the same time scale on the horizontal axis.

1. What relationship between temperature and carbon dioxide concentration is revealed by this graph?

2. What kinds of additional information would you like to have to investigate further the possible correlation between CO_2 concentration in the atmosphere and temperature?

G **REFLECT & DISCUSS** What do you predict will happen to Earth's atmospheric temperature in the future? How do the graphs in parts **D**, **E**, and **F** help you to answer this question?

Name: _____   Course/Section: _____   Date: _____

Earth is a combination of several dynamic systems. To get a feel for the dimensions of some of Earth's primary layers, let's create a scaled image of Earth from outer space to Earth's center and investigate how density plays a role in this structure.

A We are given several distances expressed in km that we need to scale so that we can represent distances within Earth as much smaller distances on our page. Our scale is that 100 mm on the page represents the 6371 km of Earth.

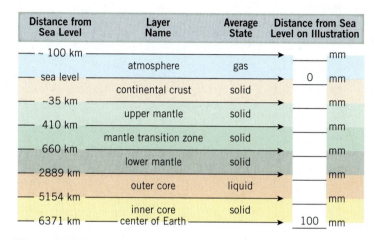

Distance from Sea Level	Layer Name	Average State	Distance from Sea Level on Illustration
~ 100 km			____ mm
	atmosphere	gas	
sea level			0 mm
	continental crust	solid	
~35 km			____ mm
	upper mantle	solid	
410 km			____ mm
	mantle transition zone	solid	
660 km			____ mm
	lower mantle	solid	
2889 km			____ mm
	outer core	liquid	
5154 km			____ mm
	inner core	solid	
6371 km — center of Earth			100 mm

Figure A1.7.1

1. We need a scaling factor that we can multiply the "distances from sea level" in km shown in **Fig. A1.7.1** to find the appropriate map distance in mm. Using our skills with proportions, we notice that 100 mm on the map is to 6371 km in Earth as our unknown conversion factor is to 1 km. Using your knowledge of proportions (and referring to the coverage of proportions earlier in the chapter text if necessary), determine the conversion factor:
   _____ mm/km

2. Use that scaling factor to compute the values in the right column of **Fig. A1.7.1**. Two values are provided for you.

3. Use a sharp pencil to carefully mark the "distances from sea level" from the right column of **Fig. A1.7.1** onto the left side of the millimeter scale on **Fig. A1.7.2**.

4. Use a drafting compass to draw concentric quarter-circle arcs from each of the pencil marks you just made on **Fig. A1.7.2**. The sharp pivot end of the compass should be held in the small circle at the 100 mm mark at the center of Earth.

5. Label each of the major layers of Earth's interior on **Fig. A1.7.2A**.

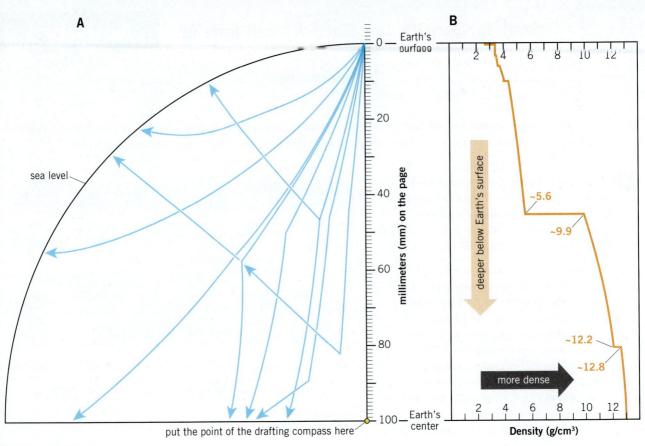

A   **B**

sea level

millimeters (mm) on the page

0 — Earth's surface

20

40

60

80

100 — Earth's center

put the point of the drafting compass here

deeper below Earth's surface

~5.6

~9.9

~12.2

~12.8

more dense

Density (g/cm³)

Figure A1.7.2

B Seismologists who study Earth's deep interior by modeling how earthquake energy propagates through Earth's interior have developed ways of recognizing the places where that energy is reflected off of boundaries where the upper layer is less dense than the lower layer. This phenomenon is like the reflection of a beam of light off of a reflective surface. The larger the difference between layer densities across a boundary, the more prominent the reflection of earthquake energy will be.

1. **Figure A1.7.2B** shows a plot of the approximate variation in density with depth in Earth's interior arranged so that its vertical scale is the same as the radius of the section through Earth in **Fig. A1.6.2A**. Circle the largest jump in density in **Fig. A1.7.2B**.

2. At what depth does the largest density jump occur in Earth's interior as depicted in **Fig. A.1.7.2B**? _____ km

3. Across the boundary between which two layers does this large density jump occur?

    less dense layer: _____   more dense layer: _____

C **REFLECT & DISCUSS** Rays indicating the paths of seismic waves through Earth due to a single earthquake that occurred near the top of **Fig. A1.7.2A** are shown in light blue on that figure. Describe how the shapes of the light blue ray paths relate (if at all) to the various boundaries you drew in section **A4** and to the boundaries indicated in the density plot in **Fig. A1.7.2B**.

D **REFLECT & DISCUSS** As Earth was forming in its earliest history and probably when the collision that resulted in the formation of the Moon occurred more than 4 billion years ago, Earth was a blob of molten rock with just a thin solid film on its outer surface. Given your experience with materials of different densities, what can you infer about the relative densities of Earth's atmosphere, ocean, rocky crust and mantle, and iron-rich core?

Plate Tectonics

Pre-Lab Video 2

http://goo.gl/NrcXgB

Contributing Authors

Edward A. Hay • *De Anza College*

Cherukupalli E. Nehru • *Brooklyn College (CUNY)*

C. Gil Wiswall • *West Chester University of Pennsylvania*

▲ The triangular Sinai Peninsula is part of a microplate between the African/Nubian Plate to the left (west) and the Arabian Plate to the right (east).

BIG IDEAS

Earth's solid outermost layer is the lithosphere, which includes the crust and the uppermost part of the upper mantle. The lithosphere is divided into plates that move relative to each other, and we can detect those motions using GPS and other technologies. Interactions between lithospheric plates along their boundary zones produce earthquakes, volcanoes, mountain ranges, mid-ocean ridges, and deep ocean trenches. Plate tectonics is the study of the motion of lithospheric plates and the geologic effects of those motions.

FOCUS YOUR INQUIRY

Think About It   How do we detect and measure plate motion?

ACTIVITY 2.1   Reference Frames and Motion Vectors (p. 41, 55)

ACTIVITY 2.2   Measuring Plate Motion Using GPS (p. 44, 57)

ACTIVITY 2.3   Hot Spots and Plate Motions (p. 48, 59)

Think About It   How does our knowledge of how materials deform help us to understand plate tectonics?

ACTIVITY 2.4   How Earth's Materials Deform (p. 50, 61)

Think About It   How does rock magnetism help us date the oceanic crust and measure sea-floor spreading?

ACTIVITY 2.5   Paleomagnetic Stripes and Sea-Floor Spreading (p. 53, 63)

ACTIVITY 2.6   Atlantic Sea-Floor Spreading (p. 53, 65)

Think About It   How do earthquakes help us locate and understand plate boundaries?

ACTIVITY 2.7   Using Earthquakes to Identify Plate Boundaries (p. 54, 67)

Introduction

We use modern technology to directly measure the present-day motion of various parts of Earth's outer surface relative to other parts of that surface—continental Africa moving away from continental South America, for example. Whether continents move relative to each other is no longer in dispute—they do. The relative motion of various parts of Earth's surface causes or contributes to many geological phenomena that are important to us, such as earthquakes, volcanism, elevation of mountain ranges and subsidence of basins, concentrations of valuable minerals, occurrence of energy resources, circulation of water in ocean basins, variations in climate, and changes in global sea level. The synthesis of knowledge related to the motion of big pieces of Earth's outer surface relative to each other over time is called **plate tectonics**.

Plate tectonics provides a context within which we can understand how different geological events and processes relate to one another. But at its core, plate tectonics is a description of the motion of plates. In a series of laboratory activities, we are going to explore some aspects of plate tectonics, emphasizing the motion of plates and some of the effects of that motion.

Earth's Outer Layers, Defined by Seismology

Pioneering studies by seismologists (geoscientists who use earthquake waves to study Earth's structure) in the late 1800s and early 1900s demonstrated that Earth has a thin, solid **crust** on top of a thick, solid **mantle**. The layers of Earth composed of rock—the crust and mantle—occur over a core composed mostly of iron with a liquid outer core and a solid inner core. The boundary between the crust and uppermost mantle was discovered in 1909 by Croatian seismologist Andrija Mohorovičić in whose honor it is called the **Moho** (pronounced MOE-hoe—pronouncing Mohorovičić is more complicated).

Earth has two fundamentally different types of crust above the Moho: oceanic and continental. **Oceanic crust** and **continental crust** differ in composition, thickness, density, average rock type, and the way they form. We will learn more about the composition of continental and oceanic crust later when we learn about minerals and rock. Oceanic crust is found beneath the deeper seafloor of major ocean basins, covering around 63% of Earth's surface. Continental crust underlies the part of Earth's surface from a couple of km below sea level to the top of Earth's highest peak.

The mantle is below the Moho and extends to a depth of ~ 2890 km or almost half-way to Earth's center. The mantle is composed of rock whose density is significantly greater than that of the crust. The average density of the crust is less than ~ 2.9 g/cm^3 whereas the density of the mantle ranges from ~ 3.3 to 5.6 g/cm^3. The iron-rich core below the mantle is much denser than the mantle.

Earth's Outer Layers, Defined by Strength

Geoscientists established the broad layering of Earth's deep interior by the end of the 1930s. As early as the 1920s, geoscientists had begun to realize that the great heat present in the core combined with the heat produced by radioactivity in the crust and mantle is likely to affect the strength or weakness of different parts of the mantle. (We use terms like *strong* and *weak* to indicate generally how difficult or easy it is to deform the material. A rock specimen you might encounter in lab is considered strong in comparison to wet clay or the inside of a banana.)

Based on laboratory experiments and theoretical modeling, the temperature at the center of Earth is interpreted to be approximately 6000°C, or about the same as the surface temperature of the Sun. Thankfully, the temperature at the ground surface where we live is much cooler, so there is a significant change in temperature with depth in Earth—a significant **geothermal gradient**. The relatively cold crust is solid and tends to break (i.e., to fracture and fault) when it is deformed. In contrast, solid rock that is heated to more than half of its melting temperature becomes softer, and very hot rock can actually flow in the solid state.

We know that liquid molten rock (**magma**) can flow. We can watch it flow in online videos of magma flowing from Kilauea Volcano in Hawaii. It seems strange that very hot **solid** rock can also flow, slowly changing its shape without fracturing or faulting. A convenient example of a solid that can flow even though it is not in the liquid state is Silly Putty™. If you mold Silly Putty into a sphere, you can bounce it on the floor just like an elastic rubber ball, and the putty will not deform during the short time it takes to impact and rebound. But if you set the ball of Silly Putty down, it will flow slowly under the force of gravity until it forms a flattened dome over the course of a half hour or less.

Earth's outermost ~ 100 km is a solid layer called the **lithosphere** that includes the crust and the uppermost part of the upper mantle (**Fig. 2.1**). The lithosphere is the relatively strong, cooler outer layer of Earth with a temperature ranging from below 0°C in some places at the ground surface to perhaps ~ 1300°C at the base of the lithosphere. Beneath the lithosphere is a hotter, much weaker layer in the mantle called the **asthenosphere**. The asthenosphere is hot enough to contain a tiny bit of magma, and that makes it particularly weak. The asthenosphere is almost entirely solid, but is weak and able to flow like Silly Putty.

https://goo.gl/8CLX0h

Lithospheric Plates

Many or most geoscientists prior to the mid-1900s assumed that Earth's solid outer layer was essentially continuous, like the unbroken shell of an egg. Global maps of earthquakes, active volcanoes, the mid-ocean ridge system, and deep ocean trenches produced in the 1950s and early 1960s led to the realization that the lithosphere is not a continuous layer but rather is a collection of pieces geoscientists call **lithospheric plates** or simply **plates** (**Fig. 2.1**). Lithospheric plates move horizontally over the

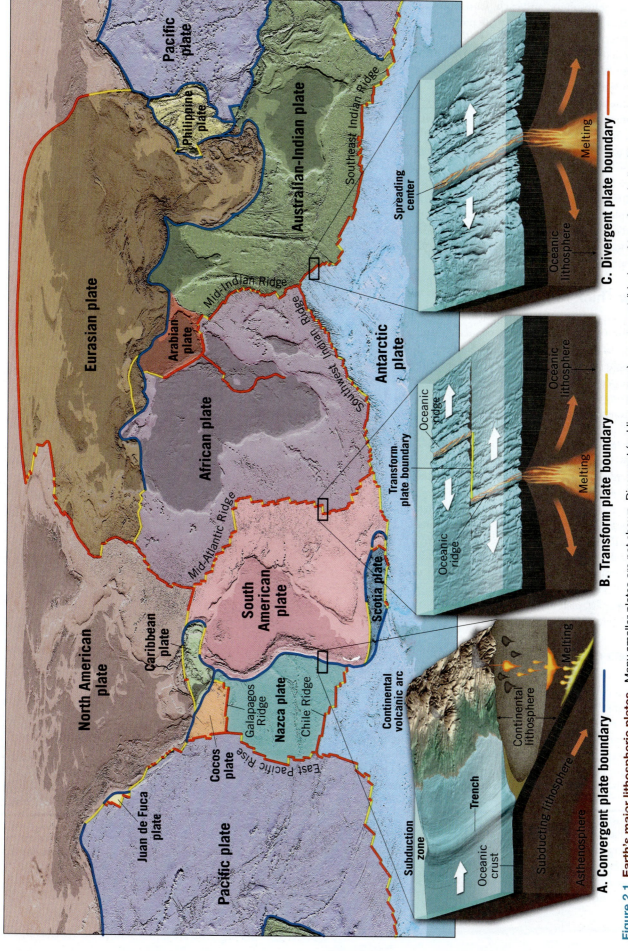

Figure 2.1 Earth's major lithospheric plates. Many smaller plates are not shown. Divergent (red line on map), convergent (blue), and transform (yellow) boundaries are illustrated in the block diagrams.

A. Convergent plate boundary ▬

B. Transform plate boundary ▬

C. Divergent plate boundary ▬

Pacific plate

Philippine plate

Australian-Indian plate

Southeast Indian Ridge

Eurasian plate

Arabian plate

Mid-Indian Ridge

Southwest Indian Ridge

African plate

Antarctic plate

Mid-Atlantic Ridge

North American plate

Caribbean plate

South American plate

Scotia plate

Galapagos Ridge

Nazca plate

Chile Ridge

Cocos plate

East Pacific Rise

Juan de Fuca plate

Pacific plate

Spreading center

Oceanic lithosphere

Melting

Oceanic ridge

Transform plate boundary

Oceanic ridge

Oceanic lithosphere

Melting

Melting

Subduction zone

Continental volcanic arc

Trench

Oceanic crust

Continental lithosphere

Melting

Subducting lithosphere

Asthenosphere

weak asthenosphere, and each plate moves relative to the other plates. The rate of plate motion is very slow—on the order of millimeters to centimeters per year along many plate-boundary segments—but the displacement across plate boundaries can be tens of kilometers over a million-year timespan.

The upper part of the lithosphere in a particular location might be composed of oceanic crust or continental crust, or perhaps some sort of transitional crust. We sometimes use the term **continental lithosphere** for lithosphere that includes continental crust (**Fig. 2A**), and **oceanic lithosphere** for lithosphere containing oceanic crust (**Figs. 2B and C**). The Pacific, Nazca, and Juan de Fuca Plates are examples of plates that are entirely or almost entirely composed of oceanic lithosphere. Most plates, like the North American Plate, include both oceanic and continental lithosphere.

How Thick Are Plates?

It is not easy to determine the thickness of the lithosphere, because the lithosphere-asthenosphere boundary is difficult to find by analyzing seismic data. The deepest part of the lithosphere and the shallowest part of the asthenosphere are both part of the upper mantle and share the same general composition. The primary difference between the two is temperature and the consequent difference in strength.

The thinnest oceanic lithosphere is along the plate boundary along a mid-ocean ridge, where the entire lithosphere is only as thick as the ~5−7 km thickness of the oceanic crust. The oceanic lithosphere thickens as it cools while moving away from the mid-ocean ridge plate boundary, ultimately reaching a thickness of perhaps 100 km before it sinks into the mantle in the subduction process. (We will learn more about subduction later.) The thickness of continental lithosphere has been even more difficult to interpret. It might be only as thick as the thinned continental crust (~20 km) in places where the continent is rifting apart, or it might be ~150−250 km thick under the oldest parts of the major continents.

Types of Plate Boundaries

In the late 1960s, geoscientists used a simplified model of plate tectonics in which Earth's surface is entirely covered by more than a dozen large lithospheric plates and some smaller microplates. Each of these plates was considered to be essentially rigid, meaning that there was little or no deformation except in the immediate vicinity of plate boundaries. Our current understanding is a bit more complex but includes many of the broad features of earlier models.

Divergent Plate Boundaries. The boundary between two plates that are moving away from each other is called a **divergent boundary** (**Fig. 2.1C**). The more common type of divergent boundary involves oceanic lithosphere along mid-ocean ridges. Partial melting of the asthenosphere below a mid-ocean ridge generates magma that rises in

small blobs, eventually making it to a **magma chamber** in the lower oceanic crust. As the two plates move apart perpendicular to the ridge crest, also known as the **ridge axis**, **tension** cracks develop along the axis that are filled in with magma. (If you grab a thin stick of chalk or a piece of string cheese and pull it apart, it will break along a tension crack oriented roughly perpendicular to the direction you pulled.) Some of the magma that intrudes the cracks continues on to the seafloor, resulting in volcanism along the axial valley of the mid-ocean ridge. Newly crystallized volcanic rock is added to the trailing edges of both plates along a mid-ocean ridge, forming new oceanic lithosphere. We call this process **sea-floor spreading**. Studies of oceanic crust formed along mid-ocean ridges indicate that about as much new crust is added to both sides of the boundary over long periods of time, although differences do occur.

Divergent boundaries in **continental lithosphere**—lithosphere with continental crust—are known as **continental rift zones**. Continued continental rifting eventually results in the generation of a new ocean basin with a mid-ocean ridge. For example, the East Africa Rift Zone might evolve into an ocean basin, just as the Red Sea and Gulf of Aden developed as the Arabian Plate rifted away from the African Plate (**Fig. 2.2**).

Convergent Plate Boundaries. The boundary between two plates that are moving toward each other is called a **convergent boundary** (**Fig. 2.1A**). If the convergent

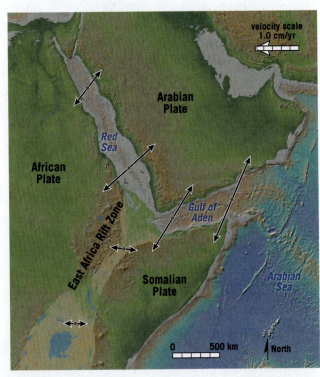

Figure 2.2 New ocean basins and continental rift. The Red Sea and Gulf of Aden are new ocean basins formed as the Arabian Plate diverged from the African Plate. The East Africa Rift is a divergent boundary that is separating the Somalian Plate from the African Plate. The arrows indicate plate motion relative to the divergent boundaries.

boundary involves oceanic lithosphere on one side of the boundary, a process called **subduction** will occur in which the oceanic lithosphere slips down beneath the other plate and sinks into the mantle (**Fig. 2.1A**). Earthquakes occur within the subducting slab of oceanic lithosphere as well as in the overriding plate, and magma begins to rise from the area above the subducting slab as it reaches depths of ~100 km. The magma feeds the chains of active volcanoes observed above subduction zones in places like the Cascades, the Andes, the Aleutians, and Japan. The oceanic lithosphere is mostly mantle material, and the oceanic crust is derived from the partial melting of the mantle, so subduction is a process in which oceanic lithosphere is recycled into the mantle. Where continental lithosphere converges with continental lithosphere, the crust is deformed into a mountain range like the Appalachians and the Himalaya.

Transform Plate Boundaries. The third principal type of plate boundary occurs where the edges of two adjacent plates are parallel to the direction in which one of the plates is moving relative to the other plate. This is called a **transform boundary**. Most transform boundaries are made up of **transform faults**—vertical or near-vertical faults oriented parallel to the direction one plate is moving relative to the other plate. A **fault** is a surface (or thin zone) along which rock bodies slip past one another, remaining in contact with each other along the fault. A vertical fault is like the surface between your favorite book and the book next to it on a tightly packed bookshelf. When you pull your favorite book horizontally away from the bookshelf, its cover slides along the cover of the adjacent book. So the "fault" between the books is vertical, and the slip direction between the books along the fault is horizontal. The books remain in contact with each other until you pull your favorite book entirely out of the bookshelf.

Transform faults are common between spreading segments along mid-ocean ridges (**Fig. 2.1B**). Other transform boundaries are more diffuse as along the western edge of the North American Plate where the relative motion between the North American and Pacific Plates is distributed across a broad zone involving several active structures. Faults within transform boundaries in continental crust can generate large earthquakes that are particularly damaging because they usually originate at shallow depths within ~15 km of the ground surface. The San Andreas Fault in California is an important part of this broader transform boundary zone (**Fig. 2.3**). Earthquakes on transform faults located along mid-ocean ridges are often even shallower because the oceanic crust is thinner, but these earthquakes usually result in little or no damage because most occur far from population centers.

Calculating Rates—The Math You Need

Several of the activities in this laboratory require you to calculate rates. You can review and learn more about calculating rates and do some practice problems at The

Math You Need, When You Need It website. This site includes math tutorials for students in introductory geoscience courses:

http://serc.carleton.edu/mathyouneed/rates/ index.html

http://goo.gl/ZnOer5

ACTIVITY 2.1

Reference Frames and Motion Vectors, (p. 55)

Think About It How do we detect and measure plate motion?

Objective Learn about reference frames, displacement, and velocity vectors related to plate motions.

Before You Begin Read the Introduction and the section below: Where Are Plates Going?

Where Are Plates Going?

Plate tectonics is, at its core, a description of the motion of lithospheric plates. In order to talk about plate motion, we need to know a few things about how motion is described in general. We can think about the direction something is moving in. We can think about how fast it is moving. We can think about whether its direction or velocity changes over time. There are other characteristics we can think about, such as whether it is rotating, but for the moment, let's limit ourselves to thinking about simple horizontal motion at a constant rate.

Compass Direction

You probably have a general idea of what we mean by the words *north*, *south*, *east*, and *west*. North is the direction toward the north spin axis of Earth, currently located in the arctic between Canada and Russia. South is toward the south spin axis in Antarctica. Looking north, east is to your right and west is to your left (**Fig. 2.4**). The conventional way of describing direction beyond these four cardinal compass directions is to measure an angle on a horizontal plane from the north direction to the direction you are interested in describing. The method we will use is called the **azimuth method** in which north is defined as having an azimuth of 0°, and the azimuth angle (or simply the **azimuth** or **compass bearing**) increases in a clockwise rotation from north. (A clockwise rotation is the same direction that the fingers on your left hand curl when they are at rest and your thumb is pointing up, so we sometimes call this a left-handed or negative rotation.) East has an azimuth of 90°, south is 180°, west is 270°, and as we swing back toward North, we approach 360°. The full circle of azimuths spans 360°.

Four arrows are shown in **Fig. 2.4**. If you were asked, "In what direction is arrow *A* pointing," the answer would be that arrow *A* is pointing toward azimuth 34°,

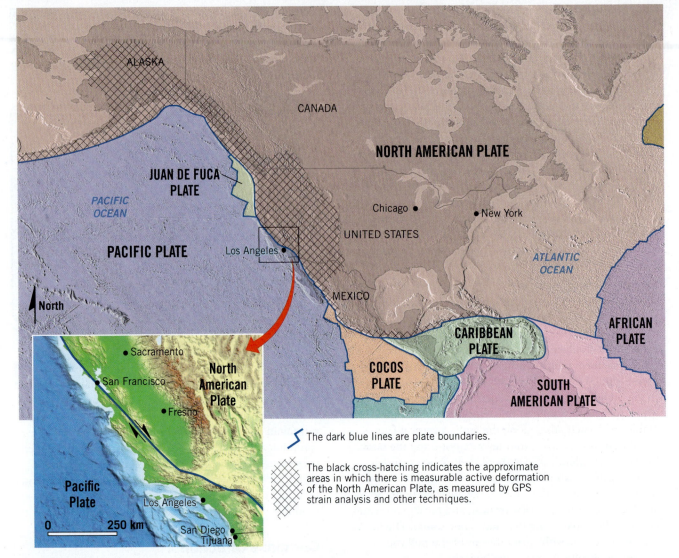

Figure 2.3 **Tectonic setting of North America.** The western part of North America (crosshatch) is actively deforming. The San Andreas Fault (blue curve in the detail map at lower left) is a major part of the transform boundary between the Pacific Plate and the North American Plate.

or approximately toward the northeast. Arrow *B* is pointing toward 150°, *C* toward 255°, and arrow *D* is pointing toward azimuth 315°.

Dust off the Pythagorean Theorem

A long time ago when you were much younger, a math teacher introduced you to a simple bit of mathematics called the *Pythagorean equation* or *Pythagorean theorem*. It's OK if you don't remember, and there's no need to deny that you have ever heard of it because you are going to learn what you need to know right now. Think about a triangle in which two sides form a right angle—that is, the two sides are 90° or *perpendicular* to each other (**Fig. 2.5**). The length of the green side at the bottom of the triangle in **Fig. 2.5** is represented by a variable we will call *a*, the length of the blue side is *b*, and we want to know the length of the orange side, which we will call *c*. (You might recall that the orange side represents the *hypotenuse* of

the right triangle.) About 2500 years ago, Pythagoras proved that

$$a^2 + b^2 = c^2,$$

which is known as the **Pythagorean equation**. We don't want to know what c^2 is, but rather we want to know what c is: the length of the hypotenuse. We find c^2 by taking the *square root* of the quantity $(a^2 + b^2)$, or

$$c = \sqrt{a^2 + b^2}$$

Nobody computes square roots by hand anymore. Just find a calculator or an app that can do the computational work for you.

Example. Imagine that we have a right triangle in which one side has a length of $a = 2$ and the length of the other side is $b = 3$. We want to know the hypotenuse length, c, and we are going to use the Pythagorean equation to find it. The order of mathematical operations in this case is to

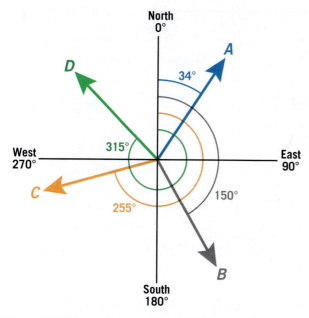

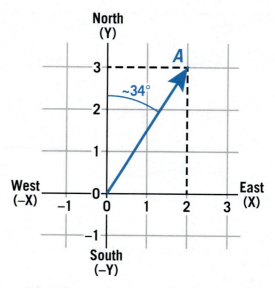

Figure 2.4 Measuring compass directions using the azimuth method. Azimuth is measured in a horizontal plane with a clockwise rotation from north. The azimuth of direction A is 34°, and the azimuths of directions B, C, and D are as shown.

Figure 2.6 2-dimensional vector. The coordinate system we will use to define vectors will be a horizontal plane with axes aligned east–west and north–south. Directions are measured clockwise from north using the azimuth method.

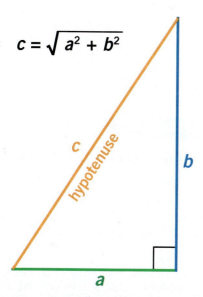

$$c = \sqrt{a^2 + b^2}$$

Figure 2.5 Length of a hypotenuse. Given a right triangle with sides of length a and b, we use a form of the Pythagorean equation to find the length of the hypotenuse, c.

square a and b first, then add them together, and then take the square root of the sum. Squaring a number is just multiplying the number by itself once, so

$$a^2 = a \times a = 2 \times 2 = 4, \text{and}$$
$$b^2 = b \times b = 3 \times 3 = 9.$$

Using a form of the Pythagorean Theorem, we find that

$$c = \sqrt{a^2 + b^2} = \sqrt{4 + 9} = \sqrt{13} \approx 3.6$$

So the length of the hypotenuse is approximately equal to 3.6. We are going to need this little bit of math.

2-D Vectors

A **vector** is a mathematical representation of something that has a **magnitude** and **direction**. We can think of a vector as an arrow extending from the origin of an X-Y coordinate system at point {0, 0} to another point, say {2, 3} (**Fig. 2.6**). The first number listed in the curly brackets is the X coordinate, and the second is the Y coordinate. We say that the coordinates of vector A are {2, 3}, meaning that the vector extends from {0, 0} to {2, 3} (**Fig. 2.6**).

The way we will define our vector coordinate system in this lab is to align the positive Y axis with north and the positive X axis with east. That allows us to use the azimuth method to describe the direction of vector A (**Fig. 2.6**). We measure the azimuth clockwise from north to vector A using a protractor and find the azimuth to be ~34°. (We can also use trigonometry to determine the angle more exactly, but measurement with a protractor will do for now.) The length or magnitude of vector A can be determined using the Pythagorean equation.

$$\text{magnitude of } A = \sqrt{2^2 + 3^2} \approx 3.6$$

Velocity is a vector quantity because it has a magnitude (the **speed**) and a direction in which the object is moving. Displacement is another vector quantity because it can be expressed as the vector from a starting location to a different location.

Motion, but Relative to What?

Every time we experience **displacement** (a change from an initial position to another position) or a **velocity** (a displacement during a time interval), our description

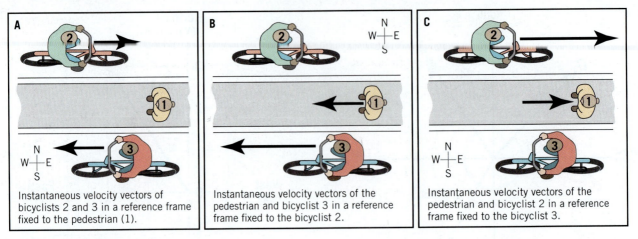

A	B	C
Instantaneous velocity vectors of bicyclists 2 and 3 in a reference frame fixed to the pedestrian (1).	Instantaneous velocity vectors of the pedestrian and bicyclist 3 in a reference frame fixed to the bicyclist 2.	Instantaneous velocity vectors of the pedestrian and bicyclist 2 in a reference frame fixed to the bicyclist 3.

Figure 2.7 Visualizing reference frames. A person (1) standing on a median in a road and two bicyclists (2 and 3) with arrows representing instantaneous velocities.

depends on the **reference frame** in which we experienced it. Any report of a displacement or velocity begs the question, "Relative to what?" That is, what was the reference frame in which the velocity was measured?

Imagine standing on a median strip in a roadway watching as a bicyclist passes you on your left and another passes on your right (**Fig. 2.7**). From your perspective—that is, in a reference frame that is fixed to you—both riders are moving, but you are not (**Fig. 2.7A**). In the reference frame experienced by rider 2, you and rider 3 are both moving west but at different speeds (**Fig. 2.7B**). Rider 3's speed relative to rider 2 is equal to the combined speeds of both riders as you measure them in your reference frame. In a similar way, rider 3 sees you and rider 2 moving east at different rates (**Fig. 2.7C**). The speed of rider 2 as observed from rider 3 is the same as the speed of rider 3 as observed from rider 2, but the directions are opposite.

We work with several reference frames in the study of plate tectonics. Sometimes we are interested in how one plate moves as observed from another plate. In that case, the reference frame is fixed to the observer's plate. Sometimes we want to describe the motion of the plates on both sides of a plate boundary, and so we use a reference frame that is fixed to that boundary (e.g., **Fig. 2.2**). Sometimes we want to know how an individual plate is moving relative to a reference frame that is not fixed to a plate or a plate boundary but rather is fixed to something else that is external to the lithosphere. A **hotspot** in the mantle below the lithosphere that generates a trail of volcanoes as a plate moves over it can be used to define an external reference frame to describe plate motions. (We will develop the hotspot idea later in this lab.)

One important kind of external reference frame that is used to express the motion of an individual plate relative to the rest of the Earth is called a "no-net-rotation" (NNR) reference frame. The details of the **NNR reference frame** are complicated, but a simplified explanation is all we will need. We can just think of the NNR reference frame as a way of measuring displacements and velocities in a reference frame that is centered on Earth but is not attached to any of the plates or their boundaries. You can think of

NNR as a view of plate motion from a vantage point in the deep mantle.

Instantaneous versus Finite Displacements and Velocities

Plates have been moving for billions of years. When geoscientists consider a displacement between two plates that occurs over a year, a few years, or even a few thousand years, we consider that an **instantaneous displacement**, and the rate of displacement would be an **instantaneous velocity**. In contrast, if geoscientists are working with displacements or velocities that involve millions of years or more, we consider those **finite displacements** or **finite velocities**.

ACTIVITY 2.2

Measuring Plate Motion Using GPS, (p. 57)

Think About It How do we detect and measure plate motion?

Objective Use velocity data from NASA for a GPS station located near where you live along with a Plate Motion Calculator from UNAVCO to determine the direction and rate that your plate is moving near your home.

Before You Begin Read the following sections: GPS—Global Positioning System and Using GPS to Study Lithospheric Plate Motion.

Plan Ahead This activity requires that data be gathered from websites hosted by NASA-JPL and UNAVCO. If you will not be able to access the web during your lab session, you will need to collect data from these websites before coming to lab. You might also need to bring a calculator that can compute a square root so that you can use the Pythagorean Theorem to compute a velocity.

GPS—Global Positioning System

The United States has established a constellation of more than two dozen navigation satellites and their ground stations that we call the Global Positioning System or **GPS**. (Russia, the European Union, India, Japan, and China also have satellite navigation systems in operation or development. These systems, along with GPS, are jointly called the Global Navigation Satellite System or **GNSS**. For simplicity, we will simply refer to GPS in this lab.) This technology allows us to determine the location of a GPS receiver that might be embedded in a phone, a car, or other devices. In addition to its other civilian and military applications, GPS is used in **geodesy**—the science of measuring changes in Earth's size and shape and the position of different parts of Earth's surface over time. We will use geodetic GPS data in this lab to help us determine the motion of lithospheric plates.

Each GPS satellite orbits Earth every 12 hours at a height of ~20,183 km. These satellites are arranged in 6 different orbital planes with at least 4 satellites in each plane. The orbital geometry was designed to ensure that between 8 and 12 satellites are always visible above the horizon by any GPS receiver anywhere on Earth. The positions of these satellites are constantly monitored by 16 sites around the globe and are precisely known. Each GPS satellite transmits radio signals that are received by GPS antennas on Earth. It takes the signals from at least four GPS satellites for a GPS receiver on Earth to compute its location through a process of **trilateration**.

Geoscientists have established networks of fixed GPS sites throughout the world. By "fixed," we mean that the antenna is attached to a building or a rigid stand that is firmly attached to the ground and that position data from that site are automatically recorded for years or tens of years. One such network of 1,100 GPS sites in the United States was developed as part of the EarthScope Project, and is called the Plate Boundary Observatory or **PBO** (**http://www.unavco.org/projects/major-projects/pbo/pbo.html**). The GPS antennas at PBO stations are pinned to Earth's crust via a stainless steel mount with four to five legs welded together that extend many meters into the ground (**Fig. 2.8**). Data from GPS satellites are collected at PBO stations every 15 seconds. Some stations collect these data at a rate of five times every second, and this wealth of data allows us to measure the displacement of Earth's surface during earthquakes. The position of each geodetic GPS site is measured to an accuracy of less than a centimeter, and the change in position is measured in millimeters per year.

Over a decade or more of measuring the position of a fixed GPS site, we can detect the slowly changing position of that part of the crust. The motion of a single GPS site does not provide enough information to characterize the motion of the entire plate that the site is a part of. We resolve a plate's motion by analyzing data from multiple GPS sites located within the plate away from GPS sites located within the plate away from plate boundaries or other areas in which the crust is actively deforming.

Figure 2.8 Example of a GPS station. Plate Boundary Observatory Station P150 on Martis Peak, California (lat 39.29238°, long −120.03386°). The antenna is under the gray dome to the left and is mounted on stainless steel pipes that extend ~15 m into the ground. Solar panels supply the energy for the electronics that acquire the data and transmit them elsewhere for analysis.

GPS stations acquire location data that are resolved into north–south, east–west, and up–down displacements from the original station position. These data are plotted on graphs with displacement along the vertical axis and time on the horizontal axis. Displacement-versus-time graphs are examples of **time-series** plots, in which we plot the variation of something as a function of time (**Fig. 2.9**). Displacements toward the north are considered positive displacements, and displacements toward the south are negative. East displacements are positive; west displacements are negative. The slope of the best-fit line or curve over a period of years is used to determine the speed and direction of site motion (**Fig. 2.9**).

Let's use some real data to derive the velocity of a GPS station in an NNR reference frame, using resources provided online by NASA. We are going to investigate the motion of a GPS station called SYDN in Sydney, Australia, located at latitude −33.780874256° and longitude 151.150381378°. First, we navigate to a website **http://sideshow/jpl.nasa.gov/post/series.html**) where GPS time-series data computed by NASA's Jet Propulsion Lab (JPL) are available. We look on the Google map to find the green dot on the east coast of Australia at Sydney. The yellow line extending to the north-northeast of the green dot indicates the direction and velocity the GPS station (and the crust it is attached to) is moving in the NNR reference frame. When we click on the green dot at Sydney, a small white box with some graphs appears. Double-click on that box, and it will expand to show the time-series plots, as shown in **Fig. 2.10**.

The vertical axis in the time-series plot shown in **Fig. 2.10A**, marked *Latitude*, indicates the north–south motion of SYDN. The time series has a positive slope, so the motion is toward north. The rate at which SYDN moves north is the same as the slope of the time series—change in position divided by the time over which that change

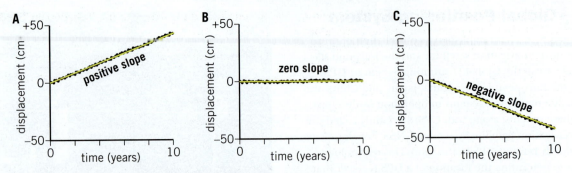

Figure 2.9 Interpreting time-series plots. Examples of time-series plots of displacement from some initial location. Positive displacements are toward the north; negative displacements are toward the south. **A.** Displacement is toward the north at an approximately constant rate. **B.** No north–south displacement. **C.** Displacement is toward the south at an approximately constant rate.

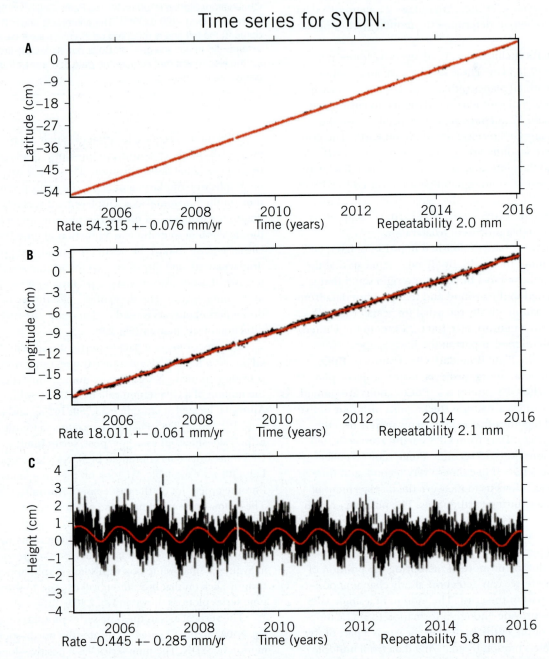

Figure 2.10 Time-series plots for GPS station SYDN. These plots indicate the displacement and average velocity of GPS site SYDN in Sidney, Australia, in an NNR reference frame. **A.** "Latitude" reflects displacement in a north–south direction. **B.** "Longitude" reflects displacement in an east–west direction. **C.** "Height" reflects vertical displacement. (Produced by NASA-JPL at CalTech and accessed online via **http://sideshow.jpl.nasa.gov/post/series.html** on January 17, 2016.)

occurred. The slope computed at JPL is printed just below the graph: 54.315 millimeters per year with an uncertainty of +/−0.076 mm/year. The vertical axis marked *Longitude* in **Fig. 2.10B** indicates the east–west motion of SYDN. The slope of the data is positive, so SYDN is moving east at a rate of 18.011 mm/year +/−0.061 mm/year. The time-series plot in **Fig. 2.10C** indicates the up–down motion of SYDN and shows an annual variation in elevation that is related to seasonal wetting and drying that causes the land surface to rise and fall a little bit every year. The average trend of the line is slightly downward at a rate of −0.445 mm/year +/−0.285 mm/year.

We can plot the north–south and east–west velocities as vectors to help us visualize the direction and speed that SYDN is moving in the NNR reference frame (**Figs. 2.6** and **2.11**). (Refer to the earlier section about *2-D vectors* if you need to.) Each square in the plot represents 10 mm/year. The north velocity vector (brown) is 54.315 mm/yr. long and points north because the north velocity is a positive number (**Fig. 2.11**). The east velocity vector (blue) is 18.011 mm/yr. long and points east because it too is a positive number. We find the total horizontal velocity vector of GPS site SYDN in the NNR reference frame by adding the north velocity vector and the east velocity vector together. The result is the black vector in **Fig. 2.11**, which has coordinates {18.011, 54.315} (e.g., **Fig. 2.6**).

Using the protractor that is included along the margins of the plot, we can see that the azimuth of the total-horizontal-velocity vector is about 18°. Its length can

be computed using a form of the Pythagorean Theorem (e.g., **Fig. 2.5**).

Total horizontal velocity
$$= \sqrt{(N - S \text{ velocity})^2 + (E - W \text{ velocity})^2}$$

In this example,

Total horizontal velocity
$$= \sqrt{(54.315)^2 + (18.011)^2} = 57.223 \text{ mm/yr.}$$

The average horizontal velocity of GPS station SYDN relative to an NNR reference frame is 57.223 mm/yr. moving toward azimuth 18.3°.

Using GPS to Study Lithospheric Plate Motion

Availability of GPS data has allowed us to investigate the initial idea that large lithospheric plates are essentially rigid and undeforming except in the vicinity of plate boundaries. The North American Plate is a good example of a lithospheric plate that is *not* rigid from boundary to boundary. While the North American Plate is approximately rigid from its eastern boundary along the Mid-Atlantic Ridge to the Rocky Mountains in the western part of the continent, GPS and other data indicate that the plate deforms between the Rocky Mountains and the western boundary along the Pacific margin (**Figs. 2.3** and **2.12**). The slow deformation of the western North American Plate is at least partly responsible for the varied landscapes, volcanism, earthquakes, and economic mineral deposits that make this region so interesting.

GPS data also allow us to measure vertical deformation in areas of Canada, the United States, and elsewhere that were covered by heavy continental ice sheets over the past couple of million years that have since melted. The crust rebounds upward and sometimes a little bit outward like an elastic cushion when a well-fed cat jumps off the sofa.

The **NAM08 reference frame** is fixed to the stable interior of the North American Plate away from boundary deformation and areas undergoing glacial rebound and allows us to depict the motion of parts of the lithosphere beyond that approximately rigid core (**Fig. 2.12**). What we see in **Fig. 2.12** is more interesting and varied than the idea that lithospheric plates are essentially rigid from edge to edge.

Data recorded at GPS sites every day are helping geoscientists develop a more detailed picture of where and how fast the various bits and pieces of the lithosphere are moving. We are gaining a better understanding of which areas are behaving like classic lithospheric plates or microplates and which are broad deformation zones between plates. This is an exciting time in the geosciences as vast new GPS and earthquake datasets accumulate that will provide the raw material for developing new and better ideas about our planet.

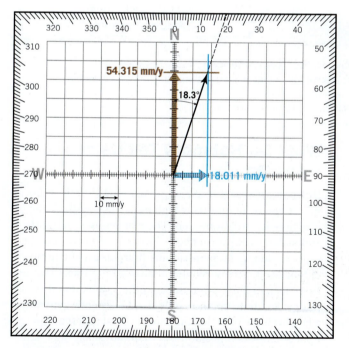

Figure 2.11 Instantaneous velocity vectors for GPS station SYDN. Plate motion plot of the north–south and east–west instantaneous velocity vectors for GPS station SYDN, which is moving toward 18.3° azimuth in an NNR reference frame.

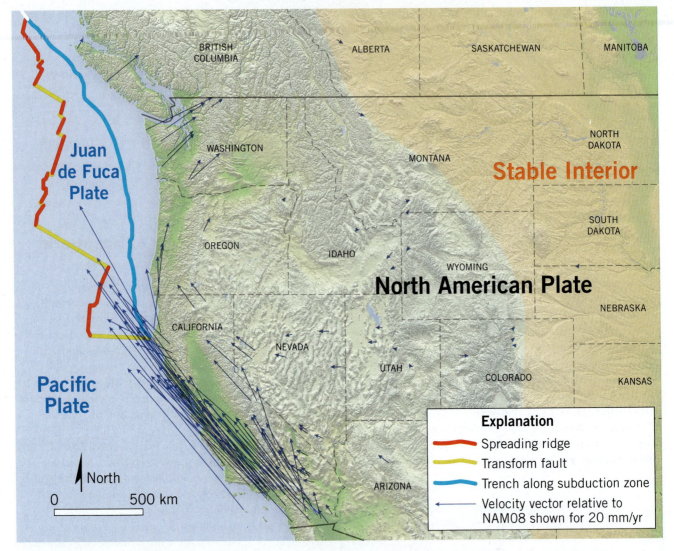

Figure 2.12 GPS velocities in western North America relative to the North American Plate. Selected GPS velocities (blue arrows) in western North America in the NAM08 reference frame that is fixed to the stable interior of central and eastern North America. (From the UNAVCO GPS Velocity Viewer and accessed online via **http://www.unavco.org/software/visualization/GPS-Velocity-Viewer/GPS-Velocity-Viewer.html** on June 26, 2016.)

ACTIVITY 2.3

Hot Spots and Plate Motions, (p. 59)

> **Think About It** How do we detect and measure plate motion?

Objective Determine the rate and direction of plate motion relative to a hotspot.

Before You Begin Read the following section: What Do Hotspots Tell Us About Plate Motion?

Plan Ahead This activity requires that data be gathered from a website hosted by NASA-JPL. If you will not be able to access the web during your lab session, you will need to collect the data you need before coming to lab. You might also need to bring a calculator that can take a square root so that you can use the Pythagorean Theorem to compute velocities.

What Do Hotspots Tell Us about Plate Motion?

Hotspots are places in the upper mantle below the lithosphere where an unusually high percentage of the asthenosphere is partially molten. The magma rises buoyantly and makes its way through the lithosphere to erupt at the surface. Many geoscientists think hotspots are the result of long-lived, narrow **plumes** of buoyant hot rock that rise from deep in Earth's mantle, but others think that they are just places in the upper mantle that persistently generate a large volume of melt. Some sort of local difference in upper-mantle composition—perhaps unusually high water content—could be responsible for a melt anomaly rather than (or in addition to) a concentration of hot rock. Improving our understanding of hotspots continues to be an area of active research. However they form, hotspots and the chains of volcanoes that form as a plate moves over them can provide a way of determining plate motion relative to the deeper mantle.

The Hawaiian Hotspot and Pacific Plate Motion

As a lithospheric plate moves over a mantle hotspot, magma rises through the lithosphere and a volcano develops on Earth's surface. Continued plate motion causes the newly formed volcano to drift away from the hotspot, depriving the volcano of its magma supply and ending its eruptive activity. New volcanoes develop over the hotspot over time, leading to the development of a trail of volcanoes whose ages increase in the direction that the plate moves relative to the mantle hotspot. A chain of active and inactive volcanoes interpreted to be related to a hotspot is called a **hotspot trail**. The Hawaiian Islands and Emperor Seamount Chain (**Fig. 2.13**) are such a line of volcanoes that formed over the Hawaiian hotspot near the center of the Pacific plate over the last ~70 million years. The current location of the Hawaiian hotspot is interpreted to be south of Kilauea Volcano in the vicinity of Lo'ihi Seamount (around latitude 18.92°N, longitude 155.27°W).

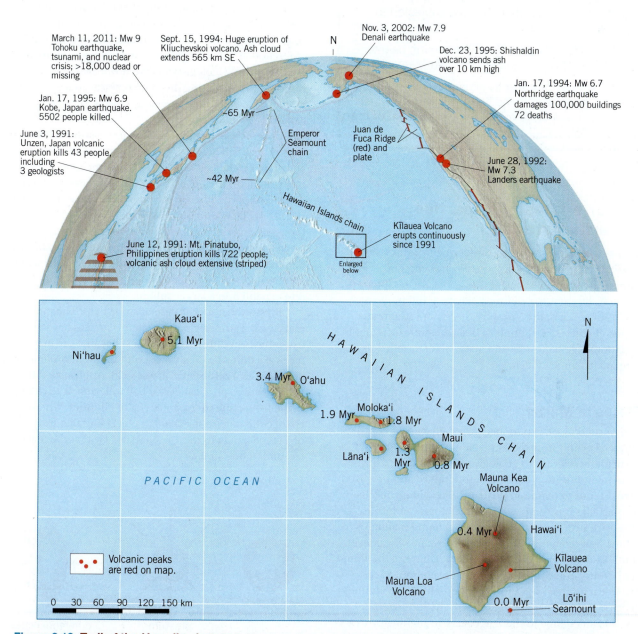

Figure 2.13 Trail of the Hawaiian hotspot. Top map shows the northern Pacific Ocean, adjacent landmasses, some notable geologic disasters, and the Hawaiian Islands and Emperor Seamount Chains. Lower map shows details of the eastern Hawaiian Islands Chain, including locations of major volcanoes. Myr means "million years".

How Earth's Materials Deform,
(p. 61)

Think About It How does our knowledge of how materials deform help us to understand plate tectonics?

Objective Investigate how the properties of a solid are different when it is hot than when it is cold. Learn about some of the ways that materials behave when they are subjected to stress or pressure.

Before You Begin Read the section below: Exploring Basic Properties of Earth Materials.

Plan Ahead This activity uses Silly Putty, small plastic bags, and some hot water and ice water (or a refrigerator) to adjust the temperature of the Silly Putty.

Exploring Basic Properties of Earth Materials

A century ago, geoscientists considering hypotheses about possible changes in the position of continents and ocean basins suffered from a lack of data about Earth's interior and how different types of rock behave at different conditions of temperature, pressure, stress, and strain rate. Improvements in our supply of basic scientific facts and in our knowledge of how Earth's materials deform led us to the plate tectonic model.

States of Matter

The world that we experience on a daily basis is composed of matter in three fundamental states that are particularly relevant to geology: solid, liquid, and gas. Under conditions of constant temperature and pressure, a **solid** tends to maintain its shape and volume because the particles it is composed of—the atoms, molecules, ions—are strongly bound together. A **liquid** maintains its volume but its shape conforms to the shape of whatever contains the liquid. The particles that compose the liquid are bound closely together but are able to move relative to each other. A **gas** will expand to fill any container because the particles in a gas are not bound together. A solid can change to the liquid state by **melting** or to a gas state by **sublimation**. A liquid can change to the solid state by **freezing** or to the gas state by **vaporization**. A gas can change to the liquid state by **condensation** or to the solid state by **deposition**.

The only liquid layer in Earth's interior is the outer core, which is thousands of kilometers below the level of lithospheric plates. The mantle and crust are almost entirely solid rock composed of crystalline mineral grains with small local pockets of liquid magma along convergent and divergent boundaries, around hot spots, and perhaps along the asthenosphere-lithosphere boundary. The asthenosphere is probably at least 99% solid on average and is probably at least 80% solid even under mid-ocean ridges and in the wedge above subducting slabs of oceanic lithosphere.

Force, Pressure, Stress, Strain

Isaac Newton's second law of motion defines **force** as the product of **mass** times **acceleration**. Force can be thought of as a vector quantity, meaning that it has a magnitude and is applied in a specific direction. Force is an abstract concept; what we actually experience is a system or field of forces that act on a surface. Force acting over an area is called **stress**. Geologists analyze how Earth's lithosphere responds to stress using the same physical concepts that describe the behavior of everyday objects, such as a rubber ball, subjected to stress.

Imagine a system of forces (that is, a **stress field**) acting on a rubber ball in which the magnitude of the force at any point on the ball's outer surface is the same as at every other point on the surface, and each of those force vectors is directed perpendicular to the ball's surface at that point (**Fig. 2.14**). That kind of stress field, in which all of the force vectors are of equal magnitude and act perpendicular to a physical surface, is called a **pressure** field. Increasing pressure causes a spherical elastic ball to contract into a smaller sphere, and decreasing the pressure would cause it to expand into a larger sphere.

Pressure is a special case of stress field. The more general case is a stress field in which the stresses have different magnitudes in different directions. Because the magnitudes are different on surfaces with different orientations, this is called a **differential stress** field. Applying a differential stress to a spherical elastic ball will cause it to change shape from a sphere to some sort of ellipsoid. When we change the state of stress on an object and that causes the object to change its size or shape, we say that the object has undergone **strain**.

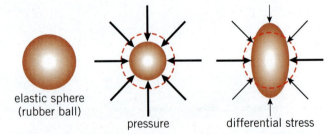

elastic sphere (rubber ball)

pressure   differential stress

Figure 2.14 Pressure and differential stress. Arrows represent force vectors within either a pressure or differential stress field. Length and thickness of arrow indicates relative magnitude of force vector—thicker, longer arrow indicates greater force. An elastic sphere is compressed with no change in shape by increasing pressure but changes shape when subjected to a differential stress.

Rheology: Elasticity, Viscosity

Different materials respond differently to stress. Squeezing a rubber ball has a different effect than squeezing a bag of cookie dough. The way that a material strains in response to stress is called the material's **rheology**, and two of the many possible rheologies are important in helping us understand plate motion: elasticity and viscosity. An **elastic** solid will change its size or shape when we squeeze it but will return to its original size or shape when we stop squeezing it. An elastic ball will distort when it hits the floor but will return to its original size and shape when it bounces up off the floor. A **viscous** material will also strain in response to a change in differential stress, but that strain will be permanent even if the stress field reverts to the original. Viscous materials flow in response to a differential stress. Matter in the solid, liquid, and gas states can all be characterized as **fluids**. (In common English usage outside of science, the words "liquid" and "fluid" are used as if they have the same meaning. In science, "liquid" is one of the states of matter, and any material that flows under a differential stress is considered a fluid.)

Elasticity and viscosity are two of the simpler rheologies we might consider, but actual rock masses tend to have more complicated rheologies that combine properties of elasticity, viscosity, and other rheologies. Earth's crust and mantle behave as elastic solids over short time intervals, such as the time needed for an elastic wave from an earthquake to pass through them. The mantle below the lithosphere behaves as a viscous fluid over longer time intervals and is able to flow while remaining in a solid state. Simplifying, we might say that the elastic lithospheric plates move over the viscous asthenosphere, as well as *through* the asthenosphere where the plates subduct.

Brittle and Ductile Deformation

The temperature and the pressure increase from the surface of the solid Earth downward toward the core. At lower temperatures and pressures in the lithosphere, rock deforms elastically or by **brittle** processes like fracturing and faulting, especially when the deformation occurs relatively quickly (**Fig. 2.15**). At higher temperatures and pressures, and especially at slower deformation rates, crystalline mineral grains can change their shape without fracturing or faulting through a variety of processes we will call **ductile** processes. Ductile processes under high temperature and pressure allow a rock to flow as a fluid.

Nature does not expend any more energy to achieve a given change than is necessary. A given mineral grain or volume of rock will deform using whatever mechanism (e.g., fracturing, faulting, ductile flow) that will result in the greatest strain for a given stress.

Strength

We use the term *strength* here to broadly indicate the amount of stress a material can resist or withstand regardless of whether the material is deforming permanently

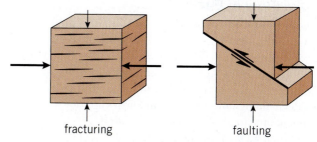

Figure 2.15 Brittle deformation: Fracturing and faulting. A brittle material subjected to a differential stress that is large enough to cause permanent deformation will develop fractures oriented roughly parallel to the greatest compressive stress direction (the longest arrows) and perpendicular to the least compressive stress direction. If that material faults, the fault surface will likely develop around 30° (and always ≤45°) to the greatest compressive stress direction and around 60° to the least compressive stress direction.

or not. For example, a fresh apple is stronger than a peeled banana. A rock's strength depends on many factors, including its mineral composition, grain size, temperature, pressure, the pressure of any gasses or liquids (water, petroleum, magma) present in the spaces between solid grains, differential stress, material properties (such as viscosity or elasticity), and strain rate. While the details are still a matter of active research, the strength of the lithosphere is thought to be significantly greater than the strength of the asthenosphere. The strength difference is significant enough to allow lithospheric plates to move horizontally across the asthenosphere and to allow subducting slabs to pass through the asthenosphere on their way into the deeper mantle.

Gravity and Buoyancy

Every particle of mass that is on or in Earth is affected by Earth's gravitational attraction. Gravity is a force that acts on every bit of mass, regardless of whether it is in the solid, liquid, or gaseous state. When a mass with a greater density is surrounded by a fluid of lesser density—imagine a ball in the air—the mass with the greater density will move downward toward the center of Earth through the less dense fluid.

What if we have a mass with lesser density surrounded by a fluid of greater density as we would if we had a blob of liquid magma surrounded by relatively soft, ductile rock? In that case, the less dense mass will tend to move upward through the more dense fluid. The upward motion is caused by what physicists call the **buoyant force**. As described in Archimedes' Principle, the magnitude of the buoyant force is the difference between the mass of the less dense material and the mass of the more dense fluid that it displaces. The rate at which the less dense material can move upward is controlled by the magnitude of the buoyant force and the viscosity of the surrounding fluid. If the surrounding fluid is very viscous, the rate of ascent will be very slow.

Heat Flow in Earth

The center of Earth has a temperature that is currently estimated to be around 6,000°C. Earth's surface temperature ranges from around −90°C to 50°C, so Earth sheds about 6,000°C of heat energy over the 6,371 km average distance from its center to the ground surface. Understanding how heat flows in Earth's interior is a fundamental topic of geoscience research.

When we increase the heat of a solid, liquid, or gas, the atoms in that material vibrate or move more rapidly. When an atom or molecule with a greater vibrational energy (more heat) is next to an atom or molecule with less vibrational energy, the particle with more energy tends to increase the energy of the particle next to it. That heat transfer continues until both particles have the same vibrational energy. This kind of direct heat transfer from a hotter body to an adjacent cooler body is called **conduction**. Conduction is most important in solids.

In most viscous fluids, increasing the fluid's temperature will decrease density, which will tend to promote buoyant upward motion. That is why hot air tends to accumulate near the ceiling and why hot water rises to the top of a pan of water being heated on a stove. The motion of heated material upward and the sinking of cooled material downward in a viscous fluid is an example of heat flow by **convection**. The most important ways that heat flows in the solid Earth are by conduction in the lithosphere and convection in the mantle.

Earth's Magnetism and Paleomagnetism

Geophysicists interested in how Earth's magnetic field changes over time were among the first to recognize that some rocks retain a record of past magnetic fields. The magnetic minerals in these rocks exhibit **remanent magnetization** that developed parallel to Earth's magnetic field at the time the magnetic minerals crystallized and that is retained unless the rock is reheated. Sedimentary particles made of magnetic minerals also seem to align themselves with the **geomagnetic field** when they are deposited. Analysis of remanent magnetization in rocks of different ages led geophysicists to the hypothesis that these rock bodies had moved relative to the magnetic poles.

If you drop a pen, it falls to the floor. The pen is under the influence of Earth's gravity—a force field that pulls every bit of mass that is in, on, or around Earth toward the center of our planet. But there is another force field around Earth that is not so obvious: the geomagnetic field. It is as though a giant bar magnet is inside Earth, giving our planet both a magnetic north pole and a magnetic south pole. Invisible flow lines of the magnetic force field arc steeply up from the south magnetic pole, curve downward so that they are horizontal at the magnetic equator, then arc steeply down into the north magnetic pole. The inclination of the field at any point on Earth's surface is related to the magnetic latitude of that point between the north and south magnetic poles. Unlike a bar magnet, the geomagnetic field changes over time because it is generated by the circulation of the liquid, iron-rich, outer core.

The strength of a magnetic field is measured in units called **teslas**. A **microtesla** is a millionth of a tesla. The small magnets we use to hold notes on refrigerator doors have a field strength of about 50,000 microteslas. The geomagnetic field strength ranges from just 25 microteslas at the equator to about 65 microteslas at the poles. Therefore, a refrigerator magnet is around three orders of magnitude (~1,000 times) stronger than the geomagnetic field. Even so, you can use the tiny magnetic needle in a compass to detect the geomagnetic field. Magnetic compass needles are not attracted to the **geographic north pole**, which is along Earth's spin axis. Instead, compass needles are attracted to the **magnetic north pole**, which is located in the Arctic Islands of northern Canada, about 700 km (~450 mi.) from the geographic north pole.

Paleomagnetism

Some iron-rich minerals, like magnetite, are naturally magnetic. The small magnetic field carried by magnetic minerals is aligned with the geomagnetic field at the time the mineral cools below its **Curie temperature**, which is 585°C for magnetite. (If the grain is reheated above the Curie temperature, this remanent magnetic field in the grain will be erased.) These magnetic grains preserve a record of the geomagnetic field at the time they last cooled below their Curie temperature. This record of a past geomagnetic field is called **paleomagnetism**. The rock that forms by crystallizing from magma along a mid-ocean ridge retains a paleomagnetic record of the geomagnetic field at the time it crystallized.

Magnetic Reversals

The circulation of liquid iron in Earth's outer core generates the geomagnetic field. A curious and useful feature of the geomagnetic field is that it flips from being north directed to south directed and back again. The north-directed field, like the current magnetic field, is said to have a **normal polarity**, and the south directed field has a **reverse polarity**. The geomagnetic field spontaneously reverses itself at irregular intervals ranging from a few thousand years to tens of millions of years in duration. The time it takes for the reversal to occur appears to be quite short—perhaps as short as a full human lifetime to as long as a thousand years. Hundreds of reversals have been documented.

The end of a compass needle that points in the direction that the geomagnetic field is flowing will point toward Earth's north magnetic pole during times of normal polarity. But during times of reversed polarity, the same end of a compass needle would point in the opposite direction, toward magnetic south. The geological effect of this reversal is that rocks containing magnetic minerals will be normally magnetized if they crystallized during a time of normal polarity, and they will be reverse magnetized if they formed during a time of reversed polarity.

Geoscientists have determined the age of all of the well-documented geomagnetic reversals. We can use the pattern of normal- and reverse-magnetized rock in an undeformed, layered sequence of volcanic or fine-grained sedimentary rock to determine the age of the rock layers based on the reversal pattern documented in the **magnetic polarity time scale** that geoscientists have developed since the 1960s.

Marine Magnetic Anomalies

Magnetic anomalies are deviations from the average strength of the magnetic field in a given area. Areas of higher than average strength are positive anomalies, and areas of less than average strength are negative anomalies (**Fig. 2.16**). During World War II, marine geophysicists began to use magnetometers towed behind ships to map the magnetic fabric of the ocean basins. (They were looking for ways to find magnetic mines and submarines.) They discovered that the oceanic crust has a striped pattern of alternating high and low magnetic intensity and that the pattern of paleomagnetic stripes is symmetric on opposite sides of mid-ocean ridges. In the early 1960s, geoscientists discovered that the symmetric pattern of paleomagnetic stripes on the seafloor was the result of two processes: the formation of oceanic crust at mid-ocean ridges while plates are moving apart and reversals of Earth's magnetic field (**Fig. 2.17**). Unexpectedly, the oceanic crust includes a magnetic recording of its own formation in the form of alternating bands of normal and reversed magnetized rock on the seafloor.

Magnetic Polarity Time Scale

Undeformed, layered sequences of volcanic rock and fine-grained sedimentary rock deposited on continental crust also contain evidence of the same magnetic field reversals that are recorded in marine magnetic anomalies. **Radiometric dating** of rock in these continental sequences has allowed us to discover when Earth's magnetic field has flipped and to establish the detailed pattern of magnetic field reversals in the past. The result is a **magnetic polarity time scale** that allows us to use the magnetic reversal pattern in a sequence of rock to date the rock. Just as we name geologic formations, some of the more important parts of the magnetic polarity time scale have names for our convenience because we refer to them so often. For example, we are living in the **Brunhes normal chron**, named for French geophysicist Bernard Brunhes who discovered that Earth's magnetic field reverses. The Brunhes normal chron began 781,000 years ago after the **Matuyama reversed chron** named for Japanese geophysicist Motonori Matuyama. In general, each of the reversals is identified by a numbering system developed by geomagnetic stratigraphers. In Activity 2.6, you will use a simpler version of a magnetic polarity time scale in which the time intervals have been color coded (instead of named) so they are easy to recognize on sight.

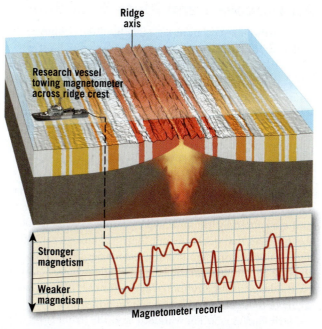

Figure 2.16 Magnetic record of sea-floor spreading. A magnetometer towed by a ship records variations in magnetic intensity, originating in the oceanic crust. A stronger magnetic field corresponds to normally magnetized rock; a weaker field indicates reverse-magnetized rock.

ACTIVITY 2.5

Paleomagnetic Stripes and Sea-Floor Spreading, (p. 63)

Think About It How does rock magnetism help us date the oceanic crust and measure sea-floor spreading?

Objective Analyze marine magnetic anomalies and infer how sea-floor spreading is related to Cascade Range volcanoes.

Before You Begin Read the section: Earth's Magnetism and Paleomagnetism.

ACTIVITY 2.6

Atlantic Sea-Floor Spreading, (p. 65)

Objective Infer how fracture zones and shapes of coastlines provide clues about how and when North America and Africa were once part of the same continent.

Before You Begin Read the section: Earth's Magnetism and Paleomagnetism.

Introduction If you worked on Activity 2.5, then you have already studied sea-floor spreading about the Gorda and Juan de Fuca Ridges off the northwest coast of the United States. This activity is an investigation of sea-floor spreading about the Mid-Atlantic Ridge.

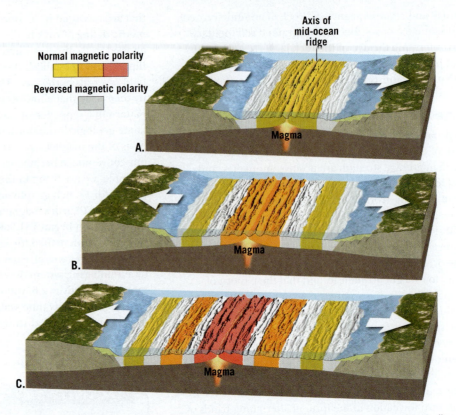

Figure 2.17 Marine magnetic anomalies record the opening of an ocean. Magnetic mineral grains in the rock are aligned with the geomagnetic field at the time the magma crystallizes along a mid-ocean ridge. While the plates slowly move apart and new crust crystallizes along their trailing edges along the mid-ocean ridge, the geomagnetic field flips its polarity. The result is that a paleomagnetic record of those geomagnetic reversals is frozen into the rock of the oceanic crust. Light gray stripes in the figure indicate reverse magnetized rock, and the yellow, orange, and red stripes indicate normally magnetized. This series of block diagrams depicts the progressive widening of an ocean basin through sea-floor spreading, from an early stage (**A**) to later stages (**B** and **C**).

ACTIVITY 2.7

Using Earthquakes to Identify Plate Boundaries, (p. 67)

> **Think About It** How do earthquakes help us locate and understand plate boundaries?

Objective Apply earthquake data from South America to define plate boundaries, identify plates, construct a cross-section of a subduction zone, and infer how volcanoes may be related to plate subduction.

Before You Begin Read the following section: Earthquakes and Plate Boundaries.

Earthquakes and Plate Boundaries

Most earthquakes occur along plate boundary zones and in subduction zones, although areas within a plate that are actively deforming produce earthquakes as well. Earthquakes tend to occur much less frequently away from boundary zones. Global maps of earthquakes produced in the 1950s and 1960s provided some of the strongest evidence that Earth's outer layer is a mosaic of plates that move relative to each other. Many of the earthquakes plotted on maps of global seismicity were located along mid-ocean ridges, deep ocean trenches, transform faults, and continental mountain ranges and rift zones (**Fig. 2.1**). Most earthquakes occur within 70 km of Earth's surface and are caused by frictional slip on faults. Essentially all earthquakes that occur deeper than ~70 km are related to subduction at convergent plate boundaries (**Fig. 2.1A**). The deepest earthquakes in the U.S. Geological Survey (USGS) earthquake catalog originated at depths of ~700 km, and many of these occurred in the subduction zones of the western Pacific near Fiji, Vanuatu, and the Santa Cruz Islands, north of New Zealand and east of Australia.

Reference Frames and Motion Vectors

Name: _____ **Course/Section:** _____ **Date:** _____

New oceanic lithosphere crystallizes from magma injected along the axis of a mid-ocean ridge. We are going to focus on a point located at latitude 45.9°N and longitude 129.9°W along the Juan de Fuca ridge between the Pacific and Juan de Fuca Plates, about 560 km west of Portland, Oregon (**Fig. A2.1.1**).

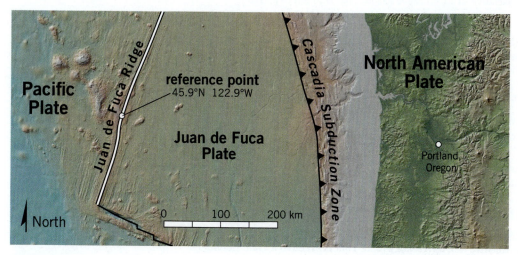

Figure A2.1.1

A **Velocity Relative to an NNR Reference Frame.** The plates move at about the same rate as that at which our fingernails grow. The Pacific plate moves at a rate of 45.7 millimeters per year (mm/yr.) toward compass azimuth 309.6° in a reference frame that is commonly used by geoscientists to define the motion of individual plates relative to the rest of the planet (a "no-net rotation," or NNR reference frame).

1. Starting at the reference point in **Fig. A2.1.2**, use a ruler to draw an arrow that is 45.7 mm long and points toward 309.6°. Label the point at the tip of the arrow **P**. This arrow represents the instantaneous velocity vector for the Pacific Plate at the reference point as that motion is observed in a NNR reference frame.

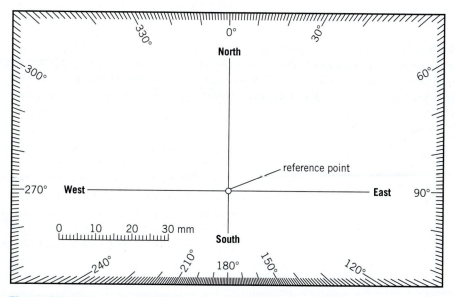

Figure A2.1.2

2. At the same reference point along the ridge axis, the Juan de Fuca Plate moves at 19.4 mm/yr. toward 42.4° relative to the NNR reference frame. That is about the rate at which your toenails grow. Starting at the reference point on **Fig. A2.1.2**, use a ruler to draw the vector—19.4 mm long pointed toward 42.4°—for the instantaneous velocity of the Juan de Fuca Plate as defined at the reference point in the NNR reference frame, and label the tip of the vector *J*.

3. We will assume that as much new lithosphere will be added to the Pacific Plate as is added to the Juan de Fuca Plate along this ridge axis for the sake of simplicity. That is, we will assume that spreading is symmetric. Find the point that is halfway between points *P* and *J*, and label that point *R* for "ridge." Draw an arrow from the reference point to *R*. That arrow represents the instantaneous velocity vector for that point along the ridge as defined in the NNR reference frame.

B Velocity Relative to a Reference Frame That Is Fixed to a Plate.

1. Draw a vector arrow from the reference point with the same length and orientation as the vector from point *P* to point *J*, and label it *P2J*. That vector represents the instantaneous velocity of the Juan de Fuca Plate relative to the Pacific Plate at the reference point, and the length of that vector is the speed.

2. How fast and toward what azimuth does the Juan de Fuca Plate move relative to the Pacific Plate at the reference point each year? Use a protractor and the north-south line to determine the azimuth as in **Fig. 2.4**.

   speed *P2J*: _____ mm/year   azimuth: _____

   That velocity is the rate at which new crust is added along the ridge at the reference point each year.

3. Draw a vector arrow from the reference point that would represent the instantaneous velocity of the Pacific Plate relative to the Juan de Fuca Plate, and label it *J2P*.

   speed *J2P*: _____ mm/year   azimuth: _____°

C Velocity Relative to a Reference Frame That Is Fixed to a Plate Boundary.

1. Draw a vector arrow from the reference point with the same length and orientation as the vector from point *R* to point *P*, and label this vector *R2P*. This is the instantaneous velocity vector of the Pacific Plate relative to the ridge axis, determined at the reference point.

   speed *R2P*: _____ mm/year   azimuth: _____°

2. Draw a vector arrow starting at the reference point that represents the instantaneous velocity of the Juan de Fuca Plate relative to the ridge axis, determined at the reference point, and call that vector *R2J*. These vectors indicate the rate at which new crust is added to the trailing edge of each individual plate at the reference point.

Important Reminder: We can measure a velocity or a displacement (the movement from an initial location to a different location) only relative to some reference frame. Before we describe a velocity or a displacement, it's important to ask "velocity relative to what?"

D **REFLECT & DISCUSS**   You watch the Sun over the course of a day. It rises in the east, passes just south of overhead at noon (if you are in the northern hemisphere above the tropics), and sets in the west. You accept that you are observing the Sun while located on the surface of a near-spherical planet and that the Sun remains about the same distance away from that planet all day long. Describe the motion of the Sun relative to Earth as observed in your reference frame that day.

Name: _____ **Course/Section:** _____ **Date:** _____

A Analyze **Figs. 2.1** and **2.3**. On what lithospheric plate do you live? (Notice that if you live in California, west of the San Andreas Fault, you are not considered to be on the North American Plate.) _____

B Go to the JPL-NASA GPS Time Series website at **http://sideshow.jpl.nasa.gov/post/series.html**. The map displays the location of each GPS station as a small green dot with a yellow line that indicates direction that the GPS station is moving in a no-net rotation (NNR) reference frame. The particular NNR reference frame used for these data is called IGS08. Use the + button in the lower right corner of the map to zoom in and to reveal more GPS stations. Find the GPS station that is the closest to where you live, click on the green circle, and when the small white box with the site velocity data opens, double-click on that box to expand it. Copy the data requested in questions 1–3, then complete the Plate Motion Plot (**Fig. A2.2.1**) for the station (see **Fig. 2.11**).

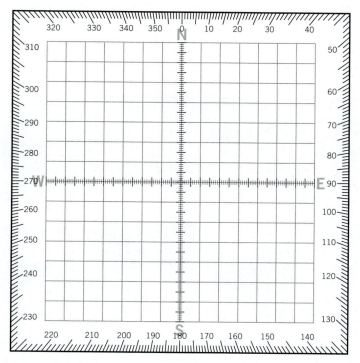

Figure A2.2.1

1. GPS station name (4 characters): _____

2. Latitude time series—the station is moving (choose one: north, south) at a rate of _____ mm/yr. with an uncertainty of _____ mm/year.

3. Longitude time series—the station is moving (choose one: east, west) at a rate of _____ mm/yr. with an uncertainty of _____ mm/year.

4. Toward what azimuth is this GPS station moving?

5. At what velocity is this GPS station moving?

6. Return to the JPL-NASA Time Series website, and click on "Geodetic Positions and Velocities" above the map. Scroll down to the name of your station, and record its current position in latitude and longitude in the following blanks. The coordinates are expressed in decimal degrees to nine places to the right of the decimal. All of those decimal places are meaningful for GPS sites that are located to the closest millimeter. North latitudes are positive numbers, and south are negative; east longitudes are positive, and west are negative.

Current latitude: _____ Current longitude: _____

7. Go to the Plate Motion Calculator hosted by UNAVCO at **https://www.unavco.org/software/geodetic-utilities/ plate-motion-calculator/plate-motion-calculator.html**. Enter the latitude and longitude of your GPS station, being sure to include the proper sign (+/−). Then enter the site name; choose MORVEL 2010 as your model; select your tectonic plate, NNR no-net-rotation for your reference frame, and HTML table w/ local E&N plate velocities as the output format. Ignore the rest of the input boxes. Then press submit. Record your results.

N velocity: _____ mm/yr.   E velocity: _____ mm/yr.   Speed: _____ mm/yr.   Azimuth: _____° clockwise from north

The Plate Motion Calculator provides the instantaneous velocity of a given point in a NNR reference frame, assuming that the plate is not deforming. How do the results from the Plate Motion Calculator compare with your earlier results (questions 2–5)?

_____

_____

The velocity of your GPS station includes the velocity of the plate it is located on as well as a component related to the deformation of the part of that plate where the GPS station is located.

C Return to the JPL-NASA Time Series website, and view the map. From the website map, draw a vector arrow from each of the green circles on the map in **Fig. A2.2.2** to show the general direction that at least one spot on Africa, Arabia, Australia, China, Europe, India, North America, Russia, and South America are currently moving relative to a NNR reference frame. In a general way, use the length of the arrows to reflect the velocities of the sites, just as the length of the yellow lines on the Time Series website indicates velocity.

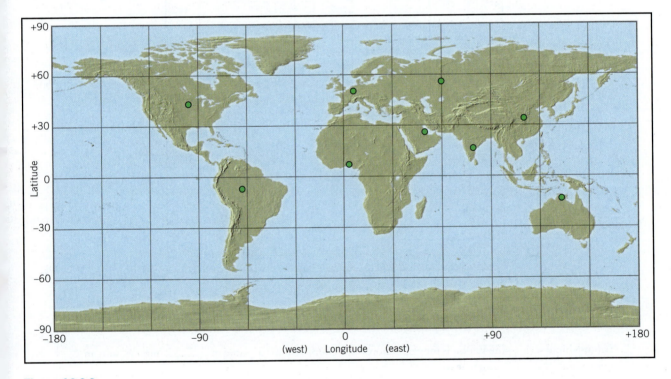

Figure A2.2.2

D **REFLECT & DISCUSS** Do you think that determining the velocity of one point on a plate is sufficient to tell where the entire plate is moving? Why or why not? *Hint:* Represent a plate using a piece of paper on a tabletop. Slide the paper a very short distance from its initial position across the tabletop, maybe with a bit of a twist of your wrist. Now consider whether you could reconstruct the motion of the piece of paper from initial to final position if you only have information about the motion of one point on the paper.

As a lithospheric plate moves over a hotspot in the upper mantle below the plate, a volcano develops directly above the hot spot. As the plate continues to move, the volcano drifts away from the hotspot and eventually becomes dormant. Meanwhile, a new volcano develops over the hotspot next to the older volcano. The result is a trail of volcanoes with one end of the line located over the hot spot and quite active and the other end distant and inactive. In between is a succession of volcanoes that are progressively older with distance from the hot spot.

A **Figure 2.13** shows the distribution of the Hawaiian Islands Chain and Emperor Seamount Chain. The numbers indicate the average age of the volcano in millions of years (Myr) obtained from isotopic dating of the basaltic igneous rock of which each island is composed.

1. If both the Emperor and Hawaiian Islands Chains developed as a result of the same mantle hotspot, what is a possible reason that the hotspot trail changes direction at ~42 Myr?

2. What was the rate of Pacific Plate motion relative to the Hawaiian hotspot as it was developing the 2,300 km long Emperor Seamount Chain from 65 Myr to 42 Myr? Express the rate in centimeters per year (cm/yr.). In what direction was the plate moving north—relative to the hotspot during that time interval?

3. What was the rate of Pacific Plate motion relative to the Hawaiian hotspot from 5.1 to 0.8 Myr, expressed in cm/year?

4. Using Loʻihi Seamount as the current location of the Hawaiian hotspot, what was the rate of Pacific Plate motion relative to the Hawaiian hotspot from 0.8 Myr to today, expressed in cm/yr?

5. Go to the JPL-NASA GPS Time Series website at **http://sideshow.jpl.nasa.gov/post/series.html**. The map locates each GPS station with a green dot with a yellow line that extends outward in the direction that the GPS station is moving relative to the NNR reference frame. GPS station HNLC is located on Oahu.
   (a) How does the current motion of HNLC on Oahu compare to the direction of Pacific Plate motion relative to the Hawaiian hotspot over the past ~42 million years?

   (b) GPS station HNLC on Oahu has the following component velocities relative to the NNR reference frame as of January 24, 2016: moving north at 3.6602 ± 0.0028 cm/yr. and moving west at 6.2665 ± 0.0030 cm/year. Use the Pythagorean Theorem to find the current speed of the Pacific Plate at Oahu, relative to the NNR reference frame. Show your work.

6. **REFLECT & DISCUSS** Based on all of your work above, explain how the direction and rate of Pacific Plate movement changed over the past ~70 million years.

B The map in **Fig. A2.3.1** shows the distribution of a trail of volcanic centers in Wyoming, Idaho, and Nevada. All of these volcanic centers are now inactive except the youngest one located in Yellowstone National Park. Hot springs, geysers, and earthquakes demonstrate that Yellowstone is still volcanically active.

1. What does this progressive chain of volcanic centers indicate about the possible origin of the active volcanism at Yellowstone? Support your answer with evidence.

2. Based on the map, what was the average speed and direction of North American Plate motion at Yellowstone, relative to the hotspot, since 13.8 Myr?

   Add an arrow (vector) and rate label to the map in **Fig. A2.3.1** to show this movement.

3. Plate Boundary Observatory GPS station P717 near the east gate of Yellowstone National Park had the following component velocities relative to the NNR reference frame as of January 24, 2016: moving south at 0.8200 ± 0.0088 cm/yr. and moving west at 1.4783 ± 0.0068 cm/year.

   (a) Use the Pythagorean Theorem to find the current speed of the North American Plate at P717 relative to the NNR reference frame. Show your work.

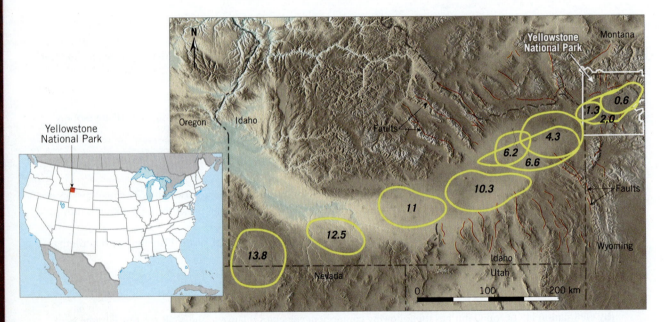

Figure A2.3.1

   (b) How do the present-day direction and speed of P717 relative to the NNR reference frame compare with your estimate of the average direction and speed of the North American Plate relative to the hotspot during the past 13.8 Myr?

4. **REFLECT & DISCUSS** How do hotspots help us understand plate tectonic processes and rates?

Name: _____ Course/Section: _____ Date: _____

A Explore how solid rock can be rigid and stiff in the lithosphere but soft and fluidlike in the asthenosphere.

1. Obtain three pieces of Silly Putty: one that is at room temperature, one that is cold (from a refrigerator or ice chest), and one that is warm (heated in a plastic bag submerged in hot water).

   (a) Take the *room-temperature* piece of Silly Putty, roll it into a sphere, and set it on the tabletop while you do steps (b) and (c) of this experiment.

   (b) Form the *cold* Silly Putty into a cube or cylinder, and place it in the middle of a wooden board (used to protect the tabletop). Do this quickly before the putty warms up.

   (1) Hit the cold putty with a quick stroke from a hammer with the intent to deform the putty. Describe how the putty deformed at a very high strain rate.

   (2) Gather up the remains of the cold putty, form it into a single mass, and pull on it with a quick motion. Describe how the cold putty deformed at a high strain rate.

   (3) Reform the putty and pull on it slowly but steadily. Describe how the cold putty deformed at the slower strain rate.

   (c) Now take the piece of *warm* Silly Putty.

   (1) Pull on it with a quick motion. Describe how the warm putty deformed at a high strain rate.

   (2) Reform the warm putty and pull on it slowly but steadily. Describe how the warm putty deformed at the slower strain rate.

   (3) Which is more difficult to deform: warm putty or cold putty?

   (d) Examine the sphere of Silly Putty that you set aside in step (a).

   (1) Describe what has happened to it.

(2) Was this change accompanied by any fractures or faults, or did it just flow? Explain.

    (e) Roll the room-temperature Silly Putty into a smooth ball, bounce it on the tabletop, and then examine the ball.

        (1) Was there any permanent deformation on the surface of the ball where it impacted the tabletop—any dents, fractures, or faults?

        (2) What rheology did the room-temperature putty display at very high strain rate?

    (f) Under what conditions of temperature (hot or cold) and strain rate (fast or slow) does Silly Putty best exhibit brittle-elastic mechanical behavior? _____

    Under what conditions does it best exhibit ductile-viscous mechanical behavior? _____

2. **REFLECT & DISCUSS** How does your research on Silly Putty help explain how rocks may behave in the lithosphere and beneath the lithosphere?

B A lava lamp contains a mixture of colored paraffin wax and another compound that makes the solid wax just a bit denser than the clear liquid (usually mineral oil or water) in the lamp. The heat is supplied by a light bulb below the sealed glass bottle of the lamp. There is a wire heating coil in the bottom of the sealed bottle to help transfer heat energy into the wax. The top and sides of the sealed glass bottle are cooler than its heated base.

1. Observe and describe the motions of the colored wax that occur over one full minute of time, starting with wax at the bottom of the lamp and its path through the lamp.

2. What causes the wax to move from the base of the lamp to the top of the lamp? (Be as specific and complete as you can.)

3. What causes the wax to move from the top of the lamp to the base of the lamp? (Be as specific and complete as you can.)

4. What heat-transfer mechanism is displayed by the rising and falling molten wax in the lamp?

C Minor partial melting in the asthenosphere causes liquid magma to form. If the molten rock is able to move at all, it will leave behind the heavier solid rock that did not melt. In what direction is the molten rock most likely to move through the viscous asthenosphere? Explain your answer.

Paleomagnetic Stripes and Sea-Floor Spreading

Activity 2.5

Name: _____ Course/Section: _____ Date: _____

A Analyze the sea-floor part of the map in **Fig. A2.5.1**, which depicts the area just off the Pacific Coast, west of California, Oregon, Washington, and southwest Canada. The colored bands are marine magnetic anomalies. Colored bands are rocks with a positive (+) magnetic anomaly, so they have normal polarity, like now. The white bands are rocks with a negative (−) magnetic anomaly, so they have reversed polarity. Different colors indicate the ages of the rocks in millions of years as shown in the magnetic polarity time scale provided.

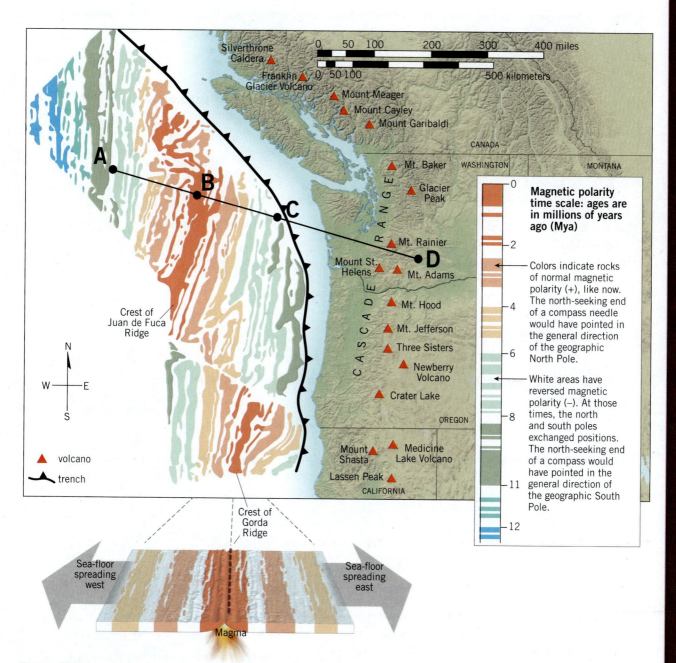

Figure A2.5.1

1. Using a pencil, draw a line on the seafloor to show where new ocean crust and lithosphere is forming now (zero millions of years old). Using **Figs. 2.1, 2.3,** and **2.12** as guides, label the segments of your line that are Juan de Fuca Ridge and Gorda Ridge (divergent plate boundaries). Then label the segments of your pencil line that are transform fault plate boundaries. Add half arrows to the transform fault boundaries to show the motion of the two plates relative to the transform fault.

2. What has been the average rate and direction of sea-floor spreading in cm per year (cm/yr.) west of the Juan de Fuca Ridge, from B to A? Show your work.

3. What has been the average rate and direction of sea-floor spreading in cm per year (cm/yr.) east of the Juan de Fuca Ridge, from B to C? Show your work.

4. Notice that rocks older than 11 million years are present west of the Juan de Fuca Ridge but not east of the ridge. What could be happening to the sea-floor rocks along line segment C-D that would explain why rocks older than 11 million years no longer exist on the seafloor east of the ridge?

5. Notice the black curve with triangular barbs just east of point C:

   (a) If you could take a submarine to view the sea floor along this line, what feature would you expect to see? (*Hint:* see **Fig. 2.1A.**)

   (b) Based on **Fig. 2.1**, what lithospheric plate is located *east* of the black barbed line at point C?

   (c) Based on **Fig. 2.1**, what lithospheric plate is located *west* of the black barbed line at point C?

6. **REFLECT & DISCUSS**   Notice the line of volcanoes that form the Cascade Range, extending from northern California to southern Canada. These are active volcanoes, meaning that they still erupt from time to time. What sequence of plate tectonic events is causing these volcanoes to form?

Name: _____ **Course/Section:** _____ **Date:** _____

A The map of the northern Atlantic Ocean Basin (**Fig. A2.6.1**) shows *isochrons* (lines of equal age) of the basaltic crust beneath the sediments that have accumulated on the seafloor. These isochrons were derived by Maria Seton and her colleagues (2012) from an analysis of marine magnetic anomalies, and their ages are based on the geomagnetic polarity time scale. *GPlates* (**http://www.gplates.org**) was used to help make this map. The red line on the map shows the location of the divergent boundary—the axis of the mid-ocean ridge—between the North American and African Plates.

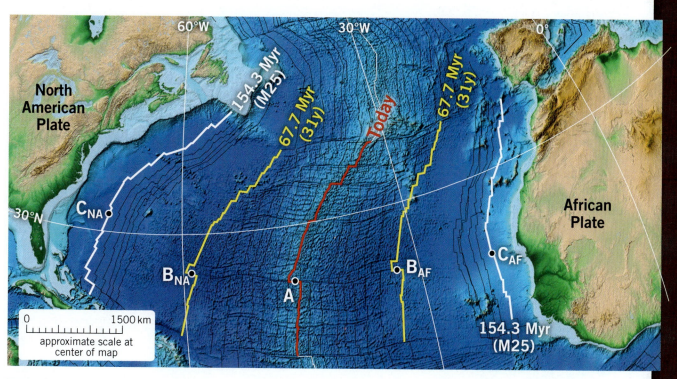

Figure A2.6.1

The approximate great-circle distances between the points identified on the map are listed here. The uncertainty in the distances provided is probably on the order of 10 km.

C_{NA} to B_{NA}	1354 km	C_{NA} to A	2617 km
B_{NA} to A	1338 km	C_{AF} to A	2599 km
B_{AF} to A	1320 km	B_{NA} to B_{AF}	2649 km
C_{AF} to B_{AF}	1279 km	C_{NA} to C_{AF}	5107 km

1. What is the average speed at which B_{NA} drifted away from the ridge at A during the past 67.7 Myr expressed in cm/yr.?
_____ cm/yr.
What is the average speed that B_{AF} drifted from A? _____ cm/yr.
Which plate moved faster relative to the ridge over the past 67.7 Myr, if either? _____

2. The area between C_{NA} and B_{NA} consists of oceanic lithosphere added along the Mid-Atlantic Ridge to the North American Plate between 154.3 and 67.7 Myr. What is the average speed at which new lithosphere was added to the North American Plate along that line? _____ cm/yr.
Do the same analysis for the lithosphere between points C_{AF} and B_{AF}. What is the average speed at which new lithosphere was added to the African Plate along that line? _____ cm/yr.
Which plate moved faster relative to the ridge between 154.3 and 67.7 Myr, if either? _____

B Given your answers in part **A**, did the North Atlantic Ocean Basin develop by adding lithosphere symmetrically along the mid-ocean ridge, or was new lithosphere added more rapidly to one side than the other? _____ If spreading was asymmetric, which plate had more lithosphere added, or did the asymmetry vary from plate to plate over time? _____

C Use the rates that you calculated above and the map scale to estimate when the coastlines of North America and Africa might have last touched before they were separated by the opening of the North Atlantic Ocean Basin. _____ _____

D **REFLECT & DISCUSS** Based on the rates you calculated above, estimate the number of meters that Africa and North America have moved apart since the United States was formed in 1776. Discuss what you did to accommodate the uncertainty in your estimate.

A Use a red colored pencil or pen to outline the location of all plate boundaries on the map in **Fig. A2.7.1**. Do your work carefully. Then label the East Pacific Ridge, Galapagos Ridge, Chile Ridge, and all of the plates. Refer to **Fig. 2.1** for help with the tectonic features.

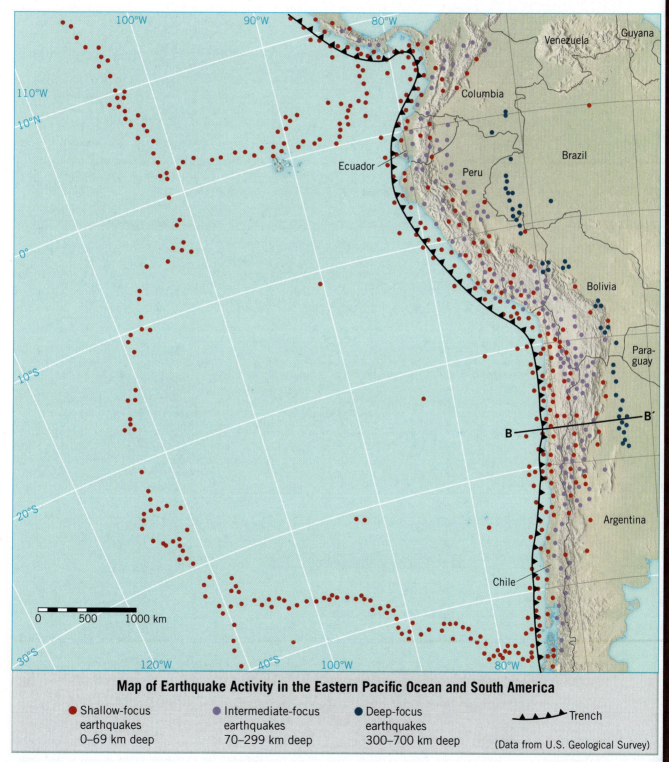

Map of Earthquake Activity in the Eastern Pacific Ocean and South America

● Shallow-focus earthquakes 0–69 km deep   ● Intermediate-focus earthquakes 70–299 km deep   ● Deep-focus earthquakes 300–700 km deep   ▲▲▲ Trench

(Data from U.S. Geological Survey)

Figure A2.7.1

B Notice line B–B' on the map in part **A** and the fact that shallow, intermediate, and deep earthquakes occur along it. [Each earthquake begins at a point beneath the surface called the *focus* (plural, *foci*).] Using data that were provided by the U.S. Geological Survey in Fig. A2.7.2, plot the locations of earthquake foci on the cross-section (Fig. A2.7.3). Volcanoes also occur at Earth's surface along line B–B'. Plot the locations of the volcanoes listed in the table by drawing a small triangle for each on the surface along the zero-depth line.

Location East or West of Trench	Depth of Earthquake (or volcano location)	Location East or West of Trench	Depth of Earthquake (or volcano location)	Location East or West of Trench	Depth of Earthquake (or volcano location)
200 km West	20 km	220 km East	30 km	410 km East	150 km
160 km West	25 km	250 km East	volcano	450 km East	50 km
60 km West	10 km	260 km East	120 km	450 km East	150 km
30 km West	25 km	300 km East	volcano	470 km East	180 km
0 (trench)	20 km	300 km East	110 km	500 km East	30 km
10 km East	40 km	330 km East	volcano	500 km East	160 km
20 km East	30 km	330 km East	40 km	500 km East	180 km
50 km East	60 km	330 km East	120 km	540 km East	30 km
51 km East	10 km	350 km East	volcano	590 km East	20 km
55 km East	30 km	390 km East	volcano	640 km East	10 km
60 km East	20 km	390 km East	40 km	710 km East	30 km
80 km East	70 km	390 km East	140 km	780 km East	530 km
100 km East	10 km	410 km East	volcano	800 km East	560 km
120 km East	80 km	410 km East	25 km	820 km East	610 km
200 km East	110 km	410 km East	110 km	880 km East	620 km

Figure A2.7.2

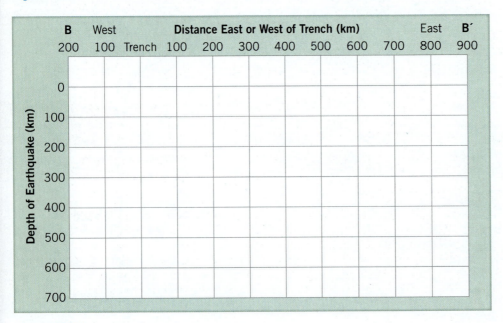

Figure A2.7.3

1. What kind of plate boundary is shown in your cross-section? _____

2. Draw a curve in the cross-section to show your interpretation of where the top surface of the subducting plate is located.

3. Label the part of your cross-section that probably represents earthquakes in the subducting slab. Then label the part of your cross-section that probably represents earthquakes in the South American Plate above the subducting Nazca Plate.

4. At what depth does magma probably originate below the volcanoes, just above the subducting plate: _____ km. How can you tell?

5. **REFLECT & DISCUSS** What is the deepest earthquake plotted on your cross-section? Why do you think that earthquakes occur at hundreds of kilometers depth along subducting slabs but not elsewhere in the mantle at that same depth?

Mineral Properties, Identification, and Uses

Pre-Lab Video 3

http://goo.gl/kKBN1l

Contributing Authors

Jane L. Boger • *SUNY, College at Geneseo*
Philip D. Boger • *SUNY, College at Geneseo*
Roseann J. Carlson • *Tidewater Community College*

Charles I. Frye • *Northwest Missouri State University*
Michael F. Hochella, Jr. • *Virginia Polytechnic Institute*

▲ Blue bands of chalcedony (microscopic quartz) surround centimeter-long quartz crystals in the middle of a geode.

BIG IDEAS

Minerals comprise rocks and are described and classified on the basis of their physical and chemical properties. Every person depends on minerals and elements refined from them. However, modern society's use-and-discard practice, in which products derived from minerals often end up in landfills, is unsustainable for industrial minerals and elements that are rare or require great risk or energy to acquire. It is better to embrace a sustainable use-and-recycle ethic.

FOCUS YOUR INQUIRY

Think About It   What are minerals and crystals, and how are they related to rocks and elements?

ACTIVITY 3.1   Mineral and Rock Inquiry (p. 70, 97)

Think About It   How and why do people study minerals?

ACTIVITY 3.2   Mineral Properties (p. 72, 98)

ACTIVITY 3.3   Determining Specific Gravity (SG) (p. 83, 100)

ACTIVITY 3.4   Mineral Identification and Uses (p. 84, 101)

Think About It   How do you personally depend on minerals and elements extracted from them? How sustainable is your personal dependency on minerals and elements extracted from them?

ACTIVITY 3.5   The Mineral Dependency Crisis (p. 93, 105)

ACTIVITY 3.6   Urban Ore (p. 95, 106)

Mineral and Rock Inquiry, (p. 97)

Think About It   What are minerals and crystals, and how are they related to rocks and elements?

Objective   Analyze photographs of four specimens. Observe the difference between an individual mineral grain and a rock. Recognize that different types of minerals may be present in the same rock. Try to recognize the boundaries of different mineral grains in a rock.

Before You Begin   Read the following sections: How Are Minerals Classified? and Mineral Structure: Crystallography.

What Are Minerals?

Most geological solids are composed of minerals. A **mineral** is a naturally occurring crystalline solid that has been formed by geologic processes, on Earth or elsewhere in the universe. Different types of minerals are distinguished from each other by their well-defined chemical composition and crystal structure, which result in a unique set of physical properties (like hardness, how it breaks, color, density) that we can use to identify the mineral.

A crystalline solid is a bit like a well-made brick wall. The wall is built up using many identical bricks, arranged in a regular pattern that is repeated throughout the wall. Elements in a **crystalline solid** occur in a regular pattern that is repeated in three dimensions billions of times in even the smallest mineral grain that we can see (**Fig. 3.1**). The regular geometric arrangement of atoms in a crystal is called a **lattice**. The 3-dimensional geometry of a lattice not only defines the internal structure of a crystal but also controls the shape of the outer surface of any crystal that grows freely into a gas or liquid without solid barriers to its growth. The flat crystal faces we observe on the mineral specimens that are on display in a museum are a direct consequence of the internal lattice structure of the mineral.

Geoscientists tend to use the word **crystal** when we are talking about the lattice structure or about mineral specimens that have nicely developed crystal faces. We use the term **mineral grain** to refer to a particle composed of a single crystal of a mineral however we find it—as a singular mineral specimen with well-developed crystal faces, as a broken piece of an individual crystal, as one crystal within an intergrown mass of crystals, or as a rounded sedimentary particle developed from a single crystal of a mineral.

Some types of mineral form because of biological activity and also develop in the geological environment independent of life processes. Examples include the mineral *aragonite* in clamshells and tiny *magnetite* crystals in the human brain. Solids that occur only because of

B. When struck with a hammer, galena breaks along flat *cleavage surfaces* (planes of weak chemical bonding within the crystal) that intersect at 90° angles to form rectangular shapes.

C. Scanning tunneling microscope (STM) image of galena showing the orderly arrangement of its lead and sulfur atoms.

x1

Sulfur atoms are blue, lead atoms are orange.
nm = nanometer = 1 millionth of a millimeter

A. Lattice structure of galena. Sulfur atoms are blue, and lead atoms are orange.

Figure 3.1 Crystal shape, cleavage, and atomic structure. Galena is lead sulfide—PbS—and is an ore mineral from which lead (Pb) and sulfur (S) are extracted. (STM image by C.M. Eggleston, University of Wyoming)

biological activity but that do not occur in the absence of life processes are not generally considered to be minerals.

Some geological solids, such as volcanic glass and the petroleum solids known as *bitumens,* do not have a crystalline structure and so are not minerals. Other geologic materials like *opal* have at least some characteristics of a mineral but do not fit the conventional definition either because of their variable composition or indefinite lattice structure. People can make synthetic crystals that have all of the physical and chemical properties of a naturally occurring mineral. Synthetic crystals are not mineral specimens because they are not formed by geologic processes. You might wonder, "Who gets to decide whether something is a mineral or not?" The International Mineralogical Association is the scientific group responsible for providing coherent guidelines for defining minerals.

Geologists have identified and named thousands of different kinds of minerals based on the mineral chemistry and structure, but minerals are often classified into smaller groups according to their importance or use. For example, some are known as **rock-forming minerals** because they are the minerals that make up most of Earth's crust. Another group, called **industrial minerals**, are the main nonfuel raw materials used to sustain industrialized societies like ours. Some industrial minerals are used in their raw form, such as quartz (quartz sand), muscovite (used in electronics, plastics, and building materials), and gemstones. Most are refined to obtain specific elements such as iron, copper, and sulfur.

Atoms, Elements, Minerals, Rocks—How Are They Related?

All minerals are composed of atoms. Understanding the basic structure and properties of different kinds of atoms is key to understanding how they combine to form minerals.

An **atom** consists of a nucleus and one or more electrons associated with that nucleus and is the smallest unit of matter that has the properties of a chemical element. There are more than 90 naturally occurring **elements**, which differ from one another by the number of protons in their nucleus. For example, every atom of the element *carbon* contains six protons in its nucleus, and every *oxygen* atom has eight protons.

Minerals are composed of atoms bonded together in a crystalline lattice. Minerals such as graphite and diamond are composed of just one type of element, but most contain two or more different elements. There are more than 5,000 different types of mineral that are currently recognized by science, which differ from one another by their chemical composition and crystal structure.

Rock is an aggregate of mineral grains. Some rocks are composed of one type of mineral, but most contain two or more types. So, atoms bond together in crystalline lattices to form the many different types of mineral, and mineral grains combine with other mineral grains to form rock.

How Are Minerals Classified?

Geologists classify minerals based on their chemical composition and lattice structure.

Mineral Chemistry

Each type of mineral has a well-defined chemical composition, and we use chemical formulas to describe those compositions. Each element has its own symbol. For example, the symbol for silicon is Si. **Ions** are charged elements or compounds, and that charge is due to an unequal number of positively charged protons and negatively charged electrons in the ion. The silicon ion has four more protons than electrons, so it is a positively charged particle. The charge on an ion is shown as a superscript after the symbol, so for silicon the ion is Si^{4+}. The chlorine (Cl) ion has one more electron than proton, so it is a negatively charged particle whose ionic symbol is Cl^-.

The chemical formula SiO_2 tells us that for every single silicon atom, there are two oxygen (O) atoms in the mineral. The formula Fe_2O_3 tells us that for every two iron (Fe) atoms there are three oxygen atoms. The formula $Ca, Mg(CO_3)_2$ tells us that calcium (Ca) and magnesium (Mg) atoms substitute for each other at the same position in the crystal lattice, and for every atom of Ca or Mg, there are two CO_3 groups containing one carbon (C) bonded to three oxygen atoms.

Minerals can be grouped into chemical classes such as the following:

- **Native elements** are minerals composed of just one type of element, such as graphite and diamond (C), copper (Cu), sulfur (S), gold (Au), silver (Ag), and mercury (Hg).
- **Silicate minerals** contain the two most abundant elements in Earth's crust bonded together: oxygen and silicon. In most silicates, silicon and oxygen are bonded with other elements such as iron, magnesium, aluminum (Al), and potassium (K). A few of the many examples of silicates include quartz (SiO_2), potassium feldspar ($KAlSi_3O_8$), and olivine [$(Mg, Fe)_2SiO_4$].
- **Oxide minerals** contain oxygen combined with one or more metals other than silicon, such as copper, iron, aluminum, magnesium, zinc (Zn), titanium (Ti), manganese (Mn), tin (Sn), chromium (Cr), uranium (U), beryllium (Be), niobium (Nb), or tantalum (Ta). Examples include hematite (Fe_2O_3), magnetite (Fe_3O_4), and corundum (Al_2O_3).
- **Hydroxide minerals** contain hydrogen and oxygen as hydroxyl ions $(OH)^-$ or water molecules (H_2O) combined with other elements. One common example is goethite [$FeO(OH)$].
- **Sulfide minerals** contain sulfur in the form of a sulfide ion (S^{2-}) bonded to a metal, such as silver, copper, iron, mercury, zinc, lead (Pb), nickel (Ni), arsenic (As), antimony (Sb), molybdenum (Mo), or cobalt (Co). Examples include pyrite (FeS_2), galena (PbS), and sphalerite (ZnS). Scratching or crushing a sulfide mineral commonly releases a faint sulfur smell.

- **Sulfate minerals** contain sulfur and oxygen as sulfate ions $(SO_4)^{2-}$ that are bonded to one or more positive ions such as calcium, copper, barium, lead, potassium, or strontium (Sr). Hydrous sulfates like gypsum $(CaSO_4 \cdot 2H_2O)$ contain water in their structure while anhydrous sulfates like barite $(BaSO_4)$ and anhydrite $(CaSO_4)$ do not.
- **Carbonate minerals** contain carbon and oxygen as carbonate ions $(CO_3)^{2-}$ that are bonded to one or more positive ions such as calcium, magnesium, iron, manganese, zinc, copper, barium, lead, or strontium. Carbonates all react to hydrochloric acid by releasing carbon dioxide (CO_2) gas. Examples of carbonates include calcite and aragonite $(CaCO_3)$ and dolomite $[Ca, Mg(CO_3)_2]$.
- **Halide minerals** contain elements belonging to the halogen group in which ions such as fluorine (F^-), chlorine (Cl^-), bromine (Br^-), or iodine (I^-) are bonded to a metal. Examples include halite (NaCl) and fluorite (CaF_2).
- **Phosphate minerals** contain phosphorus and oxygen combined in phosphate ions $(PO_4)^{3-}$ that are bonded to one or more positive ions. One common example is apatite $Ca_5(PO_4)_3(F, Cl, OH)$.

Mineral Structure: Crystallography

The scientific study of the structure and symmetry of crystals is called **crystallography**. We can think of crystallography as being concerned with the 3-dimensional geometry of a crystal lattice, but not with the specific elements that occupy those lattice positions. In other words, crystallography is concerned with structure, not chemistry. We will focus on a few parts of crystallography that might help you understand and identify common minerals you might see in the lab.

Well-built brick walls grow by the addition of bricks to the existing wall in a manner that duplicates the existing pattern of bricks. Minerals grow in approximately the same way. Atoms or small molecules (small groups of atoms bonded together) move freely in a liquid or gas around the growing crystal. The atoms or molecules bond to the surface of the crystal in a manner that extends the existing structure exactly, or nearly so. This process continues for an individual crystal until either there is no more space available to grow or the supply of necessary atoms or molecules is exhausted.

Each different type of mineral has a characteristic structure—its crystalline lattice (**Fig. 3.1A**). The shape of a mineral grain that has grown in a gas or liquid without solid barriers to its growth is the outward manifestation of that internal lattice structure. The planar, flat surfaces of a mineral grain that stopped growing when it was still in contact with gas or liquid are called **crystal faces** (**Fig. 3.2A**). The orientation of each crystal face on a mineral grain reflects the orientation of a particular plane in the crystal lattice. A crystalline lattice is present in every mineral grain whether an individual mineral grain has crystal faces or not.

Crystallographers have defined 7 crystal systems (triclinic, monoclinic, orthorhombic, tetragonal, trigonal, hexagonal, cubic), 7 lattice systems (primitive, body-centered, face-centered, 3 types of base-centered, rhombohedral), 6 crystal families (triclinic, monoclinic, orthorhombic, tetragonal, hexagonal, cubic), 32 crystal classes, and 48 crystal forms for minerals. These details are not useful to you as a novice trying to identify minerals but are worth noting in acknowledgement of the good nerdy fun that crystallographers have had over the past 400 years working to understand the internal structure of crystalline matter.

Different types of minerals can have the same crystalline structure, although different chemistry. For example, the minerals halite (NaCl) and galena (PbS) are both members of the same crystal family (cubic), system (cubic), and class. Both crystallize as cubes, as well as in other forms; for example, galena also crystallizes as a dodecahedron. Pyrite (FeS_2) is also a member of the same crystal family and system as halite and galena; however, pyrite is a member of a different crystal class. Pyrite crystallizes as a cube, a pyritohedron or an octahedron. Some of these shape names might be unfamiliar to you, but we will learn about the shapes of mineral grains in one of the sections below.

ACTIVITY 3.2

Mineral Properties, (p. 98)

Think About It How and why do people study minerals?

Objective Analyze and describe the physical and chemical properties of minerals.

Before You Begin Read the following sections: What Are a Mineral's Chemical and Physical Properties?

Plan Ahead This activity requires that you carefully cut out and use the cleavage goniometer that is included in the GeoTools Sheet 1 at the back of this manual.

What Are a Mineral's Chemical and Physical Properties?

Minerals are distinguished from one another based on their unique combination of chemical composition and internal structure. Investigating the chemical composition of an unknown mineral specimen requires specialized laboratory facilities and training, but you can observe some **chemical properties** without a geochemistry lab. For example, one family of minerals reacts when exposed to dilute hydrochloric acid. Another mineral tastes salty as a result of a chemical reaction that takes place on the surface of our tongues. Other minerals develop an oxidized or tarnished coating due to chemical interactions with water or air.

Crystal faces

Crystal faces

A.
A rock made of two large quartz crystals. Crystal growth was unobstructed, except where the two crystals touched and grew together (x1).

Top view: Crystal growth was unobstructed so crystal faces are developed (x1).

	6	1	
5			2
	4	3	

Side view: Crystal faces are not apparent among crowded intergrown crystals (x1).

B.
Rock made of many quartz crystals. Note how crystal growth was obstructed as the sides of many crystals grew together (side view), but tips of the crystals (top view) grew unobstructed into six-sided pyramids. Iron impurity gives the purple amethyst variety of quartz its color.

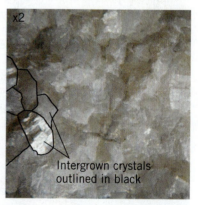

x2

C.
Crystal growth of the calcite crystals in this rock (marble) was obstructed in every direction. The crystals grew together as a dense mass of irregularly shaped crystals instead of perfect crystal forms.

Intergrown crystals outlined in black

The layers of agate are made of long intergrown quartz mineral crystals.

D.
Cut and polished surface of an agate specimen. The layers are made of tiny quartz crystals that can only be seen when magnified using a microscope.

Figure 3.2 Minerals and rocks. Rocks all have more than one mineral grain. A rock might be composed of just one type of mineral or many types of mineral. The rocks in this figure all have several grains of a single type of mineral: quartz.

The **physical properties** of a mineral are determined by its composition and structure and include the mineral's color, the luster of its surface, whether light can pass through it, how hard it is, its density, the shape of its crystals, what broken pieces of it look like, and so on. We will use physical properties that are easily observed in small specimens to identify some of the more important types of minerals in Earth's crust.

Color

We place great emphasis on color as we view the world around us. That focus on color is probably an evolutionary artifact of our need to identify threats and evaluate potential sources of food. Sunlight is a combination of all the colors in the rainbow—all of the colors in the visible spectrum of light. When we see an orange object, we are looking at something that has absorbed all of the other colors and is reflecting only orange back to our eyes.

Some minerals are always a particular color or occur in a small range of colors. Azurite is always blue, malachite is always green, and gold is always, well, gold. However, different minerals can be the same color. Specimens of fluorite and amethyst quartz can both be purple, and specimens of biotite and hornblende can both be black. Some minerals display a range of colors. For example, the microscopic quartz grains in agate (**Fig. 3.2D**) developed with different colors at different stages of the rock's development, resulting in the layered appearance of the rock. The different colors are related to different trace impurities in the quartz lattice that cause different colors of light to be absorbed. The color of some mineral grains can only be seen on a freshly broken surface because chemical weathering has changed the old outer surface in some way.

Color is an important diagnostic property of a mineral, but it is not reliable all by itself because, as discussed above, different minerals can be the same color. Color must be used along with other properties to identify an unknown mineral specimen.

Streak

The color of a fine-grained powder of a mineral is called the mineral's **streak**. We tend to presume that the color of a large mineral specimen will be the same as the color of a powder made from the same specimen, but that is not true of all minerals. All calcite has a white streak, although hand specimens of calcite exist in a variety of different colors. Pyrite has a black streak, even though it has a metallic yellow color similar to brass in hand specimens.

We investigate streak by scratching the mineral grain across a surface that is harder than the mineral. For minerals that are softer than quartz, we use tiles of unglazed porcelain, called **streak plates**: white streak plates for dark minerals and dark streak plates for light minerals (**Fig. 3.3**). Harder minerals can be crushed to a powder or dragged across a diamond file to determine their streak.

Luster

The appearance of a mineral's surface in reflected light is its *luster*. Light reflects from the outermost surface of a mineral lattice to produce a **metallic luster**, which has a mirror-like quality. Some of the many examples of minerals with a metallic luster include gold, silver, copper, pyrite, and galena. Metallic minerals are often subject to oxidation, tarnish, or other forms of chemical weathering that might alter their appearance (**Fig. 3.4**),

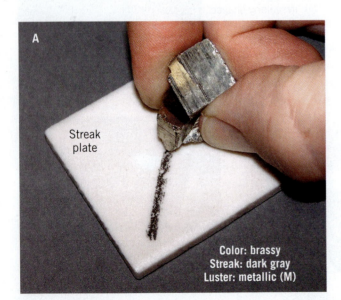

Color: brassy
Streak: dark gray
Luster: metallic (M)

Color: reddish silver
Streak: red-brown
Luster: metallic (M) to nonmetallic (NM)

Figure 3.3 Streak tests. We determine the streak of soft minerals by scratching them across a streak plate with significant force and then blowing away larger pieces of the mineral to reveal the color of the powder. Streak can also be determined by crushing or scratching part of the sample to see the color of its powdered form.

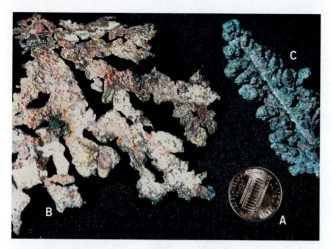

Figure 3.4 Fresh versus tarnished metallic minerals. The native elements are minerals composed of just one element, like copper. When freshly formed or broken, native copper has a reflective metallic luster like this freshly minted copper-plated coin (**A**). However, these dendritic crystals of native copper have tarnished to nonmetallic dull brown (**B**) and green (**C**) colors.

so determining the luster of these minerals might require you to find (or create) a freshly fractured surface below the altered outer surface.

Other minerals have a **nonmetallic luster**, generally because light can penetrate at least a short distance into the crystal lattice before some of it reflects back out again. Light can pass entirely through some nonmetallic hand specimens and can pass through all nonmetallic specimens if they are thin enough. Geologists use special microscopes to look at mineral and rock specimens that are ground to a thickness of just 0.03 mm (30 millionths of a meter). These carefully ground slices of minerals and rocks are mounted on glass slides and are called **thin sections**. There are many additional optical properties of nonmetallic minerals that can be observed using a geologist's petrographic microscope to examine thin sections.

A luster that seems intermediate between "metallic" and "nonmetallic" is called a **submetallic luster**. Tarnish or oxidation on a metallic mineral might impart a submetallic luster, like rust on a piece of steel.

Several other terms are used to describe the luster of nonmetallic specimens, including the following:

- **Vitreous**—resembles the luster or sheen of glass
- **Resinous**—resembles a resin like amber or dried tree sap
- **Silky**—a silk-like reflection of light from thin parallel mineral fibers
- **Pearly**—resembling the luster of a pearl
- **Earthy** (dull)—lacking reflection, like dry soil
- **Waxy**—resembles wax
- **Satin**—resembles satin cloth
- **Greasy**—looks like it is covered in a thin film of oil or grease

Diaphaneity: Transmission of Light Through a Mineral

Some minerals do not allow light to pass through them, even in the thinnest of specimens. These are called **opaque** minerals, and examples include the metallic minerals hematite, galena, magnetite, and pyrite. Nonopaque minerals are called **translucent** because they let light pass through. Some mineral specimens are so clear that you might be able to read these words through them, and they would be called **transparent** specimens.

Shape of a Mineral Grain

The shape of an individual mineral grain depends on the physical, chemical, and environmental conditions in which it grew as well as the crystalline lattice of that mineral. **Euhedral** (pronounced *you-HEE-dral*) mineral grains have well-developed crystal faces. They develop only if a crystal is unrestricted as it grows in a gas or liquid, which is not typical. It is more common for a crystal to grow in an environment that has other crystals around it, resulting in a massive network of intergrown crystals. **Subhedral** mineral grains are partly bound by well-formed crystal faces. **Anhedral** mineral grains grew into the space between existing solid mineral grains and have no apparent crystal faces on their grain boundaries.

A **crystal form** is a set of crystal faces that are related to each other by symmetry. An individual crystal might represent one closed form or two or more open forms. Some of the crystal forms we will use to describe the shape of a mineral grain are shown in **Fig. 3.5**. Many of the laboratory samples of minerals that you will analyze do not exhibit their crystal forms because they are small broken pieces of larger crystals.

The typical shape of a mineral or of aggregates of the same mineral is called the mineral's **habit**. Some common mineral habits are shown in **Fig. 3.6**.

Cleavage

The tendency of some minerals to break (*cleave*) along flat, parallel surfaces is called **cleavage**. The flat surfaces on broken pieces of galena in **Fig. 3.1B** are **cleavage planes** that coincide with surfaces of relatively weak chemical bonding in the crystal lattice. Each different set of parallel cleavage planes is referred to as a **cleavage direction** (**Fig. 3.7**). For example, muscovite has one excellent cleavage direction and splits apart like pages of a book (**Fig. 3.8**). Galena breaks into small cubes and shapes made of cubes, so it has three cleavage directions developed at right angles to one another (**Fig. 3.1**). This is called **cubic cleavage**.

Cleavage can be described as excellent, good, or poor (**Fig. 3.9**). **Excellent cleavage** results in a set of obvious, large, flat, parallel cleavage surfaces. **Good cleavage** is characterized by a set of many small, obvious, flat, parallel surfaces. **Poor cleavage** results in a set of small, flat, parallel surfaces that are difficult to detect.

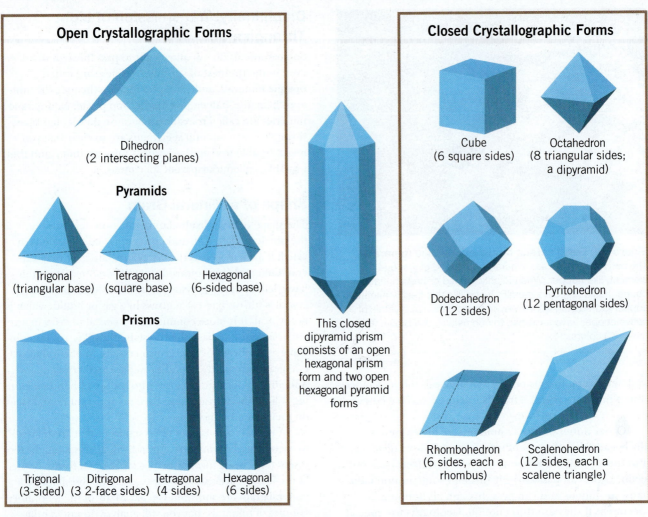

Open Crystallographic Forms

Dihedron
(2 intersecting planes)

Pyramids

Trigonal
(triangular base)

Tetragonal
(square base)

Hexagonal
(6-sided base)

Prisms

Trigonal
(3-sided)

Ditrigonal
(3 2-face sides)

Tetragonal
(4 sides)

Hexagonal
(6 sides)

This closed dipyramid prism consists of an open hexagonal prism form and two open hexagonal pyramid forms

Closed Crystallographic Forms

Cube
(6 square sides)

Octahedron
(8 triangular sides; a dipyramid)

Dodecahedron
(12 sides)

Pyritohedron
(12 pentagonal sides)

Rhombohedron
(6 sides, each a rhombus)

Scalenohedron
(12 sides, each a scalene triangle)

Figure 3.5 Crystal form. Some of the 48 crystal forms in which crystal faces are related to each other by symmetry. Open forms must be combined with other open forms to completely enclose a crystal, like a cardboard file box that needs to be combined with a top to completely enclose the box. Closed forms completely enclose a crystal by themselves.

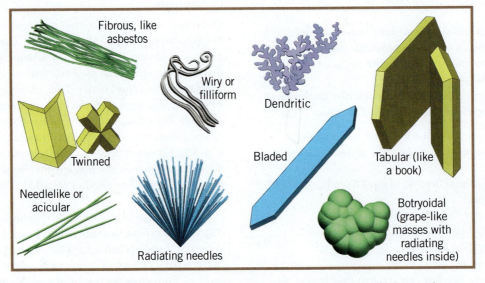

Fibrous, like asbestos

Wiry or filliform

Dendritic

Twinned

Bladed

Tabular (like a book)

Needlelike or acicular

Radiating needles

Botryoidal (grape-like masses with radiating needles inside)

Figure 3.6 Crystal habit. Some common crystal habits—the shapes in which crystals are commonly found.

Number of Cleavages and Their Directions	Name and Description of How the Mineral Breaks	Shape of Broken Pieces (cleavage directions are numbered)	Illustration of Cleavage Directions
No cleavage (fractures only)	No parallel broken surfaces; may have conchoidal fracture (like glass)	Quartz	None (no cleavage)
1 cleavage	**Basal (book) cleavage** "Books" that split apart along flat sheets	Muscovite, biotite, chlorite (micas)	
2 cleavages intersect at or near 90°	**Prismatic cleavage** Elongated forms that fracture along short *rectangular* cross-sections	Orthoclase 90° (K-spar) Plagioclase 86° & 94°, pyroxene (augite) 87° & 93°	
2 cleavages do not intersect at 90°	**Prismatic cleavage** Elongated forms that fracture along short *parallelogram* cross-sections	Amphibole (hornblende) 56° & 124°	
3 cleavages intersect at 90°	**Cubic cleavage** Shapes made of cubes and parts of cubes	Halite, galena	
3 cleavages do not intersect at 90°	**Rhombohedral cleavage** Shapes made of rhombohedrons and parts of rhombohedrons	Calcite and dolomite 75° & 105°	
4 main cleavages intersect at 71° and 109° to form octahedrons, which split along hexagon-shaped surfaces; may have secondary cleavages at 60° and 120°	**Octahedral cleavage** Shapes made of octahedrons and parts of octahedrons	Fluorite	
6 cleavages intersect at 60° and 120°	**Dodecahedral cleavage** Shapes made of dodecahedrons and parts of dodecahedrons	Sphalerite	

Figure 3.7 Cleavage in minerals.

Cleavage surface

Notice how this muscovite mica splits apart into thin, transparent, flexible sheets along its excellent cleavage surfaces

Cleavage surface

Figure 3.8 Cleavage in mica. The mica group of silicate minerals form lamellar-tabular crystals with one excellent cleavage direction. The crystals split easily into thin sheets, like pages of a book. Muscovite mica is usually clear to silvery brown in color. Biotite mica is black.

Cleavage Direction in Pyroxenes and Amphiboles.

Minerals of the pyroxene and amphibole groups generally are dark-colored nonmetallic minerals that have two good cleavage directions. However, pyroxenes can be distinguished from amphiboles on the basis of their cleavage. The two cleavages of pyroxenes intersect at ~87° and ~93°—nearly at right angles (**Fig. 3.10A**). The two cleavages of amphiboles intersect at angles of ~56° and ~124° (**Fig. 3.10B**). These angles can be measured in hand samples using the cleavage goniometer from GeoTools Sheet 1 at the back of this manual. Notice how a blue-green cleavage goniometer was used to measure angles between cleavage directions in **Fig. 3.10**.

Cleavage Direction in Feldspars.

Feldspars have two excellent-to-good cleavage directions that form cleavage prisms with uneven fracture on the ends (**Fig. 3.11**). **Alkali feldspars** (from Na-rich *albite* to K-rich *orthoclase*) have two excellent-to-good cleavage directions that are ~90° to one another. The two excellent-good cleavage directions in **plagioclase feldspars** (from albite to Ca-rich *anorthite*) are ~86° or ~94° from each other. The cleavage goniometer from GeoTools Sheet 1 can be used to help distinguish alkali feldspar from plagioclase feldspar.

Fracture

Any break in a mineral that does not occur along a cleavage plane is called **fracture**. Fracture surfaces are usually not planar, and they tend not to occur in parallel sets. Pure quartz tends to fracture like glass—along ribbed, smoothly curved surfaces called **conchoidal fractures** (**Fig. 3.12A**). Fracture can also be described as **uneven** (rough and irregular, like the milky quartz in **Fig. 3.12B**), **splintery** (like splintered wood), or **hackly** (having jagged edges, like broken metal; **Fig. 3.9D**).

A. Cleavage excellent or perfect (large, parallel, flat surfaces)

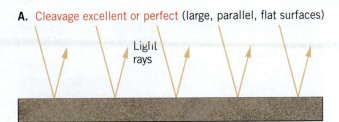

Light rays

B. Cleavage good or imperfect (small, parallel, flat, stair-like surfaces)

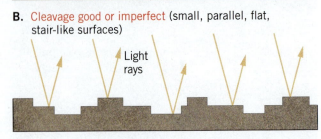

Light rays

C. Cleavage poor (a few small, flat surfaces difficult to detect)

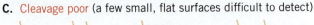
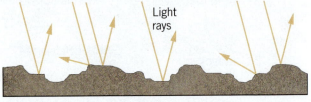

Light rays

D. Fractures (broken surfaces lacking cleavage planes)

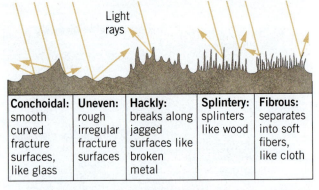

Light rays

Conchoidal:	Uneven:	Hackly:	Splintery:	Fibrous:
smooth curved fracture surfaces, like glass	rough irregular fracture surfaces	breaks along jagged surfaces like broken metal	splinters like wood	separates into soft fibers, like cloth

Figure 3.9 Recognizing cleavage and fracture. Cross-sections showing how cleavage quality is described. *Fracture* refers to any break in a mineral that does not occur along a cleavage plane. Therefore, fracture surfaces are normally not flat and don't necessarily occur in parallel sets.

Hardness

A mineral's **hardness** is a measure of its resistance to scratching. A harder substance will scratch a softer one (**Fig. 3.13**). German mineralogist Friedrich Mohs (1773–1839) developed a scale of relative mineral hardness for 10 well-known minerals in which the softest mineral (talc) has an arbitrary hardness of 1, and the hardest mineral (diamond) has an arbitrary hardness of 10. Higher-numbered minerals will scratch lower-numbered minerals. For example, diamond will scratch talc, but talc cannot scratch diamond. The **Mohs Hardness Scale** (**Fig. 3.14**) is widely used by geologists and engineers.

When identifying a mineral, you should be able to distinguish minerals that are relatively hard (6.0 or higher

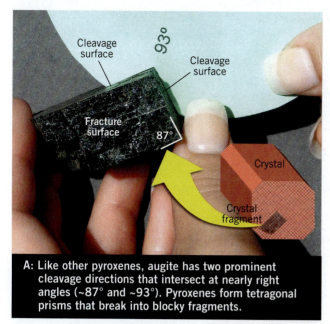

A: Like other pyroxenes, augite has two prominent cleavage directions that intersect at nearly right angles (~87° and ~93°). Pyroxenes form tetragonal prisms that break into blocky fragments.

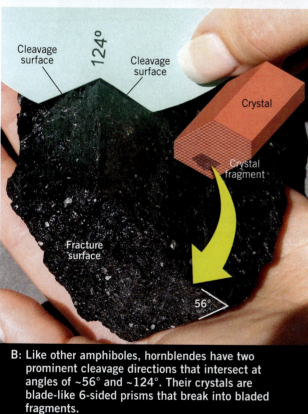

B: Like other amphiboles, hornblendes have two prominent cleavage directions that intersect at angles of ~56° and ~124°. Their crystals are blade-like 6-sided prisms that break into bladed fragments.

Figure 3.10 Cleavage in pyroxenes and amphiboles.
Pyroxenes and amphiboles are two groups of dark-colored silicate minerals with many similar properties. The main feature that distinguishes them in hand sample is the geometry of their two prominent cleavage directions.

on Mohs Scale) from minerals that are relatively soft (less than or equal to 5.5 on Mohs Scale). You can use common objects such as a glass plate, pocketknife, or steel masonry nail to make this distinction as follows:

- **Hard minerals** will scratch glass but cannot be scratched with a knife blade or masonry nail.

- **Soft minerals** will not scratch glass but can be scratched with a knife blade or masonry nail.

You can determine a mineral's hardness number on Mohs Scale by comparing the mineral to common objects shown in **Fig. 3.14** or to specimens of the minerals in Mohs Scale. Remember that the harder mineral/object is the one that scratches, and the softer mineral/object is the one that is scratched.

Density and Specific Gravity

The mass of an object (in grams, g) divided by its volume (in cubic centimeters, cm^3) is the **density** of the object. **Specific gravity** is the ratio of the density of a substance divided by the density of water, which is very close to $1 \text{ g/cm}^3 - 0.9982 \text{ g/cm}^3$ at 20°C (68°F). By dividing these two densities, the units cancel out, and so specific gravity is a dimensionless number. The value of a mineral's specific gravity is essentially the same as the value of its density but without the units. For example, the mineral quartz has a density of 2.65 g/cm^3 and its specific gravity is 2.65.

Hefting is an easy qualitative way to judge the specific gravity of one mineral relative to another. This is done by holding a piece of the first mineral in one hand and holding an equal-sized piece of the second mineral in your other hand. Feel the difference in weight between the two samples—that is, *heft* the samples. The sample that feels heavier has a higher specific gravity than the other. Most metallic minerals have higher specific gravities than nonmetallic minerals.

Other Properties

There are many additional mineral properties. Some of the more important properties that are useful in this laboratory include tenacity, reaction to acid, magnetism, and the presence of striations on crystal faces or cleavage planes.

Tenacity is the manner in which a substance resists pulling or breaking apart. Terms used to describe mineral tenacity include *brittle* (shatters like glass), *malleable* (like modeling clay or gold; can be hammered or bent permanently into new shapes), *elastic* or *flexible* (like a plastic comb—bends but returns to its original shape), and *sectile* (can be carved with a knife).

Reaction to acid differs among minerals. Cool, dilute hydrochloric acid (5–10% HCl) applied from a dropper bottle is a common "acid test." All of the *carbonate minerals* react when a drop of such dilute HCl is applied to one of their freshly exposed surfaces. Calcite ($CaCO_3$) is the most common carbonate mineral and will effervesce or "fizz" vigorously in the acid test (**Fig. 3.15**). Dolomite $[Ca, Mg(CO_3)_2]$ is another carbonate mineral, but it will fizz in dilute HCl only if the mineral is first powdered by scratching the mineral's surface with something harder than dolomite like a steel blade. If HCl is not available, then undiluted vinegar can be used for the acid test. Vinegar contains acetic acid, so the effervescence will be much less vigorous.

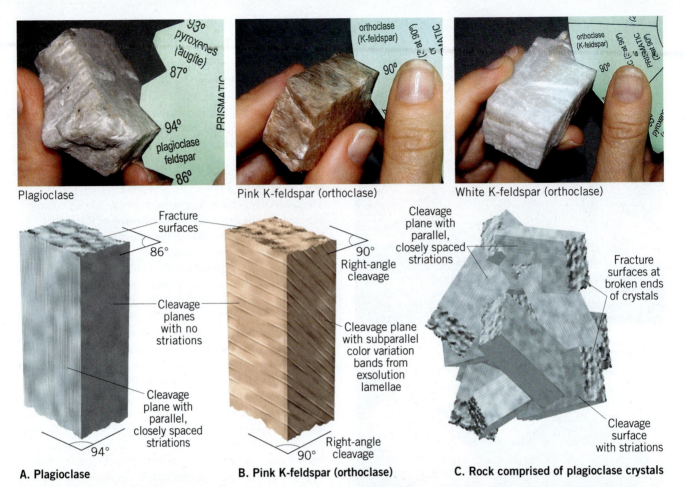

Plagioclase

Pink K-feldspar (orthoclase)

White K-feldspar (orthoclase)

A. Plagioclase

B. Pink K-feldspar (orthoclase)

C. Rock comprised of plagioclase crystals

Figure 3.11 Common feldspars. Note how the cleavage goniometer can be used to distinguish potassium feldspar (K-spar, orthoclase) from plagioclase. Orthoclase (Greek, *ortho*—right angle and *clase*—break) has perfect right-angle (90°) cleavage. Plagioclase (Greek, *plagio*—oblique angle and *clase*—break) does not. **A.** Plagioclase often exhibits *hairline striations* on some of its cleavage surfaces. They are caused by *polysynthetic twinning*. **B.** Orthoclase specimens may have intergrowths of thin, discontinuous, *exsolution lamellae*. They are alternating irregular layers of sodium feldspar and potassium feldspar. **C.** Hand sample of a rock that is an aggregate of intergrown plagioclase grains. Individual mineral grains are discernible within the rock, particularly the cleavage surfaces that have characteristic hairline striations.

A: Pure quartz is colorless, transparent, nonmetallic, and has conchoidal fracture (like glass).

B: Milky quartz forms when the quartz has microscopic fluid inclusions, usually water. It has an irregular (rough, uneven) fracture.

Figure 3.12 Fracture in quartz—SiO₂ (silicon dioxide). These hand samples are broken pieces of quartz, so no crystal faces are present. Note the absence of cleavage and the presence of conchoidal (glass-like) to uneven fracture.

Figure 3.13 **Hardness test.** You can test a mineral's hardness using a glass plate, which has a hardness of 5.5 on Mohs Hardness Scale (**Fig. 3.14**). Be sure that any sharp edges of the glass are covered with durable tape. Hold the glass plate firmly against a flat tabletop, and then forcefully try to scratch the glass with the mineral sample. A mineral that scratches the glass is considered a *hard* mineral (harder than 5.5). A mineral that does not scratch the glass is a *soft* mineral (less than or equal to 5.5).

Figure 3.15 **Acid test.** Carbonate minerals like the calcite grain in this photograph react when a drop of acid is placed on them. Calcite effervesces or fizzes vigorously.

Mohs Hardness Scale	Hardness of Some Common Objects (Harder objects scratch softer objects)
HARD 10 diamond	
9 corundum	
8 topaz	
7 quartz	
6 orthoclase feldspar	6.5 streak plate
SOFT 5 apatite	5.5 glass, masonry nail, knife blade
4 fluorite	4.5 wire nail
3 calcite	3.5 brass (wood screw, washer)
2 gypsum	2.9 copper (tubing, pipe) 2.5 fingernail
1 talc	

Figure 3.14 **Mohs Hardness Scale.** *Hard minerals* have a Mohs hardness of greater than 5.5, so they scratch glass and cannot be scratched with a steel knife blade or masonry nail. *Soft minerals* have a Mohs hardness number of 5.5 or less, so they do not scratch glass and are easily scratched by a steel knife blade or masonry nail. A mineral's hardness number can be determined by comparing it to the hardness of other common objects or mineral specimens of known hardness.

Figure 3.16 Striations in plagioclase feldspar. Fine parallel striations appear due to polysynthetic twinning (**A**) in a light sodium-rich plagioclase feldspar and (**B**) in a dark calcium-rich plagioclase feldspar.

Striations are straight "hairline" grooves on the cleavage surfaces or crystal faces of some minerals. This can be helpful in mineral identification. For example, you can use the striations of plagioclase feldspar to distinguish it from potassium feldspar. Plagioclase feldspar specimens commonly have parallel, hairline striations on cleavage faces caused by **polysynthetic twinning** in its crystal lattice (**Fig. 3.16**). Polysynthetic twinning involves thin parallel sheets of the crystal lattice that are flipped 180° from each other either during crystallization or when the crystal is deformed. In contrast, potassium feldspar often displays **exsolution lamellae** in which light-colored sodium feldspar alternates with darker potassium feldspar in irregular layers (**Fig. 3.17**).

Magnetism is a characteristic of some minerals, such as magnetite. Magnetic minerals are attracted to a magnet. Lodestone is a variety of magnetite that is itself a natural magnet that will attract steel paperclips. Hematite, maghemite, and pyrrhotite are among the other minerals that are attracted to a magnet.

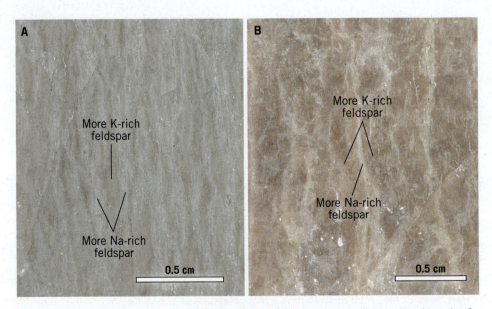

Figure 3.17 Exsolution lamellae in potassium feldspar. Irregular light-colored bands of sodium-rich feldspar alternating with darker potassium-rich feldspar in (**A**) cream-colored and (**B**) pink specimens of potassium feldspar.

Determining Specific Gravity (SG), (p. 100)

> **Think About It** How and why do people study minerals?

Objective Measure the volume and mass of minerals, calculate their specific gravities, and use the results to identify them.

Before You Begin Read the section below: Why Are Density and Specific Gravity Important?

Plan Ahead This activity requires a scale for measuring the mass of a specimen and a graduated cylinder for measuring the volume of a specimen. It also requires three single-mineral specimens that are small enough to fit into the graduated cylinder and that have different densities from one another. You might find a very thin wire useful for gently lowering the mineral specimens into the graduate cylinder, to eliminate splashes. Your instructor should supply these materials.

Why Are Density and Specific Gravity Important?

Density is a fundamental physical property, and specific gravity is a way of expressing a mineral's density using a dimensionless number. If you heft same-sized pieces of the minerals galena (lead sulfide, an ore of lead) and quartz, you can easily tell that one has a much higher specific gravity than the other. But the difference in specific gravities of different minerals is not always so obvious. In this section, you will learn how to measure the volume and mass of mineral samples, calculate their specific gravities, and use the results to identify them.

How to Determine Volume

Recall that *volume* is the amount of space that an object takes up. Most mineral samples have odd shapes, so their volumes cannot be calculated from linear measurements. Their volumes must be determined by measuring the volume of water they displace. This is done in the laboratory with a *graduated cylinder*—a slender glass or plastic vessel used to measure liquid volume (**Fig. 3.18**). Most are graduated in metric units called milliliters (mL or ml), which are thousandths of a liter. One mL (1 mL) of volume is exactly equal to one cubic centimeter (1 cm^3).

Procedures for determining the volume of a mineral sample are provided in **Fig. 3.18**. Note that when you pour water into a glass graduated cylinder, the surface of the liquid is usually a curved *meniscus*, and the volume is read at the bottom of its concave surface. In most plastic

WATER DISPLACEMENT METHOD FOR DETERMINING VOLUME OF A MINERAL SAMPLE

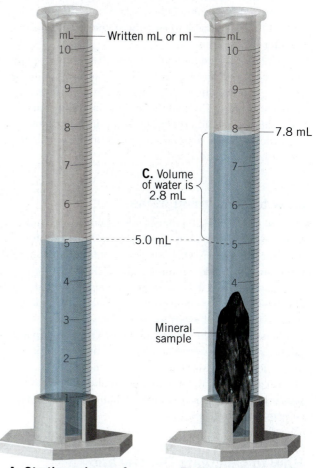

A. Starting volume of water

B. Ending volume of water

PROCEDURES

A. Place water in the bottom of a graduated cylinder. Add enough water to be able to totally immerse the mineral sample. It is also helpful to use a dropper bottle or wash bottle and bring the volume of water (before adding the mineral sample) up to an exact graduation mark like the 5.0 mL mark above. Record this starting volume of water.

B. Carefully slide the mineral sample down into the same graduated cylinder, and record the ending volume of the water (7.8 mL in the above example).

C. Subtract the starting volume of water from the ending volume of water to obtain the displaced volume of water. In the above example: 7.8 mL – 5.0 mL = 2.8 mL (2.8 mL is the same as 2.8 cm^3). This volume of displaced water is the volume of the mineral sample.

Figure 3.18 How to determine volume of a mineral sample.

graduated cylinders, however, there is no meniscus. The water level is flat and easy to read.

It is important to submerge the specimen in the water without causing a splash that changes the amount of water in the cylinder. One way of doing this is to use a thin, looped wire to ease the specimen down the cylinder slowly.

The submerged mineral specimen takes up space previously occupied by water at the bottom of the graduated cylinder, causing the water level to rise in the cylinder by an equal amount. The volume of the mineral specimen is the same as the difference between the final volume (recorded with the mineral submerged in the water) and the initial volume of water.

How to Determine Mass

Mineral grains, like all other physical bodies, have mass. **Mass** is a fundamental property of matter that supplies resistance to motion when a force is applied. The standard unit of mass is the kilogram (kg), or 1,000 grams. To determine the mass of a mineral grain, we weigh it on a scale that has been calibrated for your local laboratory conditions so that 1 kg of mass "weighs" 1 kg.

Mass and weight are not the same. One kg of quartz has the same mass everywhere in the Universe, but it weighs less on the Moon than on Earth because the Moon's gravitational attraction is only about 1/6 that of Earth. The quartz has no weight at all in Earth orbit, but its mass is unchanged.

How to Calculate Density and Specific Gravity

The amount of mass in a given volume is called **density**. A bucket filled with rock has a much greater density than an equal-sized bucket filled with air. Calculating the density of a mineral grain is a simple matter of dividing its mass by its volume.

$$\text{density} = \text{mass} \div \text{volume}$$

For example, a 37.5 gram mass of galena that occupies 5.0 cubic centimeters of volume has a density of $37.5 \text{ g} \div 5.0 \text{ cm}^3 = 7.5 \text{ g/cm}^3$.

Specific gravity (SG) is the ratio of the density of a mineral grain divided by the density of liquid water, which we will assume to be 1.0 g/cm^3.

$$\text{specific gravity of specimen} = \frac{\text{density of specimen in g/cm}^3}{1.0 \text{ g/cm}^3}$$

Dividing a density measured in g/cm^3 by another density measured in g/cm^3 yields a dimensionless number, because the "g/cm^3" units cancel in division. For example, the galena specimen whose density we just determined has a specific gravity of $7.5 \text{ g/cm}^3 \div 1.0 \text{ g/cm}^3 = 7.5$.

Calculating Density and Specific Gravity—The Math You Need

You can learn more about calculating density and specific gravity at this site featuring The Math You Need, When You Need It math tutorials for students in introductory geoscience courses:

http://serc.carleton.edu/mathyouneed/density/index.html

http://goo.gl/AwkXPZ

ACTIVITY 3.4

Mineral Identification and Uses, (p. 101)

Think About It   How and why do people study minerals?

Objective   Identify common minerals on the basis of their properties and assess how you depend on them.

Before You Begin   Read the section below: How Are Minerals Identified?

Plan Ahead   This activity requires identifying up to 20 unknown mineral specimens. Either these will be made available by your instructor, or you will need to access images of these unknowns online if actual specimens are not available. If you have access to specimens, you will also need to be able to determine their hardness and streak. Your instructor will supply some glass, a wire nail, a steel blade or masonry nail, a pure-brass screw or washer, and some pure-copper pipe or tubing for testing hardness and, for determining streak, a streak plate or some other way of creating a fine powder of the mineral specimen.

How Are Minerals Identified?

You are expected to learn how to identify common minerals on the basis of their properties and assess how you depend on them. The ability to identify minerals is one of the most fundamental skills of a geoscientist. It also is fundamental to identifying rocks, for you must first identify the minerals comprising them. Mineral identification is based on your ability to describe mineral properties using identification charts (**Figs. 3.19–3.21**) and a Mineral Database (**Fig. 3.22**). The database also lists the chemical composition and some common uses for each mineral. Some minerals like halite (table salt) and gemstones are used in their natural state. Others are valuable as *ores*—materials from which specific chemical elements or compounds can be extracted at a profit.

METALLIC AND SUBMETALLIC (M) MINERAL IDENTIFICATION

STEP 1: What is the mineral's hardness?	STEP 2: Does the mineral have cleavage?	STEP 3: What is the mineral's streak?	STEP 4: Match the mineral's physical properties to other characteristic properties below.	STEP 5: Mineral name. Find out more about it in the mineral database (Fig. 3.22).
HARD (H > 5.5) Scratches glass Not scratched by masonry nail or knife blade	Cleavage absent, poor, or not visible	Dark gray to black	Color brass gold; tarnishes brown; H 6–6.5; brittle; conchoidal to uneven fracture; forms cubes (may be striated), pyritohedrons, or octahedrons; distinguished from chalcopyrite, which is soft	pyrite
			Silvery dark gray to black; tarnishes gray or rusty yellow-brown; strongly attracted to a magnet and may be magnetized; H 6–6.5; forms octahedrons	magnetite
HARD or SOFT		Brown	Color silvery black to black; tarnishes gray to black; H 5.5–6; may be weakly attracted to a magnet; forms octahedrons	chromite
		Red to red-brown	Color steel gray, reddish-silver, to glittery bright silver (var. specular); both metallic varieties have the characteristic red-brown streak; may be attracted to a magnet; H 5–6; also occurs in nonmetallic, dull to earthy, red to red-brown forms	hematite
SOFT (H ≤ 5.5) Does not scratch glass Scratched by masonry nail or knife blade	Cleavage good to excellent	Dark gray to black	Color bright silvery gray; tarnishes dull gray; brittle: breaks into cubes and shapes made of cubes; H 2.5; forms cubes or octahedrons; feels heavy for its size because of high specific gravity	galena
		White to pale yellow-brown	Color silvery yellow-brown, silvery red, or black with submetallic to resinous luster; tarnishes brown or black; H 3.5–4.0; smells like rotten eggs when scratched, powdered, or in acid test	sphalerite
	Cleavage absent, poor, or not visible	Dark gray to black	Color bright silvery gold; tarnishes bronze brown brassy gold, or iridescent blue-green and red; H 3.5–4.0; brittle; uneven fracture; forms tetrahedrons	chalcopyrite
			Color characteristically brownish-bronze; tarnishes bright iridescent purple, blue, and/or red, giving it its nickname "peacock ore"; may be weakly attracted to a magnet; H 3; usually massive, rare as cubes or dodecahedrons	bornite
			Color opaque brassy to brown-bronze; tarnishes dull brown, may have faint iridescent colors; fracture uneven to conchoidal; no cleavage; attracted to a magnet; H 3.5–4.5; usually massive or masses of tiny crystals; resembles chalcopyrite, which is softer and not attracted to a magnet	pyrrhotite
			Color dark silvery gray to black; can be scratched with your fingernail; easily rubs off on your fingers and clothes, making them gray; H 1–2	graphite
		Yellow-brown	Metallic or silky submetallic luster, color dark brown, gray, or black; H 5–5.5; forms layers of radiating microscopic crystals and botryoidal masses	goethite
		Copper	Color copper; tarnishes dull brown or green; H 2.5–3.0; malleable and sectile; hackly fracture; usually forms dendritic masses or nuggets	copper (native copper)
		Gold	Color yellow gold; does not tarnish; malleable and sectile; H 2.5–3.0; forms odd-shaped masses, nuggets, or dendritic forms	gold (native gold)
		Silvery white	Color silvery white to gray; tarnishes gray to black; H 2.5–3.0; malleable and sectile; forms dendritic masses, nuggets, or curled wires	silver (native silver)

Figure 3.19 Identification chart for opaque minerals with metallic or submetallic luster (M) on freshly broken surfaces.

DARK TO MEDIUM-COLORED NONMETALLIC (NM) MINERAL IDENTIFICATION

STEP 1: What is the mineral's hardness?	STEP 2: What is the mineral's cleavage?	STEP 3: Compare the mineral's physical properties to other distinctive properties below.	STEP 4: Find mineral name(s) and check the mineral database for additional properties (Fig. 3.22).
HARD (H > 5.5) Scratches glass Not scratched by masonry nail or knife blade	Cleavage excellent or good	Translucent to opaque dark gray; blue-gray, or black; may have silvery iridescence; 2 cleavages at nearly 90° and with striations; H 6	plagioclase feldspar
		Translucent to opaque brown, gray, green, or red; 2 cleavages at nearly right angles; exsolution lamellae; H 6	potassium feldspar
		Green to black; vitreous luster; H 5.5–6.0; 2 cleavages at about 124° and 56° plus uneven fracture; usually forms long blades and masses of needle-like crystals	actinolite (an amphibole)
		Dark gray to black; vitreous luster; H 5.5–6.0; 2 cleavages at about 124° and 56° plus uneven fracture; forms long crystals that break into blade-like fragments	hornblendes (amphiboles)
		Dark green to black; dull to vitreous luster; H 5.5–6.0; two cleavages at nearly right angles (93° and 87°) plus uneven fracture; forms short crystals with squarish cross sections; breaks into blocky fragments	augite (a pyroxene)
	Cleavage absent, poor, or not visible	Transparent or translucent gray, brown, or purple; vitreous luster; massive or hexagonal prisms and pyramids; H 7	quartz
		Gray, black, or colored (dark red, blue, brown) hexagonal prisms with flat striated ends; H 9	corundum
		Transparent to translucent dark red to black; equant (dodecahedron) crystal form or massive; H 7	garnet
		Black or dark green; long striated prisms; H 7–7.5	tourmaline
		Olive green, transparent or translucent; no cleavage; usually has many cracks and conchoidal to uneven fracture; single crystals or masses of tiny crystals resembling green granulated sugar or aquarium gravel; the crystals have vitreous (glassy) luster	olivine
		Opaque green; poor cleavage; H 6–7	epidote
		Opaque brown prisms and cross-shaped twins; H 7	staurolite
SOFT (H ≤ 5.5) Does not scratch glass Scratched by masonry nail or knife blade	Cleavage excellent or good	Yellow-brown, brown, or black; vitreous to resinous luster (may also be submetallic); dodecahedral cleavage; H 3.5–4.0; rotten egg smell when scratched or powdered	sphalerite
		Purple cubes or octahedrons; octahedral cleavage; H 4	fluorite
		Black short opaque prisms; splits easily along 1 excellent cleavage into thin sheets; H 2.5–3	biotite (black mica)
		Green short opaque prisms; splits easily along 1 excellent cleavage into thin sheets; H 2–3	chlorite
	Cleavage absent, poor, or not visible	Deep blue; crusts, small crystals, or massive; light blue streak; H 3.5–4	azurite
		Green, green-blue, gray, yellow-green, white; platy or tabular or fibrous (chrysotile asbestos); H 2.5–3.5	serpentine minerals
		Opaque green in laminated crusts or massive; streak pale green; effervesces in dilute HCl; H 3.5–4	malachite
		Translucent or opaque dark green; can be scratched with your fingernail; feels greasy or soapy; H 1	talc
		Transparent or translucent green, brown, blue, or purple; brittle hexagonal prisms; conchoidal fracture; H 5	apatite
		Opaque earthy brick red to dull red-gray, or gray; H 1.5–5; red-brown streak; magnet may attract the gray forms	hematite

Figure 3.20 Identification chart for dark- to medium-colored minerals with nonmetallic (NM) luster on freshly broken surfaces.

LIGHT-COLORED NONMETALLIC (NM) MINERAL IDENTIFICATION

STEP 1: What is the mineral's hardness?	STEP 2: What is the mineral's cleavage?	STEP 3: Compare the mineral's physical properties to other distinctive properties below.	STEP 4: Find mineral name(s) and check the mineral database for additional properties (Fig. 3.22).
HARD (H > 5.5) Scratches glass Not scratched by masonry nail or knife blade	Cleavage excellent or good	White or pale gray; 2 good cleavages at nearly 90° plus uneven fracture; may have striations; H 6	plagioclase feldspar
		Orange, pink, pale brown, green, or white; H 6; 2 good cleavages at 90° plus uneven fracture; exsolution lamellae	potassium feldspar
		Pale brown, white, or gray; long slender prisms; 1 excellent cleavage plus fracture surfaces; H 6–7	sillimanite
		Blue, very pale green, white, or gray; bladed habit; H 4–7	kyanite
	Cleavage absent, poor, or not visible	Gray, white, or colored (dark red, blue, brown) hexagonal prisms with flat striated ends; H 9	corundum
		Colorless, white, gray, or other colors; vitreous luster; massive or hexagonal prisms and pyramids; transparent or translucent; H 7	quartz
		Colorless, white, yellow, light brown, or pastel colors; translucent; laminated or massive; cryptocrystalline; waxy luster; H 7	chalcedony (cryptocrystalline quartz)
		Pale green to yellow; transparent or translucent; H 7; no cleavage; conchoidal to uneven fracture; single crystals or masses of tiny crystals resembling green or yellow granulated sugar or aquarium gravel; vitreous (glassy) luster	olivine
SOFT (H ≤ 5.5) Does not scratch glass Scratched by masonry nail or knife blade	Cleavage excellent or good	Colorless, white, yellow, green, pink, or brown; 3 excellent cleavages; breaks into rhombohedrons; effervesces in dilute HCl; H 3	calcite
		Colorless, white, gray, cream, or pink; 3 excellent cleavages; breaks into rhombohedrons; effervesces in dilute HCl only if powdered; H 3.5–4	dolomite
		Colorless or white with tints of brown, yellow, blue, black; short tabular crystals and roses; SG 4.4–4.6; H 3–3.5; vitreous luster	barite
		Transparent, colorless to white; H 2, easily scratched with your fingernail; white streak; bladed, tabular or massive habit	gypsum (selenite)
		Colorless, white, gray, or pale green, yellow, or red; spheres of radiating needles; luster silky; H 5–5.5	natrolite (a zeolite mineral)
		Colorless, white, yellow, blue, brown, or red; cubic crystals; breaks into cubes; salty taste; H 2.5	halite
		Colorless, purple, blue, gray, green, yellow; cubes with octahedral cleavage; H 4	fluorite
		Colorless, yellow, brown, or red-brown; short prisms; splits along 1 excellent cleavage into thin flexible transparent sheets; H 2–2.5	muscovite (clear mica)
	Cleavage absent, poor, or not visible	White, gray or yellow; earthy to pearly; massive habit; H 2, easily scratched with your fingernail; white streak	gypsum (alabaster)
		White to gray; fibrous form with silky or satiny luster; H 2, easily scratched with your fingernail	gypsum (satin spar)
		Yellow crystals or earthy masses; greasy luster; H 1.5–2.5; smells like rotten eggs when powdered	sulfur (native element)
		Opaque pale blue to blue-green; conchoidal fracture; H 2–4; massive or amorphous earthy crusts; very light blue streak	chrysocolla
		Green, green-blue, gray, yellow-green, white; white streak; platy or tabular or serpentine minerals fibrous (chrysotile asbestos); H 2.5–3.5	serpentine minerals
		Opaque white, gray, green, or brown; can be scratched with fingernail; greasy or soapy feel; H 1	talc
		Opaque earthy white to very light brown masses of "white clay"; H 1–2; powdery to greasy feel	kaolinite
		Mineraloid/amophose; colorless to white, orange, yellow, blue, gray, green, or red; might have internal play of colors; conchoidal fracture	opal
		Colorless or pale green, brown, blue, white, or purple; brittle hexagonal prisms; conchoidal fracture; H 5	apatite

Figure 3.21 Identification chart for light-colored minerals with nonmetallic (NM) luster on freshly broken surfaces.

MINERAL DATABASE (Alphabetical Listing)

Mineral	Luster and Crystal System	Hardness	Streak	Distinctive Properties	Some Uses
ACTINOLITE (amphibole)	Nonmetallic (NM) Monoclinic	5.5–6	White	Color dark green or pale green; forms needles, prisms, and asbestose fibers; good cleavage at 56° and 124°; SG = 3.1	Green gem varieties are the gemstone "nephrite jade"; asbestos products
AMPHIBOLE: See HORNBLENDE and ACTINOLITE					
APATITE $Ca_5F(PO_4)_3$ calcium fluorophosphate	Nonmetallic (NM) Hexagonal	5	White	Color pale or dark green, brown, blue, white, or purple; sometimes colorless; transparent or opaque; brittle; conchoidal fracture; forms hexagonal prisms; SG = 3.1–3.4	Used mostly to make fertilizer, pesticides; transparent varieties sold as gemstones
ASBESTOS: fibrous varieties of AMPHIBOLE and SERPENTINE					
AUGITE (pyroxene) calcium ferromagnesian silicate	Nonmetallic (NM) Monoclinic	5.5–6	White to pale gray	Color dark green to brown or black; forms short, 8-sided prisms; two good cleavages that intersect at 87° and 93° (nearly right angles); SG = 3.2–3.5	Ore of lithium, used to make lithium batteries, ovenware glazes, high temperature grease, and to treat depression
AZURITE $Cu_3(CO_3)_2(OH)_2$ copper carbonate hydroxide	Nonmetallic (NM) Monoclinic	3.5–4	Light blue	Color a distinctive deep blue; forms crusts of small crystals, opaque earthy masses, or short and long prisms; brittle; effervesces in dilute HCl; SG = 3.7–3.8	Ore of copper used to make pipes, electrical wire, coins, ammunition, bronze, brass; added to vitamin pills for healthy hair and skin; gemstone
BARITE $BaSO_4$ barium sulfate	Nonmetallic (NM) Orthorhombic	3–3.5	White	Colorless to white, with tints of brown, yellow, blue, or red; forms short tabular crystals and rose-shaped masses (barite roses); brittle; cleavage good to excellent; very heavy, SG = 4.3–4.6	Ore of barium, used to harden rubber, make fluorescent lamp electrodes, and in fluids used to drill oil/gas wells
BIOTITE MICA ferromagnesian potassium, hydrous aluminum silicate $K(Mg,Fe)_3 (Al,Si_3O_{10})(OH,F)_2$	Nonmetallic (NM) Monoclinic	2.5–3	Gray-brown to white	Color black, green-black, or brown-black; cleavage excellent; forms very short prisms that split easily into very thin, flexible sheets; SG = 2.7–3.1	Used for fire-resistant tiles, rubber, paint
BORNITE Cu_5FeS_4 copper-iron sulfide	Metallic (M) Isometric	3	Dark gray to black	Color brownish bronze; tarnishes bright purple, blue, and/or red; may be weakly attracted to a magnet; H 3; cleavage absent or poor; forms dense brittle masses; rarely forms crystals	Ore of copper, used to make pipes, electrical wire, coins, ammunition, bronze, brass; added to vitamin pills for healthy hair and skin
CALCITE $CaCO_3$ calcium carbonate	Nonmetallic (NM) Hexagonal	3	White	Usually colorless, white, or yellow, but may be green, brown, or pink; opaque or transparent; excellent cleavage in 3 directions not at 90°; forms prisms, rhombohedrons, or scalenohedrons that break into rhombohedrons; effervesces in dilute HCl; SG = 2.7	Used to make antacid tablets, fertilizer, cement; ore of calcium
CHALCEDONY SiO_2 cryptocrystalline quartz	Nonmetallic (NM) No visible crystals	7	White*	Colorless, white, yellow, light brown, or other pastel colors in laminations; often translucent; conchoidal fracture; luster waxy; cryptocrystalline; SG = 2.5–2.8	Used as an abrasive; used to make glass, gemstones (agate, chrysoprase)
CHALCOPYRITE $CuFeS_2$ copper-iron sulfide	Metallic (M) Tetragonal	3.5–4	Dark gray	Color bright silvery gold; tarnishes bronze brown, brassy gold, or iridescent blue-green and red; brittle; no cleavage; forms dense masses or elongate tetrahedrons; SG = 4.1–4.3	Ore of copper, used to make pipes, electrical wire, coins, ammunition, bronze, brass; added to vitamin pills for healthy hair and skin

*Streak cannot be determined with a streak plate for minerals harder than 6.5. They scratch the streak plate.

Figure 3.22 Mineral Database. This is an alphabetical list of minerals and their properties and uses.

MINERAL DATABASE (Alphabetical Listing)

Mineral	Luster and Crystal System	Hardness	Streak	Distinctive Properties	Some Uses
CHLORITE ferromagnesian aluminum silicate $(Mg,Fe,Al)_6(Si,Al)_4O_{10}(OH)_8$	Nonmetallic (NM) Monoclinic	2–2.5	White	Color dark green; cleavage excellent; forms short prisms that split easily into thin flexible sheets; luster bright or dull; SG = 2–3	Used as a "filler" (to take up space and reduce cost) in plastics for car parts, appliances; massive pieces carved into art sculptures
CHROMITE $FeCr_2O_4$ iron-chromium oxide	Metallic (M) Isometric	5.5–6	Dark brown	Color silvery black to black; tarnishes gray to black; no cleavage; may be weakly attracted to a magnet; forms dense masses or granular masses of small crystals (octahedrons)	Ore of chromium for chrome, stainless steel, mirrors, yellow and green paint pigments and ceramic glazes, and pills for healthy metabolism and cholesterol levels
CHRYSOCOLLA $CuSiO_3 \cdot 2H_2O$ hydrated copper silicate	Nonmetallic (NM) Orthorhombic	2–4	Very light blue	Color pale blue to blue-green; opaque; forms cryptocrystalline crusts or may be massive; conchoidal fracture; luster shiny or earthy; SG = 2.0–4.0	Ore of copper, used to make pipes, electrical wire, coins, ammunition, bronze, brass; added to vitamin pills for healthy hair and skin; gemstone
COPPER (NATIVE COPPER) Cu copper	Metallic (M) Isometric	2.5–3	Copper	Color copper; tarnishes brown or green; malleable; no cleavage; forms odd-shaped masses, nuggets, or dendritic forms; SG = 8.8–9.0	Ore of copper, used to make pipes, electrical wire, coins, ammunition, bronze, brass; added to vitamin pills for healthy hair and skin
CORUNDUM Al_2O_3 aluminum oxide	Nonmetallic (NM) Hexagonal	9	White*	Gray, white, black, or colored (red, blue, brown, yellow) hexagonal prisms with flat striated ends; opaque to transparent; cleavage absent; SG = 3.9–4.1 H 9	Used for abrasive powders to polish lenses; gemstones (red ruby, blue sapphire); emery cloth
DOLOMITE $CaMg(CO_3)_2$ magnesian calcium carbonate	Nonmetallic (NM) Hexagonal	3.5–4	White	Color white, gray, creme, or pink; usually opaque; cleavage excellent in 3 directions; breaks into rhombohedrons; resembles calcite, but will effervesce in dilute HCl only if powdered; SG = 2.8–2.9	Ore of magnesium used to make paper; lightweight frames for jet engines, rockets, cell phones, laptops; pills for good brain, muscle, and skeletal health
EPIDOTE complex silicate	Nonmetallic (NM) Monoclinic	6–7	White*	Color pale or dark green to yellow-green; massive or forms striated prisms; cleavage poor; SG = 3.3–3.5	Used as a green gemstone

FELDSPAR: See PLAGIOCLASE (Na-Ca Feldspars) and POTASSIUM FELDSPAR (K-Spar)

Mineral	Luster and Crystal System	Hardness	Streak	Distinctive Properties	Some Uses
FLUORITE CaF_2 calcium fluoride	Nonmetallic (NM) Isometric	4	White	Colorless, purple, blue, gray, green, or yellow; cleavage excellent; crystals usually cubes; transparent or opaque; brittle; SG = 3.0–3.3	Ore of fluorine used in fluoride toothpaste, refrigerant gases, rocket fuel
GALENA PbS lead sulfide	Metallic (M) Isometric	2.5	Gray to dark gray	Color bright silvery gray; tarnishes dull gray; forms cubes and octahedrons; brittle; cleavage good in three directions, so breaks into cubes; SG = 7.4–7.6	Ore of lead for television glass, auto batteries, solder, ammunition; May be an ore of bismuth (an impurity) used as a lead substitute in pipe solder and fishing sinkers; May be an ore of silver (an impurity) used in jewelry, electrical circuit boards
GARNET complex silicate	Nonmetallic (NM) Isometric	7	White*	Color usually red, black, or brown, sometimes yellow, green, pink; forms dodecahedrons; cleavage absent but may have parting; brittle; translucent to opaque; SG = 3.5–4.3	Used as an abrasive; Red gemstone

*Streak cannot be determined with a streak plate for minerals harder than 6.5. They scratch the streak plate.

Figure 3.22 Mineral Database (*continued*)

Mineral	Luster and Crystal System	Hardness	Streak	Distinctive Properties	Some Uses
GOETHITE FeO(OH) iron oxide hydroxide	Metallic (M) Orthorhombic	5–5.5	Yellow-brown	Color dark brown to black; tarnishes yellow-brown; forms layers of radiating microscopic crystals; SG = 3.3–4.3	Ore of iron for iron and steel used in machines, buildings, bridges, nails, tools, file cabinets; added to pills and foods to aid hemoglobin production in red blood cells
GOLD (NATIVE GOLD) Au pure gold	Metallic (M) Isometric	2.5–3.0	Gold-yellow	Color gold to yellow-gold; does not tarnish; ductile, malleable and sectile; hackly fracture; SG = 19.3; no cleavage; forms odd-shaped masses, nuggets, and dendritic forms	Ductile and malleable metal used for jewelry; electrical circuitry in computers, cell phones, car air bags; heat shields for satellites
GRAPHITE C carbon	Metallic (M) Hexagonal	1	Dark gray	Color dark silvery gray to black; forms flakes, short hexagonal prisms, and earthy masses; greasy feel; very soft; cleavage excellent in 1 direction; SG = 2.0–2.3	Used for pencils, anodes (negative ends) of most batteries, synthetic motor oil, carbon steel, fishing rods, golf clubs
GYPSUM CaSO$_4$ · 2H$_2$O hydrated calcium sulfate	Nonmetallic (NM) Monoclinic	2	White	Colorless, white, or gray; forms tabular crystals, prisms, blades, or needles (satin spar variety); transparent to translucent; very soft; cleavage good; SG = 2.3	Plaster-of-paris, wallboard, drywall, art sculpture medium (alabaster)
HALITE NaCl sodium chloride	Nonmetallic (NM) Isometric	2.5	White	Colorless, white, yellow, blue, brown, or red; transparent to translucent; brittle; forms cubes; cleavage excellent in 3 directions, so breaks into cubes; salty taste; SG = 2.1–2.6	Table salt, road salt; used in water softeners and as a preservative; sodium ore
HEMATITE Fe$_2$O$_3$ iron oxide	Metallic (M) or Nonmetallic (NM) Hexagonal	1–6	Red to red-brown	Color silvery gray, reddish silver, black, or brick red; tarnishes red; opaque; soft (earthy) and hard (metallic) varieties have same streak; forms thin tabular crystals or massive; may be attracted to a magnet; SG = 4.9–5.3	Red ochre pigment in paint and cosmetics. ore of iron for iron and steel used in machines, buildings, bridges, nails, tools, file cabinets; added to pills and foods to aid hemoglobin production in red blood cells
HORNBLENDE (amphibole) calcium ferromagnesian aluminum silicate	Nonmetallic (NM) Monoclinic	5.5–6.0	White to pale gray	Color dark gray to black; forms prisms with good cleavage at 56° and 124°; brittle; splintery or asbestos forms; SG = 3.0–3.3	Fibrous varieties used for fire-resistant clothing, tiles, brake linings
JASPER SiO$_2$ cryptocrystalline quartz	Nonmetallic (NM) No visible crystals	7	White*	Color red-brown, or yellow; opaque; waxy luster; conchoidal fracture; cryptocrystalline; SG = 2.5–2.8	Used as an abrasive; used to make glass, gemstones
KAOLINITE Al$_4$(Si$_4$O$_{10}$)(OH)$_8$ aluminum silicate hydroxide	Nonmetallic (NM) Triclinic	1–2	White	Color white to very light brown; commonly forms earthy, microcrystalline masses; cleavage excellent but absent in hand samples; SG = 2.6	Used for pottery, clays, polishing compounds, pencil leads, paper
K-SPAR: See POTASSIUM FELDSPAR					
KYANITE Al$_2$(SiO$_4$)O aluminum silicate oxide	Nonmetallic (NM) Triclinic	4–7	White*	Color blue, pale green, white, or gray; translucent to transparent; forms blades; SG = 3.6–3.7	High temperature ceramics, spark plugs
MAGNETITE Fe$_3$O$_4$ iron oxide	Metallic (M) or Nonmetallic (NM) Isometric	6–6.5	Dark gray	Color silvery gray to black; opaque; forms octahedrons; tarnishes gray; no cleavage; attracted to a magnet and can be magnetized; SG = 5.0–5.2	Ore of iron for iron and steel used in machines, buildings, bridges, nails, tools, file cabinets; added to pills and foods to aid hemoglobin production in red blood cells

*Streak cannot be determined with a streak plate for minerals harder than 6.5. They scratch the streak plate.

Figure 3.22 **Mineral Database** (*continued*)

MINERAL DATABASE (Alphabetical Listing)

Mineral	Luster and Crystal System	Hardness	Streak	Distinctive Properties	Some Uses
MALACHITE $Cu_2CO_3(OH)_2$ copper carbonate hydroxide	Nonmetallic (NM) Monoclinic	3.5–4	Green	Color green, pale green, or gray-green; usually in crusts, laminated masses, or microcrystals; effervesces in dilute HCl; SG = 3.6–4.0	Ore of copper, used to make pipes, electrical wire, coins, ammunition, bronze, brass; added to vitamin pills for healthy hair and skin; gemstone
MICA: See BIOTITE and MUSCOVITE					
MUSCOVITE MICA potassium hydrous aluminum silicate $KAl_2(Al,Si_3O_{10})(OH,F)_2$	Nonmetallic (NM) Monoclinic	2–2.5	White	Colorless, yellow, brown, or red-brown; forms short opaque prisms; cleavage excellent in 1 direction, can be split into thin flexible transparent sheets; SG = 2.7–3.0	Computer chip substrates, electrical insulation, roof shingles, cosmetics with a satiny sheen
NATIVE COPPER: See COPPER					
NATIVE GOLD: See GOLD					
NATIVE SILVER: See SILVER					
NATIVE SULFUR: See SULFUR					
NATROLITE (ZEOLITE) $Na_2(Al_2Si_3O_{10}) \cdot 2H_2O$ hydrous sodium aluminum silicate	Nonmetallic (NM) Orthorhombic	5–5.5	White	Colorless, white, gray, or pale green, yellow, or red; forms masses of radiating needles; silky luster; SG = 2.2–2.4	Used in water softeners
OLIVINE $(Fe,Mg)_2SiO_4$ ferromagnesian silicate	Nonmetallic (NM) Orthorhombic	7	White*	Color pale or dark olive-green to yellow, or brown; forms short crystals that may resemble sand grains; conchoidal fracture; cleavage absent; brittle; SG = 3.3–3.4	Green gemstone (peridot); ore of magnesium used to make paper; lightweight frames for jet engines, cell phones, laptops; pills for good brain, muscle, and skeletal health
OPAL $SiO_2 \cdot nH_2O$ hydrated silicon dioxide	Nonmetallic (NM) Amorphous	5–5.5	White	Colorless to white, orange, yellow, brown, blue, gray, green, or red; may have play of colors (opalescence); amorphous; cleavage absent; conchoidal fracture; SG = 1.9–2.3	Gemstone
PLAGIOCLASE FELDSPAR $NaAlSi_3O_8$ to $CaAl_2Si_2O_8$ calcium-sodium aluminum silicate	Nonmetallic (NM) Triclinic	6	White	Colorless, white, gray, or black; may have iridescent play of color from within; translucent; forms striated tabular crystals or blades; cleavage good in two directions at nearly 90°; SG = 2.6–2.8	Used to make ceramics, glass, enamel, soap, false teeth, scouring powders
POTASSIUM FELDSPAR $KAlSi_3O_8$ potassium aluminum silicate	Nonmetallic (NM) Monoclinic	6	White	Color orange, brown, white, green, or pink; forms translucent prisms with subparallel exsolution lamellae; cleavage excellent in two directions at nearly 90°; SG = 2.5–2.6	Used to make ceramics, glass, enamel, soap, false teeth, scouring powders
PYRITE ("fool's gold") FeS_2 iron sulfide	Metallic (M) Isometric	6–6.5	Dark gray	Color silvery gold; tarnishes brown; H 6–6.5; cleavage absent to poor; brittle; forms opaque masses, cubes (often striated), or pyritohedrons; SG = 4.9–5.2	Ore of sulfur for matches, gunpowder, fertilizer, rubber hardening (car tires), fungicide, insecticide, paper pulp processing
PYRRHOTITE FeS iron sulfide	Metallic (M) Monoclinic	3.5–4.5	Dark gray to black	Color brassy to brown-bronze; tarnishes dull brown, sometimes with faint iridescent colors; fracture uneven to conchoidal; no cleavage; attracted to a magnet; SG = 4.6	Ore of iron and sulfur; Impure forms contain nickel and are used as nickel ore; the nickel is used to make stainless steel
PYROXENE: See AUGITE					

*Streak cannot be determined with a streak plate for minerals harder than 6.5. They scratch the streak plate.

Figure 3.22 Mineral Database (continued)

Mineral	Luster and Crystal System	Hardness	Streak	Distinctive Properties	Some Uses
QUARTZ SiO_2 silicon dioxide	Nonmetallic (NM) Hexagonal	7	White*	Usually colorless, white, or gray but uncommon varieties occur in all colors; transparent to translucent; luster greasy; no cleavage; forms hexagonal prism and pyramids; SG = 2.6–2.7 Some quartz varieties are: • var. flint (opaque black or dark gray) • var. smoky (transparent gray) • var. citrine (transparent yellow-brown) • var. amethyst (purple) • var. chert (opaque gray) • var. milky (white) • var. jasper (opaque red or yellow) • var. rock crystal (colorless) • var. rose (pink) • var. chalcedony (translucent, waxy luster)	Used as an abrasive; used to make glass, gemstones
SERPENTINE $Mg_6Si_4O_{10}(OH)_8$ magnesium silicate hydroxide	Nonmetallic (NM) Monoclinic	2–5	White	Color pale or dark green, yellow, gray; forms dull or silky masses and asbestos forms; no cleavage; SG = 2.2–2.6	Fibrous varieties used for fire-resistant clothing, tiles, brake linings
SILLIMANITE $Al_2(SiO_4)O$ aluminum silicate	Nonmetallic (NM) Orthorhombic	6–7	White	Color pale brown, white, or gray; one good cleavage plus fracture surfaces; forms slender prisms and needles; SG = 3.2	High-temperature ceramics
SILVER (NATIVE SILVER) Ag pure silver	Metallic (M) Isometric	2.5–3.0	White to silvery white	Color silvery white to gray; tarnishes dark gray to black; ductile, malleable and sectile; hackly fracture; no cleavage; forms nuggets, curled wires, and dendritic forms; SG = 10.5	Ductile and malleable metal used for jewelry and silverware; electrical circuit boards for computers and cell phones; photographic film
SPHALERITE ZnS zinc sulfide	Metallic (M) or Nonmetallic (NM) Isometric	3.5–4	White to pale yellow-brown	Color silvery yellow-brown, dark red, or black; tarnishes brown or black; dodecahedral cleavage excellent to good; smells like rotten eggs when scratched/powdered; forms misshapen tetrahedrons or dodecahedrons; SG = 3.9–4.1	Ore of zinc for brass, galvanized steel and roofing nails, skin-healing creams, pills for healthy immune system and protein production: ore of Indium (an impurity) used to make solar cells
STAUROLITE iron magnesium zinc aluminum silicate	Nonmetallic (NM) Monoclinic	7	White to gray*	Color brown to gray-brown; tarnishes dull brown; forms prisms that interpenetrate to form natural crosses; cleavage poor; SG = 3.7–3.8	Gemstone crosses called "fairy crosses"
SULFUR (NATIVE SULFUR) S sulfur	Nonmetallic (NM) Orthorhombic	1.5–2.5	Pale yellow	Color bright yellow; forms transparent to translucent crystals or earthy masses; cleavage poor; luster greasy to earthy; brittle; SG = 2.1	Used for matches, gunpowder, fertilizer, rubber hardening (car tires), fungicide, insecticide, paper pulp processing
TALC $Mg_3Si_4O_{10}(OH)_2$ hydrous magnesian silicate	Nonmetallic (NM) Monoclinic	1	White	Color white, gray, pale green, or brown; forms cryptocrystalline masses that show no cleavage; luster silky to greasy; feels greasy or soapy (talcum powder); very soft; SG = 2.7–2.8	Used as a "filler" (to take up space and reduce cost) in plastics for car parts, appliances; massive pieces carved into art sculptures
TOURMALINE complex silicate	Nonmetallic (NM) Hexagonal	7–7.5	White*	Color usually opaque black or green, but may be transparent or translucent green, red, yellow, pink or blue; forms long striated prisms with triangular cross sections; cleavage absent; SG = 3.0–3.2	Crystals used in radio transmitters; gemstone

ZEOLITE: A group of calcium or sodium hydrous aluminum silicates. See NATROLITE.

*Streak cannot be determined with a streak plate for minerals harder than 6.5. They scratch the streak plate.

Figure 3.22 **Mineral Database** (*continued*)

Mineral Identification Procedures

Obtain a set of mineral samples and analysis tools from your instructor. For each sample, fill in the Activity 3.4 tear-out worksheet using the procedures provided below.

1. Record the sample number or letter.

2. Determine and record the mineral's **luster** as metallic (M) or nonmetallic (NM)

   A. Metallic (M): mineral is opaque, looks like metal or sort of like metal

   B. Nonmetallic (NM): vitreous (glassy), resinous, silky, satin, waxy, pearly, earthy/dull, greasy

3. Determine and record the mineral's **hardness** (**Figs. 3.13** and **3.14**).

4. Determine and record the mineral's **cleavage**, if present (**Figs. 3.7–3.11**), and **fracture**, if present (**Figs. 3.9D** and **3.12**). For cleavage, try to determine number of cleavage directions (**Fig. 3.7**).

5. Determine and record the mineral's **color** on a fresh surface, and **streak** using a streak plate (**Fig. 3.3**).

   Minerals harder than 6.5 will scratch the streak plate, so no streak can be determined for them.

6. Determine and record other notable properties like these:

   A. What is the mineral's **tenacity**: brittle, elastic, malleable, or sectile?

   B. Is the specimen a natural magnet? If not, is it attracted to a magnet?

   C. Does the mineral sample display a **reaction with acid** (**Fig. 3.15**)?

   D. If crystals are visible, then what is their **crystal form** or **habit** (**Figs. 3.5** and **3.6**)?

   E. Does the mineral sample have **striations** on cleavage surfaces or crystal faces or **exsolution lamellae** (**Figs. 3.16** and **3.17**)?

   F. Estimate **specific gravity** (SG) as low, intermediate, or high.

   G. Does the mineral sample have any unique diagnostic properties like **smell** when scratched or during acid test?

7. Use mineral identification figures to identify the name of the mineral.

   A. If the mineral is opaque and metallic or submetallic, follow steps 1–5 in **Fig. 3.19**.

   B. If the mineral is dark- to medium-colored and nonmetallic, then follow steps 1–4 in **Fig. 3.20**.

   C. If the mineral is light colored and nonmetallic, then follow steps 1–4 in **Fig. 3.21**.

8. Use the Mineral Database (**Fig. 3.22**) and **Fig. 3.23** to determine and record the mineral's **chemical composition** and help you determine **how you personally depend on the mineral** and on commodities refined

from it. For more information about specific minerals or elements, you can refer to the U.S. Geological Survey's Minerals Information resources (**http://minerals.usgs.gov/minerals/pubs/commodity/**) and Mineral Commodity Summaries (**http://minerals.usgs.gov/minerals/pubs/mcs/**).

ACTIVITY 3.5

The Mineral Dependency Crisis, (p. 105)

Think About It   How do you personally depend on minerals and elements extracted from them? How sustainable is your dependency on minerals and elements extracted from them?

Objective   Evaluate our dependency on minerals.

Before You Begin   Read the following sections: What Is Mineral Dependency? and What Are Conflict Minerals?

What Is Mineral Dependency?

Did you know that some of the minerals used to make your cell phone and fluorescent light bulbs are quite rare nonrenewable resources? Many high-tech products depend on such nonrenewable mineral resources, yet many are either mined in small quantities or not mined at all within the United States. Of particular concern are minerals mined as ores for *rare earth elements*—a group of 17 elements used in products like fluorescent light bulbs, flat screen televisions, cell phones, computers, solar panels, wind turbines, hybrid cars, cameras, DVDs, rechargeable batteries, magnets, medical equipment, night-vision goggles, missile systems, and medical equipment. China currently produces nearly all of the world's supply of rare earth elements, and the United States produces almost none. This has created what is widely known as the "rare earth crisis," and a shortage of rare earth elements used to make fluorescent light bulbs has become widely known as the "phosphor crisis."

Very small concentrations of certain essential minerals can be very valuable. In some cases, the essential mineral is valuable because it is very scarce. In other cases, a valuable element might be very common but is disseminated in very low concentrations in rock rather than being concentrated in a manner that would make it economically profitable to extract.

The United States also relies on foreign supplies of many other minerals and elements extracted from them. Has the United States entered an unsustainable level of mineral dependency?

2012 U.S. NET IMPORT RELIANCE ON SELECTED NONFUEL MINERAL COMMODITIES

COMMODITY (Element, Ore, or Raw Mineral)	ORE MINERAL or RAW MINERAL	Percent Import Reliance	WHAT IS THIS COMMODITY USED FOR?
fluorine ore (F): fluorspar	fluorite	100	Fluorine is used in fluoride toothpaste, fluorocarbon refrigerant gases and fire extinguishers, and fluoropolymer plastics that coat nonstick fry pans and insulate wiring in cell phones, laptops, and airplanes.
graphite (C)	graphite	100	Graphite is used to make carbon steel, pencils, carbon fiber reinforced plastics in car bodies, and negative ends of most batteries (including those in all cell phones, power tools, computers, and hybrid/electric vehicles).
indium metal (In)	sphalerite with In as an impurity	100	Indium is used to make solar cells and liquid-crystal displays (LCDs) in cell phones, computers, and flat-screen television sets.
mica (sheet)	muscovite	100	Muscovite is used in heating elements of hair dryers and toasters, joint compound, and cosmetics with a satiny or glittery sheen.
quartz crystal (industrial)	quartz (clear transparent)	100	Crystals of cultured pure quartz are used to make quartz watches and the frequency controls and timers in every computer and cell phone.
niobium metal (Nb, "columbium")	columbite (in "coltan")	100	Niobium is used to make high-strength noncorrosive steel alloys (for jet engines, power plants) and arc welding rods, plus electrical insulation coatings in cell phones, computers, and electronic games.
tantalum metal (Ta)	tantalite (in "coltan")	100	Tantalum is used to make "tantalum capacitors" that buffer the flow of electricity between a battery and electronic parts in the circuits of cell phones, laptops, iPods, and most other electrical devices.
gallium metal (Ga)	bauxite is Ga ore	100	Gallium is used to make light-emitting diode (LED) bulbs and liquid-crystal displays (LCDs) in things like cell phones, computers, and flat-screen television sets.
vanadium metal (V)	magnetite with V as an impurity	100	Vanadium is used for cutting tools; mixed with iron to make lightweight shock-resistant steel for car axles and gears, springs, and cutting tools.
bismuth metal (Bi)	galena with Bi as an impurity	95	Bismuth is used as a nontoxic replacement for lead (in ceramic glazes, fishing sinkers, food processing equipment, plumbing, and shot for hunting) and in antidiarrheal medications.
garnet (industrial)	garnet	88	Industrial garnet is used as an abrasive in things like sandpaper and sandblasting.
zinc metal (Zn)	sphalerite is an ore of Zn	82	Zinc is used to make alloys like brass, skin-healing creams, and galvanized (rust-proof) steel and roofing nails; added to vitamin pills for a healthy immune system and to aid protein production.
silver metal (Ag)	native silver; galena with Ag as an impurity	72	Silver is used to make jewelry and silverware, photographic film, and solder on electrical circuit boards of computers and cell phones.
barium metal ore (Ba)	barite	70	Barium (Ba) is widely used to make capacitors (that store energy) and memory cells in cell phones and other portable electronic devices.
chromium metal (Cr)	chromite is an ore of Cr	66	Chromium is used to make stainless steel, yellow and green ceramic glazes and paints, and military camouflage paints; added to vitamin pills for healthy metabolism and lower cholesterol levels.
tungsten metal (W)	wolframite is W ore	49	Tungsten is a dense metal that makes cell phones and pagers vibrate (by attaching it to an electric motor spinning off center); also used for light bulb filaments, golf clubs, and tungsten carbide cutting tools.
magnesium metal (Mg)	dolomite and olivine are Mg ores	43	Magnesium is used to make strong, lightweight frames for jet engines and rockets, lightweight cell phone and laptop cases, and incendiary flares and bombs; added to vitamin pills to aid good brain and muscle function and strengthen bones.
aluminum (Al)	bauxite is Al ore	40	Aluminum is a lightweight silvery metal used to make drink cans, foil, airplanes, and solar panels.
nickel metal (Ni)	pyrrhotite contains Ni as an impurity	37	Nickel is used to make rechargeable batteries (Ni-Cd) for portable electronic devices, screw-end caps of light bulbs, and stainless steel.
copper metal (Cu)	azurite, bornite, chalcopyrite, chrysocolla, and malachite are Cu ores	36	Copper is used to make copper pipes; electrical wire for homes, businesses, electric motors, and circuit boards in cell phones and other electrical devices. Hybrid cars contain about 100 pounds (45 kg) of copper. Added to vitamin pills for healthy hair and skin.
salt	halite	32	It is used as table salt, road salt (to melt snow), in water softeners, and as a food preservative.
iron metal (Fe), steel	goethite, limonite, magnetite, and hematite are Fe ores	25	Iron and steel are used to construct machines, buildings, bridges, nails, bolts, tools, file cabinets; iron is added to vitamin pills to aid hemoglobin production in red blood cells for oxygen transport.
gypsum	gypsum	14	Gypsum is used to make plaster of paris, drywall. and for art (alabaster).
sulfur (S)	native sulfur; pyrite is a S ore	11	It is used to make matches, gunpowder, fertilizer, fungicide, insecticide, and harden rubber (car tires).
cement	calcite	10	Calcite is processed into cement, which is used to make concrete.

Figure 3.23 Selected net nonfuel mineral resource imports by the United States. Net import reliance is the total of U.S. production and imports minus the percentage of exports. (Adapted from USGS Mineral Commodity Summaries, accessed March 2016.)

U.S. Net Import Reliance on Nonfuel Mineral Resources

Commodities are natural materials that people buy and sell because they are required to sustain our wants and needs. Three classes of commodities are agricultural products, energy resources, and nonfuel mineral resources. The U.S. Geological Survey (USGS) divides the nonfuel mineral resources into two groups. **Nonmetallic mineral resources** are mostly rocks made of unrefined minerals (such as rock salt) and rocks (gravel, granite, marble). **Metallic mineral resources** are *ores* (rocks or minerals from which chemicals, usually metals, can be extracted at a profit) and chemical *elements* that have already been extracted from ore minerals.

The **USGS Mineral Resources Data System (MRDS)** is a global database of information about mineral resources. The USGS has determined that the United States was the world's largest user of nonfuel mineral resources in 2012 (about 12,000 pounds, or 11.3 metric tons, per person each year). To sustain its needs, the United States imports some of the mineral resources that it needs. **Fig. 3.23** shows the U.S. net import reliance on some selected minerals and elements refined from them as reported through 2015. The U.S. exports some of the same nonfuel mineral resources that it imports, so net import reliance is the total of U.S. production and imports minus the percentage of exports. A net import reliance of 80% means that 80% of the resource is imported. **Fig. 3.23** does not include all of the rare earth elements. Minerals and elements for which the United States is less than 5% import reliant (or a net exporter) are not shown either.

What Are Conflict Minerals?

Niobium (Nb) and tantalum (Ta) are two rare elements. Niobium is mixed with iron to make superconducting magnets for medical MRI scanners and high-strength noncorrosive steel alloys for rockets, jet engines, chemical pipelines, and nuclear power plants. Tantalum is used to make nonirritating steel for surgical steel tools and alloys with high heat tolerance for electronic microchips for small electrical devices like cell phones. Niobium is refined from the mineral columbite, and tantalum is refined from the mineral tantalite. The two minerals commonly occur together as an ore called "coltan." A primary source of coltan is the Democratic Republic of Congo (DRC) in Africa, where rebel groups use forced labor to mine the coltan and engage in extreme violence against women and children. The United States regards all coltan from the DRC as a "conflict mineral." The conflict minerals that are explicitly recognized under U.S. federal law, as of 2016, are columbite-tantalite (coltan), cassiterite, gold, wolframite, or their derivatives: tantalum, tin, and tungsten.

We can think more broadly of conflict natural resources as any natural resource produced in a manner that does not respect the basic human dignity of those who work to extract the resource and that is used to finance armed conflict. A broader definition would include

ACTIVITY 3.6

Urban Ore, (p. 106)

Think About It How do you personally depend on minerals and elements extracted from them? How sustainable is your dependency on minerals and elements extracted from them?

Objective Evaluate the prospect of recycling products and mining discarded products to extract their metals.

Before You Begin Read the section below: How Are Ores and Precious Metals Weighed?

Plan Ahead You will need to be able to connect to the Internet during your lab period to find the "New York spot gold price" in U.S. dollars today, or you will need to find that information before coming to class.

gemstones, petroleum, coal, other metals, and even biological resources like timber, rubber, cocoa, marijuana, poppy seeds, and ivory. We have a fundamental human responsibility as consumers in a global economy to take care for the welfare of our fellow humans in areas of conflict.

How Are Ores and Precious Metals Weighed?

Did you know that it is common practice in the United States to weigh gold and its ore using different systems of measurement? Within the United States, mining companies use an avoirdupois system to weigh bulk amounts of gold *ore*—rock from which gold can be extracted at a profit. If they sell the gold ore to another country, then its weight is quoted using the metric system. However, once gold is extracted from its ore, its weight is quoted everywhere in the world using the troy system.

Grams Are Metric; Ounces Are Not

A gram is a metric unit of mass equal to one thousandth of a kilogram—roughly the mass of a paper clip. There are no metric ounces. Ounces are used in the avoirdupois and troy systems of measurement. An avoirdupois ounce is 28.349523125 grams, and a troy ounce is 31.1034768 grams. Fluid/liquid ounces are units of volume, not weight.

What Is Avoirdupois?

Avoirdupois (avdp) is pronounced in English as "aver-due-pois," with the "pois" as in poison. It refers to a system of weights that is widely used in the United States and parts of Canada and Great Britain to weigh everything except precious metals, gems, and pharmaceuticals. It is based on a 28.35-gram *ounce* (oz), 16-avdp-ounce *pound* (lb), and

2,000-pound *short ton*. The concept of a pound originated in the Roman Empire, developed into a 16-ounce pound in Europe by the year 1300, and was the system of weights adopted by the 13 British Colonies that became the United States of America. However, the United States developed the 2,000-pound "short ton" that is still used today instead of the British 2,240-pound "long ton."

What Are Troy Weights?

Troy (t) refers to an old British system of weights that is now used globally to quote the weights of precious metals and gems. It is based on a 31.1-gram *troy ounce* (ozt) and a 12-troy-ounce *troy pound* (lbt). Because some gems and jewelry pieces are small, jewelers often express their weight in troy "pennyweight," abbreviated "DWT." One pennyweight is equal to 1/20th of a troy ounce (0.05 ozt) or 1.555 grams.

Unit Conversion—The Math You Need

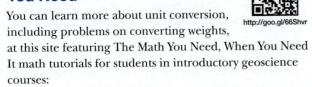

http://goo.gl/66Shvr

You can learn more about unit conversion, including problems on converting weights, at this site featuring The Math You Need, When You Need It math tutorials for students in introductory geoscience courses:

http://serc.carleton.edu/mathyouneed/units/index.html

Name: _____ Course/Section: _____ Date: _____

A All of the samples below are from Earth's crust (**Figs. A3.1.1–Fig. A3.1.4**). Record how many mineral grains you see in each sample (write 1, 2, 3, or many). Then make a numbered list of how many different kinds of minerals are in the sample and describe each one in your own words.

Figure A3.1.1 The white mass is a single mineral grain of milky quartz. How many **mineral grains** do you see in this sample? _____
List the number of different types of **minerals** in the sample and give a general description of each type of mineral:

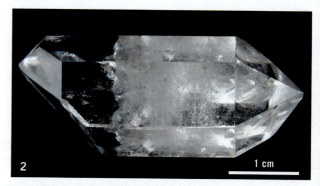

Figure A3.1.2 How many **crystals** do you see in this sample? _____
List the number of different **minerals** in the sample and give a general description of each type of mineral:

Figure A3.1.3 About how many **crystals** do you see in this sample? _____
List the number of different types of **minerals** in the sample and give a general description of each type of mineral:

Figure A3.1.4 About how many **mineral grains** do you see in this sample? _____
List the number of different types of **minerals** in the sample and give a general description of each type of mineral:

B Which of these samples seems to have crystals of a valuable chemical element? _____ What element? _____

C **REFLECT & DISCUSS** Based on your observations in this activity, how are crystals, minerals, and rocks related to—and distinguished from—each other?

Activity 3.2   Mineral Properties

Name: _____ Course/Section: _____ Date: _____

A Indicate whether the luster of each of the following materials looks metallic (M) or nonmetallic (NM):

1. a mirror: _____   2. butter: _____   3. ice: _____   4. a rusty nail: _____

B What is the streak color (i.e., color in powdered form) of each of the following substances?

1. salt: _____   2. wheat: _____   3. pencil lead: _____

C What is the crystal form or habit (**Figs. 3.5** and **3.6**) of the:

1. quartz in **Fig. 3.2A**? _____   2. native copper in **Fig. 3.4**? _____

D Look up quartz in the Mineral Database (**Fig. 3.22**) to find a list of the varieties (var.) of quartz. Then identify each quartz variety below (**Figs. A3.2.1–A3.2.4**), and write its name beneath the image.

Figure A3.2.1
var. _____

Figure A3.2.2
var. _____

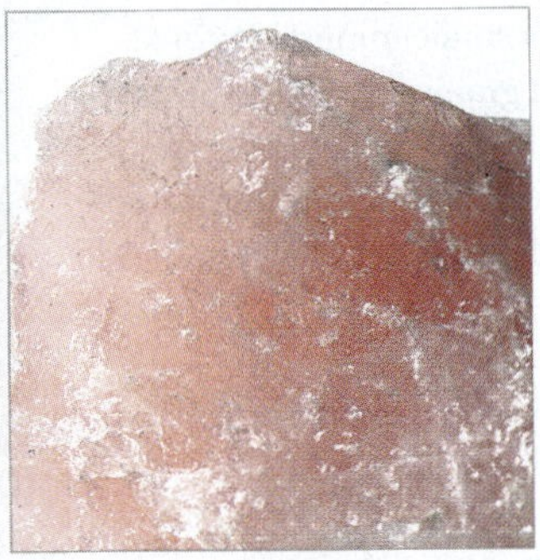

Figure A3.2.3
var. _____

Figure A3.2.4
var. _____

E A mineral can be scratched by a masonry nail or knife blade but not by a wire nail (**Fig. 3.14**).

1. Is this mineral hard or soft? _____

2. What is the hardness number of this mineral on Mohs Scale? _____

3. What mineral on Mohs Scale has such a hardness? _____

F A mineral can scratch calcite, and it can be scratched by a wire nail.

1. What is the hardness number of this mineral on Mohs Scale? _____

2. Which mineral on Mohs Scale has this hardness? _____

G The brassy, opaque, metallic mineral in **Fig. 3.3A** is the same as the mineral in **Fig. 3.13**. What is this mineral's hardness, and how can you tell?

H A mineral sample has a mass of 27 grams and takes up 10.4 cubic centimeters of space. What is the specific gravity of this mineral? Show your work.

I Analyze these two photomicrographs of ice crystals (snowflakes) by William Bentley (**Fig. A3.2.5**).

1. Both ice crystals have the same number of arms that radiate out from the center. How many arms do ice crystals have?

2. There are seven different crystal systems: triclinic, monoclinic, orthorhombic, tetragonal, trigonal, hexagonal, and cubic. Using the number of arms on a snowflake as a hint, what crystal system do ice crystals belong to? What is the basis for your interpretation?

3. Which of the crystal forms shown in **Fig. 3.5** or crystal habits shown in **Fig. 3.6** do you recognize in the ice crystals?

4. Why do you think ice crystals do not all have the same shape?

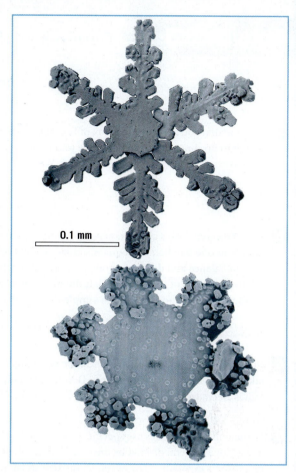

0.1 mm

Figure A3.2.5

J Analyze each crystalline household material pictured below (**Figs. A3.2.6–A3.2.8**). (Use a hand lens or microscope to observe actual samples of the materials if they are available.)

Figure A3.2.6

Figure A3.2.7

Figure A3.2.8

1. Sucrose (table sugar) displays the following crystal form(s) or habit (**Figs. 3.5** and **3.6**): _____

   How can you tell?

2. Epsomite (epsom salt) displays the following crystal form(s) or habit (**Figs. 3.5** and **3.6**): _____

   How can you tell?

3. Halite (table salt) displays the following crystal form(s) or habit (**Figs. 3.5** and **3.6**): _____

   How can you tell?

4. **REFLECT & DISCUSS** Which of these crystalline household materials (sucrose, epsomite, or halite) is not a mineral? Why not?

Name: _____   Course/Section: _____   Date: _____

A Imagine that you want to buy a box of breakfast cereal and get the most cereal for your money. You have narrowed your search to two brands of cereal that are sold in boxes of the exact same size and price. The boxes are made of opaque cardboard and have no labeling of weight. Without opening them, how can you tell which box contains the most cereal?

B Like the cereal boxes above, equal-sized samples of different minerals often have different weights. If you hold a mineral sample in one hand and an equal-sized sample of a different mineral in the other hand, then it is possible to act like a human balance and detect that one may be heavier than the other. This is called **hefting**, and it is used to estimate the relative densities of two objects. Heft the three mineral samples provided to you, and then write sample numbers/letters on the lines below to indicate the sample densities from least dense to most dense.

(Least dense)   _____   _____   _____   (Most dense)

C In more exact terms, **density** is a measure of an object's mass (in kilograms, kg) divided by its volume (how much space it takes up in cubic centimeters, cm^3). Density is usually expressed in units of g/cm^3. What is the density of a box of cereal that is 20 cm by 25 cm by 5 cm and has a mass of 0.453 kg? Show your work.

D Geoscientists sometimes use specific gravity to help identify an unknown mineral. Specific gravity is the ratio of the density of a mineral divided by the density of distilled liquid water, which we will assume to be 1 g/cm^3. Specific gravity is a dimensionless number. For example, the density of quartz is 2.6 g/cm^3, so the specific gravity of quartz is 2.6. Return to the three mineral samples that you hefted above, and do the following:

1. First (while they are still dry), determine and record the mass of each sample in grams in **Table A3.3.1**.

2. Use the water displacement method to measure and record the volume of each sample (**Fig. 3.18**). Recall that 1 fluid milliliter (mL or ml on the graduated cylinder) equals 1 cubic centimeter.

3. Calculate the specific gravity of each sample.

4. Identify each sample based on the list of specific gravities of some common minerals (**Table A3.3.2**).

SG OF SOME MINERALS	
2.1	sulfur
2.6–2.7	quartz
3.0–3.3	fluorite
3.5–4.3	garnet
4.4–4.6	barite
4.9–5.2	pyrite
7.4–7.6	galena
8.8–9.0	native copper
10.5	native silver
19.3	native gold

Sample	Mass in grams (g)	Volume in cubic cm (cm³)	Specific gravity (SG)	Mineral name

Table A3.3.1   Table A3.3.2

E REFLECT & DISCUSS   Were your data and calculations accurate enough to be useful in identifying the samples? If not, how could they be made more accurate?

Name: _____ Course/Section: _____ Date: _____

Describe the samples provided by your teacher, using the table below.

MINERAL DATA CHART

Sample Letter or Number	Luster*	Hardness	Color / Streak	Cleavage / Fracture	Other notable properties; tenacity, magnetic attraction, reaction with acid, specific gravity, smell, etc	Name (Fig. 3.19, 3.20, or 3.21) and chemical composition (Fig. 3.22)	How do you depend on this mineral or elements from it? (Fig. 3.22)

*M = metallic or submetallic, NM = nonmetallic

Table A3.4.1

MINERAL DATA CHART

Sample Letter or Number	Luster*	Hardness	Color	Streak	Cleavage	Fracture	Other notable properties; tenacity, magnetic attraction, reaction with acid, specific gravity, smell, etc	Name (Fig. 3.19, 3.20, or 3.21) and chemical composition (Fig. 3.22)	How do you depend on this mineral or elements from it? (Fig. 3.22)

*M = metallic or submetallic, NM = nonmetallic

Table A3.4.1 (*continued*)

MINERAL DATA CHART

Sample Letter or Number	Luster*	Hardness	Color	Streak	Cleavage / Fracture	Other notable properties; tenacity, magnetic attraction, reaction with acid, specific gravity, smell, etc	Name (Fig. 3.19, 3.20, or 3.21) and chemical composition (Fig. 3.22)	How do you depend on this mineral or elements from it? (Fig. 3.22)

*M = metallic or submetallic, NM = nonmetallic

Table A3.4.1 (continued)

MINERAL DATA CHART

Sample Letter or Number	Luster*	Hardness	Color	Streak	Cleavage	Fracture	Other notable properties; tenacity, magnetic attraction, reaction with acid, specific gravity, smell, etc	Name (Fig. 3.19, 3.20, or 3.21) and chemical composition (Fig. 3.22)	How do you depend on this mineral or elements from it? (Fig. 3.22)

*M = metallic or submetallic, NM = nonmetallic

Table A3.4.1 (*continued*)

The Mineral Dependency Crisis

Activity 3.5

Name: ______________________ **Course/Section:** ______________ **Date:** __________

A Refer to the list of selected net nonfuel mineral resource imports by the United States (**Fig. 3.23**).

1. Based on **Fig. 3.23**, complete the table below (**Fig. A3.5.1**).

Element used to make cell phones	**Mineral ore(s)** from which it is extracted	How is the element used to make cell phones?

Figure A3.5.1

2. Would it be possible to manufacture cell phones in the United States if a world crisis prevented it from importing minerals and elements? __________ yes __________ no What evidence from **Fig. 3.23** supports your answer?

B Refer to **Fig. 3.23** and your uses of minerals recorded in your completed Worksheet 3.4. How might the lifestyle of U.S. residents change if they could no longer import the minerals and elements on which they depend 100% on foreign suppliers?

C **REFLECT & DISCUSS** What concerns do you have about reliance on foreign sources of essential minerals? Do you have similar concerns for U.S. sources of essential minerals that are controlled by private companies based in the United States? What if the companies are based elsewhere? How do your views change, if at all, when you consider manufacturing by U.S. companies in other countries where they can more easily access essential minerals?

Name: _____ Course/Section: _____ Date: _____

A Recall that "ore" is a rock or mineral from which elements or compounds can be extracted at a profit. More than half of the gold mined in the United States is from mines in northern Nevada. These mines produced an average of 3.2 grams of gold per 2,000-pound short ton of ore in 2012.

1. Search the Internet for "New York spot gold price" in U.S. dollars (USD) per ounce, and enter it here. Note that ounces of gold are always quoted in troy ounces (ozt), but some people incorrectly report it as "oz." For example, the spot price was $1322.10 per troy ounce on July 24, 2016. What is the current spot price?

   NY spot gold price: _____

2. There are 31.1 grams (g) in 1 troy ounce (ozt). How many ozt of gold are extracted from 1 short ton of average Nevada ore? Show your work.

3. What is the gold worth (in USD) from 1 ton of the Nevada ore? Show your work.

4. It costs about USD $640 to extract 1 troy ounce of Nevada gold from the mine. So how much does it cost to mine and extract the gold from 1 short ton of the Nevada ore? Show your work.

5. Based on your answers in **3** and **4** above, what is the current average profit per ton of gold ore from northern Nevada?

B A typical smartphone contains 0.0012 grams of gold and weighs 3.951 ounces (avdp).

1. There are 16.00 ounces in 1 avoirdupois pound and 2,000 pounds in 1 short ton (avoirdupois ton). How many smartphones are there in 1 short ton? Show your work.

2. Based on your work above and the fact that there are about 0.0012 grams of gold in a typical smartphone, how many grams of gold are there in 1 short ton of smartphones? Show your work.

3. There are 31.1 grams in 1 troy ounce. How many troy ounces of gold are there in 1 short ton of smartphones? Show your work.

4. Based on the New York spot gold price in USD per troy ounce that you determined above (part **A1**), what is the current value of the gold in 1 short ton of smartphones?

C **REFLECT & DISCUSS** What materials besides cell phones could the U.S. recycle or mine as "urban ore" for metals noted in **Fig. 3.23**, and what impact would this have on the environment and the ability of the United States to sustain its need for metals and mineral ores?

Rock-Forming Processes and the Rock Cycle

Pre-Lab Video 4

http://goo.gl/NSIDMo

BIG IDEAS

*Rocks can be classified as igneous, sedimentary, or metamorphic by how they formed as indicated by their present composition and texture. The idea of a rock cycle dates back to **James Hutton** in the late 1700s. The rock cycle illustrates how different types of rock are related to each other and to the environments in which they form. The rock cycle also illustrates that rock is both a **product** of change and the **record** of geologic change over time. Embedded in this old idea of the rock cycle are the seeds of our current way of understanding Earth through its interrelated systems.*

FOCUS YOUR INQUIRY

Think About It   What is rock, and what are the three major groups of rock as classified by the way the rock forms?

ACTIVITY 4.1   Rock Inquiry (p. 108, 117)

ACTIVITY 4.2   What Are Rocks Made Of? (p. 110, 118)

- -

Think About It   How are a rock's composition and texture used to classify it as igneous, sedimentary, or metamorphic?

ACTIVITY 4.3   Rock-Forming Minerals (p. 110, 119)

ACTIVITY 4.4   What Is Rock Texture? (p. 113, 120)

- -

Think About It   How is the cycling of matter and energy evident in the three groups of rock we observe on Earth?

ACTIVITY 4.5   Rock and the Rock Cycle (p. 115, 121)

- -

Introduction

Rock is an aggregate of mineral grains that forms the solid Earth and all of the other large solid bits of matter in the Universe. What makes rock interesting is that it has a story to tell us if we learn how to interpret that story. **Igneous rock** crystallizes from **magma**—from molten rock—and we can learn things about the temperature, pressure, cooling rate, chemistry, and tectonic environment of an igneous rock by studying it. **Sedimentary rock** accumulates from material eroded or dissolved from older rock and minerals, as well as the debris of life. Sedimentary layers are very much like

▲ Light-colored igneous dikes cut the dark metamorphic rock in the Black Canyon of the Gunnison River, Colorado. Weathering and erosion of these ancient rocks produces sediment transported by the Gunnison River.

the pages of a book of Earth's history, providing information about climate, chemistry, quick events like storms and tsunamis, long events like the erosion of a mountain range, and the ebb and flow of life on Earth. **Metamorphic rock** forms from pre-existing rock or sediment that is changed by the effects of changing environmental conditions (such as temperature, pressure, stress, fluids) without melting. Every rock bears a story of its formation and change over time. So whenever you see a rock, look closely for clues about its origin and the story it has to tell.

ACTIVITY 4.1

Rock Inquiry, (p. 117)

Think About It   What is rock, and what are the three major groups of rock as classified by how the rock forms?

Objective   Analyze rock samples and infer where and how they formed.

Before You Begin   Read the section below: Rock and Naturally Occurring Glass.

Rock and Naturally Occurring Glass

Rocks contain multiple mineral grains. Recall that mineral grains form in several ways: by crystallizing from magma (i.e., from hot, liquid, melted rock material), by precipitating from water that contained ions in solution, through processes inside living organisms, through chemical reactions on the surface of other minerals, or by developing in the solid state through processes in which atoms are rearranged to form new crystal lattices.

Three Main Groups of Rocks

Geoscientists generally recognize three broad categories of rock that differ from one another based on the fundamental processes involved in rock formation (**Fig. 4.1**):

- **Igneous rock** forms as minerals crystallize during the cooling of magma, forming a mass of **intergrown crystals**. Some igneous rock crystallizes below Earth's surface, forming **intrusive igneous rock** (**Fig. 4.2**). Magma that reaches Earth's surface crystallizes to form **extrusive igneous rock**, which is also known as volcanic rock or lava.

- **Sedimentary rock** forms either from loose grains of rock, minerals, or biological debris that are collectively known as **sediments**, or from minerals precipitated from water. Sediments collect at Earth's surface in nearly horizontal layers. Layers of sediments can harden into sedimentary rock by being compacted

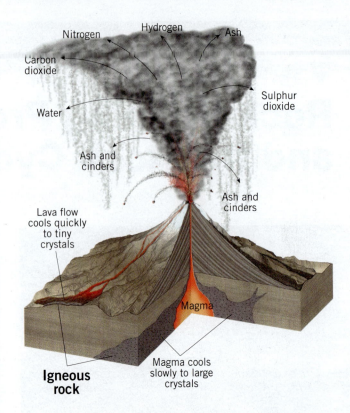

Igneous rock

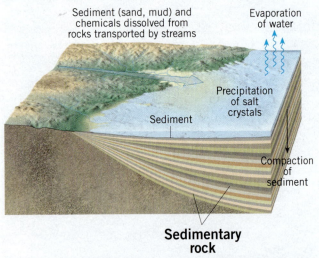

Sedimentary rock

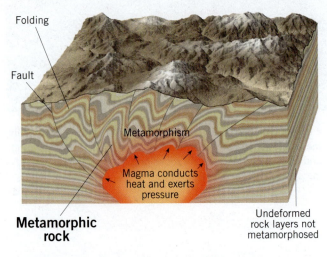

Metamorphic rock

Figure 4.1 The origin of igneous, sedimentary, and metamorphic rocks.

by the weight of new layers deposited on top of them or by being cemented together (**Fig. 4.3**). Minerals that precipitate from water tend to crystallize on whatever solid surface is available: the ground surface or seafloor, inside a hole in a rock, in the pore space between grains, or in fractures. Layering in precipitated rock follows the shape of the surface the minerals formed on (**Fig. 4.4**). Precipitated sedimentary rock is a mass of intergrown crystals.

- **Metamorphic rock** forms by changing older igneous, sedimentary, or metamorphic rock. Changes in the environment that can metamorphose pre-existing rock include changes in the gas and fluids that may be present between the solid grains within the rock, as well as changes in temperature, pressure, or stress. The result of metamorphism is a rock that is a mosaic of intergrown crystals (**Fig. 4.5**). Metamorphic processes do not involve melting because melting produces an igneous rock.

Figure 4.2 Igneous rock: Intergrown crystals.
A. Coarse intergrown crystals in the intrusive rock *granite*.
B. Fine intergrown crystals in the extrusive igneous rock *rhyolite* that has approximately the same mineralogy and chemical composition as the granite shown in **A**.

Figure 4.4 Sedimentary rock: Precipitated microcrystalline quartz. This rock is called an *agate* and is made of a microcrystalline variety of quartz called *chalcedony*. Agates usually have curved layers of chalcedony, often in a variety of colors.

Figure 4.3 Sedimentary rock: Quartz sandstone. Clastic sedimentary rock with horizontal bedding surfaces marked by the bedding-parallel fractures.

Figure 4.5 Foliated metamorphic rock. The layering developed at high temperature while the rock was subject to a differential stress. The minerals changed in the solid state, flattening in response to the greatest compressive stress, like a ball of putty flattening between your hands as you press them together.

Naturally Occurring Glass

Another naturally occurring solid, **glass**, is not an aggregate of mineral grains and is therefore not considered a rock. Glass forms when a melt cools so quickly that there is not enough time for atoms to bond together in orderly crystalline lattices. Rapid cooling traps the atoms in glass together in a disordered state without the necessary heat energy to reorganize them into ordered crystal lattices. Natural glasses that form from the very rapid cooling of magma include **obsidian** (**Fig. 4.6A**), which has a rhyolitic composition, and various glasses such as **tachylyte** that solidify from basaltic magma. Glass also forms due to meteor impact, lightning strikes, and frictional heating along faults.

Natural glass breaks just like manufactured glass breaks, forming a fracture surface that frequently has gentle curves that look like waves around the point where the fracture originated (**Fig. 4.6A**). These curved ribs on the fracture surface reminded early geologists of the whorls on a conch or gastropod shell (**Fig. 4.6B**), so this kind of fracture is called **conchoidal fracture**.

Natural glass is unstable over long time intervals and eventually changes into an aggregate of stable minerals in a process called **devitrification** (**Fig. 4.6C**). Atoms in the glass slowly organize into a crystal lattice because remaining in a disordered state within the glass requires a lot of energy. Think of the bonds between atoms as if they were springs. If the bond is just the right length, the spring is relaxed. In contrast, if the bond is too long or too short, energy is required to stretch or compress the spring. Because all of the bonds between atoms in an ordered lattice will be just the right length, a lattice has less elastic energy stored within it than a solid composed of disordered atoms does. Therefore, the disordered glass devitrifies because natural processes tend to favor whatever outcome involves the least energy—in this case, the stable minerals' ordered crystal lattice.

Safety First: Handling Volcanic Glass. Remember that obsidian specimens you might encounter in lab are pieces of broken glass. Just like any other broken glass, the edges of obsidian specimens can be very sharp. In fact, surgical scalpels made of obsidian are many times sharper than the very best steel scalpels, and 10–20 times as sharp as commercial razor blades. If you must handle an obsidian specimen in lab, do so with great care.

Figure 4.6 Natural glass.
A. The natural volcanic glass *obsidian* breaks with a conchoidal fracture, shown here as the concentric rings. **B.** Each spiral curve in the shell of this fossil gastropod is called a *whorl*. This pattern is the inspiration for the term *conchoidal fracture*, used to describe the pattern on a fractured surface of glass. **C.** The glass in this specimen of *snowflake obsidian* is slowly changing to an aggregate of new mineral grains. The gray circles in this specimen are clusters of needle-like mineral grains radiating outward from a central point, known as *spherulites*. These new minerals are growing at the expense of the glass and will ultimately replace the glass entirely.

ACTIVITY 4.2

What Are Rocks Made Of? (p. 118)

Think About It   What is rock, and what are the three major groups of rock as classified by how the rock forms?

Objective   Analyze rock samples and describe what they are made of.

Before You Begin   Read the section below: Rock Composition.

ACTIVITY 4.3

Rock-Forming Minerals, (p. 119)

Think About It   How are a rock's composition and texture used to classify it as igneous, sedimentary, or metamorphic?

Objective   Analyze and identify samples of some common rock-forming minerals.

Before You Begin   Read the section below: Rock Composition.

Rock Composition

One important way that geologists identify different types of rock they encounter in outcrops or in hand-size specimens is based on the types of minerals they contain. Most rock in Earth's crust is composed of a fairly small number of rock-forming minerals (Fig. 4.7). With practice, geologists learn to identify many of the major rock-forming minerals on sight. To view the grains in a hand-size sample of rock, start by looking at the rock closely. If you cannot see or identify the grains, then try using a magnifying glass or hand lens. Most geologists use a 10x hand lens, meaning that objects viewed through the lens appear 10 times larger than their true size. Geologists use a variety of optical or electron microscope techniques for minerals that are too small to identify even with a hand lens and can perform laboratory analyses that are beyond the scope of this course.

SOME COMMON ROCK-FORMING MINERALS

Mineral	Description	Common in Igneous Rocks	Common in Sedimentary Rocks	Common in Metamorphic Rocks
augite (pyroxene)	Very dark green to brown or dark gray, hard mineral (hardness 5.5–6.0) with two cleavages about 90 degees apart.	✓		
biotite (mica)	Glossy black mineral that easily splits into thin transparent sheets along its excellent cleavage. Hardness 2.5–3.0.	✓		✓
calcite	Usually colorless, yellow, white, or amber. Breaks along three excellent cleavages (none at 90 degrees) to form rhombohedrons (leaning blocks). Hardness 3. Reacts with dilute hydrochloric acid (HCl).		✓	✓
chlorite	Green mica-like mineral that splits into thin glossy transparent sheets along its excellent cleavage. Hardness 2.0–2.5. Occurs in large crystals or fine-grained masses.		✓	✓
dolomite	Pink, white, gray, yellow, brown, colorless. Perfect rhombohedral cleavage. Effervesces weakly in HCl if powdered. Curved, saddle-like crystals.		✓	✓
garnet	Red to black rounded crystals with no cleavage. Very hard (hardness 7).		✓	✓
gypsum	Colorless, white, or gray mineral. Easily scratched (hardness 2.0), even with a fingernail.		✓	
halite	Colorless, white, yellow, gray cubes that break into cubic shapes because they have three excellent cleavages 90 degrees apart. Brittle. Hardness 2.5.		✓	
hornblende (amphibole)	Dark gray to black, hard mineral (hardness 5.5–6.0). Breaks along glossy cleavage surfaces about 56 and 124 degrees apart.	✓		✓
kaolinite	Earthy white, gray, or very light brown clayey masses that leave powder on your fingers. Very fine grained. No visible crystals. Hardness 1–2.		✓	
muscovite (mica)	Colorless, brown, yellow, or white minerals that easily split into transparent thin sheets along its excellent cleavage. Hardness 2.0–2.5.	✓	✓	✓
olivine	Pale to dark olive green or yellow mineral with no cleavage. Very hard (hardness 7). Crystals may resemble sand grains. Brittle.	✓		✓
plagioclase (feldspar)	Usually white to pastel gray but may be colorless or black with iridescent play of colors. Exhibits fracture surfaces and two good cleavages. Cleavage surfaces may have thin striations. Hardness 6.	✓	✓	✓
orthoclase (feldspar)	Usually pink-orange or pale brown, may be white. Usually has internal discontinuous streaks (exsolution lamellae). Exhibits fracture surfaces and two good cleavages. Hardness 6.	✓	✓	✓
quartz	Usually transparent to translucent gray or milky white, may be colorless. No cleavage. Breaks along uneven fractures or curved conchoidal fractures (like glass). Very hard (hardness 7).	✓	✓	✓

Figure 4.7 Some common rock-forming minerals.

Composition of Igneous Rock

Mineral composition is essential information for classifying igneous rock. **Granite** and **rhyolite** share a common mineral composition, which is dominated by quartz, alkali (potassium or sodium) feldspar, and sodium plagioclase feldspar with perhaps 5% biotite and lesser quantities of other minerals. Of the major minerals in granite and rhyolite, all but biotite are light colored (gray, white, pink), and so the resulting rock has a light tone (**Fig. 4.2**). Granite and rhyolite are compositional twins, but granite crystallizes slowly below Earth's surface and rhyolite is a volcanic rock that crystallizes rapidly at Earth's surface. The different environments of crystallization result in different textures for intrusive and extrusive igneous rock, which we will learn about in the next section.

Gabbro and **basalt** are also compositional twins but are composed of dark-toned minerals that contain iron or magnesium, such as olivine and pyroxene. Another mineral in gabbro and basalt is calcium-rich plagioclase feldspar, which is often clear to dark toned, so the resulting rock is dark toned or **mafic**. The intrusive rock *gabbro* and the extrusive rock *basalt* have no quartz.

Between the light-toned and dark-toned igneous rock types are intermediate-toned **diorite** and **andesite**, whose composition is dominated by plagioclase feldspar, amphibole, biotite, and minor alkali feldspar and quartz. Andesite is an extrusive rock named after the volcanoes of the Andes Mountains in western South America, and diorite is its intrusive compositional twin.

Composition of Sedimentary Rock

How sedimentary rocks form is a main factor influencing their composition. The types of sedimentary rock that result from precipitation are classified based on their mineral content and, to some extent, where they precipitated. Examples of sedimentary rock composed of microcrystalline minerals precipitated from water include onyx, agate, and chert (microcrystalline quartz or chalcedony as in **Fig. 4.4**), travertine (calcite), and alabaster (gypsum or calcite). Precipitated sedimentary rock can be found around geysers and fumaroles, in caves, on the floor of lakes and oceans, or on the ground surface because of evaporation. Evaporation can lead to the precipitation of rock salt, gypsum rock, dolostone, and other sedimentary deposits that we broadly classify as evaporites.

The most common types of sedimentary rock result from deposition of sediment composed of mineral grains, rock fragments, and perhaps even fossils (**Fig. 4.8**). A **fossil** is a naturally occurring artifact of ancient life, such as a bone or a shell (**Fig. 4.9**). Many sedimentary rocks are dominated by silicate minerals like quartz, clay, and feldspar. The composition of this type of sedimentary rock is largely determined by the composition of the area that weathered and eroded to produce the sediment that now forms the sedimentary rock, and by how far the sediment traveled before deposition. Only the most durable and stable minerals, like quartz and clay minerals, tend to be present in sediment that has traveled a long distance from its source area.

Figure 4.8 Recent beach sand. Sand on a quartz-dominated beach along the Atlantic Ocean in New Jersey. Broken shell fragments are composed of carbonate minerals.

Figure 4.9 Fossil-rich sandstone. The white and gray clasts are fossils cemented together in a quartz sandstone.

The **carbonate** sedimentary rock **limestone** is composed predominately of the calcium-carbonate mineral **calcite** and typically contains fossils. The other common carbonate sedimentary rock, **dolostone**, is composed mostly of the calcium-magnesium carbonate mineral **dolomite**. Limestone effervesces vigorously when a drop of dilute hydrochloric acid (HCl) is placed on it because it is composed of calcite. Dolostone also fizzes in HCl, but you might need to grind up a powder of the dolostone to get it to fizz.

Composition of Metamorphic Rocks

The rock that existed before the metamorphism of a particular volume of rock is called the **parent rock** or **protolith**. The composition of the resulting metamorphic rock is determined by the composition of the protolith and by the full history of the temperature, pressure, differential stress, fluids, and gases that were involved in the metamorphism. Some of these changes are simple matters: quartz sandstone metamorphosing into **quartzite** through recrystallization of quartz, or limestone metamorphosing into **marble** through the recrystallization of calcite. Clay-rich mudstone or claystone might become **slate** because of differential stress and the solid-state change of clay minerals into tiny mica minerals. Other changes are complicated: minerals such as garnet, kyanite, sillimanite, and staurolite grow during metamorphism from atoms scavenged from other minerals or from liquids flowing through the rock.

Composition of Other Geological Solids

Some geological solids are not composed primarily of mineral grains. For example, **coal** is composed mostly of the altered remains of plant material. The names of different types of natural volcanic glass are based on the chemical composition of the glass. **Obsidian** is volcanic glass that has the chemical composition of the volcanic rock **rhyolite** (**Figs. 4.2B** and **4.6A**), and **tachylyte** is chemically similar to the volcanic rock **basalt**.

ACTIVITY 4.4

What Is Rock Texture? (p. 120)

Think About It How are a rock's composition and texture used to classify it as igneous, sedimentary, or metamorphic?

Objective Determine textures of rocks and classify them based on their composition and texture.

Before You Begin Read the section below: Rock Texture.

Rock Texture

Another important property we use to identify rock types is their **texture**. A rock's texture involves the size, shape, distribution, and layering of mineral grains within a rock. Rock texture does *not* involve how the outer surface of a particular rock specimen feels. For the purposes of this lab, we will focus on a limited set of rock textures that will help you to gain a basic understanding of the main groups of rock.

Clastic Versus Crystalline Texture

The grains of most rock types in Earth's crust and mantle form a mass of **intergrown crystals**, indicating that they probably formed together by one of three processes:

- Cooling from a melt (**Fig. 4.2**).
- Alteration in the solid state by metamorphic processes (**Fig. 4.5**).
- Precipitation from water that was saturated with ions that combined to form minerals (**Fig. 4.4**).

All metamorphic, most igneous, and some sedimentary rock have this **crystalline texture**.

Many types of sedimentary rock that are quite common at Earth's surface are formed by deposition of the eroded bits and pieces of other rocks along with the remains of organisms and their debris (**Fig. 4.8**). These particles are called **clasts** (from the Greek *klastós*, meaning broken). Sedimentary rock that forms from the deposition of clasts is called **clastic rock** and includes familiar rock types such as sandstone, shale, and many types of limestone. **Bioclastic** sedimentary rocks have fossils as an important constituent of the rock (**Fig. 4.9**). The composition of **volcaniclastic** rock is dominated by volcanic debris—typically material that was blasted out of the volcano into the air before the particles fell to the ground or that was carried down the volcano slope in a debris flow (lahar) or pyroclastic flow. A **siliciclastic** rock is composed largely of silicate minerals like quartz, feldspar, and clay (**Fig. 4.10**). See also **Fig. 4.3**.

Grain Size

We have already noticed that different types of rock might have different grain sizes. Granite is a type of igneous rock with coarse, intergrown crystals that grew slowly in a magma body below Earth's surface (**Fig. 4.2A**). Most of the mineral grains in granite are greater than 1 mm in diameter and are easy to see if you know what to look for. Rhyolite is an extrusive or volcanic rock with the same mineral composition as granite and is a mass of intergrown crystals. Rhyolite is a fine-grained igneous rock whose minerals are mostly less than 1 mm in diameter because they developed quickly from a rapidly cooling magma at Earth's surface (**Fig. 4.2B**). In igneous rock, grain size is a clue about how fast the magma cooled and, consequently, about where the rock formed. Coarse-grained igneous rock crystallized slowly below Earth's surface, and fine-grained igneous rock crystallized rapidly at the surface.

We will consider a rock with an average grain size of less than ~1 mm to be a **fine-grained** rock and a rock with an average grain size of more than ~1 mm to be

Figure 4.10 Quartz sandstone.

considered a **coarse-grained** rock. Some igneous rock has two distinctly different grain sizes, which is usually interpreted to indicate that the magma cooled slowly for a while (producing the larger crystals, called **phenocrysts**) and then cooled much more quickly (producing smaller crystals called the **groundmass**) until all of the magma crystallized. An igneous rock with two distinct grain sizes is called a **porphyritic** rock.

Clastic sedimentary rock is classified primarily by its grain size (**Fig. 4.11**) and secondarily by its composition. **Clay**-sized grains are so small that you cannot feel any grittiness as you rub them between your fingers or even when it is on your tongue. (If that seems unappealing, you should know that some clay minerals are used as food additives.) **Silt** is a bit bigger (1/16 to 1/256 mm) but is still so small that it feels only faintly gritty between your fingertips. Silt feels gritty on your tongue. **Sand**-sized grains (1/16 to 2 mm) definitely feel gritty between your fingers. **Gravel**-sized grains (>2 mm) include granules, pebbles, cobbles, and boulders. Sedimentary grains that are smaller than gravel are almost always composed of a single mineral grain whereas gravel-sized grains are usually rock fragments containing more than one mineral grain.

Layering

In contrast to the random orientation of grains in most igneous rocks, some metamorphic rocks have a **layered texture**. The layering that develops in a metamorphic rock during metamorphism is called **foliation** (**Fig. 4.5**). Foliation develops when a rock is more than about half of its melting temperature and is subjected to differential stress—a system of forces that is stronger in one direction than in the others. The combination of differential stress at a high enough temperature allows minerals to gradually

change their shape in the solid state without melting. They can flatten and stretch into approximately parallel layers of minerals. Parallel mica minerals in a foliated metamorphic rock can reflect light like the layered scales on a fish. In some cases, rock layers can be observed at more than one scale: over a region, in an outcrop, or in a hand sample.

Most sedimentary rocks also have a layered texture, which developed during either deposition or precipitation. Sedimentary rock generally has layers made of either clastic grains (gravel, sand, silt, clay, shells, plant fragments) or precipitated crystals of minerals like gypsum, halite, or calcite. The layering in precipitated rock follows the shape of the surface on which the minerals were precipitated, which can be quite irregular (**Fig. 4.4**). Clastic sediment is deposited on **bedding surfaces** that tend to be approximately horizontal over broad areas. On a more local scale, the shape of sedimentary bedding surfaces might be wavy due to ripples caused by the currents that transported the sediments. The resulting sedimentary rock is layered along bedding surfaces (**Figs. 4.3** and **4.10**). Sedimentary grains that are flat or elongated like a pencil will tend to become oriented approximately parallel to the bedding planes, so the layered texture extends in scale from tiny grains up to bedding surfaces that can extend over broad areas. The sedimentary layering exposed in the Grand Canyon is an example of bedding surfaces that extend for significant distances.

Vesicular Texture

Extrusive igneous rock with a **vesicular** texture has round or oval holes, called **vesicles**, that resemble the holes in a sponge or Swiss cheese (**Fig. 4.12**). The holes formed around gas bubbles that were trapped in the lava as it cooled. Some volcanic rocks have only scattered vesicles, but **pumice** is a frothy material that consists almost entirely of bubbles of volcanic glass. Because pumice is composed of volcanic glass, it is used as an abrasive in polishes and as a cosmetic exfoliant to soften skin (pumice stones, Lava™ soap). Of course, the manufacturers don't tell you that you are actually scraping your skin with tiny shards of broken glass.

Gravel (granules)

Sand

Clay

Silt

1 cm

Figure 4.11 Sediment sizes.

Vesicles

1 cm

Figure 4.12 Vesicles. The small holes in the volcanic rock (basalt) are called *vesicles* and were formed by gas trapped in the rock as it crystallized from magma.

Folded Texture

Folding of rock can occur across a range of conditions from the ground surface to the mantle and can affect many geological materials from near-surface materials (soil, sedimentary beds, ice, rock salt) to sedimentary rock and metamorphic rock. Shortening or shear of any ductile rock can result in folding. (Putting your hands together and sliding them past each other is an example of shear.) While folded foliation is quite common in metamorphic rock, folding is only of secondary importance in trying to decide whether a rock is of one type or another. Metamorphism is not required for folding.

ACTIVITY 4.5

Rock and the Rock Cycle, (p. 121)

Think About It How is the cycling of matter and energy evident in the three groups of rock we observe on Earth?

Objective Analyze and classify rocks, infer how they formed, and predict how they may change according to the rock cycle.

Before You Begin Read the following section: The Rock Cycle.

The Rock Cycle

In the late 1700s, James Hutton recognized cycles of geological events by observing outcrops and sedimentary processes in Scotland and by reading about volcanic eruptions and other geologic phenomena that occur elsewhere. Hutton was part of the Scottish Enlightenment along with Adam Smith, David Hume, Joseph Black, John Clerk, James Watt, and others. His work in geology greatly influenced Charles Lyell, who became Charles Darwin's geological mentor. Many geologists consider Hutton to be the founder of modern geology, and he is almost certainly the originator of the idea of a **rock cycle**. The rock cycle is a conceptual model of how matter and energy flow through Earth's systems over time. Our simplified diagram of the rock cycle (**Fig. 4.13**) includes some of the major processes and connections we use to understand the major rock groups.

Igneous Processes. An idealized path of rock cycling and redistribution of matter is illustrated along the broad purple arrows in **Fig. 4.13**, starting with igneous processes. Heat from deep within Earth flows through the mantle and crust on its way out to space. Matter also flows in the mantle and into the crust. Concentrations of magma are generated in a few places through partial melting of rock in the crust or upper-mantle asthenosphere (bottom of **Fig. 4.13**). Magma cools and minerals crystallize to form igneous rock, either at or below Earth's surface.

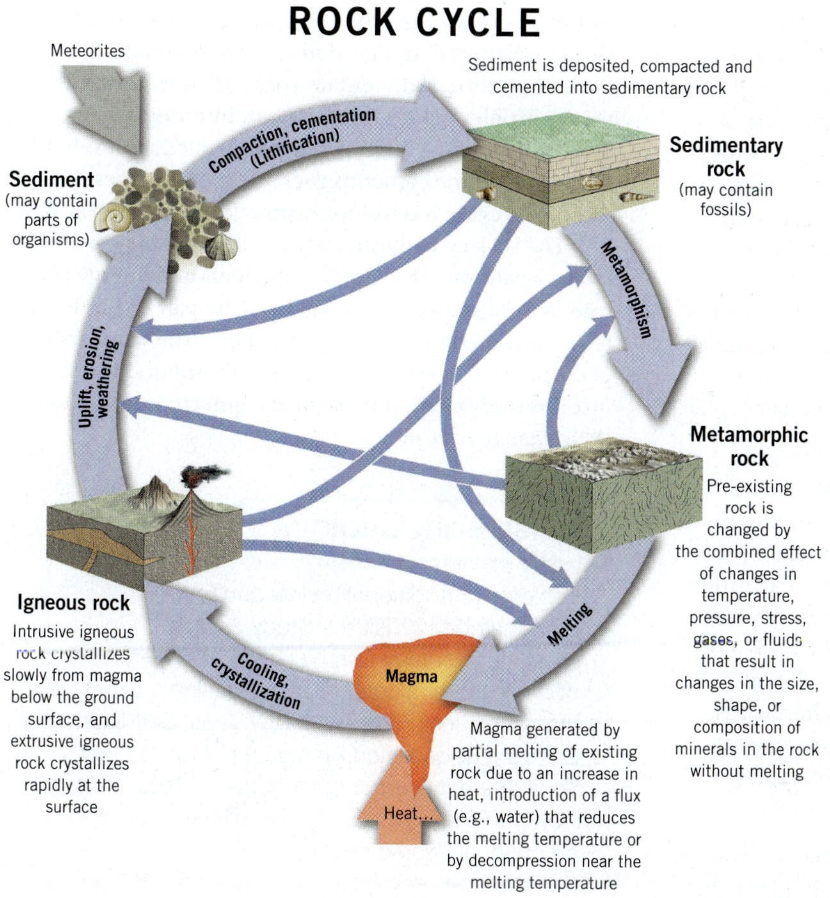

ROCK CYCLE

Figure 4.13 The Rock Cycle.
A conceptual model of how the major rock groups are formed through natural processes and environmental change. Igneous rock crystallizes from cooling magma, which formed by melting older rock. Sediments form due to weathering and erosion of older rock. Sedimentary rock forms from the deposition, compaction, and cementation of sediment or precipitation of ions in solution. Metamorphic rock forms from the alteration of older rock as it is subjected to changed environmental conditions, such as increased temperature, changes in pressure and stress, and flow of different liquids and gases through the rock.

Figure 4.14 Rock samples for analysis, classification, and evaluation. Four rock types that can be distinguished from one another based on their texture. You will work on a description of each of these samples in part B of Activity 4.5.

Sedimentary Processes. Where igneous rock (or any other type of rock) is exposed at Earth's surface, it is subjected to a variety of environmental conditions that promote change. The rock undergoes **physical weathering** and **erosion**, which produces sediment that can be transported and deposited in sedimentary layers. **Chemical weathering** of the rock produces ions carried in solution. Ions join with other ions and precipitate as solid minerals in a sedimentary layer, in open cracks or holes in rock, or as cement in the tiny spaces between solid grains. The sedimentary layers can then become **lithified** or hardened into sedimentary rock by being buried by newer layers, compacted, and cemented together. **Diagenesis** includes chemical, physical, or biological changes to sedimentary rock after it is lithified under conditions that are too mild to be considered metamorphic.

Metamorphic Processes. If rock of any kind is subject to significant changes in its environment (such as exposure to reactive gases or fluids or increases in temperature, pressure, or stress), then change might occur to the size, shape, or composition of its minerals. Layering called **foliation** might develop because of stress applied at high temperature. These changes make the rock more stable under the new conditions, and the result is a rock that is different from the original rock. In other words, the changes result in a metamorphic rock. Metamorphic processes extend up to, but do not include, complete melting of the rock.

Multiple Pathways Through the Rock Cycle. Of course, not all rocks undergo change along such a simple path.

The arrows in **Fig. 4.13** show that one rock group can be transformed to either of the other two groups *or* recycled within its own group. For example, igneous rock can be (1) weathered and eroded to form sediment that is lithified to form sedimentary rock, (2) transformed to metamorphic rock by intense heat, intense pressure, and/or hot fluids, or (3) remelted, cooled, and solidified back into another igneous rock. As rock changes through these processes, it develops distinctive textures (**Fig. 4.14**).

The rock cycle illustrates part of Earth's **biogeochemical cycles**—pathways along which elements and energy play a role in the atmosphere, biosphere (the part of Earth's systems associated with life), hydrosphere (the water system), pedosphere (the soil system), and lithosphere. Geoscience involves study of all of these interconnected systems and their change over time.

MasteringGeology™

Looking for additional review and test prep materials? Visit the Study Area in MasteringGeology to enhance your understanding of this chapter's content by accessing a variety of resources, including Pre-Lab Videos, Self-Study Quizzes, Geoscience Animations, Mobile Field Trips, *Project Condor* Quadcopter videos, *In the News* articles, glossary flashcards, web links, and an optional Pearson eText.

Name: _____ Course/Section: _____ Date: _____

A REFLECT & DISCUSS Describe the rock below (**Fig. A4.1.1**), where it may have formed, and how it may have formed.

Enlarged x4

x1 (actual size)

Figure A4.1.1

B REFLECT & DISCUSS Describe the rock below (**Fig. A4.1.2**), where it may have formed, and how it may have formed.

Enlarged x2

Figure A4.1.2

C REFLECT & DISCUSS Describe the rock below (**Fig. A4.1.3**), where it may have formed, and how it may have formed.

1 cm

Figure A4.1.3

What Are Rocks Made Of?

Name: _____ Course/Section: _____ Date: _____

A What are some visible differences between intergrown mineral grains and detrital or clastic mineral grains?

B **REFLECT & DISCUSS** Rocks are made of the materials listed below and described within the chapter. Under each sample in Fig. A4.2.1, write the name of every kind of material it contains from the list. Be prepared to compare your observations with the observations of others in your class.

Intergrown Crystals	Clasts (detrital minerals)	Gravel	Silt & Clay
Fossils (bioclasts)	Clasts (rock fragments)	Sand	Glass

Figure A4.2.1

Name: _____ Course/Section: _____ Date: _____

A **REFLECT & DISCUSS** Refer to **Fig. 4.7** and identify each rock-forming mineral below (**Fig. A4.3.1**). Write its name below the picture. Be prepared to compare your observations with the observations of others in your class.

Activity 4.4   What Is Rock Texture?

Name: _____   Course/Section: _____   Date: _____

Under each speciman in **Fig. A4.4.1**, list the textures evident in the rock using the terms below, and write whether you think it is an igneous, sedimentary, or metamorphic rock. Why did you make that interpretation?

Glassy	Fine grained	Intergrown crystals (crystalline)	Bioclastic	Layered (bedding)
Vesicular	Coarse grained	Clastic (gravely, sandy, silt/clay)	Layered (folded)	Layered (foliated)

Figure A4.4.1

Name: _____ Course/Section: _____ Date: _____

A On the rock cycle below (**Fig. A4.5.1**), color arrows *orange* if they indicate a process leading to formation of igneous rock, *brown* if they indicate a process leading to formation of sedimentary rock, and *green* if they indicate a process leading to formation of metamorphic rock. Place check marks in the table to indicate what rock group(s) is(are) characterized by each of the processes and rock properties.

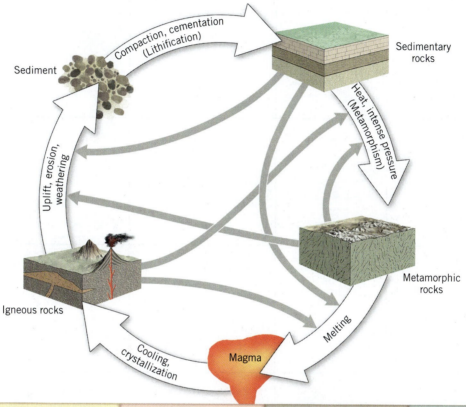

Processes and Rock Properties	Igneous	Sedimentary	Metamorphic
lithification of sediment			
intense heating (but no melting)			
crystals precipitate from water			
solidification of magma/lava			
melting of rock			
compaction of sediment			
cementation of grains			
folding of rock			
crystalline			
foliated			
common fossils			

Figure A4.5.1

B **Fig. 4.14** has photographs of four rocks, labeled **a** through **d**. For each photograph, record the following information in the chart (**Fig. A4.5.2**) on the next page:

1. In Column 2 (blue), list the rock properties that you can observe in the sample.

2. In Column 3 (pink), classify the rock as igneous, sedimentary, or metamorphic.

3. In Column 4 (yellow), describe, as well as you can, how the rock might have formed.

4. On the rock cycle diagram on the previous page (**Fig. A4.5.1**), write the figure number of the photograph/rock sample to show where it fits in the rock cycle model.

5. In Column 5 (green), predict from the rock cycle (**Fig. 4.13**) three different changes that the rock could undergo next if left in a natural setting.

Sample	ROCK PROPERTIES (grain types, textures)	ROCK CLASSIFICATION (igneous, sedimentary, metamorphic)	HOW DID THE ROCK FORM?	WHAT ARE THREE CHANGES THE ROCK COULD UNDERGO? (according to the rock cycle, Figure 4.13)
Figure 4.14A				1. 2. 3.
Figure 4.14B				1. 2. 3.
Figure 4.14C				1. 2. 3.
Figure 4.14D				1. 2. 3.

Figure A4.5.2

C **REFLECT & DISCUSS**   Starting with sedimentary rock, describe a series of processes that could transform the rock into each of the other two rock groups and back into a sedimentary rock.

Igneous Rocks and Processes

Pre-Lab Video 5

http://goo.gl/ElzuX1

Contributing Authors

Harold E. Andrews • *Wellesley College*
James R. Besancon • *Wellesley College*

Claude E. Bolze • *Tulsa Community College*
Margaret D. Thompson • *Wellesley College*

▲ Dramatic night eruption of Tungurahua Volcano (~5,023 m) in the Andes Mountains of central Equador (~1.4701°S, 78.4444°W). Tungurahua is a stratovolcano that erupts andesite-dacite rock and is related to the subduction of the Nazca Plate under South America.

BIG IDEAS

Igneous rock forms wherever magma cools and crystallizes to a solid state. The composition and texture of igneous rock samples can be used to classify them and infer their origin. Lava and igneous rock-forming processes can be observed at volcanoes, which occur along lithospheric plate boundaries, at hot spots, and in places where deformation of the crust leads to local melting. Volcanism can create new land surface in places like Hawai'i and can provide new minerals to the land surface. Volcanoes can also pose hazards to humans.

FOCUS YOUR INQUIRY

Think About It   What does igneous rock look like? How can it be classified into groups?

Think About It   What are igneous rock textures? How is texture used to classify and interpret igneous rock?

Think About It   What is igneous rock composed of? How is composition used to classify and interpret igneous rock?

Think About It   How are rock composition and texture used to classify, name, and interpret igneous rock?

Think About It   How can the shapes of igneous rock bodies be used to classify them and infer their origin?

Igneous Rock Inquiry, (p. 139)

Think About It What does igneous rock look like? How can it be classified into groups?

Objective Analyze and describe samples of igneous rock, and then infer how it can be classified into groups.

Before You Begin Read the section below: Introduction.

Introduction

Volcanic eruptions are among Earth's most astounding displays of energy. The eruption of almost 600 volcanoes has been observed and recorded in the past few thousand years, including 205 that have erupted between January 2000 and spring 2016 according to the Smithsonian Institution's Global Volcanism Program. Kilauea Volcano has been in continuous eruption since January 1983 on the flank of the tallest mountain on Earth: the shield volcano that forms the island of Hawai'i. **Shield volcanoes** have broad bases and gently sloping sides because they form from lavas that flow easily across the surface (**Fig. 5.1A**). In contrast, **stratovolcanoes** have steep sides and relatively narrow bases, and their major eruptions can be violent, explosive events (**Fig. 5.1B**). Large explosive volcanic eruptions can devastate the area around the volcano and spread volcanic dust and aerosols around the entire Earth.

We can't control or prevent volcanic eruptions, but we can learn about the igneous processes responsible for volcanism so that the worst of the hazards posed by volcanoes can be understood and avoided. To a geoscientist, a volcano is a very special window into Earth's interior. It is the surface manifestation of chemical and physical processes involving the partial melting of rock deep below the surface and the buoyant rise of molten rock, or **magma**, through the crust.

Igneous rocks are aggregates of intergrown mineral grains that crystallized from magma. Magma that reaches the surface crystallizes to form **extrusive igneous rock**, also known as volcanic rock. We use the word **tephra** broadly for the material that is blasted out of a volcano during explosive eruptions whereas we use another broad term, **lava**, for the rock formed from magma that flows from a volcano. Magma that does not reach the surface before it crystallizes forms **intrusive igneous rock**. At Yosemite National Park and throughout the Sierra Nevada Mountains of California, an enormous amount of intrusive igneous rock is exposed at the ground surface because of uplift and erosion of the mountain range long after the rock crystallized (**Fig. 5.2**).

Geoscientists have learned a great deal about igneous systems, and the plate tectonic model has provided a coherent general context for our investigations of igneous processes. But the fact remains that magma is generated deep below Earth's surface in the crust and upper mantle, so many igneous processes are difficult to study. In places

Figure 5.1 Shield and stratovolcanoes. A. Mauna Kea shield volcano as seen from the Mauna Loa Observatory, Hawai'i. The peak is ~4205 meters above sea level and more than 10 km above its base on the sea floor. **B.** Ash and steam erupting from Augustine stratovolcano, Alaska, on January 24, 2006. Peak elevation is ~1260 meters above sea level. (Photo by Michelle Coombs of the USGS Alaska Volcano Observatory.)

Figure 5.2 Sierra Nevada batholith. View of the granitic rock of the Sierra Nevada as seen from Glacier Point and looking toward Yosemite Falls, Yosemite National Park. A batholith is a continuous body of granitic rock that is exposed at the ground surface over an area of more than 100 km^2.

where magma reaches the surface, scientific and technical challenges and the threat of physical danger from heat, toxic gas, and explosive blasts combine to make active volcanoes a very exciting place to work. There is still much to discover.

Classifying Igneous Rocks

Geologists use four broad types of information to classify igneous rock: texture, mineral composition, chemical composition, and the geological context in which the rock occurs. Grain size is an important textural characteristic because it provides us with information about whether the rock might have crystallized on or below the ground surface. If the grain size is large enough, we can identify the minerals and estimate the volume of each type of mineral in the rock. Mineral identification might be done visually at an outcrop or with a hand sample, but geologists usually examine thin sections of a rock using a specialized (petrographic) microscope. We might use laboratory methods to determine the chemistry of the rock, which is a necessary step for fine-grained specimens. If it is known, the geological setting in which the rock occurs can be very useful to identification.

We will focus on classification that does not require the use of petrographic microscopes or geochemical laboratories during our work on igneous rocks and processes.

We don't have time in an introductory course to learn how to use those resources properly even if they are available. We will classify igneous rocks based on texture and the composition of the common minerals we can identify in hand specimens.

ACTIVITY 5.2

Crystalline Textures of Igneous Rock, (p. 140)

Think About It What are igneous rock textures? How is texture used to classify and interpret igneous rock?

Objective Review a crystallization experiment, infer how rate of cooling affects crystal size, and then apply your knowledge to interpret a rock with porphyritic texture.

Before You Begin Read the following section: Textures of Igneous Rocks.

Textures of Igneous Rocks

Recall that a rock's texture involves the size, shape, distribution, and layering of mineral grains within a rock. Grain size is an important clue to an igneous rock's origin. Under conditions that are otherwise identical, larger grains are thought to require a longer time to crystallize than smaller grains, so it is useful to try to specify what we mean by "large" and "small." The British Geological Survey (BGS) studied this question while compiling a guide to classifying igneous rocks for BGS geologists. They decided that grains smaller than 0.25 millimeter in diameter were effectively too small to identify without magnification, and this was adopted as the boundary size they use in classifying igneous rock textures.

Phaneritic and Aphanitic Textures

Igneous rock whose average grain size is greater than ~0.25 mm is called **phaneritic**—a word derived from the Greek word *phaneros* or "visible." A phaneritic igneous rock has grains that are large enough to be distinguished without magnification. Igneous rock whose average grain size is less than ~0.25 mm is called **aphanitic** from a Greek root *aphanes* meaning "invisible."

Figure 5.3 Porphyritic rock. A. Phenocrysts of olivine in an aphanitic groundmass of basalt with vesicles where gas was trapped during crystallization. **B.** Phenocrysts of orthoclase feldspar in a phaneritic groundmass of granite.

Porphyritic Textures

Igneous rocks are composed of intergrown mineral grains that crystallized from a melt. Many are composed of grains that are all about the same size, but some rocks have much larger grains embedded among the average, smaller grains. In this case, the smaller grains comprise the **groundmass** of the rock, and the larger grains are called **phenocrysts**. A rock that has two distinctly different sizes of grains—a groundmass with phenocrysts—is called a **porphyritic** rock or a **porphyry** (**Fig. 5.3**). Some porphyritic rocks have an aphanitic groundmass whereas others have a phaneritic groundmass. It is quite common for extrusive (volcanic) igneous rock to have an aphanitic-porphyritic texture, which is usually interpreted to indicate that some grains had already begun to grow in the magma before it erupted.

Glassy Texture

Magma sometimes cools so quickly that it is unable to nucleate and grow crystals and instead solidifies into a glass. **Obsidian** is a glass with the same chemical composition as

ACTIVITY 5.3

Glassy and Vesicular Textures of Igneous Rock, (p. 141)

> **Think About It** What are igneous rock textures? How is texture used to classify and interpret igneous rock?

Objective Experiment with molten sugar to produce glassy and vesicular textures, and then apply your knowledge to interpret rock samples.

Before You Begin Read the section: Textures of Igneous Rocks.

Plan Ahead This activity involves a simple experiment. Depending on your laboratory setup, you will either watch a video of the experiment or do the experiment in person. If you do the experiment, you will need a small metal sauce pan with a handle or a ~500 mL Pyrex™ beaker and tongs, water (~50 mL), safety goggles, aluminum foil, hand lens, sugar (~50 mL, 1/8 cup), and a hot plate. To compare the experimental results with rock, you will refer to rock specimens (or photos of rock specimens) provided by your teacher.

rhyolite and granite (**Fig. 5.4**) and is found in association with felsic or intermediate volcanic rocks. In places where there is basaltic volcanism, a glass called **tachylyte** can form when the magma is cooled very rapidly. In Hawai'i, a tachylytic glass can form as a crust on rapidly cooling lava (**Fig. 5.5**) and when the lava flows into the ocean.

Safety First: Handling Volcanic Glass. Remember that obsidian specimens you might encounter in lab are pieces of broken glass. Just like any other broken glass, the edges of obsidian specimens can be very sharp. If you must handle an obsidian specimen in lab, do so with great care.

Vesicular Textures

A **vesicle** is a round or oblong hole in an igneous rock or natural glass formed by gas trapped in the magma as it was solidifying. Vesicles occur in a variety of sizes, and can

Figure 5.4 Volcanic glass. This fresh specimen of obsidian has no mineral grains, but we can use laboratory techniques to determine its chemical composition. Obsidian has approximately the same composition as a granite or rhyolite.

be isolated within otherwise continuous rock or can be a dominant feature of the rock. The basalt in **Fig. 5.6A** has scattered vesicles ranging from smaller than a millimeter to more than a centimeter across, constituting perhaps 5–10% of the rock volume. The **scoria** in **Fig. 5.6B** is dominated by vesicles. This rock type is familiar to many people who have gas barbeques lined with "lava rock"; it is mined commercially from volcanic cinder cones. The **pumice** in **Fig. 5.6C** is a glassy foam of vesicles and has so little bulk density that it floats. It floats until all of the interconnected vesicles fill with water, and then it sinks like any other rock.

Pyroclastic Textures

Volcanic processes that generate igneous rock include surface flows of lava, emplacement of lava domes (**Fig. 5.7**), pyroclastic flows, and ejection of material from a volcano into the air and settling onto the ground (Fig. 5.1).Particles ejected from a volcano during an explosive eruption are called **tephra** and are said to be **pyroclastic** or "firebroken" deposits (**Fig. 5.8**). Tephra are grouped into several grain-size classes (**Fig. 5.9**) ranging from dust upward in size to blocks and bombs. Pyroclastic rock can display flow banding and can incorporate exotic chunks of the volcano, biologic remains, and (with recent eruptions) cultural materials. A deposit associated with a pyroclastic flow is called an **ignimbrite**, and the general name for a deposit formed from material that settled out of the air from a volcanic eruption is a **tuff**—sounds like the word "tough" and rhymes with "puff" (**Fig. 5.10**).

Xenoliths

A **xenolith** is a piece of older rock that has become incorporated in a magma either by becoming separated from the wall of a magma chamber and floating into the melt or by being plucked from the source area as the magma began to rise to where it crystallized. The bright green peridotite in **Fig. 5.11A** is from the pegmatite mines on the

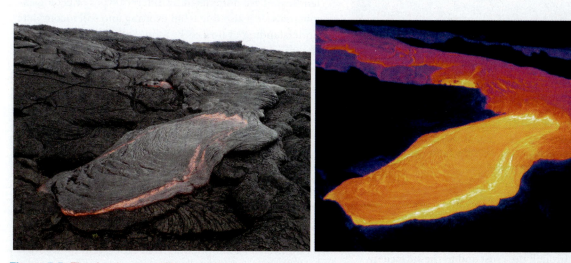

Figure 5.5 Flowing lava on Kilauea volcano, Hawai'i. A normal digital photograph of an active lava flow (left) compared with a thermal image of the same flow that illustrates the temperature of the lava. The hottest temperatures are greater than 1020°C (see temperature scale at right). After it solidifies, parts of the lava surface will probably have a glassy covering due to very rapid cooling. (Photo taken February 12, 2016. Courtesy of the USGS Hawaiian Volcano Observatory)

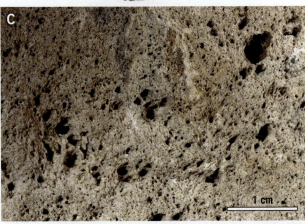

Figure 5.6 **Vesicular texture.** **A.** The oval holes are vesicles where gas was trapped during crystallization of this olivine basalt. **B.** Abundant vesicles in a scoria or cinder. The solid material between the vesicles is largely composed of mineral grains, although there might be glass as well. **C.** Small vesicles in pumice, which is mostly glass, with the same general chemical composition as a rhyolite.

San Carlos Apache Reservation in southeast Arizona. The peridotite xenoliths originated in the mantle and were brought rapidly through the crust and to the surface by rising basaltic magma. A **mafic enclave** (**Fig. 5.11B** is considered by some to be a xenolith. (The word **mafic** is used to describe magma that is enriched in magnesium and iron and that crystallizes to form dark-toned minerals and

igneous rock. In contrast, the word **felsic** describes magma that forms igneous rock with more light-toned silicate minerals. More detail is provided in the next section.) Other geoscientists interpret mafic enclaves to be blobs of mafic magma that were incorporated without complete mixing in felsic magma before crystallization.

Pegmatites

Very coarse-grained igneous rocks that seem to be associated with the final phases of crystallization of an intrusive igneous body are called **pegmatites**. Water appears to play an important role in forming many pegmatites. Perhaps the most common occurrence of pegmatites is in filling cracks through other granitic rocks or the surrounding **country rock** (the rock the granitic magma originally intruded), although isolated pods of pegmatite have also been found. The average grain size within pegmatites is commonly more than a centimeter (**Fig. 5.12**) and sometimes much more. In fact, some of the largest crystals ever found are associated with pegmatites, including tourmaline crystals the size of telephone poles. Pegmatites have been mined for a variety of industrial minerals, as well as for gem-quality crystals.

Composition of Common Igneous Rocks

It is probable that every student who has ever taken a geology course has noticed that geology involves a lot of specialized words and names for various geological things. As many as 1,500 different names have been used by geologists for various types of igneous rocks. This super-abundance of names is partly due to the fact that geology developed during the past few centuries in different countries that have different languages and in which the landscapes are dominated by different types of geological materials. It is also due to an evolving and still incomplete understanding of how rocks form. Still, the most widely

ACTIVITY 5.4

Minerals That Form Igneous Rocks, (p. 142)

Think About It   What is igneous rock composed of? How is composition used to classify and interpret igneous rock?

Objective   Identify samples of eight minerals that form most igneous rock types and categorize them as mafic or felsic.

Before You Begin   Read the following sections: Felsic Minerals and Mafic Minerals.

Figure 5.7 Lava dome. The growing lava-dome complex inside the crater of Mt. St. Helens, Washington, as it appeared on September 12, 2006. This dome is composed of dacite lava that is pushing slowly up into the crater. Dacite is intermediate in composition between rhyolite and andesite and is very viscous—more like Silly Putty than like warm syrup. The distance from one side of the crater rim to the other is about 2 km. (Photograph by Willie Scott of the USGS Cascade Volcano Observatory.)

Figure 5.8 Tephra. A "breadcrust bomb" that was blasted out of Augustine Volcano. (Photographed by Michelle Coombs, USGS.) The photo in the upper right corner is a scanning electron micrograph of a glassy, vesicular ash grain erupted from Augustine on January 13, 2006. The scale bar is 6 millionths of a meter long. (The SEM image was acquired at the University of Alaska Fairbanks by Pavel Izbekov. Both images provided courtesy of the USGS Alaska Volcano Observatory.)

Clast size in mm	Pyroclast name	Rock name
64	bomb, block	pyroclastic breccia
	lapillus	lapilli tuff
2		
0.25 (1/4)	coarse ash	coarse ash tuff
	medium ash	medium ash tuff
0.0625 (1/16)		
	fine ash or dust	fine ash tuff or dust tuff

Figure 5.9 Grain size of tephra.

used classification method for igneous rock, developed by geologists working with the International Union of Geological Sciences (IUGS), includes around 300 names for different types of igneous rocks of which less than 100 are applied to common rock types.

Do not despair! In this lab, we use a very limited set of terms to consider a few important igneous rock types to provide a brief introduction to a very broad subject.

Figure 5.10 Pyroclastic rock. Lapilli-sized pyroclast and coarse ash in a tuff with a rhyolitic composition.

Figure 5.11 Inclusions in igneous rock. A. Xenoliths of peridotite brought up from the mantle during a volcanic eruption embedded in a basaltic lava. The green minerals are olivine and pyroxene, and the dark mineral grains are a mafic mineral called spinel. **B.** A small blob of mafic material (a mafic enclave) in a granitic rock from the Sierra Nevada batholith, California. This is commonly interpreted to be a bit of mafic magma that did not mix with the more felsic magma before both crystallized.

Geoscientists who have compiled thousands of published rock analyses have observed that igneous rock exists across a broad spectrum of different compositions rather than in compositional clusters that are easy to differentiate from one another. We will subdivide this broad spectrum,

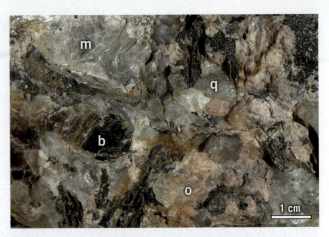

Figure 5.12 Pegmatite. A pegmatite with coarse biotite (b), muscovite (m), quartz (q), and orthoclase feldspar (o) collected from a dike near Boulder, Colorado.

generally going from light-toned to dark-toned rock: felsic, intermediate, mafic, and ultramafic rock. Within most of these classes, we will consider representative intrusive and extrusive rock types as defined by the IUGS classification system.

Felsic Minerals

The adjective **felsic** is a mnemonic derived from *fel*dspar and *fel*dspathoid + *si*lica + *c,* where the "c" is needed to turn the whole thing into an adjective. (**Feldspathoids** are a relatively uncommon group of minerals found in igneous rock that has an unusually small amount of silica.) Felsic refers both to a set of minerals—feldspars, feldspathoids, quartz, and muscovite—and to igneous rock whose composition is dominated by those minerals (**Fig. 5.13**). A felsic rock is a light-toned rock with less than ~25% dark minerals.

Feldspars are a family of minerals that are said to be the most common minerals in Earth's crust. They can be distinguished from quartz in igneous rocks by the flat cleavage faces in feldspars. Two subgroups of feldspars are used in classifying igneous rocks: alkali feldspars and plagioclase feldspars. **Alkali feldspars** form a series of minerals that vary from one another based on the percentages of sodium and potassium in their crystal lattices, ranging from sodium-rich albite ($NaAlSi_3O_8$) to potassium-rich minerals such as orthoclase and microcline ($KAlSi_3O_8$). Albite tends to be white-gray or colorless in igneous rocks whereas color in potassium feldspar ranges from white to shades of pink to brownish red. Potassium feldspar is often called **K-spar** because the chemical symbol for potassium is K. The mnemonic "pink potassium feldspar" sometimes helps, and it is the K-spar that gives many granites and rhyolites their pink color.

Plagioclase feldspars also form a compositional series, ranging from sodium-rich *albite* to calcium-rich *anorthite* ($CaAl_2Si_2O_8$) with four varieties of feldspar in between. The more calcium-rich plagioclase feldspars in igneous rocks tend to be colorless to white to light gray or even black. Even when it is colorless or light color, the

Figure 5.13 Felsic minerals. Feldspar, quartz, and muscovite are all silicate minerals found in felsic or intermediate rock. Some potassium-rich feldspar and calcium-rich plagioclase feldspar can be dark in tone, but the rest of the minerals tend to be light in tone.

sodium-rich   calcium-rich
plagioclase feldspar

quartz

sodium-rich   potassium-rich
alkali feldspar

muscovite

calcium-plagioclase mineral's translucency transmits the darkness of the surrounding mafic minerals in a way that can make the plagioclase appear darker than it is, like glass on a dark tabletop. Some plagioclase grains display fine, parallel, straight striations that are not found in alkali feldspars.

Felsic igneous rock such as **granite** and **rhyolite** have plagioclase feldspar that is enriched in sodium compared with calcium. In contrast, the mafic rocks **gabbro** and **basalt** have plagioclase feldspar that has more calcium than sodium. **Intermediate** igneous rock such as **diorite** and **andesite** are, as the name implies, intermediate in plagioclase composition—slightly enriched in sodium over calcium.

Quartz tends to look like gray or colorless bits of broken glass filling in the space between other mineral grains in igneous rock because it is one of the last minerals to crystallize as the magma cools. Among the rocks we study in this lab, quartz is only abundant in granite and rhyolite where it comprises between 16% and 57% of the rock volume. Quartz has no cleavage—it fractures like glass—is translucent to transparent, and is very hard (H = 7).

Mafic Minerals

The adjective **mafic** is also a mnemonic, combining *ma*gnesium + *f*errous/*f*erric (iron-rich) + *ic*, with the "ic" needed to convert it into an adjective. Common mafic mineral groups include biotite, pyroxene, amphibole, olivine, and iron-titanium minerals like magnetite and

ilmenite (**Fig. 5.14**). Mafic igneous rocks contain more than ~35% mafic minerals, and **ultramafic** rocks contain at least 90% mafic minerals. A mafic rock is generally a dark-toned rock.

The oceanic crust is composed of mafic rock, usually with a layer of sediments or sedimentary rock on top. There is also mafic rock *in* or *on* continental crust crystallized from magma generated by partial melting of the upper mantle above a mantle hot spot or in a few other ways. In contrast, **ultramafic rock** is normally associated with the upper mantle rather than the crust, so it is only found in unusual settings at Earth's surface: in xenoliths, in a few unusual places where a slice of oceanic crust has been faulted onto continental crust, in unusual extrusive rock called *komatiite* that erupted early in Earth's history and rarely thereafter, and in mountain ranges formed after an ocean basin is closed by subduction. As the oceanic crust subducts, the continental crust on opposite sides of the ocean converge and eventually collide to form a mountain range like the Appalachians or the Himalayas. Bits of ultramafic rock are sometimes found along these continental collision zones.

Not all dark-colored minerals are mafic minerals, and not all mafic minerals are dark gray, green, brown, or black. Actually, some felsic minerals like calcium-rich plagioclase feldspar and potassium-rich alkali feldspar can have a dark tone, and some types of the mafic mineral olivine can be a light green (**Figs. 5.3A** and **5.11A**). It might be better to avoid using the word *mafic* as a synonym for *dark* and focus on its compositional meaning—minerals that contain magnesium or iron.

amphibole

pyroxene

biotite

olivine

Figure 5.14 Mafic minerals.
Amphibole, pyroxene, biotite, and olivine are all iron- or magnesium-bearing silicate minerals found in mafic rock and tend to be dark in tone.

ACTIVITY 5.5

Estimate the Percentage of Mafic Minerals, (p. 143)

Think About It What is igneous rock composed of? How is composition used to classify and interpret igneous rock?

Objective Visually estimate the percentage of mafic minerals in a rock—the Color Index—using printed standards as a guide.

Before You Begin Read the following section: Color Index for Mafic Minerals.

ACTIVITY 5.6

Estimate Mineral Composition of a Phaneritic Rock by Point Counting, (p. 144–145)

Objective Use a form of point counting to estimate a rock's composition, and investigate how many points are needed for a reliable estimate.

Before You Begin Read the following section: Estimating Abundance of Different Minerals in a Rock.

Color Index for Mafic Minerals

The IUGS classification of igneous rocks employs a **Color Index** (CI) to describe the percentage of mafic minerals by volume in a rock. The mafic minerals used in the CI include biotite, amphibole, pyroxene, olivine, garnet, opaque minerals like magnetite and ilmenite, and a few other dark-toned minerals. A higher CI corresponds to a higher percentage of mafic minerals, resulting in a darker rock.

CI	Type	Rock Type
~5–20%	felsic	granite, rhyolite
25–50%	intermediate	diorite, andesite
~35–65%	mafic	gabbro, basalt
~90–100%	ultramafic	peridotite

The CI can be challenging to use at an outcrop or with a hand specimen. Visual estimation of the percentage of mafic minerals is often quite imprecise, even when the mafic minerals in a specimen are all very dark and the felsic minerals are all very light. When some of the felsic minerals appear to be dark or some of the mafic minerals are a bright color, confusion can sometimes descend upon us. Anyone who has had difficulty separating "light" from "dark" laundry will understand the problem. The most reliable way of determining the CI of an igneous rock is by using a thin section and a petrographic microscope in which you would either point count the felsic and mafic minerals as described below or by performing a digital image analysis of the thin section.

Estimating Abundance of Different Minerals in a Rock

Experienced geologists can make a reasonably good initial identification of many rocks by quickly inspecting the rock. This ability is developed by studying many different rock types and building a mental library of attributes. When examining an igneous rock, a geologist looks at:

- Tone (Is it generally light or dark or somewhere in-between?)
- Crystallinity (Is it a mass of intergrown crystals or glassy, or does it look like the minerals were deposited rather than formed together?)
- Grain size (Is the groundmass aphanitic or phaneritic?), and, if possible
- Mineral composition.

If the rock is phaneritic so that a geologist might be able to visually identify the minerals, she asks herself a series of questions about the rock:

- Does it have pink(ish) feldspars (it might be a granite)?
- Does it have a substantial amount of quartz (it might be a granite), a little bit (it might be a diorite), or none at all?
- Does it seem to be mostly or entirely composed of mafic minerals (it might be a gabbro or ultramafic rock)?

If the rock has an aphanitic groundmass, a geologist might base an initial assessment on color or other obvious textural information but will likely wait until seeing a thin section of the rock under a microscope before being confident about identifying the rock.

Point Counting. A geologist needs to go beyond this initial interpretation to have a reliable identification of rock type. One way to begin is to use a rock saw to cut a flat surface on the rock, polish the surface to remove the saw marks, and take a digital photo of the polished surface. Armed with the digital photo, we can take several approaches. The oldest approach is to conduct a point count to examine a statistically meaningful number of grains in the rock (see Activity 5.6). This is intended to relieve us from having to identify *every* grain that we can see in a rock. A grid of some sort is superimposed on the photograph, and the mineral under each grid point is identified. Examples of grids that have been used by geologists include a random array of dots and a rectangular or triangular grid of lines where the mineral is identified at every node point where the lines intersect. Point counting is often performed using a thin section of the rock viewed with a petrographic microscope rather than a photograph of a polished surface.

The average percentage of a rock composed by a particular mineral like quartz can be determined by performing several point counts using the same grid located over different parts of the photograph. Imagine that you complete several point counts and determine that the **average** (or **mean**) composition of the rock is 15% quartz. By itself, that number is not very useful because it does not include an estimate of the uncertainty in your data,

and that is where the **sample standard deviation** can be helpful. About 68% of a normally distributed population (e.g., a very large number of quartz composition measurements) is within 1 standard deviation above or below the average, and about 95% is within 2 standard deviations of the average. So if the average is 15% quartz and the sample standard deviation of our several measurements is 3%, we would be fairly confident that the quartz composition is $15 \pm 3\%$ and much more confident that it is $15 \pm 6\%$.

Finding a Sample Standard Deviation. Imagine we have a set of five numbers, perhaps representing measurements we have made (7, 8, 4, 6, 9). We will represent the number of values in this set with an n, so for this set, $n = 5$. We will represent the *average* or **mean** of these numbers with an x that has a bar over it ($\bar{x}$) and calculate the average by adding all the values in our set together and dividing by the number of values in the set.

$$\text{mean} = \bar{x} = (7 + 8 + 4 + 6 + 9)/5 = 6.8$$

In this case, the mean or average of our set of five numbers is 6.8.

The **sample standard deviation** is easy to calculate in steps.

Step 1. Subtract the mean from each of the values in the set, square each of those results, and add those squared results together. We'll call the final result of this step a.

$$a = [(7 - 6.8)^2 + (8 - 6.8)^2 + (4 - 6.8)^2 + (6 - 6.8)^2 + (9 - 6.8)^2] = 14.8$$

In this case, $a = 14.8$.

Step 2. Divide result a by the number of values in the set ($n = 5$) minus 1 ($n - 1 = 4$). We'll call that result b.

$$b = [a/(n - 1)] = (14.8/4) = 3.7$$

In this case, $b = 3.7$.

Step 3. The sample standard deviation is the square root of b. If we represent the sample standard deviation with an s,

$$s = \sqrt{b} = \sqrt{3.7} = \sim 1.9$$

In this case, the sample standard deviation of our set of five numbers is around 1.9. Many calculators can compute a sample standard deviation, and online calculators for simple statistics are available.

Compositional Twins

The IUGS classification of common intrusive and extrusive igneous rocks emphasizes the relative proportion of quartz, alkali feldspar, plagioclase feldspar, feldspathoids, and mafic minerals. The IUGS system has one set of terms for "plutonic" rocks—intrusive igneous rocks—and a parallel set of terms for "volcanic rocks"—extrusive igneous rocks. The result is that a given intrusive igneous rock such as **granite** has the same range of mineral composition as an extrusive igneous rock. Granite has the same mineral composition as **rhyolite**, so we call granite and rhyolite **compositional twins**. Other compositional twins include

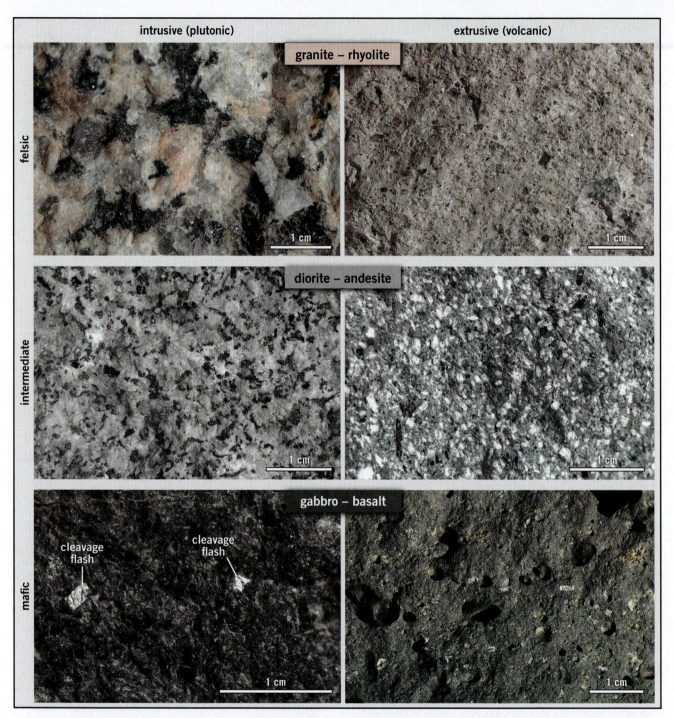

Figure 5.15 Compositional twins. Three sets of igneous rocks that are compositional twins. The intrusive igneous rocks granite, diorite, and gabbro are phaneritic with a groundmass of visible grains. The flashes from cleavage faces indicate the grain size in the gabbro specimen. The extrusive igneous rocks rhyolite, andesite, and basalt have aphanitic groundmasses and small phenocrysts. Granite and rhyolite have the same mineral composition and so are compositional twins. Diorite and andesite are also compositional twins as are gabbro and basalt.

diorite-andesite and gabbro-basalt among others that we will not work with in this lab (**Fig. 5.15**).

These compositional twins differ only in texture and mode of formation. Intrusive rock has a phaneritic texture of coarser grains and crystallizes slowly below Earth's surface whereas extrusive igneous rock has an aphanitic groundmass and solidifies quickly at Earth's surface. Below a rhyolitic dome or volcano is likely to be a mass of granite formed from a related magma.

Granite and Rhyolite. With a groundmass of grains that are large enough to be distinguished by people with good eyesight without magnification, **granite** is a phaneritic, intrusive igneous rock. **Rhyolite** is an aphanitic, extrusive igneous rock with a groundmass of very small grains. Granite and rhyolite are both **felsic rock** types that contain 20% or less mafic minerals. Quartz is as much as ~60% of the felsic component of the rock—the part that does not include mafic minerals. The other major felsic components

Generalized Sketch of the Major-Mineral Composition of Common Igneous Rocks

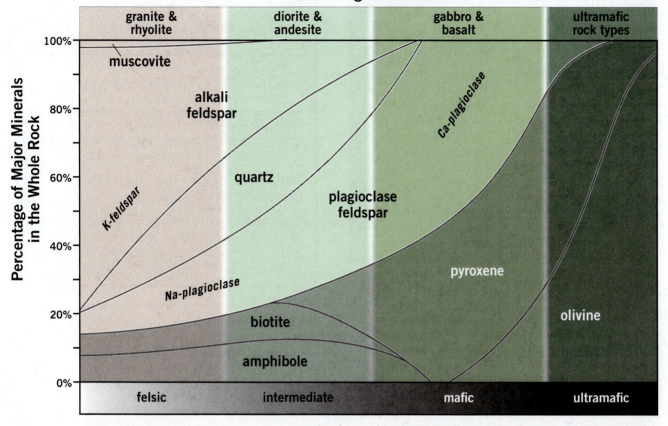

Figure 5.16 Composition of common igneous rock types. The graph at the top provides a generalized view of the major minerals in granite/rhyolite, diorite/andesite, gabbro/basalt, and ultramafic rock. The table at the bottom provides the ranges in mineral (modal) composition for these rock types, according to the IUGS system of classifying igneous rocks.

Major Minerals	Percentage of the Total Rock Volume			
	granite–rhyolite	diorite–andesite	gabbro–basalt	ultramafics
quartz	16–57%	0–15%	0–13%	—
alkali feldspar	11–68%	0–26%	0–23%	—
plagioclase feldspar	3–49%	26–75%	18–65%	≤10%
mafic and minor minerals	5–20%	25–50%	35–65%	≥90%

are alkali feldspar and plagioclase feldspar according to the current definition developed by the IUGS (Fig. 5.16). The alkali feldspar in granite tends to be potassium rich and can include grains that are white, pale to salmon pink, and even brownish red. Granite and rhyolite commonly have a pinkish tone due to the pink potassium feldspar. The plagioclase feldspar in these rock types is sodium rich and tends to be colorless to white. Common mafic minerals are biotite and the amphibole hornblende.

Diorite and Andesite. Students sometimes refer to the phaneritic rock **diorite** and its aphanitic twin, **andesite**, as the "salt-and-pepper rocks" because they are composed of white and dark minerals with a bit more salt than pepper. Diorite and andesite are **intermediate rock** types with between 25–50% mafic minerals. Mafic minerals include biotite, amphibole, and pyroxene.

Light-colored feldspars dominate the felsic composition of these rocks, which can have up to 5% of quartz. The feldspars in diorite are colorless to white, and the plagioclase feldspars have more sodium than calcium. The Sierra Nevada batholith that is so beautifully exposed in Yosemite Valley includes both diorite and granite, as well as other granitic rock types (Fig. 5.17). The North American Wall on the north side of the valley gets its name from a large mass of diorite within the granite that forms most of El Capitan. Felsic and intermediate igneous rocks commonly form in continental crust above a subducting slab (Fig. 5.18). However, these rocks can form anywhere in continental crust where we find unusual heat, availability of water to reduce the melting point of (i.e., to **flux**) hot crustal rocks, or deformation of the crust that causes a local reduction in pressure. If a mass of rock is hot enough to be on the verge of partially

Figure 5.17 Granite and diorite of El Capitan's North American Wall. The dark diorite of this vertical face on the north side of Yosemite Valley on the rock exposure called El Capitan resembles a map of North America, so climbers refer to this as the North American Wall. The lighter-tone rock is granite. The wall rises more than 750 meters (~2,500 feet) above the rubble at its base.

ACTIVITY 5.7

Analysis and Interpretation of Igneous Rock, (p. 146–147)

Think About It How are rock composition and texture used to classify, name, and interpret igneous rock?

Objective Analyze composition and texture of igneous rock specimens, and then infer how they formed.

Before You Begin Read the section: Compositional Twins.

Plan Ahead This activity involves identification of specimens of igneous rock and (perhaps) glass. Your teacher will provide you with rock specimens or photos of specimens.

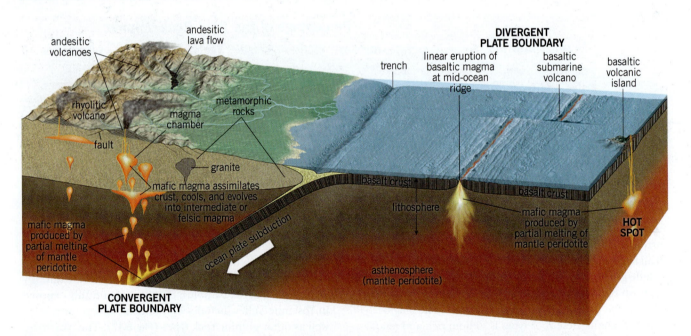

Figure 5.18 Some tectonic settings where igneous rock forms. Mafic magma is generated by the partial melting of the upper mantle. Along mid-ocean ridges, mafic magma crystallizes to form basalt at the surface and gabbro in the lower oceanic crust, adding material to the trailing edges of both plates. Partial melting of the mantle above or around mantle hot spots generates mafic magma that can rise to or through the crust where it might cause local partial melting of the crust. Reduction of melting temperature by water from the top of a subducting slab can cause melting above the slab. Rising mafic magma can crystallize on or below Earth's surface and can cause local partial melting in the crust that generates more felsic magma. The felsic magma can then rise and crystallize to form intrusive (granite, diorite) or extrusive (rhyolite, andesite) igneous rocks.

ACTIVITY 5.8

Geologic History of Southeastern Pennsylvania, (p. 148)

Think About It How can the shapes of igneous rock bodies be used to classify them and infer their origin?

Objective Analyze bodies of igneous rock in southeastern Pennsylvania using a geologic map and infer their origin.

Before You Begin Read the sections: Gabbro and Basalt and Some Igneous Rock Bodies.

melting, a reduction in pressure can cause melting to begin in a process called **decompression melting**.

Gabbro and Basalt. Partial melting of the upper mantle yields a mafic magma that crystallizes to form the phaneritic rock **gabbro** below Earth's surface or **basalt** if the magma rises to erupt at the surface (**Fig. 5.18**). Basalt is said to be the most common igneous rock on Earth because it forms the upper part of the oceanic crust that covers about 63% of the planet. The lower part of the oceanic crust is gabbro. Basalt is also found in many other locations where mafic magma has been able to make it to the surface, including above mantle hot spots and at other locations within continental crust. Gabbro and basalt are **mafic rock** types that have between 35–65% mafic minerals dominated by pyroxene with olivine or amphibole in lesser amounts. The felsic part of these rocks is dominated by calcium-rich plagioclase feldspar, which tends to be translucent and colorless, the white of thin milk, or gray. The dark tone of the major minerals in gabbro can make it difficult to discern the grain size of the groundmass, but rotating a gabbro specimen in the light will generally yield flashes of reflected light off cleavage faces of the pyroxenes or feldspars, and those flashes indicate the size of the mineral grain (see **Fig. 5.15**).

Ultramafic Rocks. Only exposed at Earth's surface because of unusual tectonic conditions, ultramafic rocks like *dunite, peridotite,* and *pyroxenite* are composed primarily of olivine and various pyroxenes. Their rarity at the surface is because these are rock types associated with the upper mantle. **Peridotite** is composed 40% or more of olivine, and rock that is more than 90% olivine is called **dunite**. **Pyroxenite** is 40% or more pyroxene. Extrusive ultramafic rocks, such as *komatiites,* are extremely rare and formed primarily during the first half of Earth's history when the flow of heat from Earth's interior is likely to have been greater than it is today.

Some Igneous Rock Bodies

Outcrops of igneous rock include some of Earth's most dramatic landforms. Whether these igneous rock bodies formed through intrusive or extrusive processes, they all began as magma deep beneath Earth's surface. Silicate magma is less dense than the rock that surrounds it in the uppermost mantle and lower crust, in the source area where the magma forms and accumulates. The rock surrounding deep magma bodies is very hot and is weaker than the same rock would be at lower temperature. The less dense magma rises buoyantly in the crust like the bubbles in a lava lamp, initially by pushing aside the cooler and less dense rock and later, nearer Earth's surface, by rising along cracks, faults, or other weak spots in the crust. We say that the magma **intrudes** the crust. Some of the magma reaches the surface to become volcanic rock, but much crystallizes below the surface to form intrusive igneous rock.

Intrusions have different sizes and shapes. The largest intrusive granitic bodies are called **batholiths** (**Fig. 5.19A**) that form when smaller bodies of magma called **plutons** accumulate to form a larger granitic body. A pluton might be 1 kilometer to a few tens of kilometers in diameter. Formation of a batholith one pluton at a time is similar to individual blobs of "lava" in a lava lamp rising to the top of the lamp to form a larger mass. Batholiths form in continental crust at subduction zones below the ridge of active volcanoes. For example, there is likely a granite-diorite batholith forming under the active volcanoes of the Cascade Mountains from northern California to British Columbia. The Southern California, Sierra Nevada, Idaho, and Coast Range batholiths of western North America are exposed at the ground surface today because of uplift and erosion (**Figs. 5.2** and **5.17**).

Smaller intrusions called **dikes** fill cracks that cut across bedding planes or metamorphic foliation in the older rock that was intruded, which geologists call the **country rock** (**Fig. 5.19B**). Many dikes are approximately planar, and some extend for kilometers. Along mid-ocean ridges during sea-floor spreading, planar dikes intrude slightly older planar dikes below the seafloor, creating **sheeted dikes** in which essentially the entire crust at that level is composed of dikes. In that setting, older dikes *are* the country rock.

Planar dikes are often found radiating out from the center of a volcano like the spokes of a bicycle wheel (**Fig. 5.19A**). These **radial dikes** are what is left of the conduits through which magma reached the flanks of the volcano to feed flank or fissure eruptions. The cylindrical conduit that brings magma to the vent of a volcano is called a **pipe**, and the eroded remnant of a pipe that remains after the volcano is no longer active is called a **neck**. Another type of dike related to volcanoes forms cylinders or cones that are concentric with the volcano. These are called **ring dikes**, and they tend to form as the

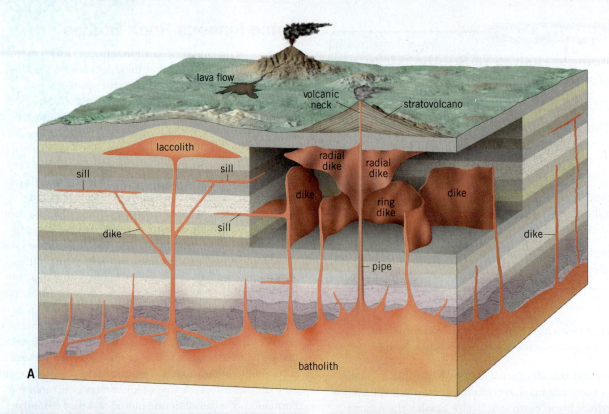

Figure 5.19 Intrusive and extrusive igneous rock bodies. A. This diagram shows the relationships between some of the more common igneous rock bodies that can be seen on the landscape and, in the case of intrusive igneous rock, in outcrops after being exposed by erosion. **B.** A pink dike cuts across pre-existing layers in an older metamorphic rock. **C.** The brownish layer between the dashed lines is an igneous sill that intruded parallel to sedimentary beds in the Raton Basin, southern Colorado. The sedimentary layers include sandstone, shale, and coal.

crust cools and contracts around the pipe and magma chamber after eruption.

An intrusion that fills a crack that grew along bedding or foliation planes in the country rock is called a **sill** (**Fig. 5.19C**). If the sill is emplaced close enough to Earth's surface, the rock above the sill might be pushed up to form a dome with the sill taking on a mushroom shape called a **laccolith**. Magma with a lower viscosity—more like water—that reaches the ground surface might flow away from the fissure or vent to form a lava flow (**Fig. 5.5**). If the magma has a higher viscosity—more like Silly Putty—it might form a lava dome like the one in the crater of Mt. St. Helens in southern Washington (**Fig. 5.7**).

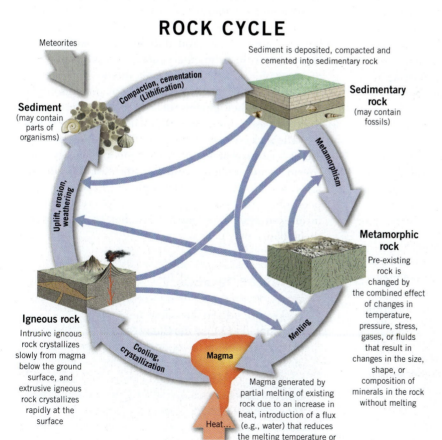

A What are some visible differences between intergrown mineral grains and detrital or clastic mineral grains?

B **REFLECT & DISCUSS**   Rocks are made of the materials listed below and described within the chapter. Under each sample in **Fig. A4.2.1**, write the name of every kind of material it contains from the list. Be prepared to compare your observations with the observations of others in your class.

Intergrown Crystals	Clasts (detrital minerals)	Gravel	Silt & Clay
Fossils (bioclasts)	Clasts (rock fragments)	Sand	Glass

1

2

3

4

5

6

7

8

9

Figure A4.2.1

Igneous Rocks and Processes

Pre-Lab Video 5

http://goo.gl/EIzuX1

Contributing Authors

Harold E. Andrews • *Wellesley College*

James R. Besancon • *Wellesley College*

Claude E. Bolze • *Tulsa Community College*

Margaret D. Thompson • *Wellesley College*

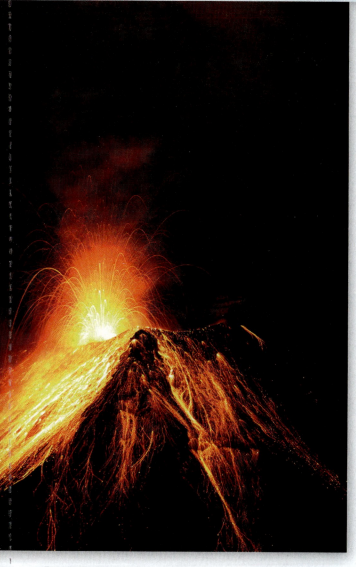

▲ Dramatic night eruption of Tungurahua Volcano (~5,023 m) in the Andes Mountains of central Equador (~1.4701°S, 78.4444°W). Tungurahua is a stratovolcano that erupts andesite-dacite rock and is related to the subduction of the Nazca Plate under South America.

BIG IDEAS

Igneous rock forms wherever magma cools and crystallizes to a solid state. The composition and texture of igneous rock samples can be used to classify them and infer their origin. Lava and igneous rock-forming processes can be observed at volcanoes, which occur along lithospheric plate boundaries, at hot spots, and in places where deformation of the crust leads to local melting. Volcanism can create new land surface in places like Hawai'i and can provide new minerals to the land surface. Volcanoes can also pose hazards to humans.

FOCUS YOUR INQUIRY

Think About It What does igneous rock look like? How can it be classified into groups?

ACTIVITY 5.1 Igneous Rock Inquiry (p. 124, 139)

Think About It What are igneous rock textures? How is texture used to classify and interpret igneous rock?

ACTIVITY 5.2 Crystalline Textures of Igneous Rock (p. 125, 140)

ACTIVITY 5.3 Glassy and Vesicular Textures of Igneous Rock (p. 126, 141)

Think About It What is igneous rock composed of? How is composition used to classify and interpret igneous rock?

ACTIVITY 5.4 Minerals That Form Igneous Rock (p. 128, 142)

ACTIVITY 5.5 Estimate the Percentage of Mafic Minerals (p. 132, 143)

ACTIVITY 5.6 Estimate Mineral Composition of a Phaneritic Rock by Point Counting (p. 132, 144)

Think About It How are rock composition and texture used to classify, name, and interpret igneous rock?

ACTIVITY 5.7 Analysis and Interpretation of Igneous Rock (p. 136, 146)

Think About It How can the shapes of igneous rock bodies be used to classify them and infer their origin?

ACTIVITY 5.8 Geologic History of Southeastern Pennsylvania (p. 137, 148)

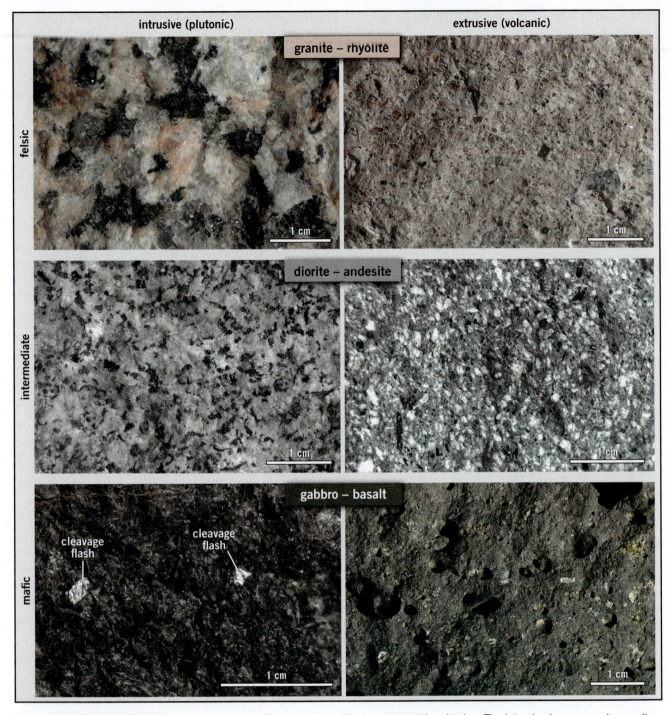

Activity 5.1

Name: _____ Course/Section: _____ Date: _____

A Analyze the igneous rocks in **Fig. A5.1.1** (and actual rock samples of them if available). They are depicted at approximately the same size as they are in nature. Beneath each picture, describe the rock's **color**, **composition** (what it is made of), and **texture** (the size, shape, and arrangement of its parts) as well as you can, using your current knowledge.

Figure A5.1.1

B **REFLECT & DISCUSS** Describe how you would classify the rocks in part A into groups. Be prepared to discuss your classification with your classmates.

Activity 5.2   Crystalline Textures of Igneous Rock

Name: _____   Course/Section: _____   Date: _____

Two students are doing an experiment to find out if crystal size in igneous rocks can be related to the speed of cooling a magma. They did not have equipment to melt rock, so they used thymol to model pieces of rock. Thymol melts easily at low temperature on a hot plate, and it cools and recrystallizes quickly. Thymol is a transparent, crystalline organic substance derived from the herb thyme and is used in antiseptics and disinfectants. Thymol gives off a very strong pungent odor that can irritate skin and eyes and cause headaches. Therefore, the students used a spoon to handle the thymol and did all of their work under a fume hood with supervision from their teacher. One of the students placed some thymol in a small Pyrex beaker and melted it completely under a fume hood to model the formation of magma. The other student poured one-half of the molten thymol into a cold petri dish and the other half into a hot petri dish of the same size.

A The results of the student's experiment are shown in **Fig. A5.2.1**. Notice that the images are enlarged. Beside each image below, measure and record the actual size range of the crystals (in mm) that formed.

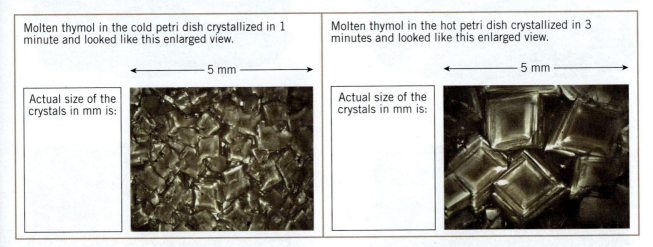

Molten thymol in the cold petri dish crystallized in 1 minute and looked like this enlarged view.

← 5 mm →

Actual size of the crystals in mm is:

Molten thymol in the hot petri dish crystallized in 3 minutes and looked like this enlarged view.

← 5 mm →

Actual size of the crystals in mm is:

Figure A5.2.1

B Igneous rocks that are made of crystals too small to see with your naked eyes or hand lens are said to have an **aphanitic** texture (from the Greek word for unseen). Those made of visible crystals are said to have a **phaneritic** texture (crystals ~1–10 mm) or **pegmatitic** texture (crystals greater than 1 cm). Which of these three igneous rock textures probably represents the most rapid cooling of magma/lava?

C **REFLECT & DISCUSS**   The rock shown in **Fig. A5.2.2** has a **porphyritic** texture, which means that it contains two sizes of crystals. The large white plagioclase crystals are called **phenocrysts** and sit in a green-gray **groundmass** of more abundant, smaller (aphanitic) crystals. Based on your work in part A, explain how this texture may have formed. More than one potentially correct answer is possible.

Figure A5.2.2

D In your collection of numbered igneous rock samples, record the sample numbers with these textures:

Sample(s) with porphyritic texture:	Sample(s) with phaneritic texture:
Sample(s) with pegmatitic texture:	Sample(s) with aphanitic texture:

Glassy and Vesicular Textures of Igneous Rock

Activity 5.3

Name: ______________________ **Course/Section:** ______________________ **Date:** ______________

Place equal parts of sugar (sucrose, $C_{12}H_{22}O_{11}$) and water in the pan/beaker and heat on medium high. Do not touch the hot plate, beaker/pan, or boiling sugar because it is very hot! Notice that steam is given off after the sugar dissolves and the solution boils. After a few minutes, there will be no more steam, and the remaining molten sugar will have a very thick consistency. At this point (and before the sugar begins to burn), pour the thick molten sugar onto a piece of aluminum foil on a flat table. DO NOT TOUCH the molten sugar, but lift a corner of the foil to observe how it flows and behaves until it hardens over a time of perhaps 2–3 minutes.

A Viscosity is a measure of how much a fluid resists flow. Water has low viscosity. Honey is more viscous than water. How did the viscosity of the sugar solution change as the water boiled off?

B What happened to the viscosity of the molten sugar as it cooled on the aluminum foil?

C When the molten sugar has cooled to a solid state, break it in half and observe its texture. Look about the room where you are now seated and name two objects that have this same texture.

D Now observe the texture of the cooled solid mass of sugar with a hand lens. Notice that there are some tiny holes within it formed by bubbles of gas. Geoscientists call these holes **vesicles**, and rocks containing vesicles are said to have a **vesicular** texture. What prevented the gas bubbles from escaping to the atmosphere?

E **REFLECT & DISCUSS** When a sugar solution is permitted to slowly evaporate, sugar crystals form. The process of crystallization depends on the ability of atoms to move about in the solution and bond together in an orderly array. What two things may have prevented crystals from forming in the molten sugar as it cooled on the aluminum foil in this experiment?

F In your collection of numbered igneous rock samples, do any of the samples have the texture that you just observed in part C? If yes, which one(s)?

G In your collection of numbered igneous rock samples, do any of the samples have the texture that you just observed in part D? If yes, which one(s)?

Activity 5.4 Minerals That Form Igneous Rock

Name: ______________________ **Course/Section:** ______________________ **Date:** ______________

A Using information from the section "Composition of Common Igneous Rocks" to identify the minerals in Fig. A5.4.1. Below each picture, write the **name of the mineral**, its **chemical composition** (chemical formula or chemical name), and **a description** that would allow you to recognize it in the future. All samples are shown at approximately their actual size.

augite (pyroxene)	biotite mica	hornblende (amphibole)	muscovite mica
olivine	plagioclase feldspar	potassium-rich alkali feldspar	quartz

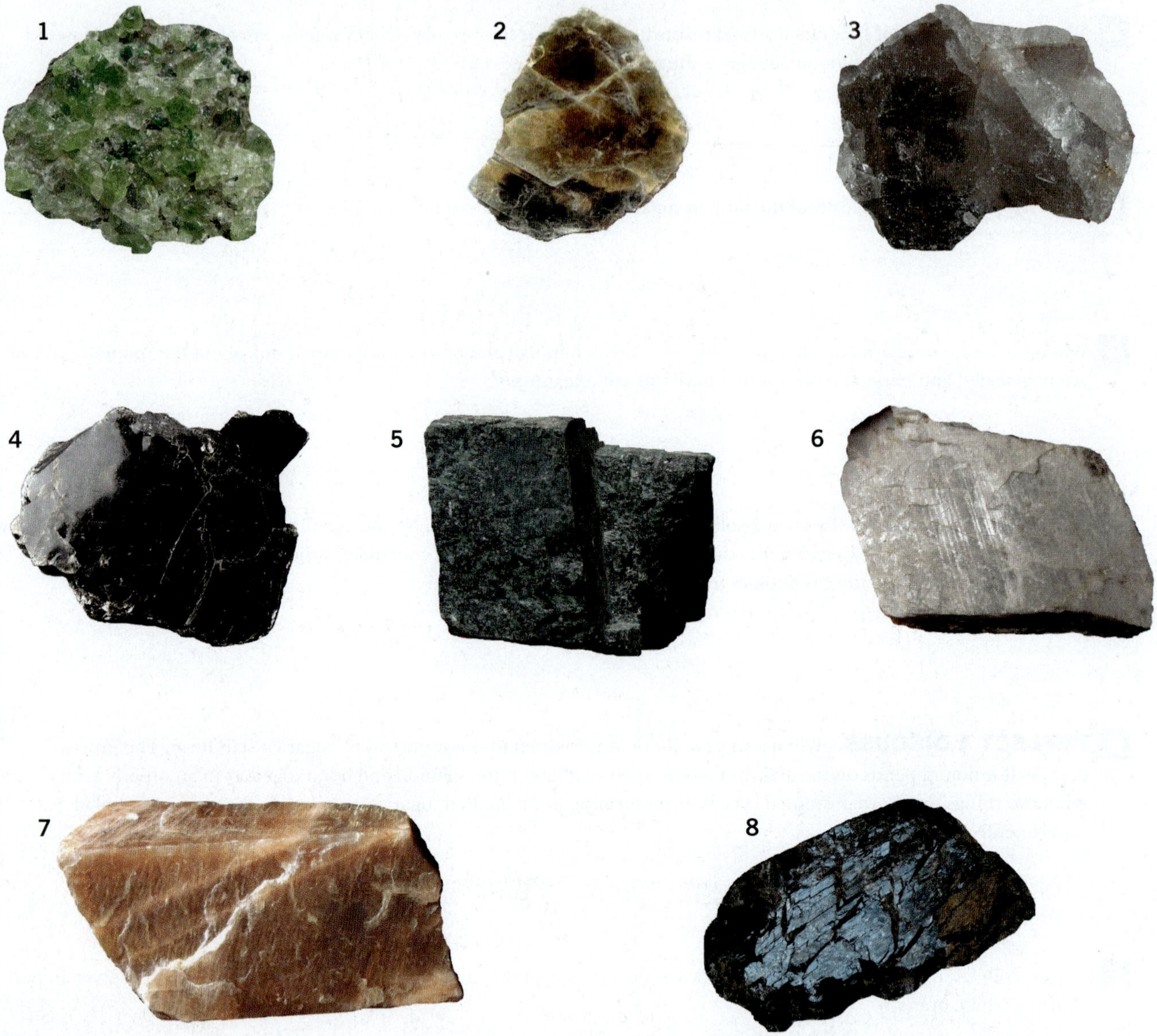

Figure A5.4.1

B **REFLECT & DISCUSS** Which specific minerals are mafic and which ones are felsic? How do you know?

Name: _____ Course/Section: _____ Date: _____

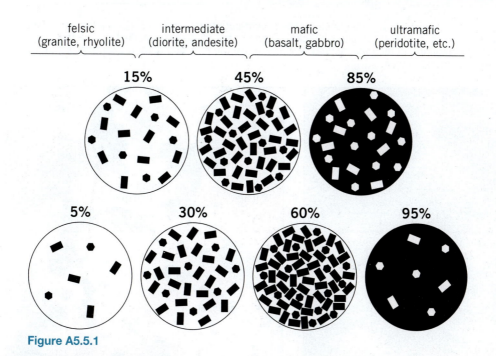

| felsic (granite, rhyolite) | intermediate (diorite, andesite) | mafic (basalt, gabbro) | ultramafic (peridotite, etc.) |

15% 45% 85%

5% 30% 60% 95%

Figure A5.5.1

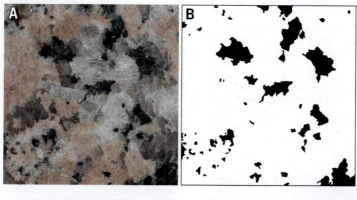

The mafic minerals in photo A are shown as black areas in map B. Use the scale above to estimate the percentage of mafic minerals in this rock.

mafic %: _____

Based on your estimate, describe this rock as felsic, intermediate, mafic, or ultramafic.

_____

Now do the same analysis for the three rocks shown below.

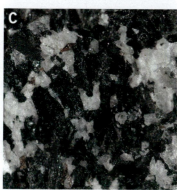

mafic %: _____
rock description:

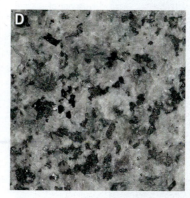

mafic %: _____
rock description:

mafic %: _____
rock description:

_____ _____ _____

Figure A5.5.2

Name: _____ Course/Section: _____ Date: _____

Geologists sometimes classify rocks using a technique called **point-counting** to estimate the relative abundance of different minerals in a rock. A phaneritic igneous rock is shown in the photograph marked A. The four major minerals in the rock are identified in the map marked B. You will use the map and a point-counting technique adapted for this lab to classify the rock.

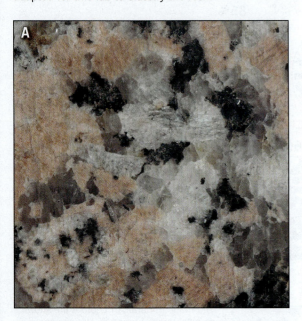

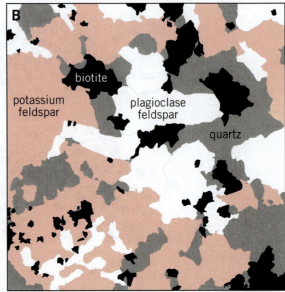

The basic idea is to identify the mineral found at each of several points on the rock. We draw a square grid with 5 vertical and 5 horizontal lines on some opaque paper. Each of the 25 spots where the grid lines cross will be called a node. At each **node**, we will create a small hole in the grid so that we can see the map of the rock through the hole.

We count the number of node points that are filled with each of the four major minerals (that is, with each of the four colors on the map), and use that number to estimate the total volume of each major mineral in the rock. You can see more than one mineral through the hole at some node points, so either pick the mineral that fills most of the hole, or if two minerals each fill about half the hole, count each as 0.5.

Figure A5.6.1

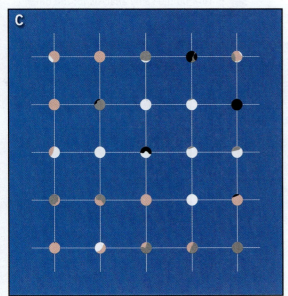

Figure A5.6.2

Point Count, Grid C

Number of
nodes
filled with
the mineral

(_____ x 4) = _____% potassium feldspar
(pink)

(_____ x 4) = _____% plagioclase feldspar
(white)

(_____ x 4) = _____% quartz (gray)

(_____ x 4) = _____% biotite (black)

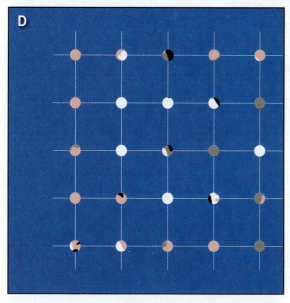

D

How reliable was that sample of 25 points as a way to estimate the modal composition of the rock? Let's see.

The grid has been shifted in three different ways, so we can sample three different sets of points. Repeat the point count for each of these three grids.

Point Count, Grid D

Number of nodes
  filled with the
    mineral

(_____ x 4) = _____% potassium feldspar (pink)

(_____ x 4) = _____% plagioclase feldspar (white)

(_____ x 4) = _____% quartz (gray)

(_____ x 4) = _____% biotite (black)

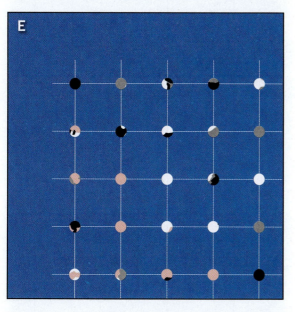

E

Point Count, Grid E

(_____ x 4) = _____% potassium feldspar (pink)

(_____ x 4) = _____% plagioclase feldspar (white)

(_____ x 4) = _____% quartz (gray)

(_____ x 4) = _____% biotite (black)

Point Count, Grid F

(_____ x 4) = _____% potassium feldspar (pink)

(_____ x 4) = _____% plagioclase feldspar (white)

(_____ x 4) = _____% quartz (gray)

(_____ x 4) = _____% biotite (black)

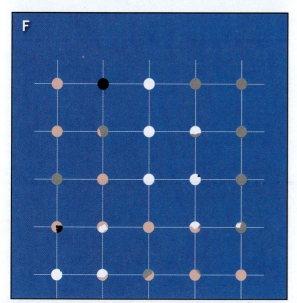

F

Use the data from the point counts of grids D, E, and F to complete the following table.

	average	standard deviation
potassium feldspar		
plagioclase feldspar		
quartz		
biotite		

Do the results of your point count of grid C on the previous page fall within one standard deviation of the average of grids D, E, and F for each of the major minerals? Explain…

What type of phaneritic igneous rock is shown in photograph A?

Figure A5.6.2 (continued)

Activity 5.7 Analysis and Interpretation of Igneous Rock

Name: ______________________ **Course/Section:** ______________ **Date:** ____________

IGNEOUS ROCKS WORKSHEET

Sample number or letter	Texture(s) present	Minerals present and their percentage of abundance.	Estimate the percentage of mafic minerals.	Rock names (Figures 5.4, 5.6, 5.15, 5.16).	Describe a geological environment where this rock might have formed (intrusive vs. extrusive, etc).

IGNEOUS ROCKS WORKSHEET

Sample number or letter	Texture(s) present	Minerals present and their percentage of abundance.	Estimate the percentage of mafic minerals.	Rock names (Figures 5.4, 5.6, 5.15, 5.16).	Describe a geological environment where this rock might have formed (intrusive vs. extrusive, etc).

Name: _____   Course/Section: _____   Date: _____

Review **Fig. 5.19**. Then study the portion of a geologic map of Pennsylvania in **Fig. A5.8.1**. The green-colored areas are exposures of 200–220 million-year-old Mesozoic sand and mud that were deposited in lakes, streams, and fields of a long, narrow valley. The red-colored areas are bodies of basalt about 190 million years old. Paleozoic and Precambrian rocks are more than 252 million years old and colored pale brown.

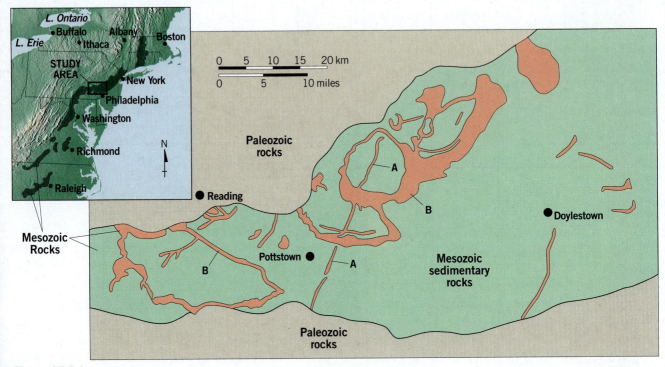

Figure A5.8.1

A   Based on their geometries (as viewed from above in map view), what kind of igneous bodies on the map are labeled A?

B   Based on their geometries (as viewed from above in map view), what kind of igneous bodies on the map are labeled B (more than one answer is possible)?

C   **REFLECT & DISCUSS**   If you could have seen the landscape that existed in this part of Pennsylvania about 200 million years ago (when the bodies of igneous rock were lava), what else would you have seen on the landscape besides valleys, streams, lakes, and fields? Explain your reasoning.

Sedimentary Processes, Rocks, and Environments

Pre-Lab Video 6

http://goo.gl/JIK8cy

Contributing Authors

Harold Andrews • *Wellesley College*
James R. Besancon • *Wellesley College*

Pamela J.W. Gore • *Georgia Perimeter College*
Margaret D. Thompson • *Wellesley College*

▲ Sedimentary layers can contain environmental, climatological, biological, and tectonic information about Earth's history. Large dinosaurs walked across this mud flat—part of the Sousa Formation in northeastern Brazil (6.73424°S, 38.26198°W)—during the Early Cretaceous, around 130 Myr ago.

BIG IDEAS

Sediments include rock fragments, mineral grains eroded from rocks, minerals precipitated from water, and the byproducts of life. Sedimentary rocks form by precipitation and by the deposition, burial, compaction, and cementation of solid sedimentary grains. Layers of sediments and sedimentary rocks are like pages of a book. Their fossils, grains, and sedimentary structures tell us about Earth's history and past environments and ecosystems.

FOCUS YOUR INQUIRY

Think About It   What do sedimentary rocks look like? How can they be classified into groups?

ACTIVITY 6.1   Sedimentary Rock Inquiry (p. 150, 169)

Think About It   How does sediment form and change?

ACTIVITY 6.2   Sediment from Source to Sink (p. 150, 170)

ACTIVITY 6.3   Clastic Sediment (p. 150, 173)

ACTIVITY 6.4   Bioclastic Sediment and Coal (p. 152, 175)

ACTIVITY 6.5   Sediment Analysis, Classification, and Interpretation (p. 152, 177)

Think About It   How do geologists describe, classify, and identify sedimentary rocks?

ACTIVITY 6.6   Hand Sample Analysis and Interpretation (p. 156, 178)

Think About It   What can sedimentary rock tell us about Earth's history and past environments and ecosystems?

ACTIVITY 6.7   Grand Canyon Outcrop Analysis and Interpretation (p. 163, 182)

ACTIVITY 6.8   Using the Present to Imagine the Past—Dogs and Dinosaurs (p. 163, 183)

ACTIVITY 6.9   Using the Present to Imagine the Past—Cape Cod to Kansas (p. 163, 184)

ACTIVITY 6.1

Sedimentary Rock Inquiry, (p. 169)

Think About It What do sedimentary rocks look like? How can they be classified into groups?

Objective Analyze and describe samples of sedimentary rocks, and then infer how they can be classified into groups.

Before You Begin Read the following sections: Introduction through Turning Clastic Sediment into Rock.

ACTIVITY 6.2

Sediment from Source to Sink, (p. 170)

Think About It How does sediment form and change?

Objective Investigate sediment forming in Yosemite Valley, California, and consider its evolution as it moves down the Merced River.

Before You Begin Read the following section: Clastic Sediment and Clastic Sedimentary Rock.

ACTIVITY 6.3

Clastic Sediment, (p. 173)

Objective Analyze clastic sediment and infer the environment in which sedimentary grains formed.

Plan Ahead You will need (or your teacher will need to supply you with) two specimens of granite or diorite, a magnifying glass or hand lens, and some sand paper.

Before You Begin Read the following section: Clastic Sediment and Clastic Sedimentary Rock.

Introduction

Natural processes related to weathering, erosion, transportation of eroded grains, deposition, precipitation, and life provide the building blocks of sedimentary rocks. Sedimentary rock consists of sedimentary grains and, in some instances, minerals that are precipitated either at Earth's surface or in near-surface cavities like caves or pore spaces between grains. The minerals precipitate from ions that are in solution in water or, in other words, that are in

aqueous solution. **Sediments** include mineral grains and rock fragments eroded from older rock, tephra erupted from volcanoes, solid parts secreted by organisms (e.g., bones, teeth, shells, or the skeletal remains of tiny organisms), carbon-based materials like kerogen or coal, and solids like the tiny egglike ooids that are precipitated from water independent of life processes.

Weathering, Erosion, Transportation, Deposition

Earth is a web of interconnected dynamic systems that are responsible for geologic, hydrologic, atmospheric, and biologic change over time. The environment around a given volume of rock can also change. Imagine igneous rock formed kilometers down in the Earth where the rock was at equilibrium with its surroundings. That rock is later uplifted and exposed at the ground surface, where it encounters significantly lower temperature and pressure in a much wetter environment. The minerals in that exposure of igneous rock would not be in equilibrium with their new environment.

Over time, materials at Earth's surface undergo a series of physical or chemical changes to bring them into equilibrium with their environment. We call this process **weathering**. If the products of weathering—sediments or **ions** in aqueous solution—move from where they originated, we say that their **source area** has undergone **erosion** (**Fig. 6.1**). Eroded sediment is available to be transported by wind, liquid water, ice, or organisms to where it is deposited. The processes of **erosion**, **transportation**, and **deposition** can be repeated many times before a sedimentary grain is finally buried and incorporated into a sedimentary rock.

Nature is a great recycler, and the Earth-surface environment is where much of the recycling happens through interactions with the atmosphere, water, and life. If you leave a steel garden tool outside, the steel will alter through a chemical process of **oxidation** to rust. A marble statue installed outdoors in a city park will begin to disintegrate almost immediately. The thermal stress of heating by day and cooling by night can promote weathering as can acidic rain and smog. As a result of weathering, the statue will shed small grains of calcite, and some calcium and carbonate ions will be carried away in aqueous solution. Weathering agents such as lichen, plants, thermal stress, expansion cracks, and the freezing and thawing of water in those cracks slowly break apart the jagged face of a granite peak in the mountains (**Fig. 6.1**). The results are sedimentary grains and ions in aqueous solution. The biosphere also participates in this process of generating sediment and ions in solution as all of the life forms on Earth eventually die, and the matter that composed them is redistributed for other uses.

Why Are There Different Types of Sedimentary Rock?

The composition of sediments and ions in solution is related to (but not necessarily the same as) the composition of the source area that weathered and was eroded to produce them. The composition of sediment can evolve as additional chemical or physical changes occur while it

Figure 6.1 **Weathering and erosion of a rock outcrop.** This rock outcrop was subjected to chemical, biological, and physical weathering that led to the erosion of sediments ranging from microscopic clay to large angular boulders. The recently eroded sediments form a talus slope of angular boulders below the outcrop.

is in transit from the source area. For example, a granite outcrop might generate sediment that includes grains of quartz, feldspar, mica, hornblende, some minor minerals, and rock fragments (**Fig. 6.2**). As the grains are transported by wind, water, or ice, the rock fragments will break into their individual minerals. Feldspar and hornblende tend to break down into clay minerals and various ions in aqueous solution. The ions are then available to form new minerals or to be used by organisms. The sediment derived from the granite is likely to have changed during its journey so that many kilometers away from the source area, it is composed mostly of quartz and clay.

Sediment on the seafloor can be a combination of material transported from an adjacent continent or island, minerals precipitated from the water, and the debris of life: shells, bones, teeth, fecal pellets, and other solids (**Fig. 6.3**). Sediment in the deep ocean related to continents or islands includes submarine landslide deposits (turbidites), windborne fine sand and silt as well as volcanic ash or dust. Away from continents in mid-ocean, most of the sediment consists of the remains of organisms that live in the water column or on the seafloor, mostly in the form of carbonate and silicate skeletal material, including shells. Carbonate minerals include *aragonite, calcite,* and *dolomite.* Carbonate minerals dissolve when they are deeper than about 5000 meters below sea level, so only silicate sediment persists on the seafloor in the deepest ocean.

Clastic Sediment and Clastic Sedimentary Rock

A **clast** is a sedimentary grain that has been transported as a solid grain from its source, which might be an eroded older rock, a volcano, or an organism. The word *clast* is derived from the Greek word for "broken

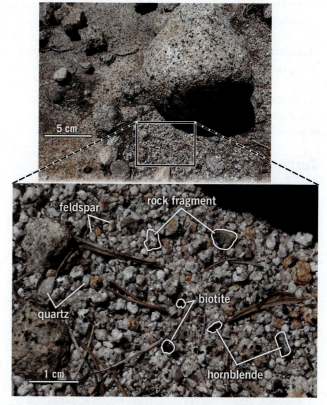

Figure 6.2 **Granitic sediment.** Sediment grains eroded from granitic rock include rock fragments, quartz, feldspar, biotite, hornblende, and probably clay.

in pieces"—*klastós.* Sediments composed of clasts are called **clastic sediments,** and the resulting sedimentary rock is called **clastic rock.** Hard parts derived from organisms (shells, teeth, bones, the skeletons of invertebrate organisms) are called **bioclasts.** The adjective **bioclastic**

Figure 6.3 Bioclastic sediment. Beach sand from Munson Island in the Florida Keys with abundant shell fragments. Virtually all of the natural material on this beach other than plant debris is composed of the carbonate minerals calcite or aragonite.

is used to describe sediment or rock in which bioclasts are abundant (**Figs. 6.3** and **6.4**). **Volcaniclastic** rock is formed of **tephra** that originated in the eruption of a volcano, transported as part of a volcanic mudflow or **lahar**, a **pyroclastic flow**, or settling out of the air or water after eruption.

The two principal types of sedimentary rocks are broadly described as siliciclastic and carbonate rock. **Siliciclastic rock** is composed primarily of silicate minerals like quartz, clay, feldspar, and mica. **Carbonate rock** is composed primarily of calcite (**limestone**) or dolomite (**dolostone**). Limestone typically contains bioclasts, although precipitated carbonate is common as a cement between the grains.

Turning Clastic Sediment into Rock

How does sand become sandstone? The general progression from a loose assortment of sedimentary grains to a hard consolidated rock can be as simple as cementing the grains together. **Cementation** can happen at Earth's surface if the water in the pore spaces between grains is saturated with ions that can combine and precipitate as

Figure 6.4 Bioclastic rock. The sediments that form this fossiliferous limestone once formed a small part of the seafloor. The fossils are all broken and transported solid bits of marine organisms.

new mineral material. The most common cements in sedimentary rock are calcite and microcrystalline quartz, but metal oxides and other precipitated minerals can also bind clastic grains together. **Coquina** is a rock composed almost entirely of broken shell fragments cemented together by calcite. Coquina looks a bit like oatmeal with broken shells in place of oats (**Fig. 6.5**) and forms at Earth's surface in or near surf zones in marine or lake environments.

Sedimentary beds are subjected to increased pressure as younger sediments bury them. The increased pressure has the effect of compacting the grains together more tightly and reducing the pore space between the grains. Freshly deposited sand might have a porosity of approximately 40%, meaning that about 60% of a given volume of sediment is solid grains and 40% is the gas or liquid between the grains. After burial and under the influence of increased pressure, the sand grains shift positions into an arrangement in which they are closer together. **Compaction** can reduce the porosity by more than half. Additional porosity reduction can occur if the grains start to deform or if the pore space starts to fill in with precipitated cement. So sediment is converted into a sedimentary rock—is **lithified**—by cementation that can be enhanced by compaction.

ACTIVITY 6.4

Bioclastic Sediment and Coal, (p. 175)

> **Think About It** How does sediment form and change?

Objective Analyze characteristics of bioclastic rock and coal, and infer how they form.

Plan Ahead You will need (or your teacher will need to supply you with) a dropper bottle containing dilute hydrochloric acid, a piece of chalk, some seashells that can be broken, some charcoal, two plastic bags, a hammer, a coal specimen, and a magnifying glass or hand lens.

Before You Begin Read the sections: Turning Clastic Sediment into Rock, Precipitated Sedimentary Rock, and Carbon in Sedimentary Rock.

ACTIVITY 6.5

Sediment Analysis, Classification, and Interpretation, (p. 177)

Objective Describe and classify samples of sediment in terms of texture and composition, and then infer environments in which they formed.

Before You Begin Read the sections: Precipitated Sedimentary Rock and Clastic Sediment and Clastic Sedimentary Rock.

Figure 6.5 Coquina. Broken shell material cemented with calcite to form the bioclastic rock *coquina*.

Precipitated Sedimentary Rock

Sedimentary rock sometimes results from the precipitation of ions from water. Minerals precipitate when the concentration of their ionic components becomes too great for the ions to remain in aqueous solution. Some limestone includes precipitated calcite like that found in **ooids** (**Fig. 6.6**). Many, if not all, ooids are thought to form in water that is agitated by waves or currents that frequently rotate the grains so that the precipitated calcite coats the grains evenly. **Travertine** is formed by precipitation of calcite at a spring, in a cave, or even around a geyser or hot spring (**Fig. 6.7**). **Tufa** is a less dense form of travertine that is formed by calcite precipitation around a spring. Silica can precipitate as microcrystalline quartz around the vent of a geyser or hot spring as the hot silica-saturated water cools at the surface, forming a sedimentary rock called **geyserite** or **siliceous sinter** (**Fig. 6.8**). Silica can also precipitate to form **agate**, **onyx**, or **opal**.

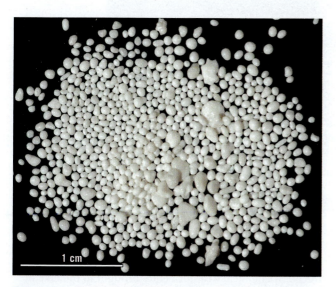

Figure 6.6 Carbonate ooids. Sand-sized grains of calcite precipitated in the shallow marine environment where the water is saturated with calcium and carbonate ions and the waves or current keeps the surface sediment moving.

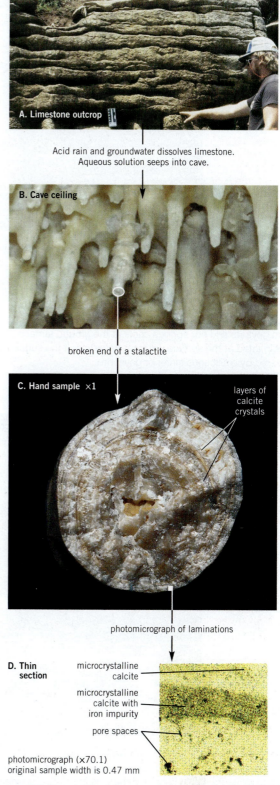

A. Limestone outcrop

Acid rain and groundwater dissolves limestone. Aqueous solution seeps into cave.

B. Cave ceiling

broken end of a stalactite

C. Hand sample ×1

layers of calcite crystals

photomicrograph of laminations

D. Thin section

microcrystalline calcite

microcrystalline calcite with iron impurity

pore spaces

photomicrograph (×70.1) original sample width is 0.47 mm

Figure 6.7 Formation of travertine. A. Limestone is dissolved by acidic water at or near Earth's surface. **B.** The resulting aqueous solution containing calcium and bicarbonate ions seeps into caves. As the solution drips from the roof of a cave, calcite precipitates to form icicle-shaped stalactites on the cave ceiling. **C.** Broken end of a stalactite reveals that it is a mass of intergrown calcite crystals. **D.** Thin section photograph reveals that the concentric laminations of the stalactite are caused by variations in iron impurity and porosity of the calcite layers.

Figure 6.8 Geysers causing mineral precipitation at Yellowstone. Ions combine to form new minerals as the water from geysers and hot springs cools at the ground surface. If the subsurface rock is a carbonate such as limestone, the minerals precipitated at the surface will form travertine. If the subsurface rock is a silicate such as quartz sandstone, the surface will be covered by a siliceous sinter or geyserite.

Figure 6.9 Playa lake in Death Valley, California. The white material at the bottom of this fault-bounded basin is composed of a variety of evaporate minerals including gypsum and halite (see the close up in the upper right corner). Groundwater emerges in springs along the basin floor, and the ions in solution precipitate at the ground surface as the water evaporates.

Precipitated minerals called **evaporites** form in sedimentary layers within playa lakes or in restricted marine basins where evaporation exceeds the rate that water enters the basin. Small basins filled with seawater can become isolated from the main ocean, perhaps during periods of time when world sea level declines. A **playa lake** within a desert basin can also become salty. Ions produced by chemical weathering are brought to the lake by groundwater or streams (**Fig. 6.9**). In playa lakes as well as in restricted marine basins, the concentration of ions in solution increases to the point where the ions must combine to form new mineral grains. Carbonates (usually calcite or aragonite) begin to precipitate when about half of the seawater in a restricted marine basin evaporates, followed by gypsum or anhydrite, halite, and finally some potassium and magnesium-bearing salts (Fig. 6.10). Gypsum ($CaSO_4 \cdot 2H_2O$) can form directly by precipitation or through the addition of water to anhydrite ($CaSO_4$). Gypsum, anhydrite, and halite are the most abundant evaporite minerals.

Carbon in Sedimentary Rock

The most obvious thing about Earth as observed from space is that it is mostly covered in water. Probably the next most obvious thing is that the observable ground surface is mostly covered by life. Life virtually saturates Earth from the deepest ocean trench up through the lower atmosphere and across the surface of the continents and seafloors. An amazing amount of microbial life exists below Earth's surface: in caves, in open cracks and pore spaces, in environments of extreme temperature and pressure, and even inside some mineral grains. These organisms and their environments are the focus of current research intended to improve our understanding of the role carbon plays throughout the Earth in living systems and in the overall biogeochemistry of our planet.

Plant material accumulates faster than it decomposes in poorly drained areas like swamps or bogs. Oxygen from the atmosphere that would normally facilitate organic decomposition is limited by the wet conditions, and so the carbon-rich plant debris can accumulate slowly. This mass of partially decayed vegetation is called **peat**. Peat has been harvested from peat bogs as a fuel source for thousands of years, but burning peat produces more particulate and greenhouse-gas pollution than other fossil fuels while generating less heat.

When peat beds are buried by other sediments, they are subjected to increased pressure and temperature. This can lead to their alteration into a series of increasingly dense, concentrated solids that are collectively called **coal** (Fig. 6.11). The lowest rank or grade of coal is **lignite**, which is a crumbly black material that is essentially compressed peat. With additional pressure supplied by deeper burial, coal becomes denser and dryer as more water is forced out by compaction. These changes result in **sub-bituminous** and **bituminous coal** (Fig. 6.11A, B), which are harder than lignite and have a dull black to somewhat shiny luster. Additional alteration resulting from increased pressure and temperature results in **anthracite coal**, which is a brittle, black material that has a very shiny luster like glass (Fig. 6.11C). Coal gradually changes from lignite to anthracite with increasing pressure and temperature, and sufficient time for change to occur.

Is coal a sedimentary rock? The fact that it takes an increase in pressure and temperature to produce

A. Rock salt (×1)

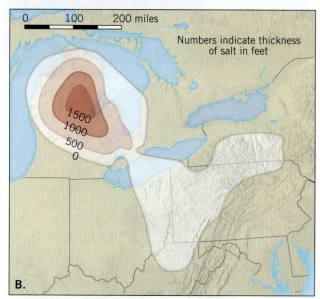

B.

Figure 6.10 Rock salt, a mass of intergrown crystals precipitated from aqueous solution. A. Hand sample from mines deep below Lake Erie shows how crystals grew together to make the rock salt. **B.** Map showing the thickness and distribution of rock salt deposits formed about 400 million years ago when a portion of the ocean was trapped and evaporated, millions of years before the Great Lakes existed.

bituminous and anthracite coal makes them candidates for consideration as metamorphic rock except for two observations. First, the anthracite coal seams in Pennsylvania are interbedded with sandstone and conglomerate that are still considered sedimentary rock (Fig. 6.11A). Second, coal is composed of carbon compounds related to plant material but is *not* composed primarily of mineral grains. If we define rock as a naturally occurring aggregate of mineral grains, coal is not a rock although it is certainly sedimentary in origin. Coal geologists describe the building blocks of coal as *macerals* in the same sense that *minerals* are the building blocks of rock. Macerals are chemically altered plant fragments

Figure 6.11 Coal. A. Bituminous coal bed overlain by quartz sandstone in the Raton Basin, southern Colorado. **B.** Specimen of dull black bituminous coal. **C.** Specimen of shiny, dense anthracite coal.

from which some of the hydrogen atoms have been stripped away from the carbon atoms to which they were originally bound.

Another way that carbon is present in sedimentary deposits is in **kerogen**. Kerogen is a solid mixture of carbon-bearing chemical compounds that originated in the proteins and carbohydrates of living organisms. After death, these biomolecules begin to break down, and their component pieces can recombine to form new molecules called polymers. These polymers are deposited along with other sediments and then buried. The increase in pressure and temperature caused by progressively deeper burial leads to the formation of kerogen from the carbon-rich polymers. Kerogen is disseminated throughout the rock rather than concentrated in layers. Further heating of some forms of kerogen generates natural gas and

ACTIVITY 6.6

Hand Sample Analysis and Interpretation, (p. 178)

Think About It How do geologists describe, classify, and identify sedimentary rocks?

Objective Be able to describe, classify, and identify hand samples of sedimentary rocks.

Before You Begin Read the following sections: Primary Properties of Sedimentary Rock and Classification of Sedimentary Rock.

Figure 6.12 **Oil in sandstone.** Part of an oil-saturated sandstone core that was extracted from an exploratory oil well. A test sample was taken for analysis by petroleum geoscientists, leaving a hole in the core.

Figure 6.13 **Arkose.** This coarse sandstone specimen from the Fountain Formation in central Colorado has abundant sand-sized feldspar, quartz, and rock fragments in a matrix that probably includes clay minerals and oxides. The red color is from iron oxides produced by weathering of iron-bearing minerals and precipitation of the resulting oxide minerals as grain coatings and cement.

liquid oil (**Fig. 6.12**). Dark shales contain a relatively large amount of carbon, although generally less than 1% of the shale's total mass.

Primary Properties of Sedimentary Rock

The characteristics of sedimentary rock on which we will focus in this lab include composition, texture, sedimentary structures, and environment of deposition. The properties that allow us to classify a sedimentary rock in hand specimen (without a thin section, petrographic microscope, or electron microscope) are its composition and texture.

Composition

Geologists classify most sedimentary rocks into two main groups based on their mineral composition: silicates and carbonates. The silicate minerals that are most important in sedimentary rocks are *quartz, feldspar,* and *clay* (**Fig. 6.13**). The clay minerals in sedimentary rock include *kaolinite, illite,* and *montmorillonite.* **Rock fragments** containing more than one mineral grain are also important constituents of sedimentary rocks. Minor minerals include the mica minerals (muscovite and biotite) and chlorite as well as a collection of minerals known as **heavy minerals** because they are more dense than either quartz or feldspar. The heavy minerals in sedimentary rocks include augite, hornblende, garnet, tourmaline, and zircon. Clastic sedimentary rock composed of silicate minerals is often called **siliciclastic rock**.

The carbonate minerals *aragonite, calcite,* and *dolomite* are the main components of **carbonate rock** (**Fig. 6.14**). The composition of aragonite and calcite is the same—$CaCO_3$—but the lattice structure of their crystals is different.

Figure 6.14 **Calcite and dolomite. A.** Calcite crystals grew into the void space in a limestone, precipitating from calcium and carbonate ions transported in ground water. **B.** Pink dolomite crystals, which have crystal faces that are distinctive because they are slightly curved.

The common marine algae *Penicillus* and other common forms of green algae secrete fine grains of aragonite that can form a major constituent of carbonate muds. Corals and most molluscs also secrete aragonite whereas other marine organisms secrete calcite to form their hard parts. The mineral *dolomite* [$CaMg(CO_3)_2$] usually forms through alteration of calcite after deposition and burial as water containing magnesium ions interacts with calcite. Dolomite can also form from direct precipitation from seawater that contains magnesium, calcium, and carbonate ions in solution, but this is uncommon. Sedimentary rocks composed of carbonate minerals are usually called **limestone**, although rocks composed predominantly of dolomite are often called **dolomite**. A few geologists proposed renaming this rock **dolostone** to avoid confusing the rock with its principal mineral, but this radical idea has not been universally adopted by the geological community. In this lab, we use *dolomite* for the mineral name and *dolostone* for the sedimentary rock name.

Limestone fizzes or effervesces vigorously when you put a drop of hydrochloric acid on it because limestone is composed of calcite. A dolostone will fizz weakly in acid. Making a powder of the dolostone will increase the fizzing a little bit. Acid does *not* cause siliciclastic rock to fizz, with two exceptions. Siliciclastic rock with fossils made of calcite or cement composed of calcite will fizz where the acid touches the cement or a fossil.

Texture

The size, shape, sorting, and packing of grains and the relative orientation of elongate or flat grains define the **texture** of a sedimentary rock. Textural descriptions can include the 3-dimensional arrangement of grains within the rock.

Grain Size. Sedimentary grains are classified by their diameter, using a simple scale devised by Chester K. Wentworth that has four major size classes: **clay**, **silt**, **sand**, and **gravel** (**Fig. 6.15**). Each of these categories is further subdivided. For example, gravel is subdivided into **granules** (2–4 mm), **pebbles** (4–64 mm), **cobbles** (64–256 mm), and **boulders** (>256 mm). Sand and silt are subdivided into *very fine*, *fine*, *medium*, and *coarse*, and sand has a *very coarse* class from 1–2 mm in diameter.

Wentworth called the smallest grain size "clay." While it is true that individual grains of clay minerals are in the "clay" size class, it is also true that quartz, feldspar, calcite, aragonite, pyrite, iron oxides, and heavy minerals like zircon, monazite, rutile, and tourmaline can all be small enough to be included in the clay-size class. Glaciers are particularly effective at grinding mineral grains into tiny fragments.

Figure 6.15 provides an accurate summary of the major divisions in the Wentworth scale, but it does not provide an intuitive feel for these size classes. The average height of a capital letter in this sentence is about 2 mm, so any sedimentary grain that is broader than that is **gravel**. **Sand** is smaller than gravel, and even the finest sand feels gritty when you rub it between your fingers. Sand does not feel good in your swimsuit. **Silt** is smaller than sand, but you might still be able to feel a little bit of grittiness of the coarsest silt when you rub it between your fingers. You might not be able to feel the finest silt between your fingers; however, anyone who has had a bit of silt deposited on his lunch on a windy day will tell you that you can sense the grittiness of silt if it gets in your mouth. In contrast, you would not be able to sense the grittiness of **clay**-sized grains.

Grain Sorting. Sediments are typically deposited in layers that are nearly horizontal. When one of these layers can be differentiated from others, it is often called a *sedimentary bed*. A **bed** is a sedimentary layer that is 1 cm or more in thickness, and a thinner layer is called a **lamina**. If the grains in a sedimentary bed are all approximately the same size, the grains are said to be **well sorted** (**Fig. 6.16A**). (The word *sort* is used here in the same sense that it is used in the laundry. All the white cotton garments go together in one bin, all the polyester disco-revival jump suits in another bin, and so on.) If a sedimentary bed has a wide variety of grain sizes throughout the bed, it is said to be **poorly sorted**.

We use qualitative assessments of sorting in this lab, but if we wanted to be more quantitative, we would use the standard deviation of the grain sizes in a representative sample as a measure of sorting. A very well-sorted sample would have a standard deviation close to 0 because all grains are approximately the same size, and a poorly-sorted sample would have a standard deviation that is greater (and perhaps much greater) than 1. A well-sorted deposit has more pore space between the grains, so it is easier for gas and liquids to flow through a sand or sandstone if it is well sorted—that is, as long as the space between the grains has not become filled with cement.

Grain Shape. A sedimentary grain might be rounded like a ball or as angular as a freshly broken brick. Geologists generally apply a qualitative assessment of grain

Grain-Size Class	Grain Diameter		
	Exponential Notation	Integers and Ratios	Decimal Notation
gravel	$\geq 2^1$ mm	$\geq$ 2 mm	$\geq$ 2 mm
sand	2^{-4} to 2^1 mm	1/16 to 2 mm	0.0625 to 2 mm
silt	2^{-4} to 2^{-8} mm	1/256 to 1/16 mm	0.0039 to 0.0625 mm
clay	$\leq 2^{-8}$ mm	$\leq$ 1/256 mm	$\leq$ 0.0039 mm

Figure 6.15 Wentworth grain-size scale.

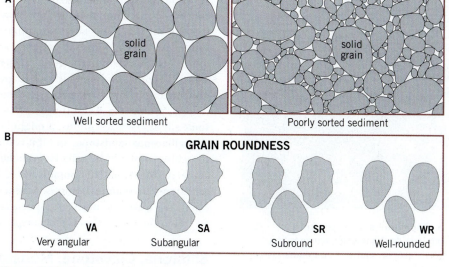

Figure 6.16 **Sorting and roundness.**

rounding and sphericity using charts like **Fig. 6.16B**. The differences in grain shape have several potential causes. One is the nature of the grain itself. For example, grains of minerals like quartz that are hard and lack cleavage tend to become rounded during transport whereas minerals like feldspar and calcite that have cleavage tend to break along their planes of weakness into smaller and smaller angular pieces. Fragments of rocks that do not have any layering, like granitic rock, tend to be rounded into more spheroidal shapes during transport. In contrast, fragments of rocks with strong internal layering, like slates, are smoothed into flatter shapes that reflect that internal structure.

The transport history of a grain is another important factor influencing the shape and roundness of a sedimentary grain. How far did the grain travel from its source? Sedimentary grains that are subjected to transport over a greater distance tend to become rounded. Did the grain travel in a turbulent, high-energy environment like a mountain stream, or was it subject to pounding surf along a beach? If so, it probably became rounded rather quickly.

Matrix. Sedimentary beds that are poorly sorted have larger grains in a matrix of smaller grains. We use the word **matrix** for the smaller grains in sedimentary rock in much the same way that we use the word *groundmass* for igneous rock to describe the little mineral grains that are around the bigger grains. It is common to have larger fossils embedded in a sandy matrix (a fossiliferous sandstone) or gravel-sized grains in a sandy matrix (a conglomerate or conglomeratic sandstone).

Classification of Sedimentary Rock

Geologists have argued about what names to apply to different types of rocks since the beginning of the science in the 1700s. Recent international efforts to better define and streamline the names of igneous and metamorphic

rock types have found broad if not universal acceptance. With sedimentary rock types, there is a widely used set of terms for naming the most common rock types and a parallel set of terms preferred by specialists such as sedimentary petrologists and sedimentologists. A specialist might use the word **arenite** to describe a sandstone, **rudite** to describe a conglomerate, **argillite** for a clay-rich rock, or **micrite** for a calcite mudstone. In this lab, we will use the common terms that are in general use because they are mostly based on well-understood English words.

Conglomerate and Breccia

A sedimentary rock composed mostly of gravel-sized grains (≥ 2 mm) with smaller grains in the matrix is called a **conglomerate** (**Fig. 6.17**). If most of the gravel is the same grain size, we can use the name of that grain size as a modifier as in *pebble conglomerate* or *boulder conglomerate*. The gravel clasts in many conglomerates are

Figure 6.17 **Poorly-sorted conglomerate.** The large clasts appear to be supported in a matrix of smaller sand and gravel-sized clasts. The gravel that became this conglomerate was probably deposited in an ancient stream.

Figure 6.18 Breccia. The gravel-sized clasts in this conglomerate are mostly angular from which we infer that the source area for these clasts is nearby. A conglomerate of angular clasts is a breccia.

rounded, but conglomerates whose clasts are angular can be called a **breccia** (pronounced *BREH*-chee-uh; **Fig. 6.18**). Sedimentary breccia contains gravel clasts that were not transported far from their source.

Sandstone

Sedimentary rock composed mostly of sand-sized grains is called a **sandstone** (**Fig. 6.19**). The composition of most sandstones includes some combination of quartz, feldspar, clay, and rock fragments, although sand of other compositions is possible. There are olivine sands on Papakolea

Beach on the island of Hawai'i, gypsum sands at White Sands National Monument in New Mexico, and calcite sands in the Bahamas. The composition of the sand grains can be used as a modifier as in **quartz sandstone** if it is mostly composed of quartz grains (**Fig. 6.19**), **lithic sandstone** if it contains abundant rock fragments, or **arkosic sandstone** (or simply **arkose**) if it contains abundant feldspar grains (**Fig. 6.13**). If the sandstone also contains some silt or clay-sized grains, it might be called a **muddy** or **argillaceous sandstone**, and if it contains some granules or pebbles, it might be called a **granular** or **pebbly sandstone**. A sandstone that contains fossils is a **fossiliferous sandstone**. Sandstones composed primarily or entirely of calcite are called *calcarenites* by specialists, but in this lab, we will just consider them to be a type of limestone.

Siltstone, Claystone, Mudstone

A sedimentary rock that is composed mostly of silt-sized grains is called a **siltstone**. The siltstone in **Fig. 6.20** has such a small grain size that it preserves tiny details of a fossil leaf. One important source of silt in siltstone is **loess**—silt-sized wind-blown dust that might have originated as fine glacial, volcaniclastic, or desert sediment. (*Loess* is pronounced with the vowel sound in the word "could.") If a sedimentary rock is composed mostly of clay-sized grains, it is a **claystone**. Silt- and clay-sized particles are difficult to discriminate from one another because they are so small. Unless there is some practical reason to determine the relative amount of silt or clay in a sedimentary rock, any rock that contains a substantial amount of silt and clay is called a **mudstone**.

Figure 6.19 Quartz sandstone. The black arrow is a field geologist's mark that points toward younger layers.

Figure 6.20 Fossil leaf in siltstone. Notice the fine details of the leaf structure that are still recognizable in this fossil.

Shale

An important type of mudstone tends to break along irregular, approximately parallel surfaces, which is a characteristic called **fissility**. A mudstone that displays fissility is called a **shale** (**Fig. 6.21A**). Clay minerals are silicates whose lattice has a layered structure like the pages in a book, and so they form platelike mineral grains. These flat minerals are more or less randomly oriented in a wet mud and are only slightly better aligned in a mudstone (**Fig. 6.21B**). Mudstone is subjected to increasing stress as more weight is deposited on top of it during burial, which has the effect of forcing many of the flat clay minerals to rotate so that they are more nearly parallel to the bedding planes (**Fig. 6.21C**). The result is a denser rock in which the platy clay minerals are more or less aligned with each other. If the sedimentary rock above the shale is eroded away so that the shale is exposed at the ground surface, the shale will expand a

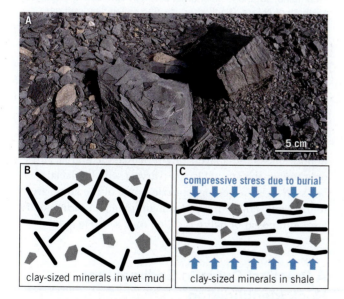

clay-sized minerals in wet mud

compressive stress due to burial

clay-sized minerals in shale

Figure 6.21 Shale. A. Fresh specimens of the Eagle Ford Shale in central Texas, showing fissility or parting along bedding planes. **B.** Platelike clay minerals (black lines) are randomly oriented in a wet mud, which includes small grains of other minerals like quartz (gray polygons). **C.** Clay minerals rotate as a result of burial and compaction, so platy clay minerals are approximately aligned in shale. The alignment of platy minerals is the reason for the rock's fissility.

little bit and will tend to break along irregular surfaces that parallel the average orientation of the clay minerals.

Limestone

A general term for a sedimentary rock composed primarily of calcium carbonate—the minerals calcite or aragonite—is limestone. It is most easily distinguished from other sedimentary rock types by placing a drop of dilute hydrochloric acid on it because limestone will fizz energetically. In contrast, a siliciclastic sedimentary rock will not fizz at all unless it contains some bits of carbonate fossil material or has calcite cement. Most limestone is cemented together with calcite, so detecting individual grains in hand specimens is often difficult. In thin sections viewed under a petrographic microscope, the various components of a limestone become apparent. Changes at low temperatures and pressures after deposition and burial (i.e., **diagenetic changes**) can be extensive, including mineralogical changes from aragonite to calcite or from calcite to dolomite.

The grains in limestone include skeletal grains that were originally secreted by living organisms (e.g., shells, grains formed by algae, microfossil skeletons, teeth, bones), ooids, fecal pellets, bits of carbonate sediment or older carbonate rock that have been eroded and redeposited, noncarbonate minerals (e.g., quartz, clay, evaporates), and whatever else might be deposited in a lake or ocean basin as the calcium carbonate sediments accumulate. Fossils that are larger than a millimeter or two are often easy to recognize as fossils, although determining which organism contributed the fossil usually requires additional background in paleontology. Sand-sized carbonate grains are called **peloids**. Many are fecal pellets of carbonate mud. **Ooids** are distinctive sand-sized grains formed by precipitation of aragonite around a central nucleus (**Fig. 6.6**). Some ooids have a concentric internal structure whereas others have a radial structure. **Pisolites** are round carbonate grains with the same internal structure as ooids and that are at least 2 mm in diameter.

Carbonate petrologists specialize in studying limestone and dolostone and have an extensive set of terms to describe these rocks. In this lab, we will employ a simple set of terms that is widely used by geologists. A **micrite** (or **microcrystalline limestone**) is a very fine-grained limestone that probably originated as a carbonate mud. Coarser-grained limestone is simply called *limestone*, although we tend to use modifiers to indicate any components that are observed in the rock. For example, a limestone with ooids is called **oolitic limestone**. A limestone with fossils is called **fossiliferous limestone** (**Fig. 6.3**). If we know that the fossils are nearly all of one type of organism, we might use that organism's name as a modifier, as in **crinoidal limestone** (**Fig. 6.22**). If the rock contains coral, it is called **coralline limestone**. If the rock is composed almost entirely of gravel-sized (≥2 mm) fossil shells cemented together without much or any matrix, it is called **coquina** (**Fig. 6.6**). A limestone with abundant clay minerals might be called a **muddy** or **argillaceous limestone**. And if the rock includes angular clasts of older limestone, it is called a **limestone breccia**.

Figure 6.22 Crinoidal limestone. A crinoid is also known as a sea lily and is an animal that is fixed to the seafloor by a stalk when it is alive. After death, the stacked disks that form the stalk tend to separate and collect on the seafloor.

Figure 6.23 Chert. A. Chert nodule in limestone, Arbuckle Mountains, southern Oklahoma. **B.** Chert specimens displaying a variety of colors. Black chert is often called flint, and red chert is jasper.

Other Sedimentary Rock Types

Processes involving precipitation, evaporation, and chemical alteration produce a variety of other sedimentary rock types. Here we examine the processes that form chert, rock salt, and rock gypsum.

Chert is a very fine-grained sedimentary rock that occurs as nodules or beds (**Fig. 6.23**) and is composed of quartz or chalcedony. Chert might be mistaken for a very fine-grained limestone, but chert does not fizz in acid as limestone does. Some chert forms through the accumulation of sponge spicules or the silica-rich skeletons of tiny single-cell organisms called **radiolaria** or **diatoms**. After deposition, the structure of these siliceous bioclasts can be altered or obliterated by physical/chemical processes that occur in sedimentary strata after deposition, known as **diagenetic processes**. Chert might also form through the alteration of clay minerals (montmorillonite) or glass in volcanic rock. The red chert called *jasper* that is found in layers within ancient banded iron formations is particularly interesting. Earth's atmosphere did not contain nearly as much oxygen ~1.9–2.5 billion years ago as it does today, so sedimentary grains of iron could be transported and deposited without quickly turning to rust. The jasper in banded iron formations is interpreted to have precipitated directly from silicate ions in seawater alternating with layers of hematite and other iron minerals. These ancient sedimentary deposits are important sources of iron for the manufacture of steel.

Rock salt and **rock gypsum** are composed of evaporite minerals that precipitate out of surface water as it evaporates. Precipitated sediments formed by evaporation of seawater include some carbonate minerals, gypsum, halite, and other salts. The set of minerals that precipitate in playa lakes and around desert springs includes gypsum and halite as well as a slightly different set of minerals than is precipitated from seawater. A sedimentary layer composed of gypsum is simply called **rock gypsum**. The Castillo Formation of west Texas has layers of gypsum separated by thin carbon-rich layers (**Fig. 6.24**). Each gypsum layer has been interpreted to be the result of one year's deposition, so these layers might be like the annual rings on a tree or the annual layers (varves) seen in glacial lakes. In a classic geologic puzzle, some of the layers are folded while adjacent layers remain planar. A sedimentary layer composed of halite is called **rock salt** (**Fig. 6.10**).

A Practical Guide to Sedimentary Rock Identification

How might a typical geologist decide what type of sedimentary rock confronts her in an outcrop or in a hand specimen? Geologists tend to ask a series of questions whose answers indicate a pretty good general classification for an unknown sedimentary rock. Some of those

Figure 6.24 Layered gypsum. The Castile Formation of west Texas features an enigmatic layered gypsum rock in which individual layers or sets of layers are folded, even though the layers on both sides are planar. Each pair of light and dark layers is interpreted to represent one year of deposition.

questions are incorporated in the worksheet for Activity 6.6, and a worked example based on **Fig. 6.3** is provided on the first of those worksheets.

- Is it black and turns your hands black without any obvious mineral grains? It is probably **lignite coal** (if it is very crumbly), **bituminous coal** (if it is mostly dull but not very crumbly), or **anthracite coal** (if it is hard, brittle, and shiny).
- Does the matrix fizz when a small drop of dilute hydrochloric acid (HCl) is put on it? Don't be fooled by fizzing calcite cement or individual fossils—the question is whether the grains that comprise most of the rock react to HCl. If it fizzes energetically, the rock is some sort of **limestone**. If it reacts weakly, it might be a **dolostone**.
- Is the matrix formed of microscopic grains? If so and it fizzes in acid, it might be **microcrystalline limestone** (if it is hard) or **chalk** (if it is soft and crumbles easily). If it does not fizz in acid, it might be a rock composed of microscopic silica, such as **chert**, **opal**, **onyx**, or **agate**. The acid test will differentiate between a carbonate and a silicate.
- Is the matrix formed by a mass of intergrown crystals that you can see without a microscope? If so it might be **travertine** (if it fizzes in HCl), **rock salt** (if it tastes salty), or **rock gypsum** or **alabaster** (if it is soft).
- If you can see the grains that constitute the rock and they are *not* a mass of intergrown crystals, what is the grain-size class of most of the rock? Is the rock mostly sand-sized particles (it's **sandstone**), gravel-sized particles (it's **conglomerate**, or it's **breccia** if the grains are angular), or finer than sand-sized particles (it's **mudstone**, **shale**, **siltstone**, or **claystone**).
- What is the composition of the grains you can see? That will help you differentiate between different types of sandstone and limestone.

Now that you have answered these questions about a particular rock specimen, you probably have at least one provisional name for the rock. Microscopic or chemical analysis might provide additional information that would cause you to classify the rock somewhat differently. It is understood that all scientific interpretations are open to reinterpretation based on new, better data that are relevant to the problem.

ACTIVITY 6.7

Grand Canyon Outcrop Analysis and Interpretation, (p. 182)

> **Think About It** What can sedimentary rock tell us about Earth's history and past environments and ecosystems?

Objective Analyze and interpret sedimentary rock from the edge of the Grand Canyon.

Before You Begin Read the following section: Sedimentary Structures and Environments.

ACTIVITY 6.8

Using the Present to Imagine the Past—Dogs to Dinosaurs, (p. 183)

Objective Infer characteristics of an ancient environment by comparing modern dog tracks in mud with fossil dinosaur tracks in sedimentary rock.

Before You Begin Read the following section: Sedimentary Structures and Environments.

ACTIVITY 6.9

Using the Present to Imagine the Past—Cape Cod to Kansas, (p. 184)

Objective Infer characteristics of an ancient environment by comparing present-day seafloor sediments with sedimentary rock formed on an ancient seafloor.

Before You Begin Read the following section: Sedimentary Structures and Environments.

Sedimentary Structures and Environments

Sedimentary rock has provided us with much of what we know about the evolution of life on Earth as well as a great deal of information about the evolution of Earth and its environment. Sedimentary rock is such a rich source of historical information because it is deposited in an orderly sequence of layers with the oldest layer on the bottom and the youngest on top. Of course, erosion and deformation can disturb this orderly record, and much of the record is difficult to access because it is buried under younger layers. We can access buried layers by drilling wells and through geophysical techniques.

The reason why we can interpret sedimentary rock so well is that we can directly observe most sedimentary processes. Geoscientists use the core idea of the **principle of uniformitarianism:** *The physical and chemical laws that govern geologic processes in the present day have been operating without change throughout the entire history of Earth.* We can use our knowledge of modern sediments and sedimentary environments to interpret their ancient counterparts. We can see how streams move sediments, how sand dunes form on continents and on continental shelves below sea level, and how life affects and is affected by the sedimentary environment. Because sedimentary processes occur at Earth-surface temperatures and pressures, we can reproduce many of these processes in the laboratory to enhance our understanding. With so much data accumulated over the years, we can only touch on a few topics in this lab. They will have to serve as a preliminary glimpse of the information we can gain from studying sediments and sedimentary rock.

Sedimentary Structures and Bed Forms

Sediment is transported by moving air, liquid water, ice, and organisms. We will leave sediment transport by glaciers and critters aside for the moment. Flowing air and liquid water can move sediment in a manner that produces **laminae** (<1 cm thick) and **planar beds** (≥1 cm thick) as well as **ripple marks** in a broad range of sizes and shapes (**Fig. 6.25**).

The size of ripples ranges from great sand dunes down to tiny ripples barely a few millimeters high. **Dunes** are ripples that are at least 5 cm high but can be many meters high. The size of a dune is related to a number of factors, including whether the sediment is being transported by air or water, the velocity of the air/water, whether the direction of flow was constant or varied, and the density and size of the grains. Dunes can form on dry land above sea level as well as below sea level in places where there is a sufficient supply of sediment and ocean current fast enough to move the sediment (**Fig. 6.26**).

The shape of ripples is controlled by how (or whether) the current direction changes over time (**Fig. 6.25**). **Symmetric ripple marks** form as a result of

currents that oscillate back and forth. **Asymmetric ripple marks** are formed by sediment moved by air or water that consistently moves in the same direction. Asymmetric ripples have a steep side that is known as the **lee slope** and a gently sloping side called the **windward** or **stoss slope**. The current of air or water that generated a given ripple flowed perpendicular to the ripple's ridgeline between the stoss and lee slopes (**Fig. 6.25**).

Each sedimentary grain moving across a current ripple bounces, rolls, or drags along up the more gently inclined stoss slope, which commonly has an inclination of just a few degrees or less relative to horizontal. When the grain reaches the top of the ridge, it is carried over the edge and falls onto the lee slope. The grains that follow then bury the grains on the lee slope, bringing their transport history to a close for now. A bed formed on the lee slope of a dune (or advancing delta) is called a **foreset bed**. Foreset beds are inclined as much as about 30° relative to horizontal in dry sand.

We see ripples when we look at most surfaces on which current moves sediment: on stream beds, along beaches, in gutters, and in places where the wind blows sand. Imagine that we are in a dry stream bed that has ripples, and we dig a vertical trench through the ripples in the same direction the water flowed. We would probably see the edges of the inclined foreset beds in the vertical wall of the trench. In this cross-sectional view, these inclined beds are called **cross beds** (**Fig. 6.25**). Sets of parallel cross beds are found sandwiched between nearly horizontal beds on the bottom and top of the set. The distance between the bottom and top of the cross-bed set is the height of the ripples or dunes that formed the cross beds. The current that resulted in the cross beds moved in the direction that a ball would roll down a given cross-bed surface, just as it would roll down the steep face of a ripple mark or sand dune. Interpreting whether a cross-bedded sequence of beds was formed in a marine or nonmarine setting usually requires environmental information, such as whether marine fossils are present in the sequence. For example, dunes can develop either on dry land or in a marine environment (**Fig. 6.26**) and might eventually be preserved as cross-bedded sedimentary rock.

Sedimentary structures are observed in a remarkable variety of types and forms that can provide hints to the environment that prevailed at the time a given sedimentary bed was deposited (**Fig. 6.25**). Raindrop impressions demonstrate a nonmarine environment. (It doesn't rain on the seafloor.) Mud cracks demonstrate drying that causes water-bearing clay minerals to shrink. Burrows and plant roots indicate what directions were "up" and "down," which is useful information when geoscientists are working with beds that have been deformed, tilted, or even overturned by later structural activity. And trackways can give us information about the organisms that made the tracks: their gait, their weight, whether they dragged a tail, and so on.

SEDIMENTARY STRUCTURES

ILLUSTRATIONS	DESCRIPTIONS	ENVIRONMENTS
	RAINDROP IMPRESSIONS: Tiny craters formed by raindrops as they impact bedding plane surfaces.	Raindrop impressions *occur on muddy land surfaces.*
	HORIZONTAL STRATA: Relatively flat *beds* (≥1 cm thick) and *laminations* (<1 cm thick).	Horizontal strata *occur where sediments settle from a standing body of water or air; or where currents travel parallel to the surface on which sediments are accumulating.*
	GRADED BED: Stratum that contains different sizes of sedimentary grains arranged from largest at the bottom of the bed to smallest at the top.	Graded beds *form when a turbulent body of water full of sediment (flood, wave, river) suddenly loses energy and calms down. Large particles settle out before small.*
	CURRENT RIPPLE MARKS: Asymmetrical ripple marks. The steep slope faces down current, and the gentle slope faces up current.	Current ripple marks *form in any environment where wind or water travels in one direction for some of the time: rivers, ocean currents, wind blowing sand dunes.*
	CROSS BEDDING: Inclined beds or laminations.	Cross bedding *forms wherever there are wind or water currents.*
	BIMODAL CROSS BEDDING: Sequence of cross bedding in which cross bedding of current ripple marks is inclined in opposite directions.	Bimodal cross bedding *forms in environments where currents of wind or water flow back and forth in opposite directions. It is common in environments with tides.*
	WAVE RIPPLE MARKS: Symmetrical ripple marks.	Wave (symmetrical) ripple marks *form in any body of water where gentle waves barely touch bottom, or where weak currents move back and forth (oscillate) in shallow water.*

Figure 6.25 Sedimentary structures.

SEDIMENTARY STRUCTURES

ILLUSTRATIONS	DESCRIPTIONS	ENVIRONMENTS

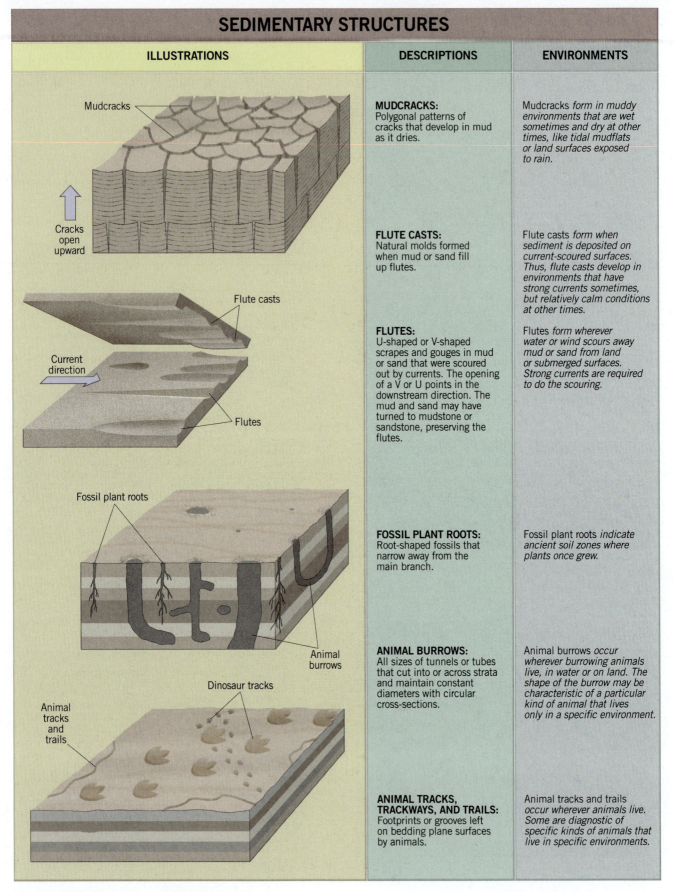

MUDCRACKS:
Polygonal patterns of cracks that develop in mud as it dries.

Mudcracks *form in muddy environments that are wet sometimes and dry at other times, like tidal mudflats or land surfaces exposed to rain.*

FLUTE CASTS:
Natural molds formed when mud or sand fill up flutes.

Flute casts *form when sediment is deposited on current-scoured surfaces. Thus, flute casts develop in environments that have strong currents sometimes, but relatively calm conditions at other times.*

FLUTES:
U-shaped or V-shaped scrapes and gouges in mud or sand that were scoured out by currents. The opening of a V or U points in the downstream direction. The mud and sand may have turned to mudstone or sandstone, preserving the flutes.

Flutes *form wherever water or wind scours away mud or sand from land or submerged surfaces. Strong currents are required to do the scouring.*

FOSSIL PLANT ROOTS:
Root-shaped fossils that narrow away from the main branch.

Fossil plant roots *indicate ancient soil zones where plants once grew.*

ANIMAL BURROWS:
All sizes of tunnels or tubes that cut into or across strata and maintain constant diameters with circular cross-sections.

Animal burrows *occur wherever burrowing animals live, in water or on land. The shape of the burrow may be characteristic of a particular kind of animal that lives only in a specific environment.*

ANIMAL TRACKS, TRACKWAYS, AND TRAILS:
Footprints or grooves left on bedding plane surfaces by animals.

Animal tracks and trails *occur wherever animals live. Some are diagnostic of specific kinds of animals that live in specific environments.*

Figure 6.25 Sedimentary structures. *(continued)*

Figure 6.26 Sand dunes. A. Quartz sand dunes in the Skardu Valley, Karakoram Range, Baltistan, around latitude 35.31°N, longitude 75.60°E. **B.** Submarine carbonate sand dunes near South Eleuthera Island, Bahamas, around latitude 24.95°N, longitude 76.42°W. (NASA photo ISS004-E-8777, March 2002.)

thousand meters below Earth's surface. Was the quartz sand in this sandstone once a part of a beach, a submarine sand bar, a sand dune in the desert, the point bar of a river, or a glacial outwash plain? Piecing together the environmental history of an area is interesting work all by itself, but it can also be essential to the exploration of water, energy, and mineral resources. If we are to make sense of environmental change in our time, it is useful to understand environmental change throughout Earth's far longer history.

Environmental reconstruction from the record contained in sedimentary rocks is a fun and challenging task. The composition and texture of sedimentary deposits is helpful but might not by themselves provide enough information for us to interpret the environment in which the sediment was deposited. For example, we think of beaches as being sandy, but sandy sediments are also deposited in deserts, by streams and elsewhere. And not all beaches are sandy—there are muddy beaches and boulder beaches as well. Factors such as the type of source area for the sediment, sediment supply, climate, topography, and vegetation can all play a role in determining the sedimentary environments present in a given place and time.

Sedimentary Environments

An important part of the interpretation of sedimentary rock involves understanding the environment in which any given sedimentary bed or sequence of beds was deposited (**Fig. 6.27**). Imagine finding a quartz sandstone in an outcrop of a core drilled from a couple of

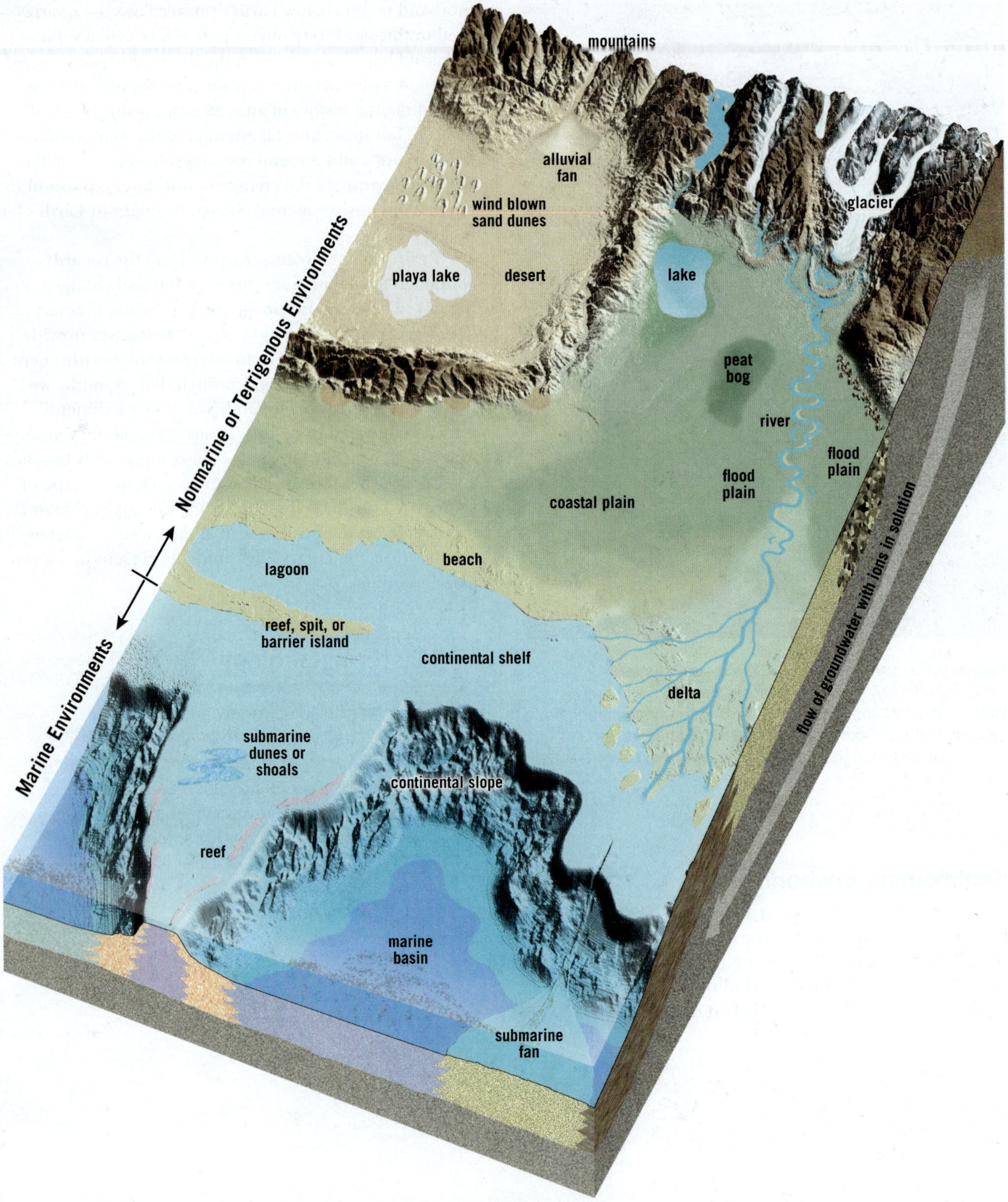
mountains
alluvial fan
wind blown sand dunes
glacier
playa lake
desert
lake
peat bog
river
flood plain
flood plain
coastal plain
Nonmarine or Terrigenous Environments
beach
lagoon
reef, spit, or barrier island
continental shelf
delta
flow of groundwater with ions in solution
Marine Environments
submarine dunes or shoals
continental slope
reef
marine basin
submarine fan

Figure 6.27 A selection of modern sedimentary environments.

Activity 6.1

Name: _____ **Course/Section:** _____ **Date:** _____

A Analyze the sedimentary rocks in **Fig. A6.1.1**, or actual rock samples of them if available. Below each photograph, describe briefly the rock's **composition** (what it is made of) and **texture** (the size, shape, and arrangement of its parts). Use your current knowledge, and complete the worksheet with your current level of ability.

1

Observation: does not fizz in acid

2

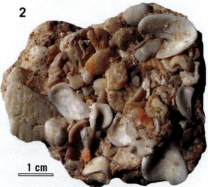

Observations: fizzes in acid; mostly shells

3

Observations: does not fizz in acid; intergrown crystals; salty

4

Observations: does not fizz in acid; very fine grained

5

Observation: fizzes in acid

6

Observations: does not fizz in acid; grains average ~0.5 mm diameter

Figure A6.1.1

B **REFLECT & DISCUSS** Reflect on your observations and descriptions of sedimentary rocks in part A. Then describe how you would classify the rocks into groups. Be prepared to discuss your classification with your classmates and teacher.

Name: _____ **Course/Section:** _____ **Date:** _____

A Look at **Fig. A6.2.1**. A rockfall from one of the steep granitic walls of Yosemite Valley (**A**) caused boulders as large as 2 meters in diameter to crash down into the forested slope below (**B**). Eventually, these sediments might end up in the steep channel of a local stream (**C**) on their way to the Merced River (**D**), which flows through the main part of Yosemite Valley.

~20 m

rockfall deposit

~200 m

~9 km

~1 m

Figure A6.2.1

1. **Rockfall deposit close to the source area** (Photo B).

   (a) List all of the grain sizes that you see or that are likely to be present in the rockfall deposit shown in photo **B** (also refer to **Fig. 6.2**). The largest sediment grain in **B** is approximately 2 meters long. Use the grain-size class names listed in **Fig. 6.15**.

(b) How would you describe the *sorting* of sedimentary grains in the rockfall deposit? Use the terms in **Fig. 6.16A**.

(c) How would you describe the *shape* of sedimentary grains you can see in the rockfall deposit? Use the terms in **Fig. 6.16B**.

(d) If the sediments in the rockfall deposit were lithified together as they currently rest without any further movement downslope, what kind of sedimentary rock would they form?

2. **Tributary stream just downslope from rockfall deposit** (Photo C).

   (a) The large sedimentary grains that can be seen in the stream channel are generally less than ~1.5 meters in diameter. Judging from the turbulence of the mountain stream, what grain sizes do you expect to be carried (suspended) in the water?

   (b) What grain sizes do you expect to be rolling, sliding, or resting on the bottom of the channel, including those that you can see?

   (c) How would you describe the *shape* of the sedimentary grains you can see in or near this stream channel (**Fig. 6.16B**)?

3. **River deposits exposed in the eroded bank of the Merced River** (Photo D).

   (a) List all of the grain sizes that you see or that are likely to be present in the Merced River bank shown in photo **D**.

   (b) How would you describe the *sorting* of sedimentary grains in the stream bank?

   (c) How would you describe the *shape* of sedimentary grains you can see in the stream bank?

(d) If the sediments in the stream bank were lithified, what kind of sedimentary rock would they form?

(e) How would you describe the change or evolution of sediments between the tributary streams and the main Merced River?

4. **Use your observations to make predictions.** The Merced River flows from Yosemite Valley at an elevation of ~1,200 meters above sea level in central Yosemite Valley to an elevation of ~250 m where the river enters Lake McClure: a reservoir in the San Joaquin Valley west of Yosemite Valley. The actual distance the river travels along its channel from Yosemite to the upper end of Lake McClure is more than 65 km. You can examine the Merced River course between latitude 37.72°N, longitude 119.63°W (central Yosemite Valley) and 37.602°N, 120.100°W (inlet to Lake McClure on the Merced River) using Google Earth.

(a) What sedimentary grain sizes are likely to be deposited in Lake McClure from the erosion of Yosemite Valley?

(b) What do you think will be the composition of most of the sedimentary grains deposited in Lake McClure from Yosemite Valley?

B **REFLECT & DISCUSS**   Based on your work, write a brief description of how the clastic sediment from Yosemite Valley might change as it travels downstream to Lake McClure. Then describe how you could use these insights to interpret clastic rocks in general.

Activity 6.3

Name: _____ Course/Section: _____ Date: _____

A Obtain two pieces of granite or diorite. Hold one in each hand and tap them together over a piece of paper. As you do this, you should notice that you are breaking tiny sedimentary grains from the larger rock samples. These broken pieces of rocks and minerals are called **clasts** (from the Greek *klastós*, meaning "broken in pieces").

1. Using a hand lens or microscope, observe the tiny clasts that you just broke from the larger rock samples. Describe what minerals make up the clasts and whether or not the clasts are fragments of mineral crystals, rock fragments, or a mixture of both.

2. Geologists commonly refer to several different kinds of clastic sediment. Circle the one that you just made.

   ▪ **pyroclastic sediment** or **tephra**—volcanic bombs, lapilli, and ash blasted into the atmosphere by volcanic eruption
   ▪ **bioclastic sediment**—broken pieces of shells, plants, and/or other parts of organisms
   ▪ **siliciclastic sediment**—broken pieces of silicate mineral crystals and/or rocks containing them

3. **Roundness** is a general description of how completely the rough edges and points of a clast have been smoothed by banging into other particles during transport (see chart in **Fig. A6.3.1**). Re-examine your clasts from part **A1** and sketch the outline of several of them. Compared to the chart, what is the roundness of the clasts that you sketched?

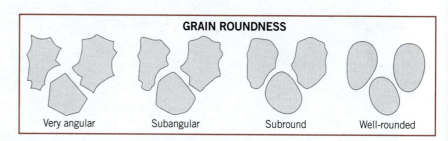

GRAIN ROUNDNESS

Very angular   Subangular   Subround   Well-rounded

A6.3.1

4. Using **Fig. 6.15** or a grain-size scale (from GeoTools 1 or 2 at the back of your manual), circle the Wentworth size classes of the clastic sediment that you made above.

   clay   silt   sand   gravel

5. Obtain a piece of quartz or garnet sandpaper and lay it flat on the table. Find a sharp corner on one of the granite/diorite samples that you used above and sketch its outline in the "before abrasion" box in **Fig. A6.3.2**. Next, rub that corner against the quartz sandpaper for about 10 seconds. Sketch its profile in the "after abrasion" box. What did this abrasion process do to the sharp corner?

Before abrasion   After abrasion

A6.3.2

173

6. The sediment that you just made by wearing down the corner of a rock clast is called **clastic sediment**. The Mississippi River carries clastic sediment that has been weathered and eroded from the landscape of much of the central United States. The river discharges its load of water and sediment through the Mississippi Delta into the Gulf of Mexico. As shown in the satellite images in **Fig. A6.3.3**, the Mississippi River passes by Lake Pontchartrain (LP) and the city of New Orleans (NO) on its way to the great "bird's foot" delta that extends into the gulf of Mexico. The detailed view of the bird's foot delta is a true-color image acquired using NASA's Landsat 7 satellite and shows the broad main channel of the Mississippi River discharging water and sediment to more than half a dozen smaller channels that extend in many different directions. The center of the larger image is around latitude 29.16°N, longitude 89.16W.

Figure A6.3.3

On the image in **Fig. A6.3.3**, write "D" to indicate all of the places where one of these smaller channels is discharging water and sediment out into the Gulf of Mexico. One example is done for you, marked with a white D near the top of the image.

B **REFLECT & DISCUSS** How do you think the roundness of sediment in the river in **Fig. A6.3.3** changes from near the source areas throughout the continent to the locations where you placed the "D"s on the image?

Name: _____ Course/Section: _____ Date: _____

A Seashells are formed through the biochemical processes of organisms, and eventually become sedimentary grains. When you find a rock with a fossil of a marine organism, you have found evidence that the rock contains sediment deposited in a marine environment. Some limestone is entirely made of shells or broken pieces of shells.

1. Obtain a seashell (e.g., a hard clam shell) and draw it below this paragraph. It may be easiest to trace it and then fill in the outline with details of what the shell looks like inside or out.

2. Next, place the shell into a plastic sandwich bag. Lightly tap the bag with a hammer to break the shell into pieces, taking care not to damage the surface that the bagged shell is on. Examine the broken pieces of shell with a hand lens. The shell fragments that you just made are called **clasts** (from the Greek *klastós*, meaning "broken"). Geologists commonly refer to several different kinds of clastic sediment. Circle the one that you just made.

   ▪ **pyroclastic sediment** or **tephra**—volcanic bombs, lapilli, and ash blasted into the atmosphere by volcanic eruption
   ▪ **bioclastic sediment**—broken pieces of shells, plants, and/or other parts of organisms
   ▪ **siliciclastic sediment**—broken pieces of silicate mineral crystals and/or rocks containing them

3. Compared to **Fig. 6.16B**, what is the roundness of your clasts? _____

4. What is the roundness of the clasts in **Fig. A6.4.1**? _____

   Explain how and in what environment the shell clasts could have attained their roundness.

Figure A6.4.1

5. Some limestone is made of shells that are calcareous (calcite or aragonite), but they are microscopic. Chalk is such a limestone. Some "chalk" used to mark sidewalks or blackboards is actually made of clay or plaster of paris rather than real chalk. Obtain a piece of chalk from your lab room or instructor. Explain how dilute hydrochloric acid can be used to help you test your chalk and find out if it is real chalk or not. Then conduct your test and report its results.

B Place a charcoal briquette into a plastic sandwich bag. Lightly hammer the bag enough to break apart the briquette, taking care not to damage the surface that the bagged briquette is on.

1. Examine the broken pieces of charcoal with a hand lens. Describe what kinds of grains you see and their texture.

2. Charcoal is made by allowing wood to smolder just enough that an impure mass of carbon remains. In the presence of oxygen, the charcoal briquette will naturally combine with oxygen to make carbon dioxide. Over a period of many years, it will all react with oxygen and chemically weather to carbon dioxide. When you burn charcoal in your grill, you are simply speeding up the process. However, if plant fragments (peat) are buried beneath layers of sediment that keep oxygen away from them, then they can slowly convert to a charcoal-like material called coal, which is stable for millions of years. Obtain a piece of coal and compare it to your charcoal. How is it different? Why?

C **REFLECT & DISCUSS**   Shells and the skeletons of microscopic organisms like plankton are made of precipitated mineral material that they secrete. Coal is made of altered plant material: leaves, grasses, wood, and so on. Lignite alters to sub-bituminous, bituminous, or anthracite coal due to increasing pressure and temperature. Do you think coal is a rock? If you do, which of the three main types of rock would you classify it as?

Sediment Analysis, Classification, and Interpretation

Name: _____ Course/Section: _____ Date: _____

A Complete parts **1** through **6** for each sample in **Fig. A6.5.1**. Refer to **Figs. 6.15**, **6.16**, and **6.27** as needed.

Sample A   1 mm

ooids

1. Grain size range in mm: _____

2. Percent of each Wentworth size class:
   clay _____ silt _____ sand _____ gravel _____

3. Grain sorting (circle):
   poor   moderate   well

4. Grain roundness (circle):
   angular   subround   well rounded

5. Sediment composition (circle):
   precipitated   siliciclastic   bioclastic   pyroclastic

6. Describe how and in what environment this sediment might have formed.

Sample B   2 mm

1. Grain size range in mm: _____

2. Percent of each Wentworth size class:
   clay _____ silt _____ sand _____ gravel _____

3. Grain sorting (circle):
   poor   moderate   well

4. Grain roundness (circle):
   angular   subround   well rounded

5. Sediment composition (circle):
   precipitated   siliciclastic   bioclastic   pyroclastic

6. Describe how and in what environment this sediment might have formed.

Sample C   10 mm

1. Grain size range in mm: _____

2. Percent of each Wentworth size class:
   clay _____ silt _____ sand _____ gravel _____

3. Grain sorting (circle):
   poor   moderate   well

4. Grain roundness (circle):
   angular   subround   well rounded

5. Sediment composition (circle):
   precipitated   siliciclastic   bioclastic   pyroclastic

6. Describe how and in what environment this sediment might have formed.

Figure A6.5.1

B **REFLECT & DISCUSS** Imagine that these sediments are rocks. Which of the samples do you think would be the least diagnostic of a specific ancient environment? Why?

Activity 6.6 Hand Sample Analysis and Interpretation

Name: ____________ Course/Section: ____________ Date: ____________

SEDIMENTARY ROCKS WORKSHEET							
Sample number or letter	Does matrix fizz in acid?	Is matrix made of microscopic grains?	Is matrix a mass of intergrown crystals, or is it clastic?	What is the grain size class of most particles (See Fig. 6.15)	What are the grains composed of? (e.g., calcite, quartz, clay, feldspar, rock fragments, fossils, ooids, evaporites, pyroclasts)	Assign a provisional rock name.	Where might this sediment have been deposited/precipitated? (See Fig. 6.27)
6.4	yes	no	clastic (bioclastic)	sand; many of the fossils are small-gravel sized	matrix (sand) is probably bioclastic calcite, abundant marine fossils made of carbonate	fossiliferous limestone	it might have been deposited on the seafloor

Figure A6.6.1

Name: ______________________________ Course/Section: ________________ Date: __________

SEDIMENTARY ROCKS WORKSHEET

Sample number or letter	Does matrix fizz in acid?	Is matrix made of microscopic grains?	Is matrix a mass of intergrown crystals, or is it clastic?	What is the grain size class of most particles (See Fig. 6.15)	What are the grains composed of? (e.g., calcite, quartz, clay, feldspar, rock fragments, fossils, ooids, evaporites, pyroclasts)	Assign a provisional rock name.	Where might this sediment have been deposited/precipitated? (See Fig. 6.27)

Figure A6.6.1 *(continued)*

Name: ______________________ Course/Section: ______________ Date: __________

SEDIMENTARY ROCKS WORKSHEET

Sample number or letter	Does matrix fizz in acid?	Is matrix made of microscopic grains?	Is matrix a mass of intergrown crystals, or is it clastic?	What is the grain size class of most particles (See Fig. 6.15)	What are the grains composed of? (e.g., calcite, quartz, clay, feldspar, rock fragments, fossils, ooids, evaporites, pyroclasts)	Assign a provisional rock name.	Where might this sediment have been deposited/precipitated? (See Fig. 6.27)

Figure A6.6.1 *(continued)*

Name: ______________________ Course/Section: ______________ Date: __________

SEDIMENTARY ROCKS WORKSHEET

Sample number or letter	Does matrix fizz in acid?	Is matrix made of microscopic grains?	Is matrix a mass of intergrown crystals, or is it clastic?	What is the grain size class of most particles (See Fig. 6.15)	What are the grains composed of? (e.g., calcite, quartz, clay, feldspar, rock fragments, fossils, ooids, evaporites, pyroclasts)	Assign a provisional rock name.	Where might this sediment have been deposited/precipitated? (See Fig. 6.27)

Figure A6.6.1 *(continued)*

Activity 6.7 — Grand Canyon Outcrop Analysis and Interpretation

Name: _____ **Course/Section:** _____ **Date:** _____

A Analyze the images in **Fig. A6.7.1**, from the South Rim of the Grand Canyon near Grand Canyon Village. The edge of the canyon here is formed by a fossiliferous limestone composed of sand-sized clastic grains, called the Kaibab Limestone, deposited about 270 million years ago during the Permian Period.

Figure A6.7.1

1. Notice that some of the beds in the outcrop are cross-bedded. Draw an arrow on the picture to show the direction that the water moved here to make this cross bedding. Refer to **Fig. 6.25** as needed.

2. Does this cross bedding indicate a steady flow of air or water, or does it indicate an oscillating (back and forth) flow? (**Fig. 6.25**) _____

B **REFLECT & DISCUSS** Describe your ideas about what the environment might have been like here about 270 million years ago (e.g., **Figs. 6.26** and **6.27**) and the evidence and logic that you used to reach your conclusion.

Using the Present to Imagine the Past—Dogs and Dinosaurs

Activity 6.8

Name: _____ Course/Section: _____ Date: _____

A Analyze photographs X and Y in **Fig. A6.8.1**.

X. Modern dog tracks in mud with mudcracks on a tidal flat, St Catherines Island, Georgia

Y. Triassic rock (about 215 m.y. old) from southeast Pennsylvania with the track of a three-toed *Coelophysis* dinosaur

Figure A6.8.1

1. How are the modern environment (Photograph X) and Triassic rock (Photograph Y) the same?

2. How are the modern environment (Photograph X) and Triassic rock (Photograph Y) different?

3. Describe the environment in which *Coelophysis* lived about 215 million years ago in what is now Pennsylvania.

B **REFLECT & DISCUSS** Use what you learned about sediment and sedimentary rocks. Develop a hypothesis about how the dinosaur footprint in Photograph Y was preserved.

A Analyze photographs X and Y in **Fig. A6.9.1** of a Kansas rock and the modern-day seafloor near Cape Cod, respectively.

X. Pennsylvanian-age rock from Kansas (290 Myr old)

Sand-sized fragments of fossil shells comprise the rock

10X close-up of thin section

Y. Modern seafloor environment, 40 m deep, ~16 km north of Cape Cod, Massachusetts.

Photo includes clastic sediment from the continent, bioclastics, and living organisms.

- 1% gravel
- 90% sand
- 9% mud

Figure A6.9.1

1. How are the modern environment (Photograph Y) and Kansas rock (Photograph X) the same?

2. How are the modern environment (Photograph Y) and Kansas rock (Photograph X) different?

3. Today, this part of Kansas is rolling hills and farm fields. Describe the environment in which the sediment in this rock sample (Photograph X) was deposited there about 290 million years ago.

B **REFLECT & DISCUSS** What would have to happen to the sediment in Photograph X to turn it into sedimentary rock?

Metamorphic Rocks, Processes, and Resources

Pre-Lab Video 7

http://goo.gl/KEp1Md

Contributing Authors

Harold E. Andrews • *Wellesley College*

James R. Besancon • *Wellesley College*

Margaret D. Thompson • *Wellesley College*

▲ Folds in foliated metamorphic rock near the terminus of the Engabreen outlet glacier of the Svartisen Ice Cap, Norway (~66.688°N, 13.755°E)

BIG IDEAS

Metamorphic rock has changed to a new and different form while in the solid state—without melting. Metamorphic change occurs because of increased heat, pressure, differential stress, or hot water. The mineralogy and texture of a metamorphic rock can be used to interpret its geologic history. This provides us with a glimpse of structural and tectonic events in the past, and helps us understand current processes occurring beyond our direct observation below Earth's surface. If the rock has undergone less intense metamorphism, we can often infer what the rock was like before metamorphism. Metamorphic rocks are widely used in the arts and construction industries and are sources of industrial minerals and energy.

FOCUS YOUR INQUIRY

Think About It   What do metamorphic rocks look like? How can they be classified into groups?

ACTIVITY 7.1   Metamorphic Rock Inquiry (p. 186, 197)

Think About It   What is metamorphic rock made of?

ACTIVITY 7.2   Minerals in Metamorphic Rock (p. 190, 198)

Think About It   How are rock composition and texture used to classify, name, and interpret metamorphic rocks?

ACTIVITY 7.3   Metamorphic Rock Analysis and Interpretation (p. 191, 199)

ACTIVITY 7.4   Hand Sample Analysis, Classification, and Origin (p. 191, 201)

Think About It   What can metamorphic rock tell us about Earth's history and the environments in which the rock formed?

ACTIVITY 7.5   Metamorphic Grades and Facies (p. 196, 204)

Introduction

The word *metamorphic* is derived from Greek and means "of changed form." *Metamorphic rock* changes from one form to another, or is metamorphosed, in response to changing environmental conditions around the rock. The change results in a rock that is more fully at equilibrium with those new conditions. Metamorphism occurs at **temperatures** above ~200°C and extends upward to the melting point of the rock (**Fig. 7.1**). Another important environmental variable is **pressure**, which is a system of forces that is has the same magnitude in all directions. Most metamorphic rock has a layering called **foliation** that develops during metamorphism because of **differential stress**, about which we will learn more in an upcoming section. Hot water (**hydrothermal fluid**) plays an important role in moving ions and in facilitating mineral changes. And **time** is important for most metamorphic processes other than lightning strikes and meteorite impacts because great change can occur to rock over long time periods.

Metamorphic change in rock is broadly similar to changes in food caused by cooking. *Heat* can be used to 'metamorphose' an egg from its fluid raw state into a firm fried egg through contact with a hot frying pan. *Pressure* within a pressure cooker can be used to cook potatoes faster. *Differential stress* can be used to crush an aluminum can into a flatter and more compact form. The chemical action of *hot water and steam* can be used to change raw vegetables into cooked forms. Inside Earth, all of these environmental conditions are more intense and capable of changing a rock from one form to another. The intensity of metamorphism varies from low-grade metamorphism at temperatures above ~200°C to high-grade metamorphism at temperatures just below the melting point of the rock. Metamorphism can change the composition, size, and shape of mineral grains as well as the texture, density, and color of the rock they are a part of.

Every metamorphic rock has a parent rock or **protolith**—the rock type that was metamorphosed. Protoliths can be any of the three main rock types: igneous rock, sedimentary rock, or even metamorphic rock subjected to another round of metamorphism. It is generally possible to interpret the protolith of low-grade metamorphic rock because changes in texture and mineralogy are less profound than in high-grade metamorphic rock. A meta-conglomerate looks a lot like the sedimentary conglomerate from which it formed. The greater the intensity of metamorphism, the more difficult it is to determine a protolith if it can be determined at all.

ACTIVITY 7.1

Metamorphic Rock Inquiry, (p. 197)

Think About It What do metamorphic rocks look like, and how can they be classified into groups?

Objective Analyze and describe samples of metamorphic rock, and then infer how they can be classified into groups.

Before You Begin Read the following sections: The Metamorphic Environment and Grain-Scale Metamorphic Processes.

The Metamorphic Environment

Metamorphism is driven by changes in the environment in which a given volume of rock exists. These environmental changes primarily involve an increase in temperature, pressure, differential stress, and exposure to hydrothermal fluids. As the environment changes, the texture or mineralogy of the rock changes so that it can approach or achieve equilibrium with the new conditions. An increase in temperature or pressure is likely to result in development of minerals that are more stable at the higher temperature or pressure. The defining characteristic of metamorphism is that the changes to the rock occur in the solid state without melting.

Temperature and Metamorphic Grade

Metamorphism occurs at temperatures above ~200°C and extends to the temperature at which the rock melts (**Fig. 7.1**). On the cooler end of this range, metamorphism and **diagenesis** (changes that occur in sedimentary rock after burial) overlap with one another. Geoscientists refer to metamorphic change at relatively low temperatures as **low-grade metamorphism**, and at relatively high temperatures, it is called **high-grade metamorphism**. Some geoscientists **include pressure** in the concept of metamorphic grade because temperature and pressure work together

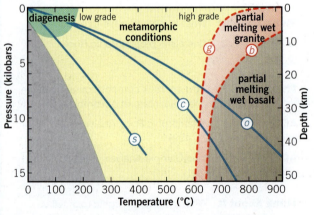

Figure 7.1 Pressure-temperature conditions for metamorphism. Metamorphic conditions (yellow area) extend from ~200°C at shallow depths to conditions that cause partial melting. P-T conditions to start partial melting of wet granite are below red dashed curve *g*; for wet basalt they are below curve *b*. Blue curves are average change in pressure and temperature in oceanic (*O*) or continental (*C*) crust away from a plate boundary, or in a subduction zone (*S*).

to produce metamorphic change, but the International Union of Geological Sciences (IUGS) recommends defining grade based solely on temperature. High temperatures result in high-grade metamorphism.

There is also a region of environmental overlap between the most extreme metamorphism and igneous processes. Compared at the same depth or pressure, different rock types begin to melt at different temperatures. For example, granite begins to melt at cooler temperatures than basalt (**Fig. 7.1**). **Partial melting** involves melting of the types of minerals that have the lowest melting point first, leaving the other minerals in a solid state. Partial melting is an igneous process that results in igneous rock when that magma cools; however, the part of the rock that did not melt is metamorphic rock. A rock that has partially melted and contains both igneous and metamorphic features is called a **migmatite**.

Pressure

In addition to temperature, pressure exerts a primary control on metamorphism. **Pressure** is a system of forces acting on an object with the same magnitude of force in all directions (**Fig. 7.2**). The magnitude of pressure is expressed either in kilobars or in pascals. One **kilobar** is approximately 1000 times the average atmospheric pressure at sea level, but it is *exactly* equal to 100 million pascals. Both the kilobar and the pascal are units of stress, which is defined as a force applied over an area. One **pascal** (Pa) is defined as equal to 1 newton of force applied over 1 square meter of area. How much is 1 newton of force? If you hold a small apple by its stem in Earth's gravitational field, the apple exerts about 1 newton of force through its stem to your fingers. [That is, 1 **newton** (N) is the amount of force needed to accelerate a 1-kilogram (kg) mass by 1 meter (m) per second squared (m/s^2): $1\ N = 1\ kg\ m/s^2$. This is nothing to worry about. Just breathe normally.] Because 1 N of force spread over a square meter of area [i.e., 1 Pa] is a very small stress, geoscientists typically work in units of a thousand

pascals (a kilopascal, or kPa), 1 million pascals (a **megapascal**, or MPa), or 1 billion pascals (a **gigapascal**, or GPa).

What is the pressure exerted by the water in the pool on a 1 square meter (m^2) area of the bottom of the pool? Water has a density of ~1000 kg per cubic meter (m^3). The bottom of a typical swimming pool is approximately 2.4 m deep. The mass of a column of water that is 1 m by 1 m by 2.4 m is (~1000 kg/m^3) × (2.4 m^3) = ~2400 kg. The acceleration of gravity at Earth's surface is typically around 9.8 m/s^2. Multiplying the mass times the acceleration gives us (2400 kg) × (9.8 m/s^2) = 23,520 kg m/s^2 = 23,520 N, or 23.52 kN. That force is applied across a 1 m^2 area at the bottom of the pool, exerting a stress of 23.52 kPa.

Changes in pressure, if applied to a material like a rubber ball that is just as strong in all directions, will result in a change in *size* but not a change in *shape*. A spherical rubber ball subjected to an increase in pressure will become a smaller spherical rubber ball. If we intend to change the shape of an object that is equally strong in all directions, we need to do something other than just changing the pressure.

Differential Stress

Many metamorphic rock types display a distinctive layering called **foliation** that develops because of differential stress during metamorphism. **Differential stress** is a system of forces that is not the same magnitude in all directions. The word *differential* refers to a difference in the magnitude of stress in different directions. A differential stress at a given point in Earth's crust will have the greatest compressive stress in one direction, and 90° away from that direction will be the least compressive stress. A differential stress applied to a rubber ball will cause it to shorten in the direction of the greatest compressive stress and maybe bulge outward in the direction of the least compressive stress (**Fig. 7.2**). In most foliated metamorphic rock, the foliation develops approximately perpendicular to the greatest compressive stress.

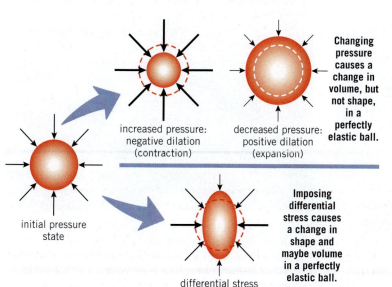

Figure 7.2 Effects of pressure and differential stress on an elastic ball. Pressure is a system of forces the same magnitude in all directions, while differential stress system involves force magnitudes that are different in different directions. The greatest shortening of an elastic ball occurs along the direction of greatest compressive stress, and the direction of least compressive stress is always 90° from the greatest stress.

initial pressure state

increased pressure: negative dilation (contraction)

decreased pressure: positive dilation (expansion)

Changing pressure causes a change in volume, but not shape, in a perfectly elastic ball.

differential stress

Imposing differential stress causes a change in shape and maybe volume in a perfectly elastic ball.

Hydrothermal Fluids

Metamorphic processes in the crust can be strongly affected by the presence of water. Earth is called "the water planet" because this remarkable substance is very common throughout the upper layer of the solid Earth and covers ~71% of Earth's surface. Liquid water is able to flow easily through very small spaces and is attracted to grain surfaces because one side of the water molecule has a slight positive charge and the other a slight negative charge. More chemicals dissolve in water than in any other solvent.

The rate at which that dissolution occurs increases with increasing temperature, so hot water (or, as geoscientists refer to it, **hydrothermal fluid**) is able to break up ionic solids into their component ions more rapidly than cold water. (That's why it is more effective to dissolve a powdered drink mix in a little bit of hot water than to try to dissolve it in ice water. Once the powder is dissolved, you can add ice and cold water to finish making the cold drink.) Hydrothermal fluids moving through rock are able to dissolve and transport ions in solution, changing the chemical composition of the rock.

Time

Enormous change can occur in rock due to the combined effects of very slow changes over a very long time. The island nation of Iceland is located along the mid-Atlantic ridge where the North American Plate moves away from the Eurasian Plate at a rate of around 18.5 $\pm$ 0.4 mm per year (about 3/4 of an inch per year). At about that same very slow rate, the entire North Atlantic Ocean Basin has developed during the past 200 million years. Planet Earth is about 23 times as old as the North Atlantic, so there has been more than enough time for very slow processes to produce dramatic changes at all scales, from plates to mineral grains. In typical metamorphic processes, the size and shape of a mineral grain can change slowly by moving one or a few atoms at a time, breaking and re-establishing chemical bonds with energy supplied by heat and stress. You may be surprised to learn that lightning strikes and meteorite impacts can produce metamorphic rocks very quickly, but those rapid changes require an enormous amount of energy.

Protolith

The chemical composition of a metamorphic rock is determined by the chemical composition of the protolith, plus or minus ions that were introduced or leached away by hydrothermal fluids. Any type of rock—sedimentary, igneous, metamorphic—can be a protolith.

Grain-Scale Metamorphic Processes

Some of the most important processes in metamorphism occur at the grain scale or smaller. In this lab, we see the end product of these processes in the textures and mineralogy of the different types of metamorphic rock. Here, we will touch on just a few of the grain-scale processes that make these rock types distinctive.

Changing Grain Size: Recrystallization

Metamorphism supplies the energy needed to change the size of mineral grains in order to reduce the total amount of energy stored in a rock. Excess energy can be stored in the form of distorted crystal lattices or as surface energy along complicated grain boundaries. Given time and sufficient temperature, the distortions in crystal lattice can be healed to recover an unstrained lattice, and the grain boundaries can adjust to minimize the surface area. This process is called **recrystallization**. These changes occur in the solid state by moving one atom (or a few atoms) at a time. The observed effect of these processes is that the average size of grains in a metamorphic rock can change, and the grain boundaries can evolve into simpler shapes. The simplest grain-boundary configuration mirrors that of a cluster of soap bubbles that are all about the same size, which is called a **granoblastic** texture. If recrystallization occurs under a low differential stress, the recrystallized grain size tends to be larger than if it occurs under a high differential stress.

A common example of recrystallization involves the change from the sedimentary rock *limestone* to the metamorphic rock *marble* (**Fig. 7.3**). Prior to metamorphism, the limestone protolith might consist of calcite grains that are too small to see. As temperature and pressure increase, the energy trapped in the small calcite grains makes that an unstable configuration. New, larger, unstrained calcite grains grow at the expense of the old, smaller, strained grains, resulting in a coarser-grained rock that has much less energy stored in its grains. Recrystallization under metamorphic conditions changes limestone into the metamorphic rock marble—a change that does not necessarily involve any change in the chemistry or mineralogy of the grain.

Figure 7.3 Marble. Marble is a fine- to coarse-grained, nonfoliated metamorphic rock with a crystalline texture formed by tightly interlocking grains of calcite or dolomite. The protolith of marble is a carbonate sedimentary rock: limestone or dolostone.

Making New Minerals: Neomineralization

Each type of mineral has a range of temperature and pressure conditions within which it is stable. **Neomineralization** involves the nucleation and growth of new minerals under metamorphic conditions. It can be thought of as a process of recycling atoms to form new minerals that are stable at the prevailing pressure-temperature (P-T) conditions. For example, the aluminum, silicon, and oxygen in a clay mineral under low P-T conditions might become part of a garnet at higher P-T conditions under which the clay mineral is not stable. New minerals formed in the solid state during metamorphism are called **neoblasts**.

Changing Rock Chemistry: Metasomatism

Recrystallization and neomineralization can occur without any change in the overall chemistry of a rock. In contrast, **metasomatism** involves the transport of atoms into and out of a volume of rock carried by hydrothermal fluids. Hence, metasomatism results in a change in rock chemistry while the rock remains in the solid state. For example, *skarn* is a metasomatic rock formed by alteration of a carbonate rock that was intruded by magma.

Dissolving Minerals Through Differential Stress

The bonds that hold the outermost atoms in a crystal lattice to the rest of the mineral along its grain boundary can be broken if a sufficient amount of differential stress is applied to that area of the lattice, in the presence of a film of water (**Fig. 7.4**). The atoms move from the high-stress region to lower-stress regions where they might precipitate to form a grain coating or fill a crack, or they might simply remain in solution and be carried out of that part of the rock. The high-stress grain boundary that lost the atoms evolves, developing a larger surface area of contact with the adjacent grain across their common boundary. If the process continues, the entire grain can be dissolved, and the total volume of the rock can be reduced.

This process of moving atoms from high-stress regions to low-stress regions in the presence of water is called **wet diffusion** or **pressure solution**. (Pressure solution is a misleading name because the process depends on *differential stress* rather than pressure.) Wet diffusion is very common in limestone at submetamorphic (diagenetic) conditions and is an important process affecting quartz and feldspar in the development of slate under low-grade metamorphic conditions.

Changing the Shape of Mineral Grains: Crystal Plasticity

The set of mechanisms by which the shape and size of a mineral grain can change in the solid state without melting and without fracturing or faulting is called **crystal plasticity**. Research into crystal plasticity provides us with two

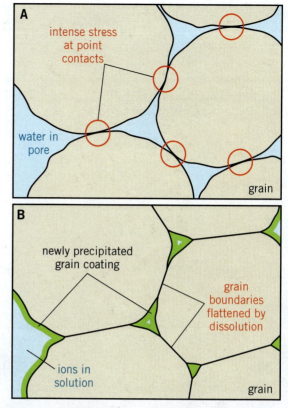

Figure 7.4 Wet diffusion (pressure solution). A. The weight of all the rock above is concentrated where grains are in contact with each other, causing intense local stress. **B.** Water coating the grains facilitates dissolution and migration of ions away from the high-stress areas to low-stress places where new minerals can precipitate. This leads to flattening of grain contacts, increasing the area of grain contact and decreasing stress along the contacts.

relevant generalizations. The first is that *the shape and size of a mineral grain can change in response to differential stress.* This change can be accomplished while maintaining the geometry of that mineral's crystal lattice. That is, the lattice is not stretched in one direction and shortened in another like a rubber ball, but rather the lattice grows in one direction and is reduced in the other through the motion of one atom (or a few atoms) at a time. The second result is that *many of the mechanisms of crystal plasticity work to produce change in a rock faster at higher temperatures.* As a result, rock becomes much weaker at temperatures that approach its melting point.

We can find deformed structures in metamorphic rock that can be very helpful in working out the geologic history of a region. These structures often occur in rock that is very hard and dense today but that looks as if it had been very soft at some earlier time. For example, the light-colored layers in the rock in **Fig. 7.5A** were likely to have been approximately planar layers before deformation but are now folded. The light and dark parts of the rock in **Fig. 7.5B** resemble soft vanilla and chocolate ice cream that has been swirled together. The strength of rock under high-grade metamorphic conditions can be much less than it is at ground-surface temperature and pressure.

Figure 7.5 Ductile deformation. Under high-temperature conditions, minerals can become quite weak and can be deformed in the solid state without melting. This solid-state flow is often called ductile deformation. **A.** This folded high-grade metamorphic rock seems to have been very soft at the time the folding occurred. **B.** Ductile fault rock (mylonite) showing wispy white areas of feldspar, red garnets, and dark foliations of biotite. (Collected from the Nanga Parbat Gneiss in the Karakoram Range of northern Pakistan.)

ACTIVITY 7.2

Minerals in Metamorphic Rock,
(p. 198)

> **Think About It** What is metamorphic rock made of?

Objective Be able to identify some of the more important minerals that are recognizable in hand samples of metamorphic rock.

Before You Begin Read the following section: Some Important Minerals in Metamorphic Rock.

Some Important Minerals in Metamorphic Rock

A very wide variety of minerals are found in metamorphic rock, including familiar minerals like calcite, quartz, various feldspar minerals, hornblende, and the mica minerals biotite and muscovite. The sheet silicates or **phyllosilicates** are an important group of minerals that help define the layering or foliation that develops because of differential stress during metamorphism. Geologists can identify the common metamorphic minerals in the following list by physical properties such as their color, shape, cleavage, luster, and hardness. (For more information about these minerals, refer to the Minerals Database in the chapter about minerals.)

- **Muscovite** is a mica mineral with one perfect cleavage breaking into very thin, flexible sheets. Individual sheets are clear and transparent. Thicker grains of muscovite composed of many sheets are light brown to silver color with a very shiny luster on the cleavage face.
- **Biotite** is a mica mineral that separates into very thin, flexible sheets because of its one perfect cleavage. Individual sheets are dark brown and transparent. Thicker biotite grains are dark brown to black color with a very shiny luster on the cleavage face.
- **Chlorite** is green to blue-green, has one perfect cleavage, and separates into thin flexible sheets.
- **Talc** is white to green color and has one perfect cleavage. It is the softest of the minerals on the Mohs hardness scale, is easily scratched with a fingernail, and has a soapy feel.
- The **serpentine minerals lizardite** and **antigorite** are both green, blue-green to white minerals with one perfect cleavage. Both are soft—lizardite is 2.5 and antigorite 2.5–3.5 on the Mohs hardness scale. The third serpentine mineral, **chrysotile**, is a fibrous asbestos mineral.

Some minerals are considered index minerals that help define metamorphic zones. The set of **metamorphic index minerals** includes the phyllosilicates **chlorite** and **biotite** as well as the following:

- **Garnet** occurs in round crystals that look a bit like soccer balls with at least 12 faces but no cleavage. Garnet is hard (H 6.5–7.5), seems heavier than average (specific gravity 3.5–4.3), is translucent to transparent, and has a vitreous (glassy or chinalike) luster. The color varies with varying composition among its six primary varieties.
- **Staurolite** crystals have a bladed form, are very hard (H 7–7.5), and are typically red-brown to brownish-black in color. Staurolite crystals are commonly twinned with blades that cross at ~60° or 90°.
- **Kyanite** has distinctive blue-gray bladed crystals with one good cleavage direction. Kyanite has a hardness of between 5 and 7—its hardness is different in different directions along the crystal.
- **Sillimanite** commonly occurs in long, slender, or fibrous crystals and is usually colorless or white although it can be yellow, brown, or green. Sillimanite has a vitreous (glassy or chinalike) luster.

ACTIVITY 7.3

Metamorphic Rock Analysis and Interpretation, (p. 199)

> **Think About It** How are rock composition and texture used to classify, name, and interpret metamorphic rocks?

Objective Be able to describe and interpret textural and compositional features of metamorphic rocks.

Before You Begin Read the following section: Metamorphic Rock Types Defined by Texture.

ACTIVITY 7.4

Hand Sample Analysis, Classification, and Origin, (p. 201)

Objective Use the texture and composition of common metamorphic rocks to interpret the rock type and possible protolith of each.

Plan Ahead Your teacher will provide you with rock (and maybe mineral) specimens or photos of specimens.

Before You Begin Read the following sections: Metamorphic Rock Types Defined by Texture and Classification Based on Composition or Context.

Metamorphic Rock Types Defined by Texture

The IUGS recognizes three major subdivisions of metamorphic rock based on texture without reference to mineral composition: **schist**, **gneiss**, and **granofels**. Schist and gneiss display a layering called **foliation** that can involve long or flat mineral grains arranged along the foliation surface, like pencils and pieces of paper on a tabletop. Foliation can also involve rock formed of one main mineral like quartz or calcite, but all of the mineral grains are flattened in the same direction. Foliation develops as a result of differential stress during metamorphism. Granofels are rocks whose mineral grains lack a strong (or any) foliation (**Fig. 7.3**).

The foliation in schists is called **schistosity** and involves the approximately parallel arrangement of long or flat mineral grains in layers throughout the rock so that the rock will break parallel to the layering on a scale of 1 cm or less. In common usage, the schists are subdivided by grain size into **slate** with its microscopic grains (**Fig. 7.6**), **phyllite** with very small (aphanitic) grains (**Fig. 7.7**), and **schist** with phaneritic grains (**Fig. 7.8**). The flat mineral grains in schist are phyllosilicates.

Foliation in gneiss is often called **gneissic layering** or **gneissic foliation**. Gneiss is characterized by parallel layers of light and dark minerals that can be discontinuous and wavy (**Fig. 7.9**) or segregated into distinct planar bands. When gneiss is whacked with a big hammer, it does not generally break along single foliation surfaces that are less than 1 cm apart. Gneiss is a phaneritic rock that often resembles a layered granite or diorite.

Foliated Texture

The Latin word for "leaf" or "sheet" is *folium*, which is the word we use to describe a single layer of aligned flat minerals that developed during metamorphism. In general,

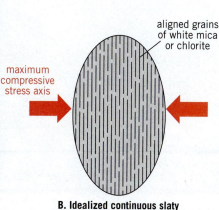

aligned grains of white mica or chlorite

maximum compressive stress axis

B. Idealized continuous slaty cleavage at high magnification

Figure 7.6 Slate. Slate is a foliated rock with dull luster, excellent slaty cleavage, and no visible grains. Slate forms from low-grade metamorphism of mudstone, shale, or claystone. Clay minerals in the protolith convert to chlorite and muscovite whose flat grains grow perpendicular to the maximum compressive stress axis. Slate can split into hard, flat sheets less than 1 cm thick.

Figure 7.7 Phyllite. Phyllite is a foliated, fine-grained rock with a satiny luster caused by tiny mica grains on its wavy foliation surfaces. Protoliths of phyllite might be mudstone (shale, claystone), slate, or other rocks rich in clay, chlorite, or mica. Phyllite forms under conditions that are between the low-grade metamorphism that produces slate and the intermediate metamorphism that produces schist.

Figure 7.8 Schist. Schist is a medium- to coarse-grained, foliated rock that displays schistosity. Protoliths of schist are quite varied and include rocks rich in clay, chlorite, or mica. Schist forms under conditions that are between the low-grade metamorphism that produces phyllite and the high-grade metamorphism that produces gneiss.

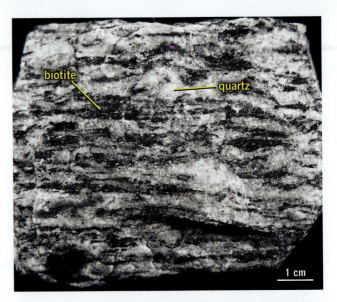

Figure 7.9 Gneiss. Gneiss is a medium- to coarse-grained metamorphic rock with *gneissic banding*—alternating layers or lenses of light and dark minerals. Many gneisses look like layered granitic rocks. Light-colored layers are rich in quartz or feldspars and alternate with dark layers of mafic minerals such as biotite and hornblende. Gneiss is a high-grade metamorphic rock that can form from many different protoliths.

metamorphic foliation develops perpendicular to the greatest compressive stress direction. Imagine pressing a ball of Silly Putty between the palms of your hands. The ball flattens perpendicular to the direction you apply the greatest compressive stress (**Fig. 7.2**). Though the mechanisms and materials are different, the effect is the same in developing the foliation in metamorphic rock.

Slate. The characteristic features of slate are its dull surface luster, its aphanitic (microscopic) grain size, its greater hardness and density compared with shale, and the fact that it can be broken into thin, very planar sheets because of its **slaty cleavage** (**Fig. 7.6**). Slate has probably been used for blackboards since the dawn of writing but is more likely to be seen these days in flooring, decorative wall tiles, the top of high-end billiard tables, and roof shingles. Slate occurs in several colors, including red slates that contain some hematite, green slates with chlorite, and gray or black slates with some graphite.

The protolith of slate is usually mudstone or shale—sedimentary rock that is composed primarily of clay minerals (**Fig. 7.10**). Some slate preserves a remnant of the original beds, called **relict bedding**, which can help demonstrate the sedimentary nature of the protolith. As temperature rises to ~100°C during burial, many of the clay minerals in the mud are converted in the solid state to another clay mineral called *illite*, which has a structure and composition that is similar to muscovite. Continued heating into the range of low-grade metamorphism causes the illite to convert to very fine-grained white mica (muscovite or paragonite) while other clays convert to chlorite.

Slaty cleavage develops perpendicular to the axis of maximum compressive stress just as the Silly Putty

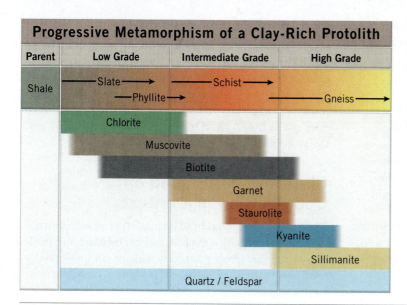

Progressive Metamorphism of a Clay-Rich Protolith

Parent	Low Grade	Intermediate Grade	High Grade
Shale	— Slate →	— Schist →	
	— Phyllite →		— Gneiss —

Chlorite

Muscovite

Biotite

Garnet

Staurolite

Kyanite

Sillimanite

Quartz / Feldspar

Figure 7.10 Metamorphism of a clay-rich protolith. A mudstone, claystone or shale subjected to increasing metamorphic grade will change to slate, phyllite, schist and even gneiss. Minerals grow in the rock during progressive metamorphism, each within a set of pressure-temperature conditions. By examining these minerals, we can infer the P-T conditions the rock was subjected to.

flattens perpendicular to the direction in which you squeeze the putty. At low metamorphic temperatures below ~350°C, microscopic grains of white mica or chlorite become aligned parallel to each other and perpendicular to the maximum compressive stress axis. The orientation of this metamorphic foliation is commonly different from the orientation of the original sedimentary bedding, so slaty cleavage usually cuts across any relict bedding.

Phyllite. The word *phyllite* (pronounced as FILL-ite) is derived from the Greek word for leaf, *phyllon*. Phyllite is intermediate in metamorphic grade and grain size between slate and schist (**Fig. 7.7**). The foliation surfaces in phyllite are typically irregular (wavy, bumpy, or corrugated) and have a silky or lustrous sheen related to very fine-grained graphite, white mica, or chlorite along the foliation surfaces. The color of phyllite is variable, including red, green, gray, and black. The protolith might be a mudstone or sandy mudstone.

Schist. A medium- to coarse-grained metamorphic rock with a well-developed schistosity that involves more than 50% of the rock's grains is called *schist*—a word derived from the Greek word *schistos* meaning "to split" (**Fig. 7.8**).

Schist is a coarser-grained rock than phyllite and reflects a higher metamorphic grade. Schistosity is a metamorphic layering that can be expressed by mica minerals (muscovite, biotite), chlorite, or amphibole and that so completely pervades the rock that it can be split along schistose layers on a scale of 1 cm or less. The intermediate-grade metamorphism of mudstone or sandy mudstone is likely to produce schist, although many types of rock can become schist through metamorphism.

Gneiss. The word *gneiss* sounds just like the word *nice*, as in "It was a nice spring day." Gneiss is a phaneritic rock that typically contains quartz and feldspar along with mica minerals (**Fig. 7.9**). The mica minerals form a schistosity that can be discontinuous, poorly developed, or widely spaced. Unlike schist, gneiss does *not* split along foliation on a scale of 1 cm or less. Gneiss often resembles a granitic rock that is layered.

Gneiss commonly has a banding of alternating light and dark layers with light layers containing quartz and feldspar whereas the dark layers have micas or mafic minerals (**Fig. 7.11**). Gneiss forms in intermediate to high-grade metamorphic conditions within a differential

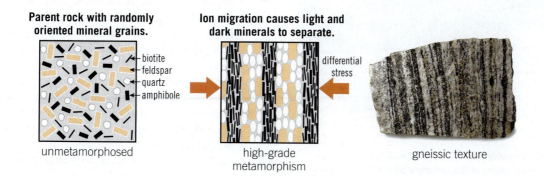

Parent rock with randomly oriented mineral grains.

- biotite
- feldspar
- quartz
- amphibole

unmetamorphosed

Ion migration causes light and dark minerals to separate.

differential stress

high-grade metamorphism

gneissic texture

Figure 7.11 Compositional layering in gneiss. A rock composed of felsic and mafic minerals can be subjected to high temperature and pressure as well as a differential stress. Pressure solution and other processes cause the felsic minerals to concentrate in bands between the bands of mafic minerals. The result is a medium- to coarse-grained rock with gneissic banding.

stress field and can form from many different protoliths. Mudstone or shale subjected to high-grade metamorphism might produce a gneiss (**Fig. 7.10**).

Granofelsic Texture

The characteristic feature of a granofels is that it is a metamorphic rock that lacks schistosity (e.g., **Fig. 7.3**). The mineral grains have an approximately random orientation in the rock. **Granofelsic texture** lacks schistosity, but there can be other types of mineralogical layering in the rock. For example, a quartzite with a granofelsic texture might have relict bedding (**Fig. 7.12**). Examples of rock types that might have a granofelsic texture include marble, quartzite, and hornfels.

1 cm

relict cross-bedding

Figure 7.12 Relict bedding in quartzite. The protolith of this quartzite was a quartz sandstone that had cross beds. The cross beds are preserved as relict beds in the quartzite, which was subjected to low-grade metamorphism.

chlorite

1 cm

pyrite porphyroblast (brassy cube)

Figure 7.13 Porphyroblastic texture. This texture is characterized by large, visible crystals that occur in a fine-grained groundmass of one or more other minerals. This medium-grained schist contains porphyroblasts of pyrite (brassy metallic cubes) in a groundmass of chlorite. The rock can be called a pyrite chlorite schist.

Porphyroblastic Texture

Just as igneous rock might have a porphyritic texture in which there are large grains in a groundmass of smaller grains, metamorphic rock might have a **porphyroblastic texture** in which large grains that grew as a result of metamorphism, called **porphyroblasts**, occur within a groundmass of finer grains (**Fig. 7.13**). If the large grains are derived from the protolith, they are called **porphyroclasts**.

Classification Based on Composition or Context

Texture defines several fundamental types of metamorphic rock including, phyllite, schist, and gneiss. Slate and phyllite are too fine grained for us to add any compositional modifiers to their names based on what we can observe in hand samples without magnification.

Schist is the root term for a medium- to coarse-grained foliated rock, and we can describe the type of schist based on the minerals we observe. Standard practice is to list the distinctive or major minerals in order of increasing abundance. So, for example, if a particular schist has a few garnets but is mostly muscovite, it would be a garnet muscovite schist (**Fig. 7.8**). **Blueschist** is a metamorphic rock formed at relatively low temperatures in subduction zones.

Some other metamorphic rock types are based on composition or other characteristics and include the following.

- **Marble** is mostly composed of carbonate minerals: calcite, aragonite, dolomite (**Fig. 7.3**). Calcite/aragonite marble fizzes energetically in dilute hydrochloric acid, and powdered dolomite marble fizzes weakly. The protolith is limestone or dolostone.
- **Quartzite** is composed of quartz, and the protolith is quartz sandstone (**Fig. 7.12** and **Fig. 7.14**). Quartzite occurs in many colors and is both hard and durable.
- **Metaconglomerate** is, as the name implies, a metamorphosed conglomerate (**Fig. 7.15**). The gravel-sized clasts can be recognized and might be flattened in response to a differential stress applied during metamorphism.
- **Metabreccia** is composed of gravel-sized angular clasts. Metabreccias can be metamorphosed sedimentary breccias, fault breccias, or volcanic breccias.
- **Amphibolite** is composed largely of black, brown, or green amphibole (hornblende), usually with plagioclase feldspar, and may contain quartz, pyroxene, garnet, biotite, or other minerals. The protolith is typically basalt or gabbro. Amphibolite might have a gneissic or granofelsic texture.
- **Serpentinite** is composed of green serpentine minerals: lizardite, antigorite, chrysotile (**Fig. 7.16**). Serpentinite is a product of the hydrothermal alteration of basalt, gabbro, or ultramafic rock.
- **Soapstone** is composed largely of talc, possibly with some minor carbonate minerals. Like serpentinite, soapstone is a product of the hydrothermal alteration of basalt, gabbro, or ultramafic rock. Soapstone is used in sculpture and to make tabletops, and is very soft.

Figure 7.14 **Quartzite.** The protolith of quartzite is quartz sandstone. Quartzite that formed under low-grade conditions might be only slightly altered, with relict bedding, flattened grain contacts, and a new coating of quartz on the older quartz sand grains. Under high-grade conditions, quartzite might undergo extensive recrystallization. Most quartzite is not obviously foliated.

Figure 7.15 **Metaconglomerate.** The protolith of metaconglomerate is a conglomerate: a sedimentary rock with gravel-sized grains. This specimen was metamorphosed with a differential stress that resulted in the flattening of the rounded gravel-sized grains approximately perpendicular to the maximum compressive stress axis.

Figure 7.16 **Serpentinite.** This is a sheared serpentinite from the Franciscan Formation in California, which was developed in a subduction zone between the Farallon and North American Plates. It is composed primarily of the serpentine minerals lizardite and antigorite formed by metamorphism of the basaltic oceanic crust.

Figure 7.17 **Hornfels.** Hornfels is a fine-grained, nonfoliated metamorphic rock having a dull luster and a microcrystalline texture that may appear smooth or sugary. It is usually very hard and dark in color, but it sometimes has a spotted appearance caused by patchy chemical reactions with the metamorphosing magma or hydrothermal fluid. Hornfels forms by contact metamorphism, and the protolith can be virtually any rock type.

- **Eclogite** is a dense rock composed of bright green omphacite (pyroxene) and red garnet. The protolith is probably basalt, gabbro, or ultramafic rock metamorphosed under high-pressure conditions, possibly in a subduction zone.

Some metamorphic rock types are based on context as much as their texture or composition and include the following.

- **Hornfels** is formed by *contact metamorphism* (which will be described in a later section) and is composed of silicate and oxide minerals with a granofelsic/granoblastic texture (**Fig. 7.17**). The grain size is approximately

the same within a given sample of hornfels, but grain size for hornfels in general is variable, tending to be small. Context is important in recognizing this rock type—it occurs next to a younger igneous body that was the heat source for the contact metamorphism.

- **Mylonite** is a type of **fault rock** formed by shear displacement along a fault under metamorphic conditions. Mylonite was subjected to enough temperature and pressure that it was able to flow even though it was still in the solid state. Mylonite has a well-developed schistosity and typically has a small average grain size and porphyroblasts.

Metamorphic Grades and Facies, (p. 204)

> **Think About It**   What can metamorphic rock tell us about Earth's history and the environments in which the rock formed?

Objective   Infer regional geologic history and the relationship of metamorphic facies to plate tectonics using index minerals, pressure-temperature diagrams, and geologic maps.

Before You Begin   Read the following sections: Local and Regional Metamorphism and Metamorphic Zones and Facies. Review Figure 7.10.

Local and Regional Metamorphism

The effects of metamorphism can be described as local or regional based on the volume of rock affected. A variety of causes can result in **local metamorphism**, including:

- Lightning (**lightning metamorphism**)
- Meteorite strikes (**impact metamorphism**)
- Lava flows and deposition of hot tephra (**pyrometamorphism**)
- Hot water (**hydrothermal metamorphism**)
- Faulting (**dislocation-** and **hot-slab metamorphism**)
- Combustion of coal, oil, or natural gas (**combustion metamorphism**)

Probably the most important cause of local metamorphism is related to the emplacement of a hot magma body, which subjects the older surrounding rock to **contact metamorphism** as the magma conducts heat into the country rock.

The lower crust exists within the pressure-temperature zone in which metamorphism occurs, resulting in **regional metamorphism** of global extent (**Fig. 7.1**). Where these deep rock bodies have been uplifted and uncovered by erosion at the ground surface, they are recognized as regionally metamorphosed rocks. Sedimentary rock that is buried deeply enough to be subject to temperatures above ~200°C is said to be affected by **burial metamorphism**.

Apart from metamorphism caused simply by location deep enough to be subject to metamorphic conditions, the primary cause of regional metamorphism is plate tectonics. All plate boundaries are areas of significant differential stress. Hydrothermal activity is an essential element of the active part of mid-ocean ridges and continental rifts where the rate of heat flow is much greater than elsewhere in the lithosphere. The high heat flow that leads to the generation of magma along rifts, in subduction zones, and above hotspots also generates the conditions for regional metamorphism.

Metamorphic Zones and Facies

Geologists have recognized at least since the time of James Hutton (1726–1797) that older rock is changed where it is intruded by magma, and that the intensity of that change decreases with increasing distance from the intrusion. The reason for this change with distance or **metamorphic gradient** is that the heat conducted from a cooling magma body into the surrounding rock decreases with distance. Metamorphic gradients reflect changing pressure-temperature (P-T) conditions that might be caused by depth of burial, closeness to an igneous body, or position within a deforming mountain range.

The change in P-T conditions across an area is reflected by a change in mineralogy, which was first described by British geologist George Barrow (1853–1932). Starting in the early 1890s, Barrow used minerals to map zones of different metamorphic intensity in the Grampian Highlands of Scotland. The most intense P-T conditions in Barrow's zones were marked by the occurrence of the mineral **sillimanite**, followed by a zone that had **kyanite**, and then the **garnet**, **biotite**, and **chlorite** zones in the less intensely metamorphosed rocks (**Fig. 7.10**). One mnemonic to remember the order of these index minerals and their associated zones is "**S**illy **k**ids **g**arner **b**ig **c**huckles." In Barrow's honor, these are called Barrovian Zones.

Starting with the idea of using groups of metamorphic minerals to help us identify the approximate P-T conditions of metamorphism, several geologists developed the idea of **metamorphic facies**. Barrow's zones were appropriate for rocks whose protolith included clay minerals (i.e., was *pelitic*) and that were subject to metamorphic conditions of increasing temperature and pressure. In contrast, metamorphic facies provide a tool intended to be useful across a broad range of rock compositions and metamorphic conditions.

The IUGS currently recognizes 10 different metamorphic facies, each of which occupies a separate area in a graph of metamorphic pressure and temperature. Each facies is defined by a specific group of metamorphic minerals. (*Facies* is one of those words that is both a singular and a plural noun.) The assemblage of minerals that defines a facies is known to be in a happy state of thermodynamic equilibrium within a general range of P-T conditions. We will use index minerals to interpret metamorphic facies in Activity 7.5.

MasteringGeology™

Looking for additional review and test prep materials? Visit the Study Area in MasteringGeology to enhance your understanding of this chapter's content by accessing a variety of resources, including Pre-Lab Videos, Self-Study Quizzes, Geoscience Animations, Mobile Field Trips, *Project Condor* Quadcopter videos, *In the News* articles, glossary flashcards, web links, and an optional Pearson eText.

Name: _____ **Course/Section:** _____ **Date:** _____

A Analyze the metamorphic rocks in **Fig. A7.1.1** and actual rock samples of them, if available. Beneath or beside each picture, describe the rock's **composition** (what it is made of) and **texture** (the size, shape, and arrangement of its parts) as well as you can, using your current knowledge and observational skills.

1

2

3

4

5

6

Figure A7.1.1

B **REFLECT & DISCUSS** Reflect on your observations and descriptions of metamorphic rocks in part **A**. Then describe how you would classify the rocks into groups. Be prepared to discuss your classification with your classmates or teacher.

Name: _____   **Course/Section:** _____   **Date:** _____

Review the descriptions of metamorphic minerals. Below each of the numbered photos in **Fig. A7.2.1**, identify the type of mineral shown. Be prepared to discuss your interpretations with your teacher or with others in your lab.

What is the slender gray-white mineral?

(Photo by Tom Mortimer)

The white mineral is quartz. What is the blue mineral?

A7.2.1

Name: _____ Course/Section: _____ Date: _____

A Analyze the samples of sedimentary limestone and metamorphic marble in **Fig. A7.3.1**.

1. These rocks are both composed of the same mineral. What is it? _____

   What test could you perform on the rocks to be sure?

2. How is the texture of these two rocks different?

limestone   marble

Figure A7.3.1

B **Figure A7.3.2** shows three metamorphic rocks that had a mudstone or shale protolith and that display excellent foliation. The three specimens represent different metamorphic grades. Higher-grade types can also be formed from protoliths other than mudstone, but we are going to think about a progression from mudstone through successively more intense metamorphism.

1. Describe the change in grain size from slate to schist.

2. How does the texture of phyllite differ from that of schist?

3. Why do you think that the micas (flat minerals) in these rocks are all parallel, or nearly so, to one another?

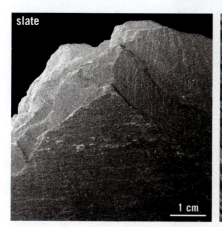 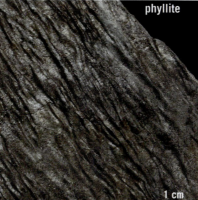 

slate   phyllite   schist

A7.3.2

C Analyze the rock sample in **Fig. 7.5A**. The white layers in this metamorphic rock—a gneiss—were originally flat, parallel layers that were later folded during metamorphism. Describe a process that could account for how this strong, dense gneiss was folded without breaking during regional metamorphism. (*Hint*: How could you bend a brittle candlestick without breaking it?)

D Analyze the foliated metamorphic rock sample in **Fig. A7.3.3**.

1. What mineral defines the foliation in this rock?

2. Notice that the rock consists mostly of muscovite but also contains scattered garnet crystals. What is the name for this kind of texture?

3. What is the name of this metamorphic rock?

4. What type(s) of rock might have been the protolith for this rock?

Figure A7.3.3

E Analyze the metamorphic rock sample in **Fig. A7.3.4**.

1. Is this rock foliated or nonfoliated (granofelsic)? What features in the photograph did you use to make your interpretation?

2. What is the name of this metamorphic rock?

3. What type(s) of rock might have been the protolith for this rock?

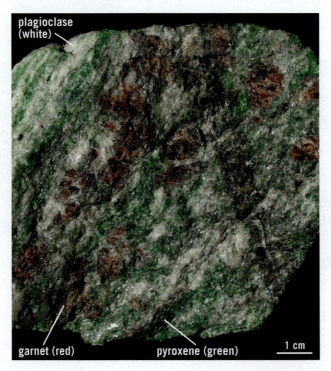

Figure A7.3.4

F **REFLECT & DISCUSS** Which one of the rocks in this activity do you think has the highest metamorphic grade? Explain your answer.

Name: _____ Course/Section: _____ Date: _____

METAMORPHIC ROCKS WORKSHEET

Sample Letter or Number	Texture	Mineral Composition or Other Distinguishing Properties	Rock Name Interpretation	Possible Protolith(s)
	☐ foliated ☐ nonfoliated			
	☐ foliated ☐ nonfoliated			
	☐ foliated ☐ nonfoliated			
	☐ foliated ☐ nonfoliated			
	☐ foliated ☐ nonfoliated			

Figure A7.4.1

Name: ______________________ Course/Section: ______________ Date: __________

METAMORPHIC ROCKS WORKSHEET

Sample Letter or Number	Texture	Mineral Composition or Other Distinguishing Properties	Rock Name Interpretation	Possible Protolith(s)
	☐ foliated ☐ nonfoliated			
	☐ foliated ☐ nonfoliated			
	☐ foliated ☐ nonfoliated			
	☐ foliated ☐ nonfoliated			
	☐ foliated ☐ nonfoliated			

Figure A7.4.1 *(continued)*

Name: _____ Course/Section: _____ Date: _____

METAMORPHIC ROCKS WORKSHEET

Sample Letter or Number	Texture	Mineral Composition or Other Distinguishing Properties	Rock Name Interpretation	Possible Protolith(s)
	☐ foliated ☐ nonfoliated			
	☐ foliated ☐ nonfoliated			
	☐ foliated ☐ nonfoliated			
	☐ foliated ☐ nonfoliated			
	☐ foliated ☐ nonfoliated			

Figure A7.4.1 (continued)

203

Name: _____ Course/Section: _____ Date: _____

A British geologist George Barrow mapped rocks in the Scottish Highlands that were metamorphosed by granitic igneous intrusions. He discovered that as he walked away from the granitic intrusive igneous rock, there was a sequence of mineral zones that generally reflected the intensity of metamorphism. He defined the following sequence of **index minerals**, which represent intensity of metamorphism along a gradient from low to high pressure-temperature (P-T) conditions:

Chlorite (lowest P-T), biotite, garnet, staurolite, kyanite, sillimanite (highest P-T)

1. Boundaries between Barrow's metamorphic zones are called **isograds**. On the geologic map (**Fig. A7.5.1**), color in the zone of *maximum* metamorphic intensity as indicated by Barrow's index minerals.

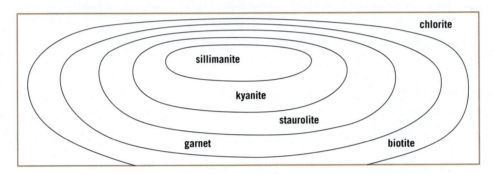

Figure A7.5.1

B Metamorphism is typically caused by increases of both pressure and temperature. Geologists represent these relationships on pressure-temperature (P-T) diagrams (or phase diagrams) showing the stability of different index minerals. The minerals andalusite, kyanite, and sillimanite shown on this phase diagram (**Fig. A7.5.2**) are *polymorphs*: minerals that have the same chemical composition but different crystalline structure and physical properties that can be used to distinguish them. Each polymorph is stable under pressure and temperature conditions that are different from the others. Note that any two of these minerals can occur together only under P-T conditions represented by the boundary lines in the diagram and that the three minerals can occur together only at the point where these three lines intersect: approximately 500°C and 4 kilobars, which normally occurs about 15 km below Earth's surface.

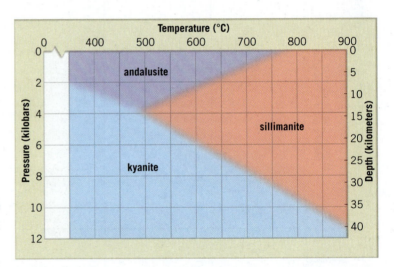

Figure A7.5.2

1. Study the mineral zones and isograds on the two maps in **Fig. A7.5.3**. Which region was metamorphosed at higher pressure? How can you tell?

2. What was the minimum temperature at which the rocks in Map **B** were metamorphosed? _____

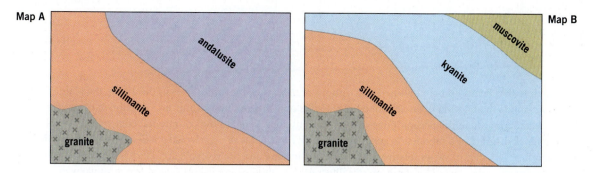

Figure A7.5.3

C Finnish geologist Pentti Eskola recognized that the volcanic rock *basalt* could be metamorphosed to different *metamorphic facies* (unique assemblages of several minerals) under changing conditions of pressure and temperature:

- Amphibolite facies (low pressure, high temperature): black hornblende amphibole, sillimanite.
- Greenschist facies (low pressure, low temperature): green actinolite amphibole and chlorite.
- Eclogite facies (high pressure, high temperature): red garnet, green pyroxene.
- Blueschist facies (high pressure, low temperature): blue amphibole (glaucophane, riebeckite) and lawsonite.

1. Write the names of these metamorphic facies in their proper places on the dotted lines marked A through D in the P-T diagram (**Fig. A7.5.4**). Notice that pressure and depth increase downward in this diagram, and temperature increases to the right.

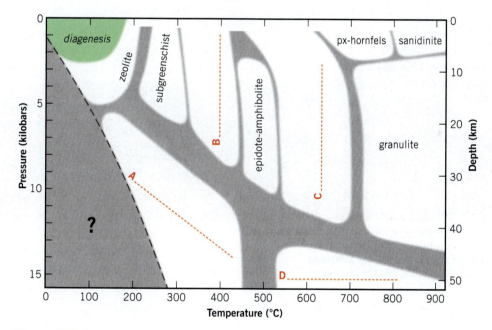

Figure A7.5.4

2. **REFLECT & DISCUSS** What is the significance of the gray area of the P-T diagram in **Fig. A7.5.4** that is marked by a large question mark? *Hint:* Where on or in Earth would you find the P-T conditions in that gray area?

3. At the time that Pentti Eskola published descriptions of these metamorphic facies in the 1920s, the plate tectonics model had not yet been developed. Geologists now realize that volcanic arcs develop at convergent plate boundaries where the oceanic edge of one plate subducts beneath the continental edge of another plate. In the block diagram of a subduction zone (**Fig. A7.5.5**), notice how the *geothermal gradient* (rate of change in temperature with depth) varies relative to the subduction zone and the volcanic arc. Place letters in the white circles that are linked to the starred locations on this illustration to show where Eskola's facies are most likely to occur: A = amphibolite, G = greenschist, E = eclogite, B = blueschist.

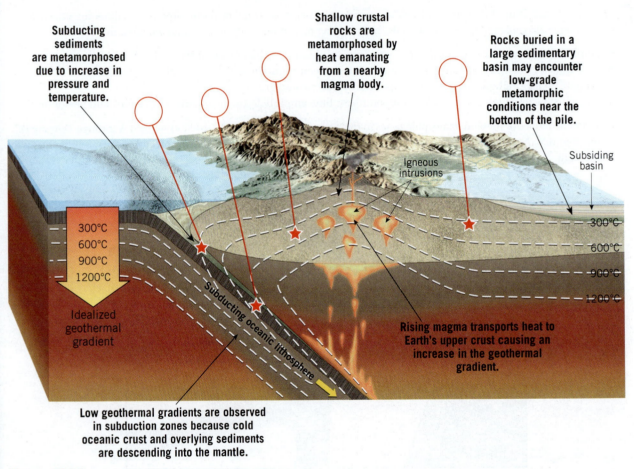

Subducting sediments are metamorphosed due to increase in pressure and temperature.

Shallow crustal rocks are metamorphosed by heat emanating from a nearby magma body.

Rocks buried in a large sedimentary basin may encounter low-grade metamorphic conditions near the bottom of the pile.

Igneous intrusions

Subsiding basin

300°C
600°C
900°C
1200°C

Idealized geothermal gradient

Subducting oceanic lithosphere

300°C
600°C
900°C
1200°C

Rising magma transports heat to Earth's upper crust causing an increase in the geothermal gradient.

Low geothermal gradients are observed in subduction zones because cold oceanic crust and overlying sediments are descending into the mantle.

Figure A7.5.5

4. **REFLECT & DISCUSS** Based on what you learned in this activity, write a brief description of how pressure, temperature, depth, and intensity of metamorphism are related.

Dating of Rocks, Fossils, and Geologic Events

Pre-Lab Video 8
http://goo.gl/k3LPxG

Contributing Authors

Jonathan Bushee • *Northern Kentucky University*

John K. Osmond • *Florida State University*

Raman J. Singh • *Northern Kentucky University*

▲ James Hutton visited this site at Salisbury Crags above Edinburgh, Scotland (55.9431°N, 3.1672°W) in the late 1700s. Hutton recognized that the layered sandstone had been intruded by molten rock, which baked and locally deformed the sandstone. The molten rock cooled to form the red dolerite shown here above the layered sandstone.

BIG IDEAS

Geologists use relative and numerical dating techniques to determine the ages of geologic features and events. Relative dating is the process of determining the order in which geological events occurred. Numerical dating is the process of determining the age of a geological material or event in years, generally through the analysis of certain radioactive isotopes and their stable daughter products that are present in minerals and rocks. The geologic time scale provides a graphic way of identifying major phases of Earth's history along with the numerical ages of the beginning and ending of those phases. Geoscientists developed the time scale over centuries of work, initially using fossils to define different parts of the time scale before we could determine numerical ages of geologic materials. Many of the boundaries are related to extinctions that are evident in the geological record of life on Earth.

FOCUS YOUR INQUIRY

Think About It How can you tell relative age relationships among the parts of geologic cross-sections exposed in outcrops?

ACTIVITY 8.1 Geologic Inquiry for Relative Dating (p. 208, 221)

Think About It How can geologic cross-sections be interpreted to establish the relative ages of rock units, contacts, and other geologic features?

ACTIVITY 8.2 Determining Sequence of Events in Geologic Cross-Sections (p. 208, 223)

Think About It How are fossils used to tell geologic time and infer Earth's history?

ACTIVITY 8.3 Using Fossils to Date Rocks and Events (p. 213, 225)

Think About It How do geologists determine the numerical age, in years, of Earth materials and events?

ACTIVITY 8.4 Numerical Dating of Rocks and Fossils (p. 216, 226)

Think About It How are relative and numerical dating techniques used to analyze outcrops and infer geologic history?

ACTIVITY 8.5 Infer Geologic History from a New Mexico Outcrop (p. 216, 227)

ACTIVITY 8.6 Investigating a Natural Cross-Section in the Grand Canyon (p. 216, 229)

Introduction

Geology is a historical science. It involves the physical, chemical, and biological laws of Nature and the mathematical description of those laws just like all other physical sciences. However, geoscience also involves deep time and the story of how the physical, chemical, and biological world has changed through time. **Deep time** refers to the vast time span that geoscientists work with contrasted with the much more limited time spans that nongeoscientists consider in their daily lives.

If we liken a 45.5-year-old's life span to the 1 mm depth of a drop of water on the deck by the side of a swimming pool, how deep would the pool have to be in order for its depth to be comparable to the age of Earth (~4.55 billion years)? We can approach this as a proportion problem: 45.5 years is to 1 mm as 4.55×10^9 years is to x mm, or

$$\frac{45.5 \text{ yr.}}{1 \text{ mm}} = \frac{4.55 \times 10^9 \text{ yr.}}{x \text{ mm}}$$

To solve a ratio of the form

$$\frac{a}{b} = \frac{c}{d}$$

we rearrange the equation to isolate the unknown quantity by itself on one side of the equation.

$$d = \frac{(b \times c)}{a}$$

so

$$x \text{ mm} = \frac{(1 \text{ mm}) \times (4.55 \times 10^9 \text{ yr.})}{45.4 \text{ yr.}} = 10^8 \text{ mm}$$

The pool would have to be 10^8 mm $= 10^5$ m $= 10^2$ km $= 100$ km deep. The deepest part of the ocean is the Challenger Deep in the Marianas Trench of the southeastern Pacific Ocean (11.329903°N, 142.199305°E) at a depth of $10,984 \pm 25$ m. So our deep-time swimming pool would have to be just over *9 times deeper than the deepest part of the ocean* at a scale of 45.5 years = 1 mm in order to represent the age of Earth.

It is our ability to date geological materials that has allowed us to know when and in what order various animals, plants, microbes, and even our distant ancestors have lived on Earth. Dating geological materials allows us to establish the history of fault motion so that we can define the recurrence intervals of damaging earthquakes and can

differentiate between active and inactive faults. Beyond helping us tell the story of the biological and geological history of Earth, our ability to measure time has also given us the ability to analyze how climate has changed over Earth's history, which is crucial knowledge to inform current discussions about the effect of geologically recent human activities on climate.

ACTIVITY 8.1

Geologic Inquiry for Relative Dating, (p. 221)

Think About It How can you tell relative age relationships among the parts of geologic cross-sections exposed in outcrops?

Objective Identify features of geologic cross-sections exposed in outcrops, infer their relative ages, and suggest rules for relative age dating.

Before You Begin Read the following sections: The Geologic Record, Principles for Determining Relative Age, and Contact Relationships.

ACTIVITY 8.2

Determining Sequence of Events in Geologic Cross-Sections, (p. 223)

Think About It How can geologic cross-sections be interpreted to establish the relative ages of rock units, contacts, and other geologic features?

Objective Apply principles of relative dating to analyze and interpret sequences of events in geologic cross-sections.

Before You Begin Read the following sections: The Geologic Record, Principles for Determining Relative Age, and Contact Relationships.

The Geologic Record

When we drill a well kilometers deep into Earth's crust, we are not simply passing through mute sediment and rock. We are encountering material that has a story to tell of its formation and subsequent history. The evidence needed to reconstruct that story is provided by the fossils, minerals, sedimentary structures, and other physical and chemical features observed in the rock (**Fig. 8.1**). A particularly clear part of the **geologic record** is contained within layers of sedimentary rocks that are stacked one atop the other like

Figure 8.1 Fossil record of life on Earth. The state fossil of Wyoming, *Knightia eocaena*, preserved in siltstone was deposited during the Eocene Epoch ~50 Myr ago as part of the Green River Formation. *Knightia* was a small, bony fish that lived in freshwater lakes.

pages in a book. As each new layer of sediment is deposited, it covers older layers and becomes the youngest layer of the geologic record at that time.

Geologists have studied sequences of rock layers wherever they are exposed in mines, quarries, riverbeds, road cuts, wells, and mountainsides throughout the world. Geologists have also followed the major rock units or formations from one place to another across regions and continents. Thus, the geologic record of rock layers is essentially a stack of stone pages in a giant natural book of Earth's history. And like some of the pages in an old book, some rock layers have been folded, torn (fractured or faulted), and even removed by geologic events.

Geologists tell time based on relative and numerical dating techniques. **Relative dating** is the process of determining the order in which geological events occurred relative to other events. For example, if you have a younger brother and an older sister, then you could describe your relative age by saying that you are younger than your sister and older than your brother. **Numerical dating** is the process of determining when a geological event occurred as measured in years. Using the example above, you could describe your numerical age just by saying how old you are in years.

Principles for Determining Relative Age

The earliest dating methods involved determining the order in which geological events occurred relative to each other. Many of these principles or laws were first articulated in print by Nicolas Steno in 1669, while others are due to the insights of William Smith, who created the first geological map of England and Wales in 1815. Although we now have the ability to determine the numerical age of many materials, determination of relative age is still an important part of the process of geological analysis. Building upon definitions in the AGI *Glossary of Geology*, the elementary principles of relative dating can be described as follows.

- **Law of Original Horizontality.** Water-laid sediments are deposited in beds (**strata**) that are horizontal or nearly horizontal and parallel or nearly parallel to Earth's surface (**Fig. 8.2**). The cross beds formed by wind or flowing water might be more steeply inclined, but the upper and lower surfaces that bound the cross beds are nearly horizontal at the time of deposition. Sedimentary beds that are no longer horizontal are likely to have been deformed after deposition.

- **Law of Original Continuity.** At the time it is deposited, a water-laid bed (**stratum**) extends laterally in all directions until it thins out because the sediment supply was exhausted or until it encounters an obstruction like the edge of a depositional basin.

- **Law of Superposition.** In any layered sequence of sedimentary or extrusive volcanic rocks that has not been tilted or overturned, the youngest layer is at the top and the oldest is at the base **Fig. 8.2**. That is, each layer is younger than the layers beneath it but older than the layers above (**Fig. 8.2**). In tilted or overturned sedimentary layers (**Fig. 8.2A**), the age of the layers increases systematically in a particular direction. The direction in which the age increases must be interpreted from other geological data.

- **Law of Inclusion.** The source rock that eroded to produce a clast that was subsequently included in a sedimentary layer is older than the sedimentary layer

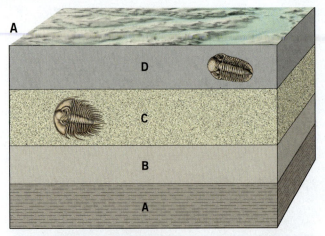

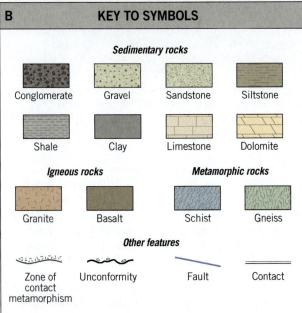

Figure 8.2 Superposition and undeformed (original) horizontal strata. Block diagram showing undisturbed horizontal sedimentary beds in superposed order—with the oldest bed (formation **A**) on the bottom of the sequence and the youngest (formation **D**) at the top. No erosional surfaces are evident within the sequence, so we can make an initial interpretation that the contacts between the formations are *conformable*, resulting from continuous deposition. The sequence of events was deposition of **A, B, C,** and **D,** in that order and stacked one atop the other.

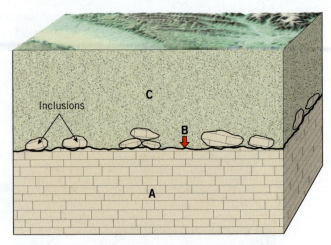

Figure 8.3 Inclusions above a disconformity. In this undeformed sequence of horizontal sedimentary layers, we use the principle of superposition to infer that limestone **A** is the oldest unit. An erosional surface **B** is observed at the top of limestone **A**, and limestone clasts eroded from limestone **A** are included in the sedimentary formation **C** that was deposited on the erosional surface **B**, forming a *disconformity* along **B**. The order of events began with deposition of **A** followed by removal by erosion of whatever strata might once have been above **A**, erosion of the top of **A** to form surface **B** along with development of limestone clasts eroded from **A**, and deposition of **C** that includes the limestone clasts.

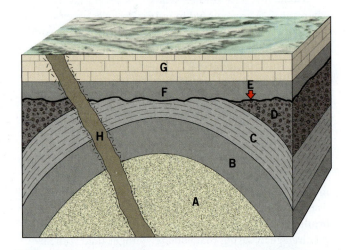

Figure 8.4 Cross-cutting dike. The igneous dike **H** is the youngest rock unit because it cuts across all of the others. We infer tentatively that **A** is the oldest formation because it is at the bottom of the sedimentary rock sequence that is cut by **H**, but this is a provisional interpretation because of the deformed nature of beds **A–D**. After formations **A–D** were deposited, they were folded, and an erosional surface (**E**) developed on the deformed strata. Formation **F** was deposited on the erosional surface to form an unconformable contact with the strata below, and formation **G** was deposited conformably on **F**. Finally, a crack filled with magma propagated across all of the formations and cooled to form an igneous dike. The contact along erosion surface **E** is interesting because it is an *angular unconformity* in most places but a *disconformity* along the hinge of the fold where the orientation of formation **C** is locally horizontal.

in which the clast is embedded (**Fig. 8.3**). In an igneous rock, the source rock of a xenolith is older than the igneous rock that crystallized around the xenolith. Note that a fossil is approximately the same age as the sedimentary layer or erosional surface it was deposited on unless the fossil was eroded from an older layer and redeposited.

- **Law of Cross-Cutting Relationships.** A dike, vein, fracture, or fault that cuts across the fabric of another rock body is younger than any other rock that it cuts across. For example, an intrusive dike is younger than the rock that it intrudes (**Fig. 8.4**). The most recent

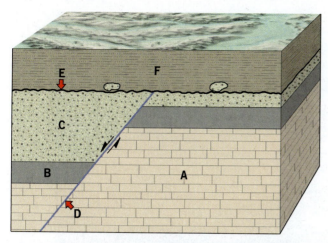

Figure 8.5 Cross-cutting fault. We use the principle of superposition to infer that formation **A** is the oldest unit in the undeformed horizontal sequence of sedimentary beds below erosional surface **E**, and **C** is the youngest in this sequence. Formations **A**, **B**, and **C** are cut by fault **D**, which does not displace the erosional surface **E** or formation **F**. This means that the fault **D** must be younger than **C** and older than **E** and **F**. The sequence of events began with deposition of formations **A**, **B**, and **C** in that order. Those formations were then cut by fault **D**. After faulting, the land surface was eroded. Siltstone **F** was deposited on the erosional surface, forming a *disconformable contact* along **E**.

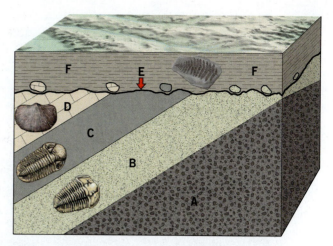

Figure 8.6 Angular unconformity. We know that formations **A–D** have been subjected to deformation because they are tilted and not in their original horizontal orientation. The fossil evidence indicates that formation **D** is younger than **C**, which is younger than **B**. The contacts between formations **A–D** are all illustrated as *conformable contacts* formed by continuous sedimentation, so we use the principle of superposition to interpret formation **A** as the oldest in this set of formations. After tilting, the upper edges of formations **A–D** were eroded to form surface **E**. Formation **F** was deposited on erosional surface **E**, forming an *angular unconformity* with the tilted formations below **E**.

motion along a fault is younger than the rock that is displaced by the fault. This is analogous to saying that a tear in a piece of paper must be younger than the piece of paper (**Fig. 8.5**).

- **Law of Fossil Assemblages.** Similar groups of fossils indicate similar geologic ages for the rocks that contain them. If sedimentary strata in two different locations contain the same assemblage of fossil species, they are approximately the same age.
- **Law of Fossil Succession.** Organisms preserved as fossils in sedimentary rock appeared, became extinct, and were succeeded by newer organisms over time. Hence, the assemblage of fossils in one sedimentary formation will be distinct from the fossil assemblage in formations above and below that formation. The relative age of a sedimentary rock formation can be determined by matching the fossil content of the formation with the fossils that geologists use to define one of the subdivisions of the geologic time scale. Finding Cretaceous fossils in a formation makes it older than Paleocene formations and younger than Jurassic formations.

Contact Relationships

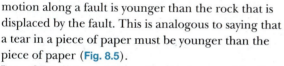

The surface between two distinct rock units is called a **contact**. Here we discuss a few types of contacts that are particularly relevant to unraveling the geologic history of an area.

Sedimentary Contacts

Nicolas Steno described types of sedimentary contacts in 1669 based on his observations in Tuscany, Italy. His interest in sedimentary rock was rooted in his interest in fossils because the nature of fossils was a matter of active debate at the time.

Conformable Contacts. Sedimentation occurs constantly on the seafloor, in lakes, and along rivers. Contacts between sedimentary layers deposited during continuous sedimentation are called **conformable contacts** (**Fig. 8.2**). Many soils developed on near-horizontal surfaces above sea level also have a conformable contact with the material below them.

Unconformable Contacts (Unconformities). James Hutton was interested in sedimentary contacts that are different than the conformable contacts that develop as a result of the ordinary sequence of continuous sedimentation. Hutton was particularly interested in **unconformities** that hinted at extraordinary events in the history of Earth. In 1787–1788, Hutton sought and found examples of interesting geological contacts in Scotland on the Isle of Arran, Inchbonny near Jedburgh, and Siccar Point that were similar to examples he had read about in descriptions written by natural scientists in Europe.

- **Angular Unconformity.** Sedimentary deposition on an erosional surface formed on older tilted sedimentary strata results in an **angular unconformity** (**Fig. 8.6**).

Upper Devonian Old Red Sandstone nonmarine strata

erosional surface

Silurian marine strata

Figure 8.7 Hutton's angular unconformity at Siccar Point. The Old Red Sandstone of Devonian age was deposited on an erosional surface on steeply inclined marine strata of Silurian age. Parts of the erosional surface are indicated with dashed lines. The entire rock sequence was later tilted a bit more as you can see by comparing the orientation of the originally horizontal upper red beds with the present-day horizon. This site is located ~4 km east of Cockburnspath, Scotland—lat 55.9315°N, long 2.3014°W.

James Hutton considered angular unconformities to be evidence of Earth's long and interesting history. Hutton visited Siccar Point with James Hall and John Playfair in 1788, where they discussed his interpretation of this remarkable outcrop (**Fig. 8.7**). They observed steeply inclined beds that must have been deposited as horizontal beds on the seafloor over a long time period. The horizontal marine strata were then deformed (folded), uplifted, and eroded. Onto this erosional surface, the younger beds of the Old Red Sandstone were deposited in horizontal layers above sea level. All agreed that this series of events must have taken a very long time to occur, although they had no way of knowing just how long. Playfair later wrote, "The mind seemed to grow giddy by looking so far into the abyss of time." We now know that the beds below the erosional surface were deposited in the Silurian Period ~435 Myr ago, and the Old Red Sandstone was deposited during the youngest part of the Devonian Period ~370 Myr ago.

- **Disconformity.** Sedimentary deposition on an erosional surface formed on undeformed older sedimentary strata, in which the younger layers are approximately parallel to the older layers below the erosional surface, results in a **disconformity** (**Fig. 8.3**).

This term is usually employed where a significant thickness of sedimentary layers has been removed by erosion prior to deposition of the younger layers atop the erosional surface. A surface on which there was simply no deposition over a significant interval of time can also be considered a disconformity.

- **Nonconformity.** As the term is typically used, a **nonconformity** is the result of sedimentary deposition on an erosional surface formed on older igneous or metamorphic rock (**Fig. 8.8**).

Igneous Contacts

Extrusive igneous rocks such as lava and pyroclastic flows cover the ground surface, often causing minor contact metamorphism to the material below (**Fig. 8.9**). An igneous flow is younger than the ground surface on which it cooled and so is younger than the rock units below that surface.

Intrusive igneous rock is younger than the rock that it intrudes (**Fig. 8.10**). Intrusive rock bodies include dikes that cut across pre-existing rock layering, such as sedimentary bedding or metamorphic foliation, and sills that are parallel to pre-existing layering. You can think of igneous dikes and sills as cracks that open in an older rock unit and that are filled with magma. Another common type of

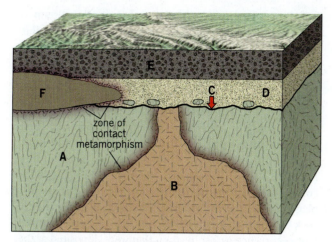

Figure 8.8 Intrusive igneous contact and nonconformity. Gneiss **A** is at the base of this rock sequence, separated from granite **B** by a zone of contact metamorphism. This suggests that a body of magma intruded **A** and then cooled to form the contact zone and granite **B**. There must have been erosion of both **A** and **B** *after* this intrusion, forming erosional surface **C** because there is no contact metamorphism between **B** and **D**. Formation **D** was deposited horizontally atop the eroded igneous and metamorphic rocks, forming a nonconformity across erosional surface **C**. After **E** was deposited, a second body of magma intruded **A, C, D,** and **E**, cooling to form **F**. Intrusive body **F** is a sill because it was emplaced parallel to the existing layering.

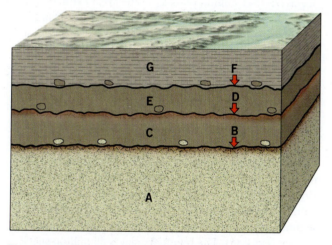

Figure 8.9 Extrusive igneous contacts and nonconformity. Lava flow **C** extends over the erosional contact **B** on top of the sandstone **A**, causing local metamorphism in **A** along the igneous contact. A second flow **E** covers the upper surface **D** of the older flow **C**. Sediment of formation **G** eventually covers the upper surface of flow **E**. That surface, **F**, is a contact between sediment deposited under low temperature and pressure conditions and an igneous rock that crystallized at high temperature and surface pressure. Hence, there is a *nonconformity* between the sedimentary layer **G** and the igneous layer **E** across surface **F**.

intrusive igneous body is a *pluton*, which was once a bit like a bubble of buoyant magma before cooling and crystallization. Igneous contacts often display distinctive features. The magma in intrusive bodies cools more rapidly along

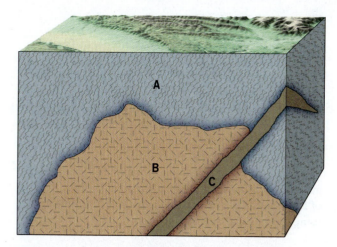

Figure 8.10 Igneous intrusions and cross-cutting. The body of granite **B** must have formed from the cooling of a body of magma that intruded the pre-existing rock unit **A**, called *country rock*. Country rock **A** has a zone of contact metamorphism adjacent to the granite. Lastly, a crack filled with magma propagated across **A** and **B**, causing local metamorphism along the intrusive igneous contact. The magma cooled to form dike **C**.

the edge of the body, and so the contact margins of the intrusion generally have a smaller grain size than the interior of the intrusion. The rock that was intruded—the country rock—is often subject to local metamorphism along the igneous contact.

Fault Contacts

A fault is a surface or zone between two rock masses along which the rock masses have slid past one another. The most recent activity along a fault is younger than the rock on either side of the fault (**Fig. 8.5**).

Some faults have long displacement histories so that the early history of the fault might be older than some of the rock units that are displaced along the fault. For example, a sedimentary layer might have been deposited over a still-active fault in the time between widely spaced large earthquakes, later to be displaced along the fault. The fault as a whole is older than that sedimentary bed, but the part of the fault that displaces the sedimentary bed is younger than that bed.

ACTIVITY 8.3

Using Fossils to Date Rocks and Events, (p. 225)

Think About It   How are fossils used to tell geologic time and infer Earth's history?

Objective   Use fossils to determine the ages of rock bodies and infer some of Earth's history.

Before You Begin   Read the following section: The Geologic Time Scale.

The Geologic Time Scale

Nicolas Steno described the geologic history of Tuscany as being divisible into six phases and indicated that this division might be of worldwide extent. This might be considered the beginning of the development of a geologic time scale. At the time Steno published his geological work in 1669, a tradition had not yet developed in the West for using observable geologic facts to decipher the history of Earth independent of the creation accounts in religious texts. Geoscientists working over the past 350 years have gained a tremendous amount of knowledge of the history of Earth and of life on Earth, organizing it chronologically using the geologic time scale.

Using Fossils to Date Sedimentary Rock

The sequence of strata that makes up the geologic record is a graveyard filled with the fossils of millions of kinds of organisms that are now extinct. Geologists know that they existed because of their fossilized remains, the traces of their activities like burrows and tracks, and sometimes the chemical traces left behind by organisms.

In the late 1700s and early 1800s, William Smith used his observations of fossils to help him correlate individual formations across large distances in Britain. He recognized that certain types of fossils were found grouped together in particular fossil assemblages and that the types of fossils changed from older strata to younger strata. That is, different fossils are found in rocks of different ages, and the same fossils tend to be found in rocks of the same age. From these observations by Smith and others arose the idea of fossil succession—the fossil record in sedimentary rock shows that species occur, become extinct, and are replaced by new species over time.

There is a practical advantage of being able to correlate sedimentary layers using their fossils. Once the sequence of major sedimentary layers had been established in a region (e.g., shale is overlain by coal, which is overlain by sandstone, which is overlain by limestone), people interested in exploiting natural resources or quarrying building stones had only to identify how a particular outcrop fit into the sedimentary section, using its unique combination of rock type and fossils. Then such people would be able to predict that the coal needed to produce energy is just below the sandstone, or the limestone needed for building stone is just above it. Names became attached to the major rock layers, and the first and last occurrence of particular fossils were also used to identify parts of the section by age. Eventually, the time-related names that we now use (e.g., Cambrian, Mesozoic, Pleistocene) became standardized into the eons, eras, periods, epochs, and stages of the modern geologic time scale.

The first geologic time scales were based on relative ages of geological units as marked by fossils. The laws of fossil succession and fossil assemblages were used to correlate sedimentary units across large distances. The divisions of relative time in the early geologic time scales were based on the appearance and disappearance of certain index fossils in the sedimentary record. An **index fossil** is a type of fossil that is particularly useful to define the age of sedimentary layers. The organisms that make the best index fossils were swimmers or floaters that were very widespread geographically, very abundant or common, readily preserved in sedimentary strata as fossils, and evolved rapidly in forms that are easy to identify. Although some index fossils are large, many are small or very small in size—microfossils. Pollen is also used as an index (whether it is fossilized or not) in sediments, sedimentary rock, and ice cores. The **range zone** of a given fossil species (or assemblage of fossil species) refers to the layers of sedimentary rock that contain fossils of that (or those) species from first occurrence to last in strata that were deposited at the time of extinction.

Numerical Dates Added to the Geologic Time Scale

Arthur Holmes first associated numerical dates to what had been an entirely relative time scale in 1915, based on developments in nuclear physics that permitted the use of radioactive isotopes to measure the age of minerals and rocks. Eventually, the numerical ages of key events in the history of life on Earth were determined. We found that the first occurrence of fossils that have hard parts, which approximately marks the base of the Cambrian Period, occurred ~541 million years ago. The great extinction that marks the end of the Paleozoic Era occurred ~252 million years ago. These events and their ages relative to each other were already well known because of the fossil record preserved in sedimentary rocks.

Most major extinctions are probably not rapid events like turning out the lights with the flip of a switch, but some seem to have been particularly abrupt by geological standards. The extinction that marks the end of the Cretaceous Period and the Mesozoic Era claimed the last of the dinosaurs (other than birds and crocodilians) as well as 70 to 90% of planktonic foraminifera species worldwide. The Cretaceous-Paleocene boundary is marked geologically by a thin clay bed that can be found in many places where sedimentary rock of that age is exposed at the ground surface (**Fig. 8.11**). The boundary clay is 66 million years old and is interpreted to have developed as a result of the collision of a ~10 km wide meteorite with Earth. The massive meteorite hit Earth on a continental shelf near the present-day northwest corner of the Yucatan Peninsula of eastern Mexico. The collision left a scar known as the Chicxulub crater, which is now buried under younger sedimentary rock. In addition to significant local and regional effects, the collision generated a tremendous volume of dust that circulated around Earth for weeks or months. The dust was enriched in iridium and included tiny glass spheres formed of the mist of molten rock blasted from the impact site as well as quartz crystals that are deformed in a distinctive way that is consistent with meteorite impact. The dust eventually settled to form the boundary clay.

Figure 8.11 Cretaceous-Paleogene boundary. The Cretaceous-Paleogene boundary layer (thin light-colored layer) as it occurs near the top of a coal seam near Trinidad in south-central Colorado. The equivalent of this layer can be found many places worldwide, so it is the result of a worldwide event. The cause is interpreted to be a ~10 km wide meteorite that hit Earth 66 Myr ago. The pencil is included to give you a sense of the size of the boundary layer and surrounding strata.

Two figures are provided to help you with the geologic time scale. **Fig. 8.12** is scaled so that the entire history of Earth is represented in two columns—one extending from Earth's formation ~4.55 billion years ago until ~2.275 billion years ago, and the other extending from ~2.275 to the present. In this illustration, the major time periods of the geologic time scale are scaled to their actual length or duration relative to each other. **Fig. 8.12** illustrates that 88% of Earth's history occurred during the Precambrian Eon before animals with hard parts appear in the fossil record. The earliest evidence of life on Earth, as currently understood by geoscience, is more than ~3.5 billion years old. The second figure (**Fig. 8.13**) is not scaled to time but rather is more of a table showing the order of geologic eras and periods along with their boundary ages and a few representative large-fossil (macrofossil) forms. They include the following:

- **Brachiopods** such as *Mucrospirifer, Platystrophia, Chonetes,* and *Strophomena* are members of a phylum of marine invertebrate animals with two symmetrical seashells of unequal size. They range throughout the Paleozoic, Mesozoic, and Cenozoic Eras, but they were most abundant in the Paleozoic Era. Only a few species exist today, so they are nearly extinct.

- **Trilobites** such as *Olenellus, Elrathia, Flexicalymene,* and *Phacops* are an extinct group of marine arthropods. Arthropods share a phylum of animals with external skeletons such as insects, spiders, and crustaceans like crabs and lobsters. Trilobites evolved into many forms and are found only in Paleozoic rocks, so they are good index fossils for the Paleozoic Era and its subdivisions.
- **Mollusks** such as *Inoceramus* and *Exogyra* are parts of a phylum of animals that includes snails, cephalopods (squid, octopuses), and bivalves (oysters, clams) with two asymmetrical shells of unequal size.
- **Plant** fossils of land plants occur in rock as old as ~475 Myr. Plant fossils are important indicators of environmental and climatic conditions.
- **Reptiles** are the class of vertebrate animals that includes lizards, snakes, turtles, and dinosaurs. **Dinosaurs** are found only in Triassic, Jurassic, and Cretaceous formations, so they are characteristic fossils for the Mesozoic Era.
- **Mammals** are a class of warm-blooded vertebrate animals that nurse their young, have hair, and are represented in the fossil record for about 195 million years. We are mammals.

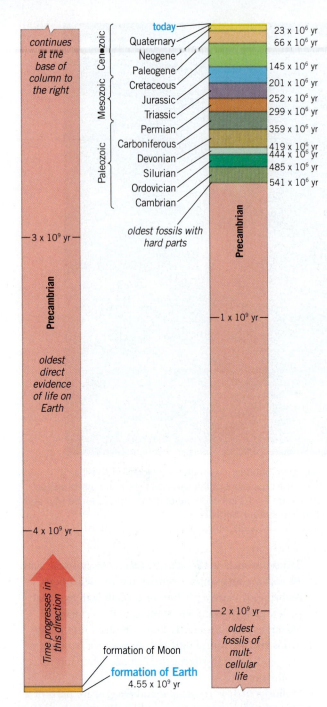

Figure 8.12 Geologic time scale. Major divisions of the geologic time scale shown in a length scale that is directly proportional to time.

- **Amphibians** are the class of vertebrate animals that includes modern frogs and salamanders.
- **Sharks** are a group of fish with teeth but with skeletons made of cartilage rather than bone. The earliest known sharks lived more than 400 million years ago.

The approximate time range of these fossil forms is indicated by the black vertical lines adjacent to the picture of the organism or its fossil in **Fig. 8.13**.

ACTIVITY 8.4

Numerical Dating of Rocks and Fossils, (p. 226)

Think About It   How do geologists determine the numerical age, in years, of Earth materials and events?

Objective   Calculate numerical ages to date Earth materials and events.

Before You Begin   Read the following section: Determining Numerical Ages by Radiometric Dating.

Plan Ahead   You might want to have access to a basic scientific calculator in working on this activity.

ACTIVITY 8.5

Infer Geologic History from a New Mexico Outcrop, (p. 227)

Think About It   How are relative and numerical dating techniques used to analyze outcrops and infer geologic history?

Objective   Apply relative and numerical dating techniques to analyze an outcrop in New Mexico and infer its geologic history.

Before You Begin   Read the following section: Determining Numerical Ages by Radiometric Dating as well as all preceding sections.

Plan Ahead   You might want to have access to a basic scientific calculator in working on this activity.

ACTIVITY 8.6

Investigating a Natural Cross-Section in the Grand Canyon, (p. 229)

Objective   Apply relative and numerical dating techniques to analyze an outcrop in the Grand Canyon and infer its geologic history.

Before You Begin   Read the following section: Determining Numerical Ages by Radiometric Dating as well as all preceding sections.

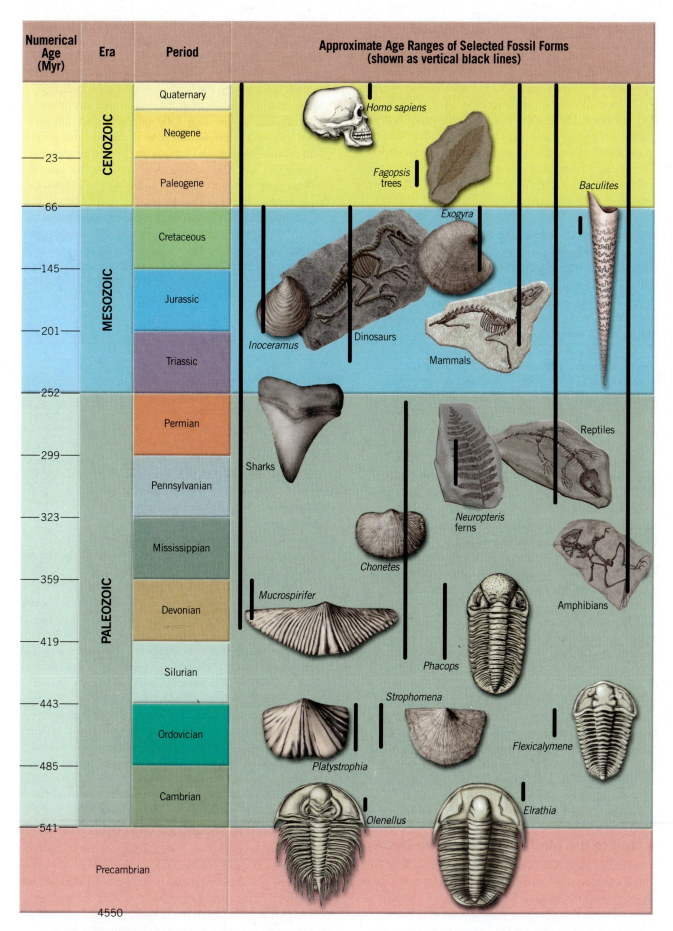

Figure 8.13 Range zones. Range zones (vertical bold black lines) of some well-known fossil forms shown within a representation of the geologic time scale.

Determining Numerical Ages by Radiometric Dating

Geological materials have been dated using radioactive isotopes since 1905 when Ernest Rutherford, Robert Strutt, and Bertram Boltwood pioneered the use of uranium isotopes to measure the time since certain minerals crystallized. Our understanding of nuclear chemistry and the processes of nuclear decay has developed steadily over the intervening time as has the accuracy of laboratory instruments used to measure the relative abundance of different isotopes in a sample.

Isotopes, Nuclides, Decay, and Half-Life

Each of the 98 naturally occurring **elements** on Earth is differentiated from the other elements by the number of **protons** in its **nucleus**. So, for example, any nucleus that contains six protons is by definition a carbon nucleus. All naturally occurring elements also have several **isotopes**, which differ from one another because they have different numbers of **neutrons** in their nucleus. The most abundant isotope of carbon is carbon-12, which has six neutrons and six protons in its nucleus ($6 + 6 = 12$) and comprises almost 99% of all carbon on Earth. The radioactive isotope carbon-14 is well known for its use in radiometric dating and has eight neutrons and six protons in its nucleus ($8 + 6 = 14$).

Rather than talk specifically about the isotopes of an element, it is convenient to refer to an atom or nucleus that has a specific number of protons and neutrons as a distinct **nuclide**. A nuclide is analogous to a specific species (e.g., *Homo sapiens*) rather than an individual of that species (e.g., my neighbor Juanita). Our current understanding is that there are 254 **stable nuclides** that do not change over time and more than 1500 unstable nuclides. An **unstable nuclide** is said to be **radioactive** and spontaneously decays through one of several mechanisms (e.g., **alpha decay**, **beta decay**, **electron capture**) into another nuclide. Potassium-40 decays to argon-40, for example.

Radioactive decay is a statistical process that involves changes in countless trillions of atoms over time. As Brent Dalrymple of the USGS noted, "as little as 0.00001 gram of potassium contains 150,000 trillion atoms!" The **half-life** of a radioactive nuclide is the amount of time it takes for one-half of the atoms in a population to decay into another nuclide. The half-life for a given radioactive nuclide is a constant. The half-lives of several radioactive nuclides that are useful in dating geologic materials are given in **Fig. 8.14**.

A Nod to the Math of Radioactive Decay

Rather than jump into the deep end right away, let's work with a situation that is analogous to radioactive decay but easier to visualize. Imagine that you inherit $1 billion dollars. You decide that your personal sense of justice demands that every year on the anniversary of your windfall inheritance, you take half of the remaining inheritance and give it away to worthy charities. Hence, the half-life of your inheritance fund is 1 year. (We will also imagine that you do not spend any of it, foolishly choose not to invest any part of it, and the government exempts you from paying gift taxes so that the total amount does not change between anniversaries.) On the first anniversary of your inheritance, you give away $500 million and are left with $500 million. On the second anniversary, you give away half of the $500 million that remains in your inheritance, leaving you with $250 million. And so on. At that rate, it will take you 30 years of selfless giving—30 half-lives—for your fund to drop to below $1.

Radioactive decay follows this same mathematical pattern. If we can measure the abundance of the radioactive parent isotope (P_t) of a given atom and the abundance of its stable daughter product (D_t) in a geological sample and we know the half-life of this decay sequence ($t_{1/2}$), we can compute the age (t) of the specimen.

$$t = \left(\frac{t_{1/2}}{-\ln(0.5)} \right) \times \ln\left(\frac{D_t}{P_t} + 1 \right)$$

The "ln" is the natural log, or in other words, the logarithm to base e ($e \approx 2.71828\ldots$). There is no need to have a panic attack because there is an "ln" function button on most inexpensive scientific calculators, and there are simple web calculators (e.g., **http://web2.0calc.com** and **www.mathgoodies.com/calculators/calc4chem.html**) and calculator apps for smartphones that can compute natural logs. The built-in calculator on an iPhone becomes a scientific calculator if you rotate the phone into a horizontal (landscape) position.

http://goo.gl/wTp2x7

http://goo.gl/vPDhgs

Let's see if the equation actually works to model our charitable activities in the example. We can look at the pattern that was explained above and conclude that on the day before the third anniversary, you would have $250 million in the fund, so on the anniversary you would give away $125 million and retain $125 million. Hence, you would retain $P_t = \$125,000,000$ and would have given away a total of $D_t = \$875,000,000$, where $P_t + D_t = \$1$ billion. The half-life ($t_{1/2}$) of your charitable giving is 1 year. Plugging the values for variables $t_{1/2}$, D_t, and P_t into the equation, the answer you would compute (t) is 3, which you had already determined using an independent process. Now that you have a mathematical model that seems to follow the pattern faithfully, you can use it for radiometric age determination.

Radiometric Dating of Geologic Materials

To determine the age of a geological material, it must contain atoms of a suitable radioactive nuclide and of its stable daughter product, and that population of atoms must have remained in the specimen without addition or loss since the decay process began in the specimen. The abundance of the parent isotope (P_t) and of the daughter isotope (D_t) in the specimen is determined using a geochemical

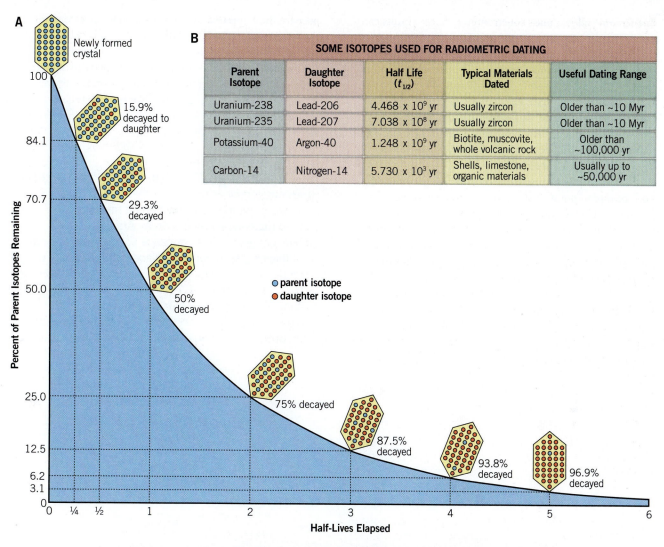

A. Radiometric decay curve showing percent of parent isotopes remaining versus half-lives elapsed.

- parent isotope
- daughter isotope

Newly formed crystal

15.9% decayed to daughter

29.3% decayed

50% decayed

75% decayed

87.5% decayed

93.8% decayed

96.9% decayed

Percent of Parent Isotopes Remaining

Half-Lives Elapsed

B.

SOME ISOTOPES USED FOR RADIOMETRIC DATING				
Parent Isotope	Daughter Isotope	Half Life ($t_{1/2}$)	Typical Materials Dated	Useful Dating Range
Uranium-238	Lead-206	4.468×10^9 yr	Usually zircon	Older than ~10 Myr
Uranium-235	Lead-207	7.038×10^8 yr	Usually zircon	Older than ~10 Myr
Potassium-40	Argon-40	1.248×10^9 yr	Biotite, muscovite, whole volcanic rock	Older than ~100,000 yr
Carbon-14	Nitrogen-14	5.730×10^3 yr	Shells, limestone, organic materials	Usually up to ~50,000 yr

C.

Half-Life of Radioactive Isotope ($t_{1/2}$)	Half-Lives Elapsed (n)		Percent of Radioactive Parent Isotope (P_t)	Percent of Stable Daughter Isotope (D_t)	Age of Specimen (t)
	As a Ratio	As a Decimal number	$P_t = \left(\frac{1}{2^n}\right)$	$D_t = \left(1 - \frac{1}{2^n}\right)$	$t = n \times t_{1/2}$
Find the half-life for the appropriate radioactive isotope in Table B above	0	0.000	100.0	0.0	$0.000 \times t_{1/2}$
	1/64	0.016	98.9	1.1	$0.016 \times t_{1/2}$
	1/32	0.031	97.9	2.1	$0.031 \times t_{1/2}$
	1/16	0.062	95.8	4.2	$0.062 \times t_{1/2}$
	1/8	0.125	91.7	8.3	$0.125 \times t_{1/2}$
	1/4	0.250	84.1	15.9	$0.250 \times t_{1/2}$
	1/2	0.500	70.7	29.3	$0.500 \times t_{1/2}$
	1	1.000	50.0	50.0	$1.000 \times t_{1/2}$
	1½	1.500	35.4	64.6	$1.500 \times t_{1/2}$
	2	2.000	25.0	75.0	$2.000 \times t_{1/2}$
	3	3.000	12.5	87.5	$3.000 \times t_{1/2}$
	4	4.000	6.2	93.8	$4.000 \times t_{1/2}$
	5	5.000	3.1	96.9	$5.000 \times t_{1/2}$

Figure 8.14 Radiometric dating. A. Radiometric decay curve. **B.** Table of some isotopes useful for radiometric dating, their decay parameters, and their useful ranges for dating. **C.** Table of points along the decay curve with corresponding equations to compute the age given the half-life ($t_{1/2}$).

instrument called a **mass spectrometer**. These results can either be used along with the half-life to directly compute the age, or a decay curve can be used to estimate the number of half-lives that must have passed in order to result in the observed isotopic abundances (**Fig. 8.14A**). The number of half-lives that have elapsed (n) can then be multiplied by the half-life of that particular decay series ($t_{1/2}$) to yield an estimate of the age of the specimen (see **Fig. 8.14B** and **8.14C**). The decay curve can be constructed ahead of time using the data in **Fig. 8.14C**.

For example, a sample of Precambrian granite contains biotite mineral crystals, so it can be dated using the potassium-40 to argon-40 decay pair. If there are three argon-40 atoms in the sample for every one potassium-40 atom, then the sample is 25.0% potassium-40 parent atoms (P_t) and 75.0% argon-40 daughter atoms (D_t). This means that two half-lives have elapsed, so the age of the biotite (and hence the granite) is 2.0 times 1.3 billion years, which equals 2.6 billion years.

This explanation of radiometric dating has necessarily simplified the process. Accurate age determinations require excellent instrumentation, experienced technicians to prepare the samples and operate the mass spectrometer, a thorough understanding of the samples and their geological environment, and the ability to analyze and interpret results based on the best practices developed during more than a century of research effort. The payoff is our ever-improving understanding of how our world has changed over time.

MasteringGeology™

Looking for additional review and test prep materials? Visit the Study Area in MasteringGeology to enhance your understanding of this chapter's content by accessing a variety of resources, including Pre-Lab Videos, Self-Study Quizzes, Geoscience Animations, Mobile Field Trips, *Project Condor* Quadcopter videos, *In the News* articles, glossary flashcards, web links, and an optional Pearson eText.

Name: _____ Course/Section: _____ Date: _____

A Analyze the piece of layer cake in **Fig. A8.1.1**. Each side of the block of cake is a vertical **cross-section** of the layers. Also notice the surfaces between the layers, where two different layers touch each other. Geologists refer to surfaces between layers or other bodies of rock as **contacts**.

1. Think about the process used to construct the layer cake from depositing the first layer to depositing the last layer. On the left edge of the cake, number the layers to show the sequence of steps in which they were deposited to make the layer cake from 1 (first step) to "n" (the number of the last step).

2. Using a pen, draw lines on the layer cake to mark all of the contacts between layers. Then place arrows along the right edge of the cake that point to each contact. Label each arrow to show its relative age from 1 for the first or oldest contact to "n"—the number corresponding to the youngest contact.

Figure A8.1.1

B **Figure A8.1.2** shows an outcrop about 5 meters thick near Sedona, Arizona. The red rock in **Fig. A8.1.2** is an ancient soil called a paleosol. The brown layer in which grass is rooted near the top of the photo is modern soil. The blocky brown-gray rock with wide fractures is an ancient lava flow. This outcrop is a natural geologic cross-section of rock layers, analogous to the cake.

1. Which layer is the oldest? How do you know?

2. Using a pen, carefully draw lines on **Fig. A8.1.2** that mark the position of:
   (a) the contact between the red ancient soil and the lava flow.
   (b) the contact between the top of the lava flow and the base of the darker brown modern soil in which grass is growing.

3. Notice the fractures that cut across the lava flow layer. Are they older or younger than the lava flow? How do you know?

4. Notice that *clasts* of the lava flow are included in the brown soil. Are they older or younger than the brown soil? How do you know?

Figure A8.1.2

C Analyze the outcrop shown in **Fig. A8.1.3** photographed by geologist Thomas McGuire. It is another natural geologic cross-section with red sandstone layers on the bottom and a tan conglomerate (gravel) rock layer on top. Notice that the red rock layers are not horizontal. They are bent like a gently folded newspaper, down in the middle.

1. Using a pen on **Fig. A8.1.3**, carefully trace two of the contacts between layers within the red sandstone as well as you can. Assuming that the red sandstone layers were originally horizontal, what might have caused them to be folded in this way?

2. On both sides of **Fig. A8.1.3**, use an arrow to indicate the location of the contact between the red sandstone and the horizontal tan conglomerate above it. This surface is an unconformity like the one shown in **Fig. 8.4**. What sequence of events may have happened to form the unconformity in **Fig. A8.1.3**?

D **REFLECT & DISCUSS**   In all of your work above, you had to interpret the relative ages of rock layers, fractures, folds, and clasts included in soil. Based on your work, write down three rules that a geologist could follow to tell the relative ages of rock layers, fractures, clasts, and folds in geologic cross-sections.

Figure A8.1.3

Name: _____ Course/Section: _____ Date: _____

A Review the key to symbols at the bottom of **Fig. A8.2.1**. On the lines provided for each cross-section in **Figs. A8.2.1** and **A8.2.2**, write letters to indicate the sequence of events from oldest (first in the sequence of events) to youngest (last in the sequence of events). Refer to **Figs. 8.2–8.10** and the principles of relative dating that are explained in the text as needed.

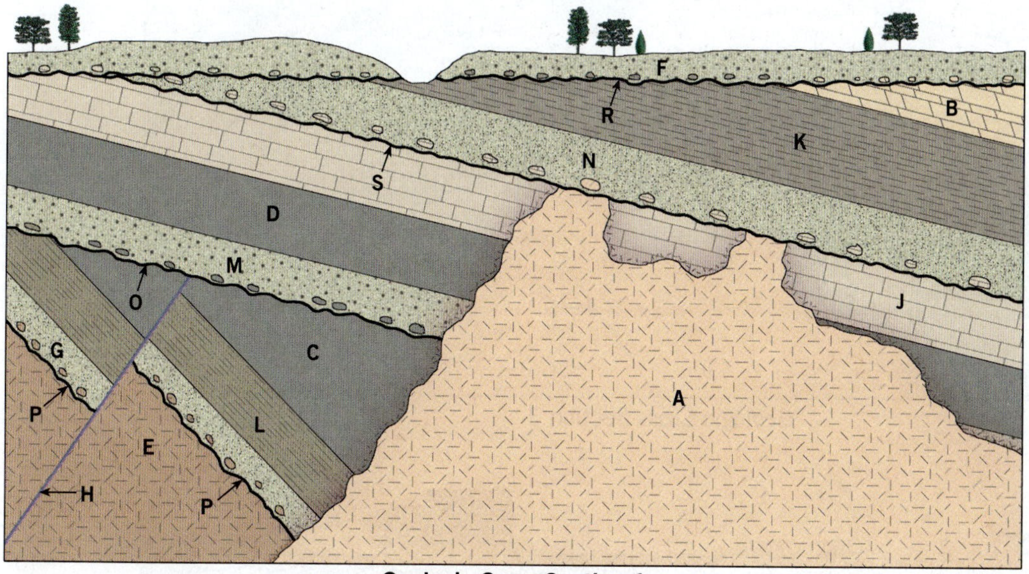

Geologic Cross Section 1

Youngest _____
_____
_____
_____
_____
_____
_____
_____
_____
_____
_____
_____
_____
_____
Oldest _____

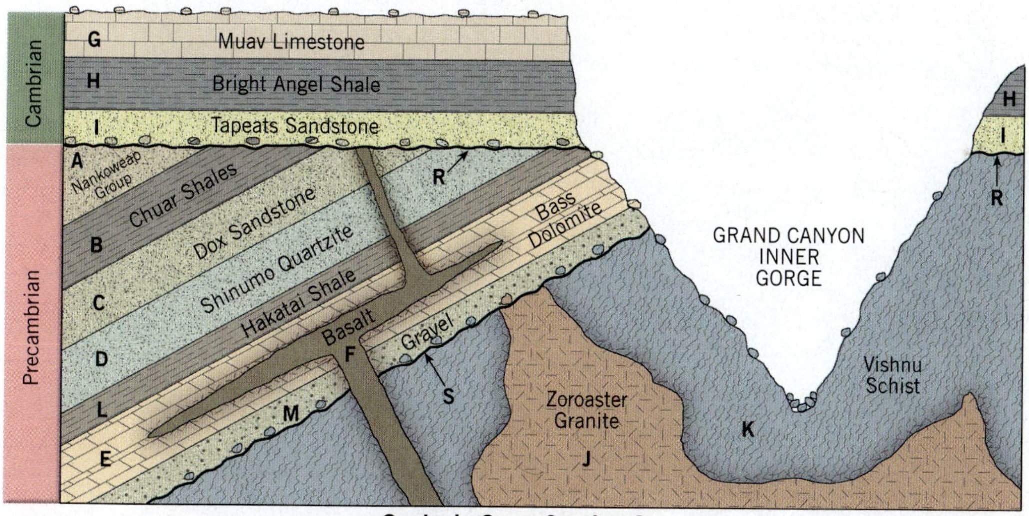

Geologic Cross Section 2

Youngest _____
_____
_____
_____
_____
_____
_____
_____
_____
_____
_____
_____
_____
Oldest _____

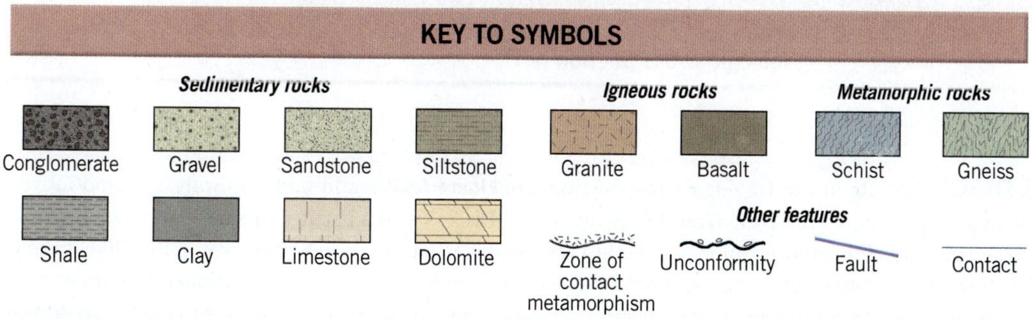

KEY TO SYMBOLS

Sedimentary rocks

Conglomerate   Gravel   Sandstone   Siltstone

Igneous rocks

Granite   Basalt

Metamorphic rocks

Schist   Gneiss

Shale   Clay   Limestone   Dolomite

Other features

Zone of contact metamorphism   Unconformity   Fault   Contact

Figure A8.2.1

223

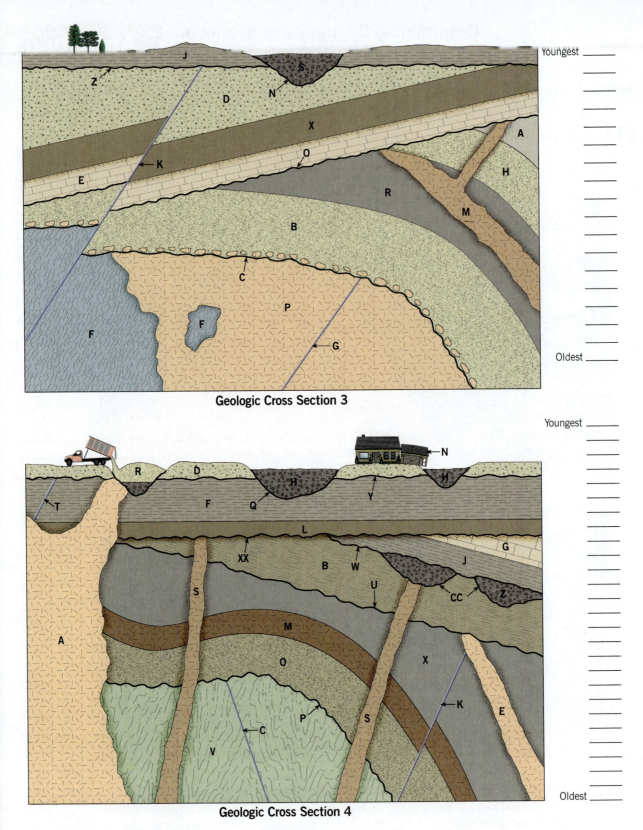

Geologic Cross Section 3

Youngest _____
_____
_____
_____
_____
_____
_____
_____
_____
_____
_____
_____
_____
_____
Oldest _____

Geologic Cross Section 4

Youngest _____
_____
_____
_____
_____
_____
_____
_____
_____
_____
_____
_____
_____
_____
_____
_____
_____
_____
_____
_____
Oldest _____

Figure A8.2.2

B **REFLECT & DISCUSS** Return to Geologic Cross-Section 2 in **Figure A8.2.1**, and notice how the Colorado River has cut down through the rocks to create the Grand Canyon Gorge. Discuss with a partner or small group what law(s) of relative dating you would need to apply if you were asked to reconstruct what the cross-section would look like if the Grand Canyon Gorge had not developed in this area. As carefully as you can, apply the law(s) and use dashed lines to draw in the contacts between named rock layers that were eroded away in Grand Canyon Gorge. Compare your completed drawing with those of other students.

Name: _____ Course/Section: _____ Date: _____

A Analyze this fossiliferous rock from New York (**Fig. A8.3.1A**).

1. What index fossils from **Fig. 8.13** are present?

2. The overlap in ranges of the index fossils indicates the age of the rock. Express that age range in terms of the period(s) of the geologic time scale, using informal modifiers like "early," "middle," or "late" as appropriate. Example: late Triassic to early Jurassic.

3. Using the information in **Fig. 8.13** as your guide, how old is this rock? Express your answer as a range in millions of years.

A

x1

B Analyze this fossiliferous sand from Delaware (**Fig. A8.3.1B**).

1. What index fossils from **Fig. 8.13** are present?

2. Express the age range of this sandstone in terms of the period(s) of the geologic time scale using informal modifiers like "early," "middle," or "late" as appropriate.

3. Using the information in **Fig. 8.13** as your guide, how old is this rock? Express your answer as a range in millions of years.

B

x1/3

C Analyze this fossiliferous rock from Ohio (**Fig. A8.3.1C**).

1. What index fossils from **Fig. 8.13** are present?

2. Express the age range of this rock in terms of the period(s) of the geologic time scale, using informal modifiers like "early," "middle," or "late" as appropriate.

3. Using the information in **Fig. 8.10** as your guide, how old is this rock? Express your answer as a range in millions of years.

C

x1

D Using **Fig. 8.13**, re-evaluate the geologic cross-section in **Fig. 8.2** based on its fossils.

1. Which two formations are separated by a disconformity?

Figure A8.3.1

2. The discontinuity indicates a gap in the rock record. What is the minimum and maximum amount of time that is not reflected in the rock record in this location?

minimum: _____ Myr   maximum: _____ Myr

E **REFLECT & DISCUSS** What geologic event occurred during the Mesozoic Era in the region where **Fig. 8.6** is located? Explain.

A A solidified lava flow containing zircon mineral crystals is present in a sequence of rock layers that are exposed in a hillside. A mass spectrometer analysis was used to count the atoms of uranium-235 and lead-207 isotopes in zircon samples from the lava flow. The analysis revealed that 71% of the atoms were uranium-235, and 29% of the atoms were lead-207. Refer to **Fig. 8.14** to help you answer the following questions.

1. About how many half-lives of the uranium-235 to lead-207 decay pair have elapsed in the zircon crystals?

   _____

2. What is the numerical age of the lava flow based on its zircon crystals? Explain how you arrived at your answer.

3. What is the age of the rock layers above the lava flow? _____

4. What is the age of the rock layers beneath the lava flow? _____

B Astronomers think that Earth probably formed at the same time as all of the other rocky materials in our solar system, including the oldest meteorites. The oldest meteorites ever found on Earth contain nearly equal amounts of both uranium-238 and lead-206. Based on **Fig. 8.14**, what is Earth's approximate age? Explain your reasoning.

C The radioactive isotope carbon-14 (C-14) is continuously replenished in organisms while they are alive. When an organism dies, it is no longer able to take in new C-14, and so the amount of C-14 decreases as it decays to its stable daughter product: nitrogen-14 (N-14).

1. The carbon in a buried peat bed has about 6% of the C-14 of modern shells. When the plants that now form the buried peat were alive, they absorbed C-14 and probably had about the same amount of C-14 as modern shells, so about 94% of the peat's original C-14 has decayed. What is a reasonable initial estimate of the age of the peat bed? Explain.

2. In sampling the peat bed, you must be careful to avoid any young plant roots or old limestone. Why?

D Zircon ($ZrSiO_4$) forms in magma as it cools into igneous rock. It is also useful for numerical dating (**Fig. 8.14**).

1. If you walk on a modern New Jersey beach, then you will walk on some zircon sand grains. Yet if you determine the numerical age of the zircons, it does not indicate a modern age (zero years) for the beach. Why?

2. Suggest a rule that geologists should follow when they date rocks based on the radiometric ages of mineral grains inside the rocks.

E **REFLECT & DISCUSS** An "authentic dinosaur bone" is being offered for sale on the Internet. The seller claims that he had it analyzed by scientists who confirmed that it is a dinosaur bone and used carbon dating to determine that it is 400 million years old. Discuss the seller's claims with a partner or in a small group. Should you be suspicious of this bone's authenticity? Explain. (See **Figs. 8.13** and **8.14**.)

Name: _____ Course/Section: _____ Date: _____

A Refer to **Fig. A8.5.1**, which shows an outcrop in a surface coal mine in northern New Mexico. Note the sill, sedimentary rocks, fault, and places where a fossil leaf was found and isotope data for zircon crystals in the sill.

1. What is the relative age of the sedimentary rocks in this rock exposure? Explain your reasoning.

2. What is the numerical age of the sill? Use the information in **Fig. 8.14**, and show how you calculated the answer.

3. Locate the fault. Approximately how much separation has occurred along this fault? _____ m
   What additional information would you like to have to make a better estimate of fault separation?

Figure A8.5.1

B Make a numbered list of the geologic events that contributed to the development of the geological features in this outcrop, starting with deposition of the sandstone (oldest event: 1) and ending with the time this picture was taken. Include the name(s) of relevant period(s) from the geologic time scale as well as the isotopic age of the sill in your writing. *Your reasoning and number of events may differ from those of other students.*

1.

2.

3.

4.

5.

6.

7.

8.

C **REFLECT & DISCUSS** Write a question that you have about the geologic history of this location. What geologic evidence would you need to answer the question?

Investigating a Natural Cross-Section in the Grand Canyon

Activity 8.6

Name: _____ Course/Section: _____ Date: _____

A **Figure A8.6.1A** is a photograph of part of the bottom of the Grand Canyon, which extends east to west across northern Arizona. The photo was taken west of Grand Canyon Village on the south rim of the Canyon, looking at the north side of the bottom of the canyon.

Carefully analyze the photograph for rock layering. The very bottom rock layers in the foreground are folded Precambrian metamorphic rock called the Vishnu Schist, which contains narrow white bodies of igneous rock. The Vishnu Schist is overlain here by relatively horizontal layers of sedimentary rock.

Figure A8.6.1

1. Using a pen or colored pencil on **Fig. A8.6.1B**, carefully trace the contact between the Vishnu Schist and the relatively horizontal sedimentary rocks above it.

2. Based on **Figs. 8.2–8.10**, what specific kind of unconformity occurs along the contact you just traced on **Fig. A8.6.1B**?

B **REFLECT & DISCUSS**   The Vishnu Schist has a numerical age of about 1700 million years. The Lower Cambrian Tapeats Sandstone was deposited on top of the unconformity that you considered in part **A**. Imagine that you find reliable information indicating that the Tapeats Sandstone was deposited at the very start of the Cambrian Period and then how about much time is represented by the rock record that is "missing" across this unconformity—missing in the sense of erosion, nondeposition, or both prior to deposition of the Tapeats Sandstone on the erosional surface at the top of the Vishnu Schist?

C **Figure A8.6.2A** is another photograph of part of the bottom of the Grand Canyon. The photo was taken east of Grand Canyon Village on the south rim of the Canyon, looking at the north side of the canyon. All of the rocks in this scene are sedimentary rocks.

1. Analyze this canyon scene. Then, using a pen or colored pencil on **Fig. A8.6.2B**, carefully trace the contact that marks a significant unconformity in the photo.

2. What kind of unconformity did you identify?

Figure A8.6.2

D **REFLECT & DISCUSS** What evidence did you apply to justify drawing an unconformity where you did on **Fig. A8.6.2B**?

Topographic Maps

Pre-Lab Video 9

https://goo.gl/ymsX24

CONTRIBUTING AUTHORS

Charles G. Higgins • *University of California*
John R. Wagner • *Clemson University*

James R. Wilson • *Weber State University*

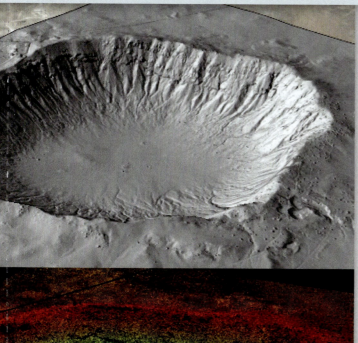

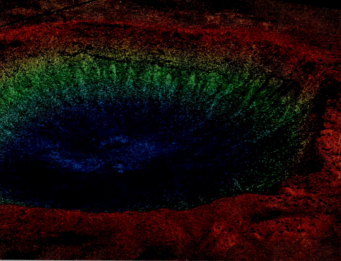

▲ At top is a hillshade map of Meteor Crater, Arizona (35.027°N, 111.023°W), based on a 3-D digital elevation model constructed using an airborne laser system (lidar). Below is a cloud of lidar points, colored by elevation from blue (lower) to red (higher).

BIG IDEAS

Topographic maps are two-dimensional representations of three-dimensional landscapes viewed from directly above. Horizontal (two-dimensional) positions of landscape features are given relative to a coordinate reference frame and represented at consistent scale throughout the map. The third dimension, elevation (height) of the landscape, is represented with contours that join points of equal elevation relative to sea level. The three-dimensional and quantitative aspect of topographic maps makes them valuable to geologists and other people who want to know the shapes and elevations of Earth's surface. They can be used in combination with orthoimages—aerial photographs that have been adjusted to the same scale as the map.

FOCUS YOUR INQUIRY

Think About It   How are specific places located using the geographic latitude–longitude coordinate system, and how does Google Earth help us to study Earth's surface?

ACTIVITY 9.1   Map and Google Earth Inquiry (p. 232, 253)

Think About It   What coordinate systems, scales, directional data, and symbols are used on maps?

ACTIVITY 9.2   Map Locations, Distances, Directions, and Symbols (p. 232, 255)

Think About It   How are topographic maps constructed and interpreted?

ACTIVITY 9.3   Topographic Map Construction (p. 244, 258)

ACTIVITY 9.4   Topographic Map and Orthoimage Interpretation (p. 244, 259)

Think About It   How is a topographic profile constructed from a topographic map, and what is its vertical exaggeration?

ACTIVITY 9.5   Relief and Gradient (Slope) Analysis (p. 250, 262)

ACTIVITY 9.6   Topographic Profile Construction (p. 250, 264)

Map and Google Earth Inquiry, (p. 253)

Think About It How are specific places located using the geographic latitude–longitude coordinate system, and how does Google Earth help us to study Earth's surface?

Objective Apply the latitude–longitude coordinate system to locate a country, quadrangle, and place of your choice, and then explore it in greater detail using Google Earth tools, layers, ruler, and historical imagery.

Before You Begin Read the following sections: Introduction and Latitude–Longitude and Quadrangle Maps.

Plan Ahead This activity requires that you use Google Earth. You will either need to have access to the web and to Google Earth during the lab period, or this activity will have to be completed outside of lab time. Google Earth is a free application, and current information about how to access and use Google Earth is available online at earth.google.com.

Map Locations, Distances, Directions, and Symbols, (p. 255)

Think About It What coordinate systems, scales, directional data, and symbols are used on maps?

Objective Identify and characterize features on topographic maps using printed information, compass bearings, scales, symbols, and three geographic systems: latitude and longitude, the U.S. Public Land Survey System (PLSS), and the Universal Transverse Mercator (UTM) system.

Before You Begin Read the following sections: Latitude–Longitude and Quadrangle Maps, Map Scales, Declination and Bearing, UTM—Universal Transverse Mercator System, Geodetic Datum, and Public Land Survey System.

Plan Ahead You will need a simple calculator or calculator app on your smartphone, a ruler, and a protractor.

Introduction

Imagine that you are seated with a friend who asks you how to get to the nearest movie theater. To find the theater, your friend must know locations, distances, and directions.

You must have a way of communicating your current location, the location of the movie theater, plus directions and distances from your current location to the movie theater. You may also include information about the topography of the route (whether it is uphill or downhill) and landmarks to watch for along the way. Your directions may be verbal or written, and they may include a map, satellite image, or a photograph taken from an aircraft. Geologists are faced with similar circumstances in their field (outdoor) work. They must characterize geologic features, the places where they occur, and their sizes, shapes, elevations, and locations in relation to other features. Satellite images and aerial photographs are used to view parts of Earth's surface from above (**Fig. 9.1A**), and this information is summarized on maps.

A **map** is a flat representation of part of Earth's surface as viewed from above and reduced in size to fit a sheet of paper or computer screen. A **planimetric map** (**Fig. 9.1B**) is a flat representation of Earth's surface that shows horizontal (two-dimensional) positions of features like streams, landmarks, roads, and political boundaries. A **topographic map** shows the same horizontal information as a planimetric map but also includes **topographic contours** to represent elevations on the ground surface. The contours are the distinguishing features of a topographic map and make it appear three dimensional. Thus, topographic maps show the shape of the landscape in addition to horizontal directions, distances, and a system for describing exact locations.

The U.S. Geological Survey (USGS) publishes topographic maps of the United States that are available from their US Topo website (**http://nationalmap.gov/ustopo/** or **http://store.usgs.gov**). GeoPDF files of current and historic USGS topographic maps can be downloaded at no cost via **http://geonames.usgs.gov/pls/topomaps/**. Canadian topographic maps are produced by the Centre for Topographic Information of Natural Resources Canada (NRCAN: **http://maps.nrcan.gc.ca**). State and provincial geological surveys as well as the national geological surveys of other countries also produce and distribute topographic maps.

Latitude–Longitude and Quadrangle Maps

Earth is nearly spherical, so any coordinate system used to map its surface must be adapted to its shape. Points on its surface can be located using a geographic coordinate system based on latitude and longitude (**Fig. 9.2**).

Latitude–Longitude Coordinate System

Earth's nearly spherical surface is divided into circles of latitude (**parallels**) that are parallel to the equator and semicircles or **meridians** of longitude that extend from pole to pole (**Fig. 9.2**). The poles represent the points where the spin axis of Earth intersects the ground surface. The plane that is perpendicular to Earth's spin axis and that passes through Earth's center is called the **equator**. All points along the equator have a latitude of 0°, and the poles have latitudes of 90° north and 90° south, respectively. In some systems, a south latitude is considered a negative latitude.

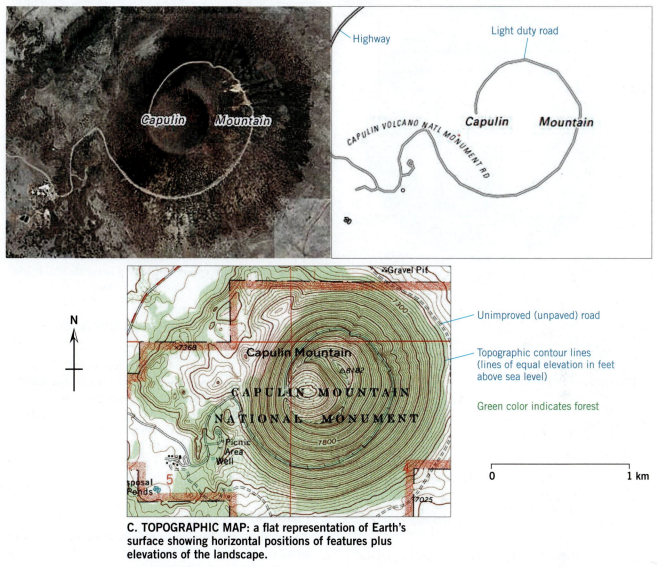

A. AERIAL PHOTOGRAPH: a flat picture or image of Earth's surface.

B. PLANIMETRIC MAP: a flat representation of Earth's surface showing horizontal positions of feature.

Highway

Light duty road

Capulin   Mountain

CAPULIN VOLCANO NATL MONUMENT RD

N

Capulin Mountain

CAPULIN MOUNTAIN

NATIONAL MONUMENT

Picnic Area
Well

Gravel Pit

Unimproved (unpaved) road

Topographic contour lines (lines of equal elevation in feet above sea level)

Green color indicates forest

0   1 km

C. TOPOGRAPHIC MAP: a flat representation of Earth's surface showing horizontal positions of features plus elevations of the landscape.

Figure 9.1 Comparison of an aerial photograph with planimetric and topographic maps.

By international agreement in 1884, the **prime meridian** was defined as passing through the center of the Airy Transit Circle Telescope at the Royal Astronomical Observatory in Greenwich, England. A century later in the age of orbital satellites, we were able to determine the exact location of Earth's center of mass, reportedly within centimeters. It then became clear that the axis of the Airy telescope was not perfectly aligned with a plane that passes through Earth's center. So in 1984, the prime meridian was redefined in light of more accurate data. The current prime meridian is the intersection of Earth's surface and a plane that passes through Earth's geodetic center and spin axis, and that is 0.001475° east of the 1884 prime meridian. A group called the *International Earth Rotation and Reference Systems Service* (**IERS**) keeps track of this for us. Points located along the prime meridian have a longitude of 0°. Longitudes are either specified by direction (east or west longitude) or by sign (positive longitudes are east of

the prime meridian and negative longitudes are west of the prime meridian).

For example, point **A** in **Fig. 9.2** is located at coordinates of 20° north latitude, 120° west longitude (20°N 120°W or 20° −120°). When it is necessary to specify a location more exactly, two ways of subdividing degrees of latitude and longitude are common. Each degree of latitude and longitude can be subdivided into 60 minutes, and each minute can be divided into 60 seconds—for example, 36° 42′ 23.4″. A more useful and less complicated way of subdividing degrees is to use decimal degrees—for example, 36.7065°.

Quadrangle Maps. Most topographic maps published by the USGS depict rectangular sections of Earth's surface, called quadrangles. A **quadrangle** is a map area that is bounded by lines of latitude at the north and south and by lines of longitude on the west and east (**Fig. 9.2**).

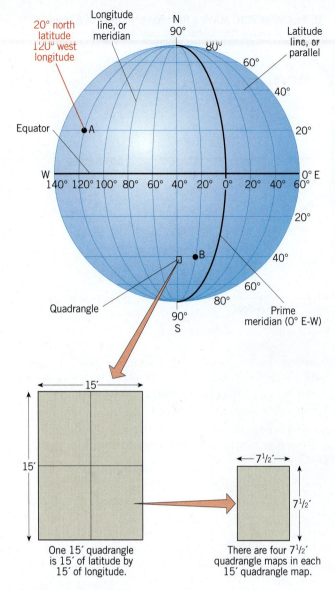

Figure 9.2 Latitude and longitude coordinate system and quadrangles. Point **A** is located at coordinates of 20° north latitude, 120° west longitude.

Quadrangle maps are published in many different sizes, but the most common USGS sizes are 15-minute and 7.5-minute quadrangle maps (**Fig. 9.2**). The numbers refer to the amount of area that the maps depict in degrees of latitude and longitude. A 15-minute topographic map represents an area that is one-quarter of a degree (15 minutes) of latitude long by one-quarter of a degree of longitude wide. A 7.5-minute quadrangle is one-eighth of a degree on a side, so four 7.5-minute quadrangle maps compose one 15-minute quadrangle map (**Fig. 9.2**).

A reduced copy of a 7.5-minute USGS topographic map is provided in **Fig. 9.3**. Notice its name (Ritter Ridge, CA) and size (7.5-Minute Series, SW ¼ of the Lancaster 15′ Quadrangle) in the upper right and lower right corners of the map, respectively. Also notice that the map has colors, patterns, and symbols (**Fig. 9.4**) that are used to depict water bodies, vegetation, roads, buildings, political boundaries, place names, and other natural and cultural

features of the landscape. The lower right corner of the map indicates that the map was originally published in 1958, and that it was photorevised in 1974. *Photorevised* means that aerial photographs were used to discover changes on the landscape, and the changes are over-printed on the maps in a standout color like purple, red, or gray. The main new features shown on this 1974 photorevised map are several major highways and the California Aqueduct that carries water south from the Sierra Nevada Mountains to the urban areas of southern California.

Map Scales

Maps are representations of an area of Earth's surface. The real sizes of everything on a map have been reduced so they fit a sheet of paper or computer screen, so maps are scale models. To understand how the real world is depicted by the map, you must refer to the map scales. Topographic maps commonly have a bar scale and might have a fractional scale printed on the map.

Bar Scales for Measuring Distances on the Map

The most obvious scales on a USGS topographic map are the **bar scales** printed in its lower margin (**Fig. 9.3**). Bar scales are rulers for measuring distances on the map. USGS topographic maps can have as many as four different bar scales: miles, feet, kilometers, and meters. The benefit of bar scales in a digital age is that they expand and contract with the map as it is reproduced or displayed at different scales.

Fractional Scale

A numerical scale based on a ratio is commonly expressed above the bar scales in the bottom margins of topographic maps that are carefully printed at scale on paper. This is called a **representative fraction scale** or simply a **fractional scale**. The typical scale of a USGS 7.5-minute topographic map is 1:24,000, which means that a length of 1 unit of some kind on the map—say, a 1 cm line—is equal to 24,000 of that same unit within the area depicted by the map, on the ground—24,000 cm or 240 meters (m). The USGS produced maps at that scale because most people in the United States use the system of measurement we inherited from the English when we were colonies. So 1 inch on a 1:24,000 scale map is the equivalent of 2000 feet on the ground in the mapped area. This English system is cumbersome to use, and even the English have abandoned it in favor of the metric system. The USGS has published topographic maps at a variety of scales, ranging from about 1:12,000 to 1:10,000,000.

A fractional scale that is printed on a map is meaningful only on the original printed copy of that map or on a duplicate that is exactly the same size as the original. That limitation makes the fractional scale much less useful than bar scales for maps that might be viewed or printed at different sizes.

Declination and Bearings

All USGS topographic quadrangle maps are bounded by meridians of longitude and parallels of latitude. So the left and right edges of a USGS quadrangle are not quite parallel to one another, but each is aligned north–south for that part of the map area. (Imagine you cut a wedge out of an apple and carefully pick and scrape the pulp of the apple off of the apple skin. When you flatten the apple skin, you will see that the left and right edges of the apple skin are not parallel except at the "equator" of the apple. In a similar way, the left and right edges of a quadrangle from the northern hemisphere are not parallel lines but are curves that converge toward the north pole.)

The orientations of **true north**, **magnetic north**, and the UTM **grid north** for the point at the center of a USGS topographic quadrangle map are indicated by three lines called **mean declination arrows** that are printed at the bottom margin of the map, as in **Fig. 9.3B**. The arrow that ends in a star (or TN) points toward true north. The arrow labeled MN pointed toward magnetic north at the time the map graphic was created. The arrow labeled GN indicates the orientation of the north line in the UTM grid based on the geodetic datum of the map.

What Is Declination?

The north and south magnetic poles are not located at the north and south poles, and their location varies over time. Hence, magnetic north (the direction toward the north magnetic pole) indicated by a compass needle is not the same as true north (the direction toward the north spin axis of Earth) at most places on Earth. The direction of Earth's magnetic field changes over time as measured at any point on Earth's surface. The difference between true north and magnetic north as measured at a point on Earth's surface on a particular day is the **magnetic declination** of that point on that day.

The mean declination arrows printed on the bottom margin of USGS topographic maps show the magnetic declination between magnetic north and true north and the declination between true north and UTM grid north (e.g., **Fig. 9.3B**). Both of these declinations were determined for a point at the center of the map area on the day on which that part of the map was finalized prior to printing.

The magnetic declination indicated on **Fig. 9.3B** is an **east declination** of 15°, so the line pointing toward magnetic north (MN) points 15° east of the line pointing toward true north (i.e., toward the star). Magnetic north is a clockwise rotation from true north for an east declination. A **west declination** means that magnetic north is a little west of true north and is found by a counterclockwise rotation from true north.

The magnetic declination for any given point changes slowly but continually. You can obtain the most recent magnetic data for your location from the Magnetic Declination Estimated Value Calculator (**http://www.ngdc .noaa.gov/geomag-web/**) maintained by the U.S. National Oceanic and Atmospheric Administration's National Centers for Environmental Information (NOAA-NCEI).

Maps showing the varying location of the north and south magnetic poles since 1590 are available from NOAA-NCEI at **http://www.ngdc.noaa.gov/geomag/GeomagneticPoles .shtml**.

What Is a Bearing?

A **bearing** is the azimuth along a line from one point to another, and an **azimuth** is an angle measured clockwise in a horizontal plane from north to the line you are interested in. For example, if you are looking due east at a fossil platypus sticking out of a hillside, the bearing of the mammal fossil would be about 90°. A general explanation of how to find the bearing from one point to another point on a map is given in the caption of **Fig. 9.5**.

Linear geologic features (faults, fractures, dikes), lines of sight and travel, and linear property boundaries are all defined on the basis of their bearings. But because a compass needle points to Earth's **magnetic north pole** rather than the true north pole, an uncorrected **compass bearing** will be inaccurate by an amount equal to the magnetic declination at that location. (The compasses used by field geoscientists can be adjusted to compensate for magnetic declination.) If magnetic north is 5° east of true north, for example, you will need to subtract 5° from the compass bearing to find the true bearing—the azimuth relative to true north—from one point to another. A magnetic declination of 7° west requires that you add 7° to the compass bearing to determine the true bearing.

How to Determine a Bearing on a Map

Imagine that you are buying a property for your dream home. The boundary of the property is marked by four metal rods driven into the ground, one at each corner of the property. The location of these rods is indicated on a map made carefully by a surveyor (**Fig. 9.5A**) as points *a*, *b*, *c*, and *d*. The property deed notes the distances between the points *and* bearings between the points. This defines the shape of the property. Notice that the northwest edge of your property lies between two metal rods located at points *a* and *b*. You can measure the distance between the points using a tape measure. How can you measure the bearing on an accurate map of the property?

First, lightly draw a line in pencil that extends from *a* through *b* and beyond until it intersects the closest meridian you can find on the map. In **Fig. 9.5A**, the edge of the map is a convenient north–south line near points *a* and *b*. Next, orient a protractor so that its 0° and 180° marks are on the edge of the map with the 0° end toward geographic north. Place the origin of the protractor at the point where your line *a–b* intersects the edge of the map. You can now read a bearing of 43°. If you were to determine the opposite bearing, from *b* through *a*, then the bearing would be pointing southwest and would be read as an azimuth of 223°.

You also can use a compass to read bearings on a map, as shown in **Fig. 9.5B**. Ignore the compass needle and use the compass as a circular protractor. Square azimuth protractors for this purpose are provided in GeoTools Sheets 3 and 4 at the back of this manual.

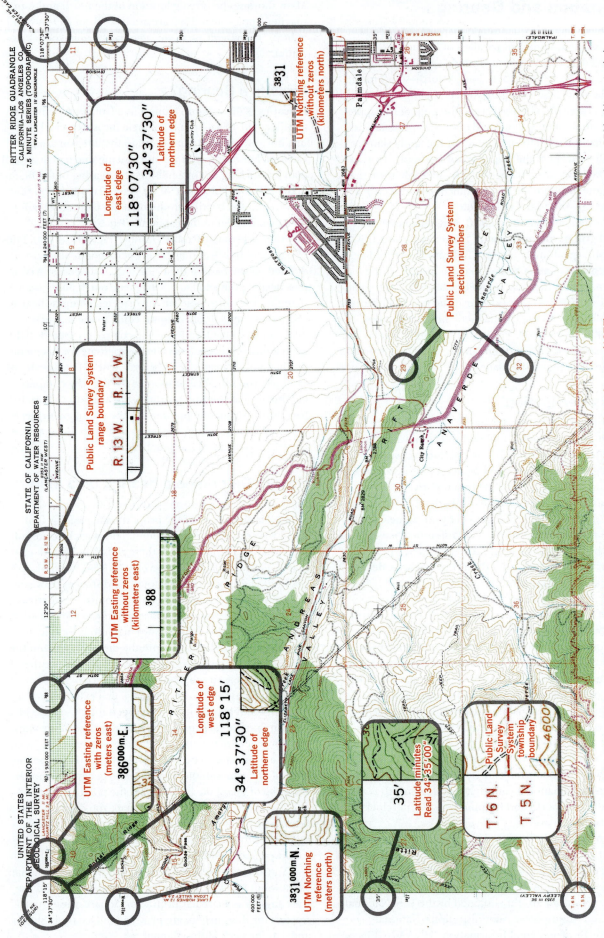

Figure 9.3A Northern part of the Ritter Ridge, California, USGS 7.5-minute topographic quadrangle map (1975). This map was reduced to about 55% of its actual size.

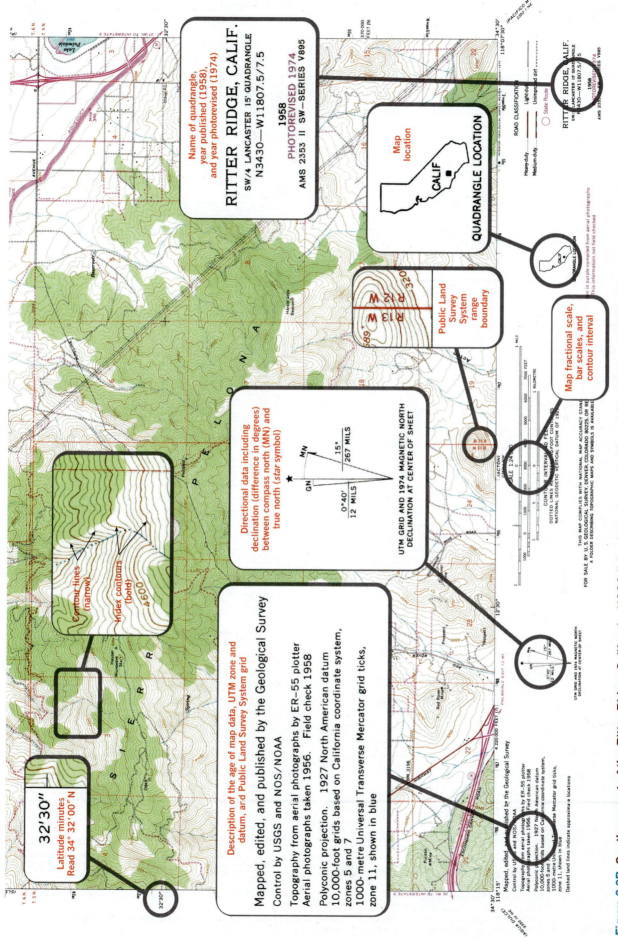

Figure 9.3B Southern part of the Ritter Ridge, California, USGS 7.5-minute topographic quadrangle map (1975). This map was reduced to about 55% of its actual size.

Control data and monuments

Vertical control

Third order or better, with tablet	BM x 16.3
Third order or better, recoverable mark	x 120.0
Bench mark at found section corner	BM 18.6
Spot elevation	x 5.3

Contours

Topographic

Intermediate	
Index	
Supplementary	
Depression	
Cut; fill	

Bathymetric

Intermediate	
Index	
Primary	
Index primary	
Supplementary	

Boundaries

National	
State or territorial	
County or equivalent	
Civil township or equivalent	
Incorporated city or equivalent	
Park, reservation, or monument	

Surface features

Levee	Levee
Sand or mud area, dunes, or shifting sand	(Sand)
Intricate surface area	(Strip mine)
Gravel beach or glacial moraine	(Gravel)
Tailings pond	(Tailings pond)

Mines and caves

Quarry or open pit mine	
Gravel, sand, clay, or borrow pit	
Mine tunnel or cave entrance	
Mine shaft	
Prospect	X
Mine dump	(Mine dump)
Tailings	(Tailings)

Vegetation

Woods	
Scrub	
Orchard	
Vineyard	
Mangrove	(Mangrove)

Glaciers and permanent snowfields

Contours and limits	
Form lines	

Marine shoreline

Topographic maps

Approximate mean high water	
Indefinite or unsurveyed	

Topographic-bathymetric maps

Mean high water	
Apparent (edge of vegetation)	

Submerged areas and bogs

Marsh or swamp	
Submerged marsh or swamp	
Wooded marsh or swamp	
Submerged wooded marsh or swamp	
Rice field	(Rice)
Land subject to inundation	Max pool 431

Coastal features

Foreshore flat	Mud
Rock or coral reef	
Rock bare or awash	*
Group of rocks bare or awash	
Exposed wreck	
Depth curve; sounding	3
Breakwater, pier, jetty, or wharf	
Seawall	

Rivers, lakes, and canals

Intermittent stream	
Intermittent river	
Disappearing stream	
Perennial stream	
Perennial river	
Small falls; small rapids	
Large falls; large rapids	
Masonry dam	
Dam with lock	
Dam carrying road	
Perennial lake; Intermittent lake or pond	
Dry lake	(Dry lake)
Narrow wash	
Wide wash	Wide wash
Canal, flume, or aquaduct with lock	
Well or spring; spring or seep	o ⦶

Buildings and related features

Building	
School; church	
Built-up area	
Racetrack	
Airport	
Landing strip	
Well (other than water); windmill	o
Tanks	
Covered reservoir	
Gaging station	
Landmark object (feature as labeled)	⊙
Campground; picnic area	
Cemetery: small; large	(Cem)

Roads and related features

Roads on Provisional edition maps are not classified as primary, secondary, or light duty. They are all symbolized as light duty roads.

Primary highway		
Secondary highway		
Light duty road		
Unimproved road		
Trail		
Dual highway		
Dual highway with median strip		

Railroads and related features

Standard gauge single track; station	
Standard gauge multiple track	
Abandoned	

Transmission lines and pipelines

Power transmission line; pole; tower	
Telephone line	Telephone
Aboveground oil or gas pipeline	
Underground oil or gas pipeline	Pipeline

Figure 9.4 Symbols used on U.S. Geological Survey topographic quadrangle maps.

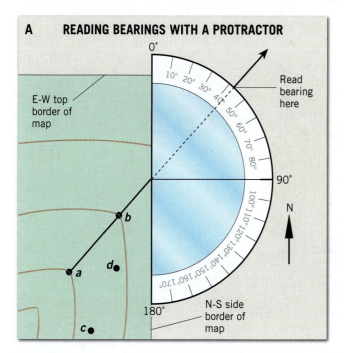

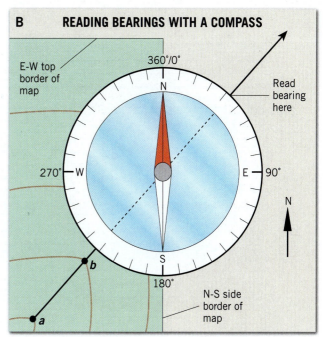

Figure 9.5 How to determine the bearing between two points on a map. A. Bearing from point *a* through *b* on a map, determined using a protractor. Center the protractor where a north–south meridian near the points intersects the line from *a* through *b*. Align the protractor with the meridian. Measure the azimuth clockwise from north to the line from *a* to *b*. In this case, the bearing is ~43°. The range of possible bearing azimuths is from 0 to 360°. **B.** You can use a compass as a protractor. Align the north–south axis of the compass ring with the meridian, and place the center pivot of the compass at the intersection of the bearing line and the meridian.

GPS—Global Positioning System

The Global Positioning System (GPS) is a technology that allows us to locate points on Earth with great accuracy. GPS has become woven into the fabric of our lives and is used for everything from keeping track of where people, cars, and even pets are located to its scientific use in geodesy—the science of measuring changes in Earth's size and shape and the position of points on its surface over time. The GPS technology operated by the U.S. government is based on a constellation of about 30 satellites whose orbits are designed so that a minimum of 6 satellites will be above the horizon at any point on Earth at any time. We access the GPS system through small GPS receivers—either handheld receivers like those used by hikers or receiver circuitry that is built into common products such as smartphones and navigation systems in cars.

The accuracy of a GPS location depends on several variables, but is generally in the range of about ±3 to 9 meters for consumer GPS receivers. The GPS systems used by surveyors and geoscientists can have uncertainties that are much less than 1 meter, but achieving that kind of accuracy is significantly more difficult and costly.

UTM—Universal Transverse Mercator System

The U.S. National Imagery and Mapping Agency (NIMA) adapted and improved on a global military navigation grid and coordinate system after World War II, and the result

is the **Universal Transverse Mercator (UTM)** coordinate system. Unlike the latitude–longitude grid that is spherical and measured in degrees, minutes, seconds, the UTM grid is rectangular and measured in meters. The full specification of a location in the UTM system requires four elements: zone, latitude band (or hemisphere), easting, and northing.

UTM Zones and Latitude Bands

The UTM grid (top of **Fig. 9.6**) is based on 60 north–south **zones**, which are strips of longitude with a width of 6°. The zones are consecutively numbered from Zone 01 (between 180° and 174° west longitude) at the left margin of the grid to Zone 60 (between 174° and 180° east longitude) at the east margin of the grid. Each zone has a north–south **central meridian** that is perpendicular to the equator (**Fig. 9.6**). A version of UTM based on the Military Grid Reference System (MGRS) divides the zones into east–west segments called **latitude bands** that are identified by different letters (**Fig. 9.6**). Latitude bands are lettered consecutively from C (between 80° and 72° south latitude) through X (between 72° and 84° north latitude) and are 8° long except for band X, which is 12° long. Letters I and O are not used because they could be confused with numbers 1 and 0.

Easting and Northing

Points located along the central meridian of any UTM zone have identical east–west coordinates, called **eastings**, of 500,000 mE in the UTM system. The abbreviation *mE*

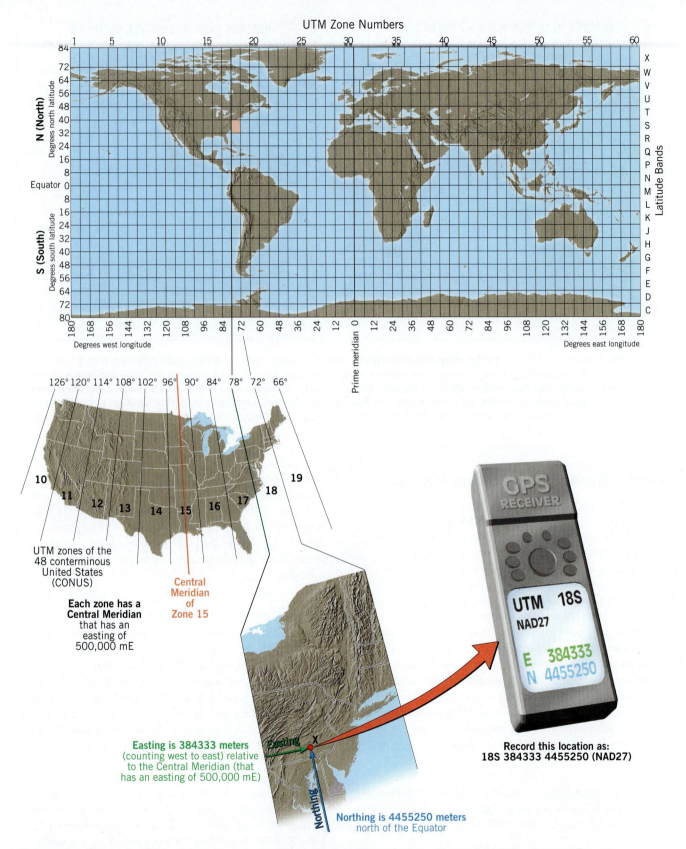

UTM Zone Numbers

Figure 9.6 UTM with GPS. A handheld Global Positioning System (GPS) receiver is set to display the Universal Transverse Mercator (UTM) grid for the *North American Datum 1927* (*NAD27*). When operated at point **X**, the receiver displays its location as a zone and designator letter (18S) plus an easting and northing. Most handheld GPS devices can select a geodetic datum from among several options, including WGS84 and NAD27. Be sure your grid system is the same as that of any map on which you may plot the GPS coordinates.

means "meters east." The eastings of other points within that zone are measured from the central meridian. For example, a point that is 10,000 m west of the central meridian has an easting of $500,000 - 10,000 = 490,000$ mE, and a point that is 20,000 m east of the central meridian has an easting of $500,000 + 20,000 = 520,000$ mE. The north–south coordinates in the UTM system are called **northings** and are measured relative to the equator along meridians. In the northern hemisphere, the northing is simply the distance in meters to the equator, and that distance is marked with the abbreviation *mN* for "meters north." In the southern hemisphere, the equator is assigned a northing of 10,000,000 mS (*mS* means "meters south"), and the northing is counted down from this number by the distance south of the equator. For example, the Sydney Opera House in Australia has a northing of approximately 6252280 mS, so it is about $10,000,000$ m $- 6,252,280$ m $= 3,747,720$ m south of the equator.

The conventional way to report UTM coordinates is to write the zone number and designator letter, then the easting, and then the northing (i.e., 18S 384333 4455250 in **Fig. 9.6**). It is also useful to note what geodetic datum was used to define the coordinates.

Geodetic Datum

Measurements of location or motion can be made only relative to some frame of reference. Geodesists and cartographers use a reference frame that is called the World Geodetic System, or WGS, which is revised from time to time. As this edition was being prepared, the current version of WGS is called **WGS84**. Older USGS maps you might encounter are commonly based on an older reference frame called the North American Datum of 1927, or **NAD27**. The coordinate origin of WGS84 is at the center of Earth, and it uses the prime meridian specified by the IERS and adopted as the international standard in 1984.

The reason it is important to notice the reference frame used to produce a map is that the same feature on Earth's surface will have a different location in WGS84 than in NAD27, whether in latitude–longitude or UTM. The difference can cause a location error of hundreds of meters. Because the difference in datum affects UTM coordinates, grid north is different within a given quadrangle map created using WGS84 compared with a map constructed using NAD27. The GPS system and Google Earth use the WGS84 datum.

Locating Points Using UTM

Study the illustration of a GPS receiver in **Fig. 9.6**. Notice that the receiver is displaying UTM coordinates (based on NAD27) for a point **X** in Zone 18S (north of the equator). Point **X** has an easting coordinate of 384333 mE, which means that it is located $500,000$ m $- 384,333$ m $= 115,667$ m to the west of the central meridian of Zone 18. Point **X** also has a northing

coordinate of 4455250 mN, which means that it is located 4,455,250 m north of the equator. Therefore, point **X** is located in southeast Pennsylvania. To plot point **X** on a 1:24,000 scale, 7.5-minute topographic quadrangle map, see **Fig. 9.7**.

Point **X** is located within the Lititz, PA 7.5-minute topographic quadrangle map (**Fig. 9.7**). Information printed on the map margin (**Fig. 9.7A**) indicates that the map has blue ticks spaced 1000 m apart along its edges that match the local UTM grid. The map area is in UTM Zone 18 as defined relative to the NAD27 datum. Notice how the ticks for northings and eastings are represented on the northwest corner of the Lititz map (**Fig. 9.7B**). One northing label is written out in full (4456000 mN) and one easting label is written out in full (384000 mE), but the other values are given in UTM shorthand for kilometers (i.e., they do not end in 000 m). Point **X** has an easting of 384333 mE within Zone 18S, so it is located 333 m east of the tick mark labeled 384000 mE along the top margin of the map. Point **X** has a northing of 4455250 mN, so it is located 250 m north of the tick mark labeled 4455 in UTM shorthand (which stands for 4455 km or 4,455,000 m). Distances east and north can be measured using a ruler and the map's graphic bar scale as a reference (333 m = 0.333 km, 250 m = 0.250 km). However, you can also use the graphic bar scale to construct a UTM grid like the one in **Fig. 9.7C**. If you construct such a grid and print it onto a transparency, then you can use it as a UTM grid overlay. (However, a grid overlay works only if it is printed at exactly the same scale as the map with which it is used.) To plot a point or determine its coordinates, place the grid overlay on top of the square kilometer in which the point is located. Then use the grid as a two-dimensional ruler for the northing and easting. Grid overlays for many different scales of UTM grids are provided in GeoTools Sheets 2–4 at the back of the manual for you to cut out and use.

Public Land Survey System

The **U.S. Public Land Survey System (PLSS)** was initiated in 1785 when the U.S. government required a way to divide public land west of the 13 original colonies into small parcels that could be transferred to ownership by private citizens. The U.S. Bureau of Land Management (BLM) regulates and maintains public land using the PLSS, which is also used as the basis for many legal surveys of private land that was once publicly owned.

PLSS Township-and-Range Grids

The PLSS is a square grid system centered on any one of dozens of **principal meridians** (north–south lines) and **base lines** (east–west lines). Once a principal meridian and base line were established, additional lines were surveyed parallel to them and 6 miles (mi.) apart. This created a grid of 6 mi. by 6 mi. squares of land (**Fig. 9.8**). The 6-mile-wide rows extending east–west parallel to the base lines

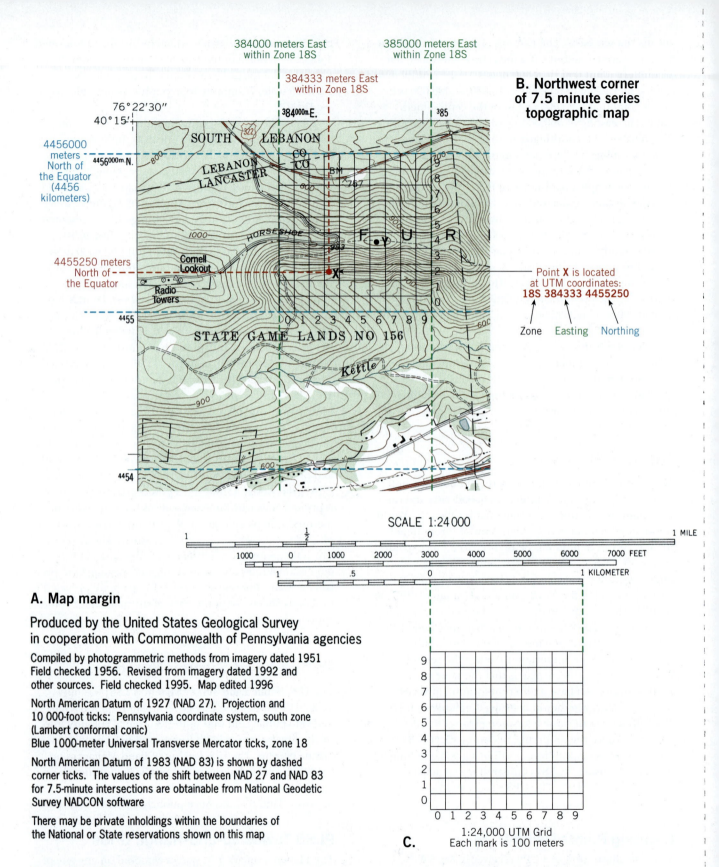

SCALE 1:24 000

**B. Northwest corner
of 7.5 minute series
topographic map**

Point **X** is located
at UTM coordinates:
18S 384333 4455250

Zone   Easting   Northing

A. Map margin

Produced by the United States Geological Survey
in cooperation with Commonwealth of Pennsylvania agencies

Compiled by photogrammetric methods from imagery dated 1951
Field checked 1956. Revised from imagery dated 1992 and
other sources. Field checked 1995. Map edited 1996

North American Datum of 1927 (NAD 27). Projection and
10 000-foot ticks: Pennsylvania coordinate system, south zone
(Lambert conformal conic)
Blue 1000-meter Universal Transverse Mercator ticks, zone 18

North American Datum of 1983 (NAD 83) is shown by dashed
corner ticks. The values of the shift between NAD 27 and NAD 83
for 7.5-minute intersections are obtainable from National Geodetic
Survey NADCON software

There may be private inholdings within the boundaries of
the National or State reservations shown on this map

C.

1:24,000 UTM Grid
Each mark is 100 meters

Figure 9.7 UTM with topographic maps. A. Map margin indicates that the map includes UTM grid data based on North
American Datum 1927 (NAD27) and represented by blue ticks spaced 1000 m apart along the map edges. **B.** Connect the blue
1000-m ticks to form a grid square. Northings (blue) are read along the N–S map edge, and eastings (green) are located along the
E–W map edge. **C.** You can construct a 1-km grid from the map's bar scale and then make a transparency of it to form a grid overlay
(see GeoTools Sheets 2 and 4 at back of manual). Place the grid overlay atop the 1-km square on the map that includes point **X**, and
determine the UTM coordinates of **X** as shown (red).

PUBLIC LAND SURVEY SYSTEM (PLSS)

A. The Township-and-Range Grid.

The grid is made of E-W *township* strips of land and N-S *range* strips of land (columns of land) surveyed relative to a *principal meridian* (N-S line) and its *base line* (E-W line). Township strips are 6 miles high and numbered T1N, T2N, and so on north of the base line and T1S, T2S, and so on south of the base line. Range strips (rows) of land are 6 miles wide and numbered R1E, R2E, and so on east of the principal meridian and R1W, R2W, and so on west of the principal meridian. Each intersection of a township strip of land with a range strip of land forms a square, called a *township*. Note the location of Township T1S, R2W.

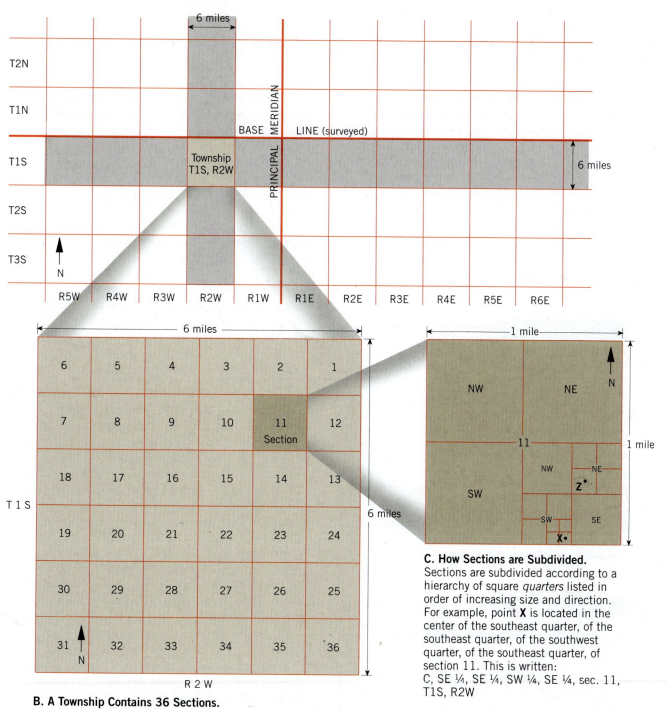

B. A Township Contains 36 Sections.

Each township is 6 miles wide by 6 miles long (36 square miles) and subdivided into 36 sections. Each section is 1 square mile (640 acres), called a *section*, and numbered as shown here.

C. How Sections are Subdivided.

Sections are subdivided according to a hierarchy of square *quarters* listed in order of increasing size and direction. For example, point **X** is located in the center of the southeast quarter, of the southeast quarter, of the southwest quarter, of the southeast quarter, of section 11. This is written:
C, SE ¼, SE ¼, SW ¼, SE ¼, sec. 11, T1S, R2W

Figure 9.8 U.S. Public Land Survey System (PLSS). This survey system is based on grids of square townships, which are identified relative to *principal meridians* (N–S lines) of longitude and *base lines* (E–W lines, surveyed perpendicular to the principal meridian) that are unique to specific states or regions.

were called **townships**, so *Township 1 North* (T1N) identifies the first row north of the base line. The 6-mile-wide columns extending north–south parallel to the principal meridians were called **ranges**, so *Range 3 West* (R3W) is the third column west of the principal meridian. Each 6 mi. by 6 mi. square is, therefore, identified by its township and range position in the PLSS grid. For example, the township in **Fig. 9.8B** is located at T1S (Township 1 South) and R2W (Range 2 West). Although each square like this is identified as both a township and a range within the PLSS grid, it is common practice to refer to each of the 6-by-6 squares as a *township*.

Defining Land Areas Using PLSS

The PLSS is designed to define the location of square or rectangular subdivisions of land. The 6 mi. by 6 mi. townships are used as political subdivisions in some states and often have place names. Each township square is divided into 36 squares, each having an area of 1 square mile (640 acres). These square-mile subdivisions of land are called **sections**.

Sections are numbered from 1 to 36, beginning in the upper right corner of the township (**Fig. 9.8B**). Sometimes section boundary lines are shown on topographic quadrangle maps, like the red-brown grid of square numbered

---

ACTIVITY 9.3

Topographic Map Construction, (p. 258)

> **Think About It** How are topographic maps constructed and interpreted?

Objective Construct topographic maps by drawing contours based on maps showing elevations of specific points and a digital terrain model.

Before You Begin Read the following section: What Are Topographic Maps?

ACTIVITY 9.4

Topographic Map and Orthoimage Interpretation, (p. 259)

Objective Interpret topographic maps and determine their effectiveness in comparison to, and combination with, U.S. Topo orthoimages.

Before You Begin Read the following section: What Are Topographic Maps?

Plan Ahead You will need a simple calculator or calculator app on your smartphone, a ruler, and a red pen or pencil.

---

sections in **Fig. 9.3**. Any tiny area or point can be located within a section by dividing the section into quarters (labeled NW, NE, SW, SE). Each of these quarters can itself be subdivided into quarters and labeled (**Fig. 9.8C**). If a point is at the center of one of these quarters, that is signified with a C; otherwise, the position within the quarter is indicated by specifying the number of feet from the two closest edges of the quarter.

What Are Topographic Maps?

Topographic maps are miniature models of Earth's three-dimensional landscape, displayed in two dimensions on a horizontal surface (**Figs. 9.1C** and **9.3**). The third dimension, elevation or height, is shown using **topographic contours**, which are lines connecting points of equal elevation (**Fig. 9.9**). Notice how the contours in **Fig. 9.9** occur where the landscape intersects horizontal planes of specific elevations: 0, 50, and 100 feet. Zero elevation is sea level, so it is along the coastline of the imaginary island. You can think of the contours for 50 and 100 ft. above sea level as the location of the coastline if sea level rose by 50 or 100 feet. An "x" or triangle is often used to mark the highest point on a hilltop with the exact elevation noted beside it. The highest point on the map in **Fig. 9.9** is above the elevation of the highest contour (100 ft.) but below 150 ft. because there is no contour for 150 feet. In this case, the exact elevation of the highest point on the island is marked by spot elevation ("x" labeled with the elevation of 108 ft.).

Topographic Map Construction

The development of topographic maps has progressed in stages over time. In the 1800s, topographic maps relied on surveyors mapping with a compass and a measuring tape or chain, estimating elevation using a barometer. Mapping with the use of planetable and alidade brought increased dimensional accuracy to the process. With the advent of aviation, aerial photography became important in the creation of topographic maps. Photographs were taken automatically as an airplane flew at a constant elevation along straight flight lines, eventually covering an entire area. The resulting mosaic of photographs was fitted together and corrected in a process called photogrammetry that identifies points of equal elevation on the landscape. Until recently, most USGS topographic maps were the product of a combination of techniques, including photogrammetry, aerial photo analysis, and field surveying (**http://nationalmap.gov/ustopo/125history.html**).

Today, topographic maps are often based on data obtained using various remote sensing technologies. The Shuttle Radar Topography Mission in 2000 collected topographic data from about latitude 60°S to 60°N using a radar altimeter. NASA knew the location of the shuttle orbiter at any given time to a high degree of accuracy, and the radar altimeter provided data on its height above the ground surface. Combining those two data sets allows us to determine the position of points on the ground surface that the shuttle passed over. Topographic data currently

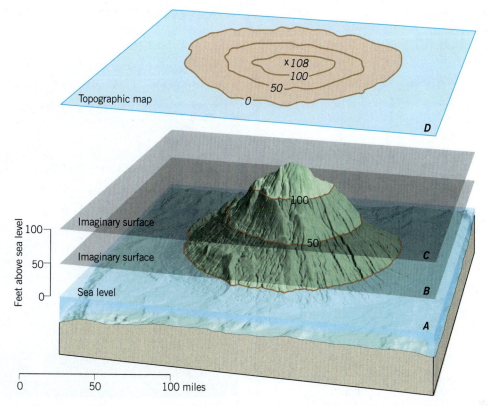

Figure 9.9 Topographic map construction. A *contour* is drawn where a horizontal plane (A, B, or C) intersects the land surface. Where sea level (plane A) intersects the land, it forms the 0-foot contour. Plane B is 50 ft. above sea level, so its intersection with the land is the 50-foot contour. Plane C is 100 ft. above sea level, so its intersection with the land is the 100-foot contour. D is the resulting topographic map of the island. It is a view looking down onto the island from above that shows the 0-, 50-, and 100-foot contours. The elevation change between any two contours is 50 ft., so the map is said to have a 50-foot *contour interval*.

available from this mission have a resolution of about 30 meters. Some NASA satellites can measure the ground surface as well. The ASTER instrument on NASA's Terra satellite acquires stereo images of Earth's surface that can be used to resolve elevation, and ICESat uses lasers to map land elevation.

The highest-resolution maps today are produced using a combination of GPS and **lidar** technologies. The word *lidar* is an acronym that stands for "light detection and ranging." Lidar uses a laser to measure distances very accurately. When a lidar system is operated from an aircraft that has a sophisticated guidance system linked to GPS, points on Earth's surface can be located with accuracies of less than 1 meter—sometimes substantially less. The image on the first page of this laboratory chapter includes a gray hillshade image made from aerial lidar data collected over Meteor Crater (or Barringer Crater), Arizona. Below that image is a representation of the **point cloud** generated by that lidar survey. Each point in the cloud represents a point on Earth's surface that was hit by the lidar laser as the aircraft flew over and whose position in space was resolved during the subsequent analysis. The points in the point cloud are colored by elevation.

Lidar data are available to the public from a variety of sources, including OpenTopography.org where research data from a variety of studies are archived. The USGS, NOAA, and other federal agencies are heavily invested in aerial lidar mapping. They plan to use a combination of

aerial lidar and interferometric synthetic aperture radar to map the entire country in the coming decades (**http://nationalmap.gov/3DEP/** and **https://coast.noaa.gov/inventory/?redirect=301ocm**). Those high-resolution location data will form the basis of the next generation of topographic maps.

US Topo Maps and Orthoimages

Historic USGS map series (**Figs. 9.1C** and **9.3**) and those of most other countries are one-page paper maps. However, the latest series of USGS topographic maps are layered digital maps called "US Topo" maps (**Fig. 9.10**). US Topo maps are PDF files that can be read using Adobe Acrobat Reader. The map layers can be turned on or off, including an aerial photograph layer called an *orthoimage*. The digital products can be downloaded free of charge, or they can be ordered as printed paper maps.

Aerial photographs are sometimes taken at angles oblique to the landscape, but topographic maps are representations of the landscape as viewed from directly above. **Orthoimages** are digitized aerial photographs or satellite images that have been orthorectified—corrected for distortions until they have the same geometry and uniform scale as a topographic map. Therefore, an orthoimage correlates exactly with its topographic map and reveals visual attributes of the landscape that are not visible on the topographic map. The topographic map, orthoimage,

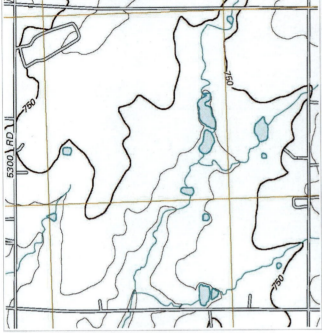

A. Topographic map base (contour lines), UTM grid lines (WGS84, Zone 15S), hydrography, and transportation features.

B. Orthoimage base with all other data layers: contour lines, UTM grid lines (WGS84, Zone 15S), geographic names and boundaries, hydrography, and transportation features.

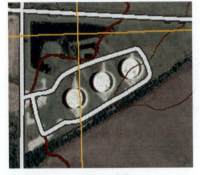

C. Enlarged portion of B.

Figure 9.10 US Topo (USGS) digital topographic maps and orthoimages of quadrangles. The latest series of USGS topographic maps are digital maps with data layers for planimetric features and an orthoimage. Choose layers displayed on a topographic map base (**A**) or combine the topographic base as one layer of data on an orthoimage base (**B**) that can be enlarged with high resolution (**C**). Use Adobe Reader (or Adobe Acrobat) to print products in GeoPDF format. Using Adobe Acrobat, you can save products as JPEG images for enhancement with any photo processing software.

and other orthorectified "layers" of data can be added or removed to give the viewer extraordinary perspectives of the landscape. All of this can be done at US Topo, courtesy of the USGS and its partners. One can display features like hydrography (water bodies), roads, and UTM grid lines on a topographic base (**Fig. 9.10A**) or display the topographic map layer on an orthoimage base (**Fig. 9.10B**). All layers can be enlarged with outstanding resolution (**Fig. 9.10C**). To learn more about obtaining and using US Topo products, watch a 6-minute USGS video (**https://www.usgs.gov/media/videos/exploring-us-topo-geopdfs**).

Interpreting Topographic Contours

Each **topographic contour** connects points on the map that have the same elevation above sea level (**Fig. 9.11**). Look at the topographic map in **Fig. 9.3** and notice the

light brown and heavy brown contours. The heavy brown contours are called **index contours** because they have elevations printed on them whereas the four lighter contours between the index contours do not have their elevations labeled. Index contours are your starting point when reading elevations on a topographic map. For example, the index contours in **Fig. 9.3** are labeled with elevations in increments of 200 feet. The map has five contours for every 200 ft. of elevation, or a **contour interval** of 40 feet. This contour interval is specified at the center of the bottom margin of the map (**Fig. 9.3**). USGS topographic contours are multiples of the contour interval above sea level. For example, if a map uses a 10-foot contour interval, then the contours represent elevations of 0 ft. (sea level), 10 ft., 20 ft., 30 ft., 40 ft., and so on. Most maps use the smallest contour interval that will allow easy readability and provide as much detail as possible.

RULES FOR MAKING AND INTERPRETING TOPOGRAPHIC CONTOURS

1. Every point on a contour has the same elevation; that is, contours connect points of equal elevation. The contour *must* go through control points that are the same elevation as the contour.

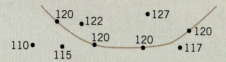

2. Interpolation is used to estimate the elevation of a point B located in line between points A and C of known elevation. To estimate the elevation of point B:

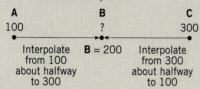

A	B	C
100	?	300

Interpolate from 100 about halfway to 300   **B = 200**   Interpolate from 300 about halfway to 100

3. Extrapolation is used to estimate the elevations of a point C located in line beyond points A and B of known elevation. To estimate the elevation of point C, use the distance between A and B as a ruler or graphic bar scale to estimate in line to elevation C.

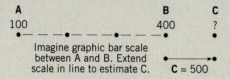

A		B	C
100		400	?

Imagine graphic bar scale between A and B. Extend scale in line to estimate C.   **C = 500**

4. Contours always separate points of higher elevation (uphill) from points of lower elevation (downhill). You must determine which direction on the map is higher and which is lower, relative to the contour in question, by checking adjacent elevations.

5. The elevation between any two adjacent contours of different elevation on a topographic map is the *contour interval*. Often every fifth contour is heavier so that you can count by five times the contour interval. These thicker contours are known as *index contours*, because they generally have elevations printed on them.

6. Contour lines do not cross each other except where an overhanging cliff is present. In such a case, the hidden contours are dashed.

7. Contours can merge to form a single contour line only where there is a vertical cliff or wall.

8. Evenly spaced contours of different elevation represent a uniform slope.

9. The closer the contours are to each other the steeper the slope. More widely-spaced contours indicate a less steep slope.

Steep   Less steep

10. A concentric series of closed contours represents a hill:

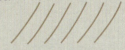

11. *Depression contours* have hachure marks on the downhill side and represent a closed depression:

See Figure 9.13

12. Contours form a V pattern when crossing streams. The apex of the V always points upstream (uphill):

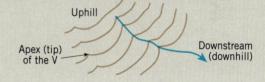

Uphill

Apex (tip) of the V   Downstream (downhill)

Figure 9.11 Rules for Making and Interpreting Topographic Contours

Additional rules for interpreting topographic contours are provided in **Fig. 9.11**, and some of the common kinds of landforms represented by contours on topographic maps are depicted in **Fig. 9.12**. Your ability to use a topographic map is based on your ability to interpret what the contours mean and to be able to visualize the topography.

Reading Elevations

If a point on the map lies on an index contour, you simply read its elevation from the label on the index contour. If the point lies on an unnumbered contour, then its elevation can be determined by counting up or down from the nearest index contour. For example, if the nearest index contour is 300 ft., your point of interest is on the fourth contour *above* it, and the contour interval is 20 ft., then you simply count up by 20s from the index contour: 320, 340, 360, 380. The point is 380 ft. above sea level. Or, if the point is three contours *below* the index contour, you count down: 280, 260, 240; the point is 240 ft. above sea level.

If a point lies between two contours, then you must estimate its elevation by interpolation (**Fig. 9.11**). For example, on a map with a 20-foot contour interval, a point between the 340- and 360-foot contours has an

A. PERSPECTIVE VIEW OF LANDSCAPE

B. TOPOGRAPHIC MAP

Hill

Saddle

Hill

Saddle

Hill

Saddle

Valley

Valley

Spur

Spur

200

200

Closed
depression
at top of
hill

Ridge

Steep
slope

Gentle
slope

Ridge

266×

200

Overhanging
cliff

Vertical
cliff

100

Ridge

Ridge

100

BM
24
×

0

N

0   1 mile

Contour interval 20 ft.

Figure 9.12 Names of landscape features observed on topographic maps. Perspective view (**A**) and topographic map (**B**) features: **valley** (low-lying land bordered by higher ground), **hill** (rounded elevation of land; mound), **ridge** (linear or elongate elevation or crest of land), **spur** (short ridge or branch of a main ridge), **saddle** (low point in a ridge or line of hills, resembling a horse saddle), **closed depression** (low point/area in a landscape from which surface water cannot drain; contours with hachure marks), **steep slope** (closely spaced contours), **gentle slope** (widely spaced contours), **vertical cliff** (merged contours), **overhanging cliff** (dashed contour that crosses a solid one; the dashed contour indicates what is under the overhanging cliff).

elevation between 340 and 360 ft. above sea level. If a point lies between a contour and the margin of the map, then you must estimate its elevation by extrapolation (**Fig. 9.11**).

Depressions

Figure 9.13 shows topographic contours in and adjacent to a depression. *Hachure marks* (short line segments pointing downhill) on some of the contours in these maps indicate the presence of a closed depression—a depression from which water cannot drain (**Fig. 9.11**). As depicted in **Fig. 9.13**, contours of the same elevation repeat on opposite sides of the rim of the depression at the top of a hill.

On the side of a hill, the contours repeat only on the downhill side of the depression.

Ridges and Valleys

Figure 9.14 shows how topographic contours represent linear ridge crests and valley bottoms. Ridges and valleys are roughly symmetrical, so individual contours repeat on each side (**Fig. 9.11**). To visualize this, picture yourself walking along an imaginary trail across the ridge or valley (dashed lines in **Fig. 9.14**). Every time you walk up the side of a hill or valley, you cross contours. Then, when you walk down the other side of the hill or valley, you recross contours of the same elevations as those crossed walking uphill.

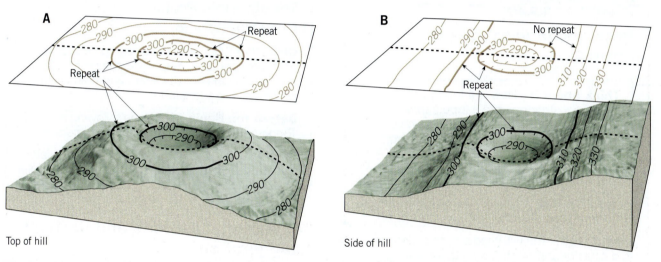

Figure 9.13 Contours for depressions. Enclosed depressions in the topography are indicated by short hachure lines on the downslope side of contours. **A.** Notice the pair of contours with the same elevation at the top of this hill, just below the rim of the central crater. **B.** Notice the pair of contours with the same elevation on the downslope side of this feature, just below the rim of the depression on the left side. There are no repeating contours on the upslope side because elevation simply increases from the bottom of the depression to the top of the slope.

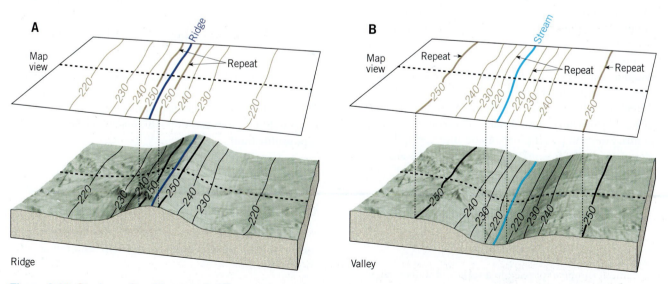

Figure 9.14 Contours for ridges and valleys. Contours with the same elevation occur on both sides of a ridgeline (**A**) and on both sides of a valley (**B**). In **A**, for example, walking along the dashed line you would cross the 220-, 230-, 240-, and 250-foot contours, go over the ridgeline, and cross the 250-, 240-, 230-, and 220-foot contours again as you walk down the other side.

Spot Elevations and Benchmarks

Elevations of specific points on topographic maps (tops of peaks, bridges, survey points, etc.) sometimes are indicated directly on the maps as **spot elevations** beside a small triangle, black dot, or an x symbol at the exact spot of the elevation indicated. The elevations of prominent hilltops, peaks, or other features are often identified. For example, the highest point on the ridge in the west central part of **Fig. 9.12B** has an elevation of 266 ft. above sea level. The notation "BM" denotes a **benchmark**, a permanent marker (usually a metal plate) placed by the USGS, BLM, or another governmental agency at the point indicated on the map (**Fig. 9.4**).

ACTIVITY 9.5

Relief and Gradient (Slope) Analysis, (p. 262)

> **Think About It** How is a topographic profile constructed from a topographic map, and what is its vertical exaggeration?

Objective Calculate relief and gradients from a topographic map and apply the gradient data to determine a driving route.

Before You Begin Read the following section: Relief and Gradient (Slope).

Plan Ahead You will need a simple calculator or calculator app on your smartphone and a ruler.

Relief and Gradient (Slope)

Recall that **relief** is the difference in elevation between two points on a landscape or map. *Regional relief* (total relief) is the difference in elevation between the highest and lowest points on a topographic map. The highest point is the top of the highest hill or mountain; the lowest point is generally where the major stream of the area leaves the map, or a coastline. A slope's topographic **gradient** is a measure of the steepness of a slope. The topographic gradient between two points is simply the difference in the elevation of those two points (relief) divided by the horizontal distance between those two points. For example, if points **A** and **B** on a map have elevations of 200 ft. and 300 ft., and the points are located 2 miles apart, then:

$$\text{gradient} = \frac{\text{difference in elevation}}{\text{horizontal distance}}$$
$$= \frac{100 \text{ ft.}}{2 \text{ mi.}} = 50 \text{ ft./mi.}$$

A gradient can be expressed in a variety of units such as ft./mi. as above, or it can be expressed as a dimensionless number if it is computed using the same units in the dividend and divisor. For example, a gradient of 50 ft. per 5280 ft. is 0.00947, or 0.947%, which is the same slope as 50 feet/mile.

Calculating Gradient (Slope) — The Math You Need

You can learn more about calculating slope (gradient) at this site featuring *The Math You Need, When You Need It* math tutorials for students in introductory geoscience courses: **http://serc.carleton.edu/mathyouneed/slope/index.html**.

ACTIVITY 9.6

Topographic Profile Construction, (p. 264)

Objective Construct a topographic profile from a topographic map using the profile box provided on the activity sheet and then calculate its vertical exaggeration.

Before You Begin Read the following section: Topographic Profiles and Vertical Exaggeration.

Plan Ahead You will need a simple calculator or calculator app on your smartphone and a ruler.

Topographic Profiles and Vertical Exaggeration

A topographic map provides an overhead (aerial) view of an area, depicting features and relief by means of its symbols and contours. Occasionally a profile of the topography is useful. A **topographic profile** depicts the shape of the ground surface on a vertical plane between two points on the ground surface, as in a side view (**Fig. 9.15**). To construct a topographic profile, follow the steps in **Fig. 9.15**.

Topographic Profiles — The Math You Need

You can learn more about constructing topographic profiles at this site featuring *The Math You Need, When You Need It* math tutorials for students in introductory geoscience courses: **http://serc.carleton.edu/mathyouneed/slope/topoprofile.html**.

How to Obtain a US TOPO Map or Orthoimage

Watch a 6-minute USGS video (**https://www.usgs.gov/media/videos/introduction-national-map**) and follow these steps:

Step 1: Navigate to the USGS Store on the web (**http://store.usgs.gov**). Select "Map Locator and Downloader."

Step 1

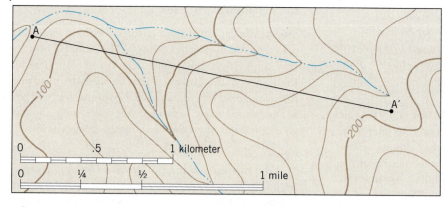

Step 2

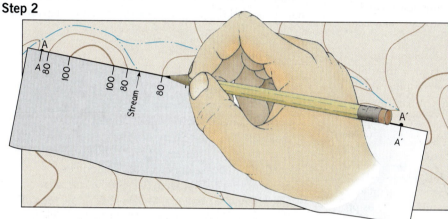

Step 3

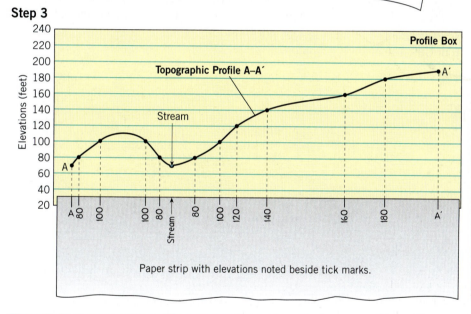

Paper strip with elevations noted beside tick marks.

Step 4 Vertical exaggeration

On most topographic profiles, the vertical scale is exaggerated (stretched) to make landscape features more obvious. One must calculate how much the vertical scale has been exaggerated in comparison to the horizontal scale.

Let's use the variable *a* to represent the total range of the vertical scale on the profile box (220 feet) in Step 3. Variable *b* is the length of the vertical scale on the profile box, measured in some convenient unit like mm.

Then find the bar scale on the map and determine its length in the same units you used for the vertical scale: feet and mm, for example. The bar scale in Step 1 is one mile in length or 5280 feet. Call that distance *c*. Variable *d* is the length of the one-mile bar scale printed on the map.

Divide *a* by *b* to find a scale ratio for the profile, in units of feet (in nature) per mm (on the map). This division yields a number of feet of elevation for every mm on the vertical scale of the profile box. Let's call that resulting number *e*.

Divide *c* by *d* to find a scale ratio for the map, in units of feet per mm. This division yields a number of feet for every mm measured horizontally, either on the map or in the profile box. Let's call that result *f*.

If you actually follow this process for the map in Step 1 and the profile in Step 3, you will find that *f* is a lot bigger than *e*. How many times bigger? Divide the bigger number (*f*) by the smaller (*e*) and you will find that the result is about 16.7, give or take a bit for uncertainty.

That means that the vertical scale on the profile box is about 16.7 times the horizontal scale. We say there is a 16.7x vertical exaggeration in the profile.

Figure 9.15 Topographic profile construction and vertical exaggeration. Shown are a topographic map (Step 1), topographic profile constructed along line **A–A'** (Steps 2 and 3), and calculation of vertical exaggeration (Step 4). **Step 1**—Select two points (**A, A'**), and the line between them (line **A–A'**), along which you want to construct a topographic profile. **Step 2**—To construct the profile, the edge of a strip of paper was placed along line **A–A'** on the topographic map. A tick mark was then placed on the edge of the paper at each point where a contour and stream intersected the edge of the paper. The elevation represented by each contour was noted on its corresponding tick mark. **Step 3**—The edge of the strip of paper (with tick marks and elevations) was placed along the bottom line of a piece of lined paper, and the lined paper was graduated for elevations (along its right margin). A black dot was placed on the profile above each tick mark at the elevation noted on the tick mark. The black dots were then connected with a smooth line to complete the topographic profile. **Step 4**—*Vertical exaggeration* of the profile was calculated by comparing the vertical scale to the horizontal scale.

Step 2: At the Map Locator and Downloader site, you will see a map of the United States.

a. Zoom in to the area for which you would like to obtain a map. Use the + and − buttons to zoom in or out, and click-hold-drag to move around in the map. Move on to the next step when you have found the area you are interested in.

b. To the right of the map, select "MARK POINTS: Click on a place to add a marker." Click on the center of the area you are interested in, and a balloon marker will be placed there. Then click on the marker and an information bubble will appear with a list of maps that are available. Download the map or maps that you want by clicking on the corresponding download link. The US Topo files are large (7–20 MB) and may take some time to download and open.

US Topo maps and images are displayed in GeoPDF® format and can be viewed only with Adobe Reader® or Adobe Acrobat®. If your computer does not have Adobe Reader (or Adobe Acrobat), then download and install it by selecting the "Get Adobe Reader" bar at the bottom of the page or go to Adobe (**http://get.adobe.com/reader/**).

Step 3: Modify the map if necessary to suit your needs. When you obtain and open a US Topo 7.5-minute quadrangle, you can add or subtract layers from them by using the menu along the left-hand side of the image. You have the option of downloading a free TerraGo Toolbar that allows you to measure distances, add comments, and merge products with Google Maps or your GPS. The older series of topographic maps are single-layer products scanned from paper maps.

Step 4: You can print all or parts of the maps and orthoimages that you display in US Topo. The GeoPDF files are very large (10–20 MB), so be patient and allow time for them to load, display, and print. To print an entire map or orthoimage on letter-size paper, be sure to set your printer to the "shrink to fit" setting. If you have a snipping tool on your computer, then you can snip the images as low-resolution JPEG files.

You can also use the *US Topo and Historical Topographic Map Collection* at **http://geonames.usgs.gov/pls/topomaps** to obtain USGS topographic maps if you already know the map name and the primary state in which the map area is located.

MasteringGeology™

Looking for additional review and test prep materials? Visit the Study Area in MasteringGeology to enhance your understanding of this chapter's content by accessing a variety of resources, including Pre-Lab Videos, Self-Study Quizzes, Geoscience Animations, Mobile Field Trips, *Project Condor* Quadcopter videos, *In the News* articles, glossary flashcards, web links, and an optional Pearson eText.

Name: _____ Course/Section: _____ Date: _____

A Imagine that a friend has asked you, "Where are you located right now?" Give three different answers below:

1.

2.

3.

B **REFLECT & DISCUSS** One of the oldest ways of describing a location on our spherical Earth is by using a coordinate system of latitude and longitude. On the map in **Fig. A9.1.1**, notice how the red location is identified as 30°N, 150°W.

1. On the map, lightly color in the country where you live.

2. Name a place of your choice to locate on the map (your current location, home, school, place of work, or other location). This will be referred to as your "home place" in the inquiry items below.

3. Place a dot on the map to show approximately where your home place is located, and then estimate its latitude and longitude as is already done for the red and blue points on the map.

   Latitude: _____ Longitude: _____

4. Notice that the lines of latitude and longitude outline somewhat rectangular areas called *quadrangles*, like the one shaded pale red on the map. A quadrangle is an area bounded by parallels of latitude and meridians of longitude. Describe the location of the red quadrangle by giving its bounding latitudes and longitudes.

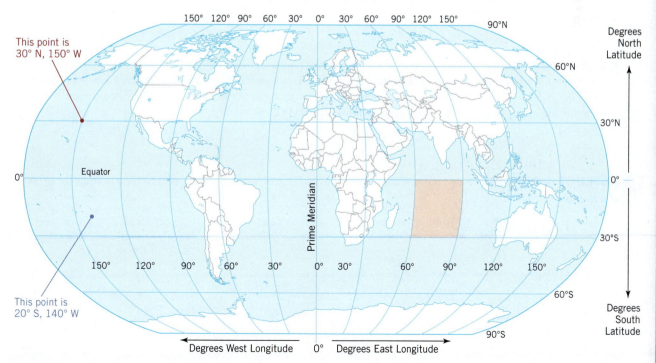

Figure A9.1.1

C Use Google Earth to locate places on Earth. If you do not have Google Earth on your web-enabled device, then you can install a free copy of the app via **http://earth.google.com**. When you have a functioning copy of Google Earth, open the app and do the following:

1. Determine the latitude, longitude, and elevation of your home place.

   (a) Start by using the Google Earth "Search" option located in the upper left-hand corner of the screen. You can type in the name (e.g., Washington Monument) or address of your home place, and then click on the "Search." Google Earth will then zoom in on the location. You can then change the zoom (in or out) using your mouse wheel, trackpad, or the plus–minus slide bar in the upper right-hand corner of the screen.

   (b) Next, place your cursor over your home place to locate its latitude and longitude more exactly. As you move the cursor, notice along the bottom right-hand edge of your screen that the latitude and longitude coordinates of the cursor location are identified. (If you do not see degrees of latitude north or south and longitude east or west, then go to the top of the Google Earth screen and choose "Tools" and then "Options." On that menu, under "Show Lat/Long," choose "Degrees, minutes, seconds." Then click on "Apply" and "OK.") Notice that more than just degrees of latitude and longitude are indicated. For finer measurements of latitude and longitude, each degree can be subdivided (like a clock) into 60 subdivisions called minutes ('), and the minutes can be divided into 60 equal subdivisions called seconds ("). Record the exact location of your home place below in degrees, minutes, and seconds.

   Latitude: _____   Longitude: _____

   (c) Notice that elevation above sea level (elev) is indicated to the right of latitude and longitude at the bottom right-hand edge of the Google Earth screen. What is the elevation of your home place? _____

2. Experiment with the Layers of Google Earth to learn more about your home place. From the menu on the left-hand side of the Google Earth screen, open "Layers." Experiment with turning layers on and off to see how it affects what Google Earth displays. For example, you may want to start by choosing Roads, Borders and Labels, Gallery, or More to see what is available to display. When you are done, describe something that you learned about your home place or the region around it by experimenting with the Layers.

3. Measure a distance in Google Earth. Move your cursor over the toolbar icons at the top of the screen until you identify the "Show Ruler" icon and then click on it to open the Ruler menu. Select the "Line" tab, and use the pull-down menu to select units of measurement. Then use your mouse to make a measurement using the ruler (by clicking on a starting location and an ending location. Describe something that you measured and note the measurement.

4. Explore historical imagery. Again, move your cursor over the toolbar icons at the top of the screen until you identify the "Show historical imagery" icon (a clock symbol), and then click on it to open the slider of dates. Use your mouse to move the slider, and observe changes in the Google Earth images. Explore the region where your home place is located. Describe something that you learned using this feature of Google Earth.

D **REFLECT & DISCUSS** Based on your knowledge of Google Earth, suggest how it could be used to study the geology of a region.

Name: _____ Course/Section: _____ Date: _____

A What are the latitude–longitude coordinates of point B in **Fig. 9.2**?

Latitude: _____ Longitude: _____

B Refer to **Fig. 9.8**, which describes the Public Land Survey System (PLSS).

1. Review **Fig. 9.8** to understand how the location of point X in **Fig. 9.8C** was determined using PLSS shorthand. What is the location of point Z in **Fig. 9.8C** in PLSS shorthand?

2. How many acres are present in the township in **Fig. 9.8B**? (*Hint*: There are 640 acres in 1 mi.^{2}.) Show your work.

3. Imagine that you wanted to purchase the NE 1/4 of the SE 1/4 of section 11 in **Fig. 9.8C**. If the property costs $500 per acre, then how much must you pay for the entire property?

C USGS 30 × 60-minute quadrangle maps have a scale of 1:100,000.

1. One inch on such maps equals about how many miles? Show your work.

2. One cm on such maps equals about how many meters? Show your work.

D The standard base map for many regional geologic studies is the 7.5-minute topographic quadrangle. Imagine that a field geologist uses a handheld 12-channel GPS receiver with a location uncertainty of ±5 m. What would be the *diameter* of the uncertainty area associated with that GPS unit if you plotted it as a circle on the 1:24,000 scale map (in mm)?

E Refer to **Fig. 9.5**.

1. What is the bearing (azimuth) from point *c* to point *d*? _____
2. What is the bearing (azimuth) from point *d* to point *c*? _____

F Refer to the Ritter Ridge, California, 7.5-minute topographic quadrangle map in **Figs. 9.3A** and **9.3B**.

1. What are the latitude–longitude coordinates of the NW corner of the map?

2. The USGS used a polyconic projection to map points on the curved surface of Earth onto the plane of this map. What is the *datum* used for this map projection? (*Hint:* Read the information at the lower left corner of the map in **Fig. 9.3B**.) _____ (The most recent version of this map uses the North American Datum of 1983, so the coordinates of a given point on the older map are slightly different from those on the current map.)

3. In what UTM zone is this map located? (*Hint:* Read the information at the lower left corner of the map in **Fig. 9.3B**.) UTM zone _____

4. In what year was this map originally made? _____ In what year was it photorevised? _____

5. (a) What was the magnetic declination at the center of the map when the map was prepared? *Hint:* Find the north arrow at the bottom of **Fig. 9.3B** and read the angle between true north (indicated with the black star) and magnetic north (MN) in degrees and minutes of arc. What was the magnetic declination angle? _____°

   (b) What is the current magnetic declination for the center of the map—34.5625°N, 118.1875°W—as indicated by the NOAA-NCEI Magnetic Declination Estimated Value Calculator? (**http://www.ngdc.noaa.gov/geomag-web/**) _____

6. What is the PLSS location of section 21 in the southwest corner of the map?

G Refer to **Fig. A9.2.1** to see the SW corner of the Ritter Ridge, California, 7.5-minute topographic quadrangle map, published by the USGS (1975).

1. What general kind of vegetation is indicated on the west (left) side of **Fig. A9.2.1**? (Refer to symbols in **Fig. 9.4**.)

2. List the different kinds of roads depicted on the map based on the symbols explained in **Fig. 9.4**.

3. Gold, copper, and titanium were mined here until the early 1900s. Notice that there are three different symbols used to indicate the kind of mining activity that occurred here. Draw the mine symbols below and identify what they mean using **Fig. 9.4**.

4. What are the UTM (NAD27) coordinates of Puritan Mine near the southwest corner of the map? (*Hint:* Cut out and use the 1:24,000 UTM Grid from Geotools Sheet 4 at the back of the manual, or use the orange lines to help you compute the coordinates using proportions. The orange lines are exactly 1000 m apart.)

5. Use a protractor to measure the bearing (azimuth) from Red Rover Mine to Puritan Mine. The black line at the left edge of the map is oriented north–south.

6. The black line at the left edge of the map is along a meridian that extends to the north and south poles. Notice that the orange lines of the UTM grid are not quite parallel and perpendicular with the black line. Grid north is the same as true north for points along the central meridian of a zone, along which the eastings are all defined as 500,000 mE. The farther away from the central meridian a point is, the larger the angle of difference between UTM grid north and true north.

(a) What is your estimate of the angle between grid north and true north in **Fig. A9.2.1**? _____

(b) Look at the north arrow in **Fig. 9.3B** and read the angle between grid north (GN) and true north (indicated with the black star) in degrees and minutes of arc. What is that angle? _____

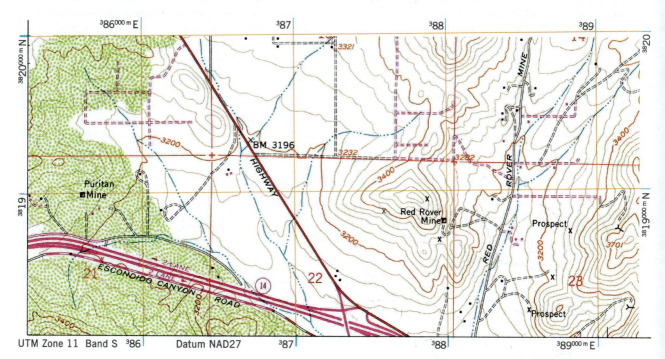

Figure A9.2.1

H **REFLECT & DISCUSS** Below **Fig. A9.2.1**, add bar scales to show how long 1 mi. and 1 km are on the map, and then explain how you determined the lengths of the bars in your bar scales.

Name: Journie Dyehouse Ferguson **Course/Section:** _____ **Date:** 10/29/2020

A Use interpolation and extrapolation to estimate and label elevations of all points in **Fig. A9.3.1** that are not labeled (see **Fig. 9.11** for help). Estimate elevations to the nearest 50 feet. Using a pencil, sketch and label contours using a contour interval of 100 feet. Notice that the 0-foot and 100-foot contours have already been drawn. Each contour must go *through* points that have the same elevation as the contour—not near or next to but *through*. Contours should be deflected upstream where they cross stream drainages. Most contours end at the edge of the map.

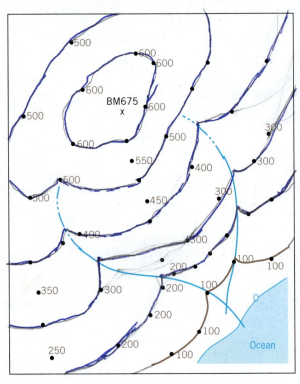

Figure A9.3.1

B Using a pencil, contour the elevations in **Fig. A9.3.2** using a contour interval of 10 feet. Refer to **Fig. 9.11** as needed.

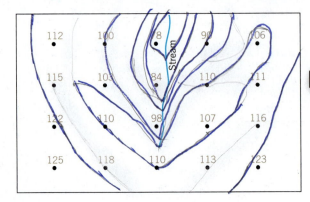

Figure A9.3.2

C Using a contour interval of 10 ft., label the elevation of every contour in **Fig. A9.3.3**. (*Hint:* Start at sea level and refer to **Figs. 9.11** and **9.12**.) The short tick marks on some contours are hachures and indicate an enclosed depression.

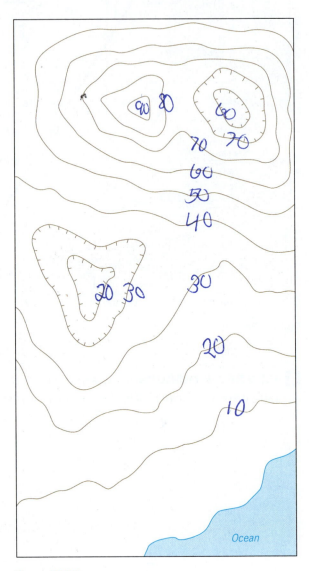

Figure A9.3.3

D **REFLECT & DISCUSS** The elevations in **Fig. 9.1C** are in ft. above sea level. What is the contour interval of the map, and how did you determine it?

Name: _____ **Course/Section:** _____ **Date:** _____

A **Figure A9.4.1** is a portion of the 1989 SP Mountain, AZ 7.5-minute topographic quadrangle map. The black 1 km by 1 km UTM grid is NAD27, Zone 12S. Also note the 1 mi. by 1 mi. PLSS grid sections with red numbers in their centers.

1. What is the point of highest elevation on the map, and how can you tell?

2. Draw a small "x" over the point of lowest elevation on the map, and label it with the elevation.
3. What is the total relief of SP Mountain measured from its base to the north of the peak up to its highest elevation? _____ ft.

4. Circle the four places within the map where streams (blue) begin. How can you tell which direction is upstream based on the contours?

5. Notice that there is an SP Mountain and an SP Crater. How can you tell which part is the crater? Color it red.

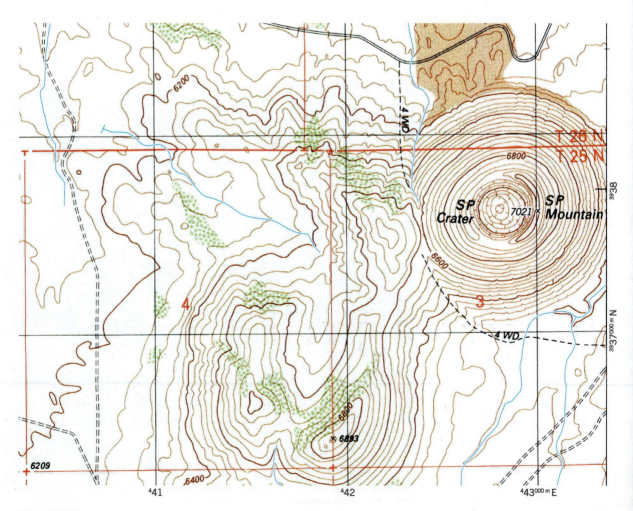

6. Describe the north–south diameter (in km and mi.) and three-dimensional shape of SP mountain, and interpret what kind of geological feature it might be. Use the UTM and PLSS grids to help with scaling.

7. Based on your answer above, what material do you think is represented by the orange-brown color (section 34)?

B **Figures A9.4.2A** and **A9.4.2B** are 2011 and 2014 US Topo Series products. The digital map is matched with an orthoimage of the same area.

1. The orthoimage (**Fig. A9.4.2B**) reveals that SP Mountain is a cinder-cone volcano with a very visible closed depression or crater at its summit. The image also reveals an older, reddish volcano that is very eroded. Draw a dashed line on the orthoimage and topographic map (**Fig. A9.4.2A**) to show the outline of this older volcano.

2. Draw a solid line on the orthoimage and map to show the outline of the crater of the older volcano. Why is it not shown with hachure lines on the map like the crater for SP Mountain?

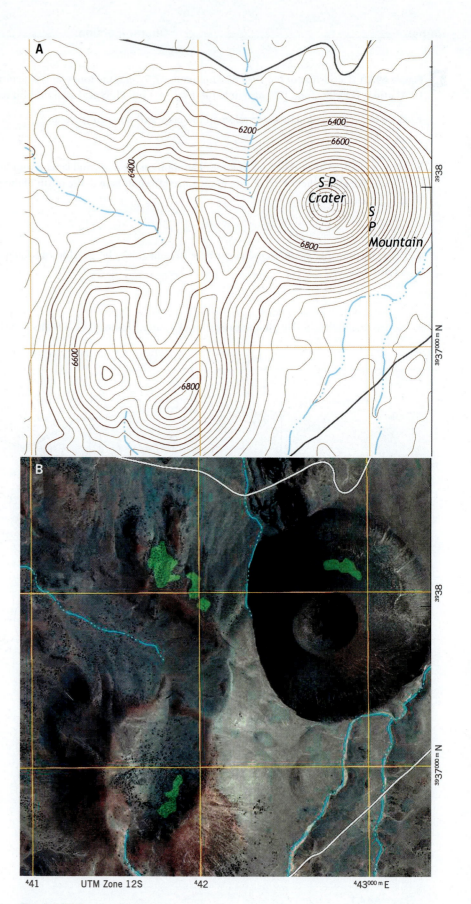

Figure A9.4.2

C Compare the 2014 topographic map (**Fig. A9.4.2A**) with the 1989 map (**Fig. A9.4.1**).

 1. How are the contours different in the 2014 map?

 2. What is one thing that changed on the ground in this area from 1989 to 2014?

D **REFLECT & DISCUSS** Which series of USGS products do you think is more useful: the 1989 paper maps series or the 2014 US Topo series of digital maps with matching orthoimages? List some advantages and disadvantages of your choice.

Advantages:

Disadvantages:

On a visit to Yosemite, you and some friends find a nice place along the Merced River for a picnic. You have a topographic contour map (**Fig. A9.5.1**) in which the index contours are labeled in feet. From your resting place at point A, you decide to gain a better vantage point by walking up to point B.

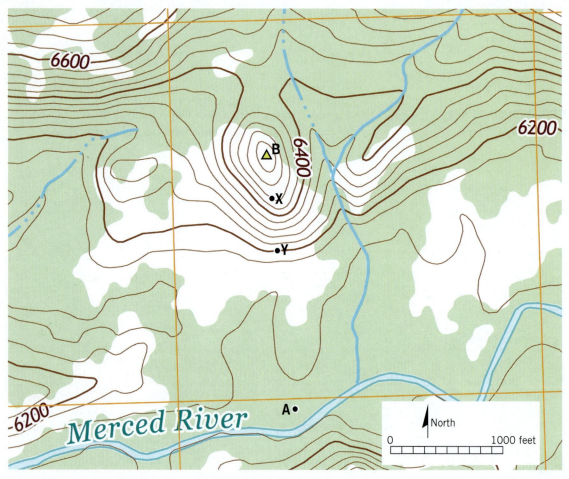

Figure A9.5.1

A Using the labeled elevations of the index contours as your guide, what is the contour interval (the elevation difference between two adjacent contours) in **Fig. A9.5.1**? _____ ft.

B You have a bar scale on the map that indicates a distance of 1000 ft. on the ground in your picnic area. How long is that bar scale on the map in mm? _____ mm

C What is the horizontal distance from A to B measured along a straight line in ft? (*Hint:* Measure the distance from A to B using a metric ruler on the map, and then use your measurement of how many mm on the map equal 1000 ft. in the field to find the answer using proportions.) _____ ft.

D What is the average *gradient* (elevation change divided by horizontal distance, both expressed in ft.) along a straight line from A to B? _____

E In fact, it gets much steeper between X and Y along the slope to B. What is the gradient between X and Y? _____

F The slope between X and Y does not appear to be the steepest slope in the area. *Circle the area on the map* that has the steepest slope over an elevation change of at least 80 feet.

G **REFLECT & DISCUSS** There is no need to walk in a straight line from A to B, so pick a path that has the gentlest overall slope to the top of the knob at point B, and use a pencil to sketch that path with a dashed curve. Share your proposed path with other students, and listen to their explanations of why they chose the paths they drew.

Name: Journie Dyhouse Ferguson   **Course/Section:** _____   **Date:** 10/29/2020

A   Construct a topographic profile for section line **A–A′** in the profile box (**Fig. A9.6.1**). The profile box represents a vertical plane along line **A–A′** through the ground surface.

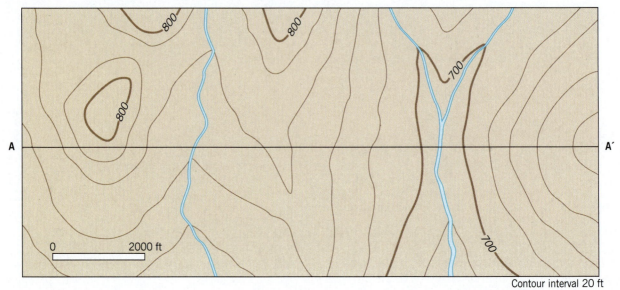

Contour interval 20 ft

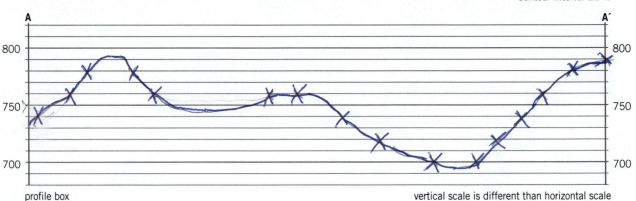

profile box   vertical scale is different than horizontal scale

Figure A9.6.1

B   What is the vertical exaggeration of the topographic profile that you constructed above? Show your work.

$$1 = \frac{a}{b} = \frac{140 ft}{4.2 cm} = 33.\overline{3} \quad \Big| \quad 2 = \frac{c}{d} = \frac{2000 ft}{2.5 cm} = 800 \quad \Big| \quad \frac{2}{1} = \frac{800}{33.3} = 24.03x$$

C   **REFLECT & DISCUSS**   Why is it important to always know the vertical exaggeration of a topographic profile?

To show how steep the slope is between the contour lines in reality

Pre-Lab Video 10

https://goo.gl/26f7MB

Geologic Structures, Maps, and Block Diagrams

CONTRIBUTING AUTHORS

Michael J. Hozik • *Stockton College of New Jersey*

Raymond W. Talkington • *Stockton College of New Jersey*

William R. Parrott, Jr. • *Stockton College of New Jersey*

▲ Piqiang Fault displaying strike-slip displacement in the Keping Shan thrust belt of Zinjiang, China (40.24°N, 77.67°E). Part of a NASA Landsat 8 OLI.

BIG IDEAS

The rock units exposed on Earth's surface provide us with clues about the sequence of events in an area's geologic history. Construction of geologic maps and cross-sections is an important part of the process of documenting and understanding the geometric relationships among rock units, revealing important details about how those rocks formed and changed over time. Earth's crust is subject to stresses that can distort rock through fracturing, faulting, folding, and other deformation mechanisms. Through specialized map symbols, geoscientists can represent rock types and structures on geologic maps and cross-sections. The wide variety of geologic maps provides society with important information about hazards, resources, geological processes, and Earth's history.

FOCUS YOUR INQUIRY

Think About It How do geologists use aerial imagery to map geologic structures on and beneath Earth's surface?

ACTIVITY 10.1 Map Contacts and Formations (p. 266, 281)

Think About It How are deformed rocks identified and classified?

ACTIVITY 10.2 Geologic Structures Inquiry (p. 272, 284)

Think About It What does a fault look like from above?

ACTIVITY 10.3 Fault Analysis Using Orthoimages (p. 272, 286)

Think About It How do geologists visualize geologic structures using geologic maps and cross-sections?

ACTIVITY 10.4 Appalachian Mountains Geologic Map (p. 275, 288)

Think About It How are 3-dimensional models used to visualize geologic structures?

ACTIVITY 10.5 Cardboard Model Analysis and Interpretation (p. 276, 289)

ACTIVITY 10.6 Block Diagram Analysis and Interpretation (p. 276, 291)

Introduction

A **geologic map** shows the distribution of different rock types across part of Earth's surface (**Fig. 10.1A**). The geology mapped across the ground surface is just a hint of the full 3-dimensional distribution of rock in the crust below the surface and can even give us a glimpse of the 4-dimensional geological evolution of the area over time. Geologists construct vertical **cross-sections** that help us to visualize the geometry of rock bodies below the ground surface (**Fig. 10.1B**), just like a vertical cut in a layer cake lets us see the internal structure of the cake.

Geologic maps are usually layered on top of an existing topographic base map, so they incorporate all of the information about the horizontal position and elevation of different points on the landscape. The geoscientists who make these maps then go one step further to indicate some geologic property associated with each of those points. The most common property is the formally defined rock unit or formation that is found at that point. Geologic maps also depict the location of contacts between different rock units: conformable and unconformable sedimentary contacts, igneous and metamorphic contacts, and fault contacts.

Different types of geologic maps might be developed for the same quadrangle to highlight different aspects of the area's geology such as water resources, soils and other recent sediments, building materials, economic minerals, structure, engineering geologic properties, landslide susceptibility, geophysical properties of the gravity or magnetic field, and so on. You can search for U.S. geologic maps and formation descriptions from the National Geologic Map Database and Geologic Names Lexicon (**http://ngmdb. usgs.gov**). This site also has links to state geologic maps.

Geologic field mapping is a fun, intellectually stimulating activity much of the time. Field geologists like to say that any day in the field is better than a good day in the office. Many challenges confront field geologists, including access restrictions on private property; various materials that cover the rock they would like to map (soil, stream sediment,

vegetation, buildings, roads, artificial fill); topographic challenges like steep cliffs; weather; and various plants and animals that are worth avoiding. However, there are usually ways to overcome the challenges to geologic mapping projects.

Geologic Mapping and Map Symbols

Between 1799 and 1815, the English civil engineer and geologist William Smith created the first true geologic maps that cover large areas, although people before that had made local maps of soils. *A Geological Map of England and Wales and Part of Scotland* by William Smith was first published in 1815 and has been called "the map that changed the world." Geologic maps help to identify the locations of Earth materials that are interesting, useful, or economically valuable. Smith found that having accurate geologic maps made it possible for him to build canals and other public works that were more durable because they were designed with the prevailing geological conditions in mind. Today, geologic mapping is conducted by a vast number of public and private organizations worldwide, including the U.S. Geological Survey (USGS) and individual state geological surveys in the United States. In order to be effective in communicating information, modern geologic maps use styles and symbols that are widely used and understood by geoscientists to express characteristics of the geology.

Orientation of Geologic Surfaces

How do the field geologists who make geologic maps describe the orientation of a geological surface such as a bedding plane, a fault plane, or the surface of a dike? First, let's not worry about the orientation of the entire bedding surface or fault plane because the orientations of most geologic surfaces vary at least a little bit with distance across the surface. Instead, we measure the orientation of *part* of the broader surface, concentrating on a small area of a rock surface that is about the size of a binder or book cover—a few tens of centimeters on a side (**Fig. 10.2**).

There are plenty of ways of describing the orientation of a surface, but the convention among field geologists is to measure the strike, dip direction, and dip angle of a surface. The **dip** (or **dip vector**) is the steepest line of descent on an inclined surface. The dip vector is the blue arrow in **Figs. 10.2A** and **10.2B**, which points along the path that a ball will begin to roll or a drop of water will flow down the surface. The **dip angle** tells us how steep the surface is and is measured in a vertical plane between the dip vector and horizontal (**Fig. 10.2B**). Dip angles vary from 0° for a horizontal surface to 90° for a vertical surface. The gray horizontal vector shown in **Fig. 10.2** points toward the **dip direction**, which is usually expressed relative to one of the eight primary compass directions—N, NE, E, SE, S, SW, W, or NW.

Strike is the azimuth of a horizontal line on the surface and is perpendicular to the dip vector and to the vector showing the direction of dip. We use the **azimuth**

ACTIVITY 10.1

Map Contacts and Formations, (p. 281)

Think About It   How do geologists use aerial imagery to map geologic structures on and beneath Earth's surface?

Objective   Map formations with geometrically simple contact surfaces using an aerial photograph and a topographic map.

Before You Begin   Read the following section: Geologic Mapping and Map Symbols.

Plan Ahead   You will need a ruler and a pencil for initial sketching of contacts on a map and cross-section.

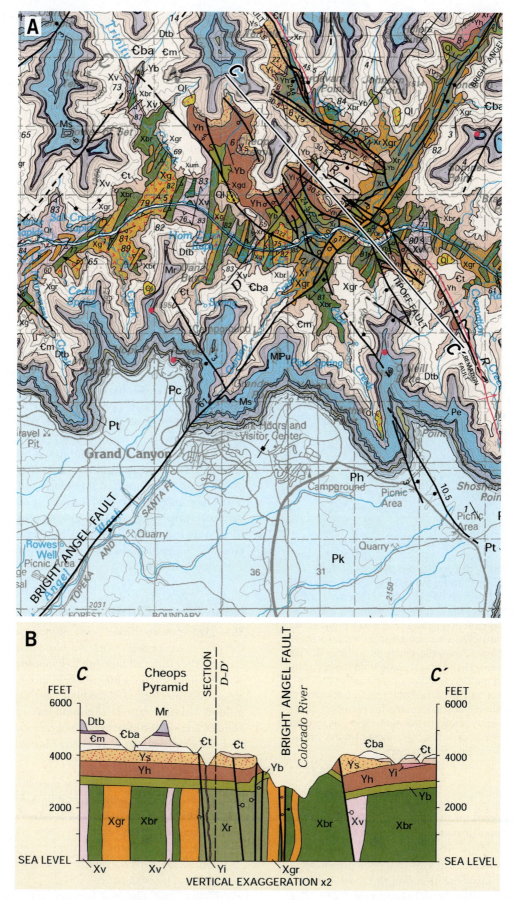

Figure 10.1 Geologic map and cross-section. Part of a geologic map of the Grand Canyon by George Billingsley, published in 2000 by the USGS (**http://pubs.usgs.gov/imap/i-2688/i-2688.pdf**). The C–C′ line of section is shown in the geologic map (**A**) and the corresponding cross-section is shown in **B**. Nearly horizontal beds overlie near-vertical Precambrian rock units along this line of section, which is also crossed by several normal faults.

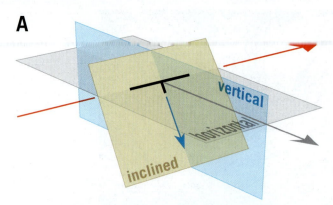

A

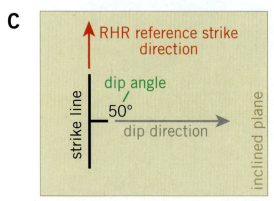

B

edge of inclined plane

horizontal vector
in dip direction

dip vector

vertical
plane

dip
angle
50°

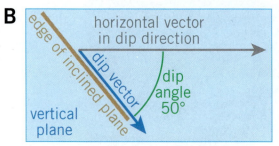

C

RHR reference strike
direction

dip angle

50°

strike line

dip direction

inclined plane

Figure 10.2 Strike, dip angle, dip direction. A.
Perspective diagram showing the strike of an inclined surface
(yellow) as the intersection of that surface with a horizontal
plane (gray). The strike is along the red lines and longer
black line. The dip direction (gray arrow) and the dip vector
(blue arrow) of the inclined surface are in a vertical plane
that is perpendicular to the yellow inclined surface. **B.** The
dip angle is measured in a vertical plane from the horizontal
vector pointing in the dip direction (gray arrow) to the dip
vector (blue arrow) on the inclined surface. **C.** Look straight
down on the inclined plane to illustrate the relationship of
the dip direction (gray arrow) to the right-hand-rule reference
strike direction (red arrow). The black symbol is the strike-
and-dip symbol used on geologic maps.

method to measure the orientation of the line segment
in which 0° is toward north and the angle increases in
a clockwise manner through 90° east, 180° south, and
270° west and back to north.

There are a number of ways that geologists have
recorded the strike orientation over the years, but we
will focus on one called the **right-hand-rule** (**RHR**) that
is commonly used by geologists who use computers to
produce their maps and help analyze structural data.

The strike of a particular area on a geological surface is a
line segment with one end directed 180° from the other
end. Which one is the strike, or might either one be the
strike? The RHR convention is to point the outstretched
fingers of your right hand parallel to the dip vector with
your thumb pointing up. Then relax your fingers. As
they curl, your fingers will point toward the **RHR strike**
or **reference strike** (**Fig. 10.2C**). The reference strike is a
right-handed (counterclockwise) rotation from the dip
direction.

If the reference strike is 42°, the full description of the
orientation of that surface is 42°, 23°SE where the strike is
listed first, the dip angle second, and (optionally) the dip
direction third. Using the RHR makes the dip direction a bit
unnecessary, but including it makes interpreting the strike-
and-dip description easier.

Formations

Formally defined rock units that can be distinguished
from one another "in the field" and are large enough
to appear on published geologic maps are called **formations**. Formations may be subdivided into *members* and
various smaller divisions. Geologists assign each formation a formal name, which is capitalized (e.g., Muav
Limestone, Temple Butte Formation). Descriptions
of the distinguishing features and "type locality" of a
formation are compiled and published in an accessible
database. The USGS maintains the *Geologic Names Lexicon*
of names and descriptions of geologic units that currently has more than 16,500 units in its web-searchable
database (**http://ngmdb.usgs.gov/Geolex/search**). The
surfaces between formations are called **formation contacts** and appear as thin black lines on geologic maps
and cross-sections.

Near-Planar Contacts and Topography

Field geologists who make geologic maps are very atten-
tive to contacts between rock units and map them as accurately as they can. Contacts are 3-dimensional surfaces, as
is the ground surface whose shape is given in topographic
maps. The intersection of a contact surface and the
ground surface is called the **ground-surface trace** of that
contact. Let's investigate how topography and the orientation of a simple planar contact combine to affect the trace
of that contact.

- If the contact is horizontal, its trace follows the topographic contours, which are all lines of equal elevation
  (**Fig. 10.3A**). Every point on a horizontal surface has
  the same elevation.
- If the contact is vertical, its trace is a straight line
  (**Fig. 10.3B**). This is a bit like looking straight down on
  a vertical piece of paper.
- If the contact dips in a direction that is opposite to the
  topographic slope, the trace is deflected in an upslope
  direction in stream drainages (**Fig. 10.3C**).

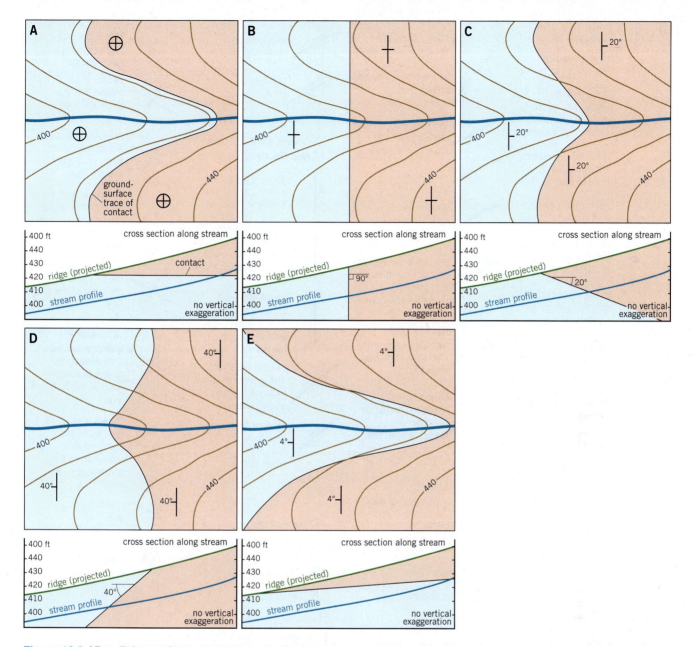

Figure 10.3 Visualizing a planar contact on geologic maps and cross-sections. A. Horizontal contact between the blue formation below and the pink formation above. The ground-surface trace of the contact is parallel to the topographic contours. **B.** Vertical contact, producing a straight trace that is not deflected in any way by the topography. **C.** Inclined contact dipping in the upslope direction, producing a trace that is deflected in the upslope direction. **D.** Inclined contact dipping in the downslope direction at a greater dip angle than the inclination of the stream channel, producing a trace that is deflected in the downslope direction. **E.** Inclined contact dipping in the downslope direction at a smaller dip angle than the inclination of the stream channel, producing a trace that is deflected upslope.

- If the contact dips in the same direction as the topographic slope and a stream channel is not as steep as the dip of the contact, the trace is deflected in a downslope-downstream (**Fig. 10.3D**).
- If the contact dips in the same direction as the topographic slope and the channel is steeper than the dip of the contact, the trace is deflected in an upslope-upstream direction in stream drainages (**Fig. 10.3E**).

Map Symbols

The USGS publication that contains the standards for geologic map symbols (**http://ngmdb.usgs.gov/fgdc_gds/ geolsymstd/fgdc-geolsym-all.pdf**) has ~150 pages of tables explaining the map symbols, colors, patterns, and other cartographic standards used on USGS geologic maps. Some of the most important symbols used in geologic maps are shown in **Fig. 10.4** and will be needed in several of the activities.

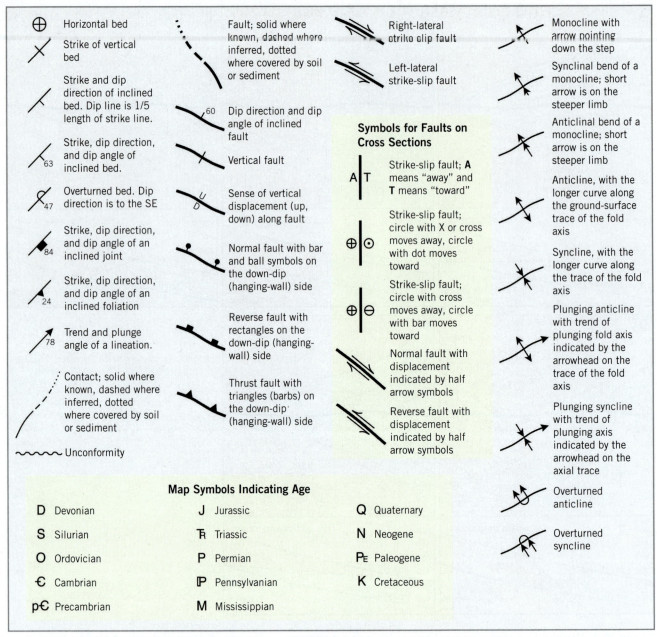

Map Symbols Indicating Age

D	Devonian	J	Jurassic	Q	Quaternary
S	Silurian	Ŧ	Triassic	N	Neogene
O	Ordovician	P	Permian	Pᴇ	Paleogene
Є	Cambrian	ℙ	Pennsylvanian	K	Cretaceous
pЄ	Precambrian	M	Mississippian		

Figure 10.4 Structural geology symbols and time-related abbreviations used on geologic maps. These map symbols agree with USGS standards for geologic maps.

How to Make a Simple Geologic Cross-Section

Geoscientists "map" below the surface using information from geophysics, boreholes, mines, and geological features exposed at the ground surface that can be projected below the ground surface. We typically represent our interpretation of the geology at depth using cross-sections. A cross-section is a kind of geological map that is projected onto a plane that extends into the Earth—usually, along a vertical plane (**Fig. 10.1B**).

The simplest geological cross-sections to construct involve undeformed horizontal sedimentary formations. If a contact is a horizontal plane, the intersection of that contact with the ground surface (if any) will form a ground-surface trace that is parallel to topographic contours. The northern edge of the light blue formation marked Pk—the Kaibab Formation—that forms the lower third of **Fig. 10.1A** is approximately parallel to topographic contours as are the contacts of several of the formations below the Kaibab Formation.

Other than undeformed horizontal beds, the easiest cross-sections to make are constructed in a vertical plane that is perpendicular to strike in an area where the strikes of all of the layers are parallel to each other as in **Fig. 10.5**. The line of section in **Fig. 10.5** is from **A** to **A′**

HOW TO CONSTRUCT A GEOLOGIC CROSS SECTION

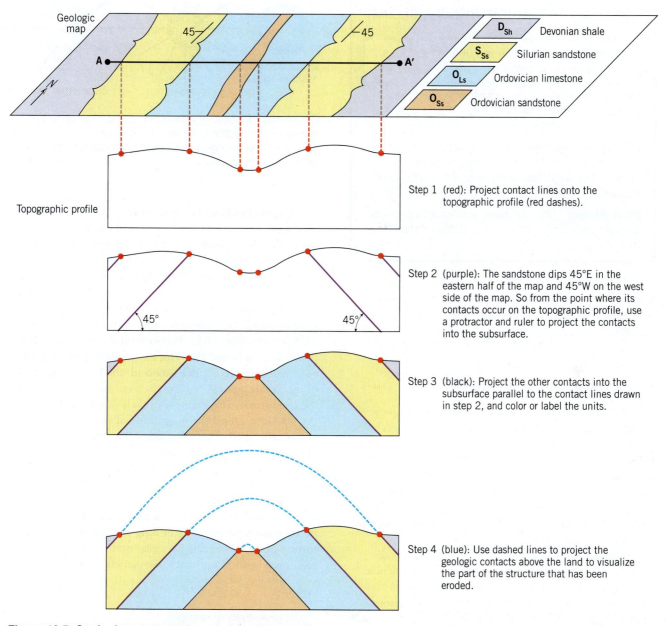

Geologic map

D_{Sh}	Devonian shale
S_{Ss}	Silurian sandstone
O_{Ls}	Ordovician limestone
O_{Ss}	Ordovician sandstone

Topographic profile

Step 1 (red): Project contact lines onto the topographic profile (red dashes).

Step 2 (purple): The sandstone dips 45°E in the eastern half of the map and 45°W on the west side of the map. So from the point where its contacts occur on the topographic profile, use a protractor and ruler to project the contacts into the subsurface.

Step 3 (black): Project the other contacts into the subsurface parallel to the contact lines drawn in step 2, and color or label the units.

Step 4 (blue): Use dashed lines to project the geologic contacts above the land to visualize the part of the structure that has been eroded.

Figure 10.5 Geologic cross-section construction. Follow the four steps in the illustration to construct a cross-section.

as shown on the geologic map at top. The profile of the ground surface along the line of section is drawn carefully in the profile box using elevation data from a topographic map.

The places where contacts cross the line of section are transferred to the profile (**Fig. 10.5**). Strike-and-dip information on the map indicates that the beds on the west side of the map dip toward the west, and beds on the map's east side dip toward the east. We project those dip data to the line of section and transfer them to the profile. We assume that the orientations of contacts that we measure at the ground surface can be projected at

least a little way below the ground surface along the cross-section. As a consequence of that assumption, the contacts are drawn on the cross-section at about the same dip angle as indicated on the geologic map. We interpret these beds to be part of a fold and extrapolate the shapes of the contacts into areas where we have no data.

The creation of geologic maps and sections is constrained by reproducible data and is an interpretive process. All geologic maps and cross-sections can be thought of as a combination of reproducible data—scientific facts—and spatial hypotheses that relate those facts to

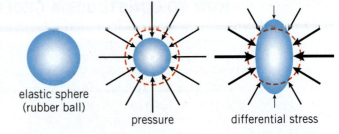

ACTIVITY 10.2

Geologic Structures Inquiry, (p. 284)

Think About It How are deformed rocks identified and classified?

Objective Classify types of rock deformation based on an analysis of deformed and underformed rocks.

Before You Begin Read the following section: How Does Rock Deform? Refer back to the section Geologic Mapping and Map Symbols as needed.

Plan Ahead You will need a ruler and a pencil for initial sketching of strike and dip in this activity.

Figure 10.6 Effects of changing pressure or differential stress on an elastic ball. Increasing pressure makes the ball smaller without changing its shape, but increasing differential stress changes its shape.

A stress field can be thought of as a system of forces acting on a mass. Imagine a spherical rubber ball drifting in the "zero gravity" environment of the International Space Station. (Actually, it is gravity that keeps the ISS in orbit around Earth, but the occupants of the ISS do not sense Earth's gravity while in orbit.) The ball experiences **pressure**, which is a special type of stress state in which the magnitude of the stress is the same in all directions (**Fig. 10.6**). Pressure on the ball is exerted by the air inside the ISS. The more typical case in geology involves a **differential stress** in which the magnitude of the greatest stress is different from the magnitude of the least stress. The greatest stress is oriented 90° from the least stress, and the magnitude of stress varies systematically in all other directions (**Fig. 10.6**). Changes in pressure will make an elastic rubber ball larger or smaller, but it takes a differential stress to change its shape—to distort the ball.

The kinds of deformation that we will be thinking about in this lab involve big volumes of rock deformed in perhaps multiple events over long periods of time. The history of the deformation can be quite complicated, and there is nothing reliable we can say about the history

one another. Geologic maps and sections are always provisional and are subject to revision as new data become available.

How Does Rock Deform?

In this lab, we will work with a few concepts of geologic mapping and structural geology to give you a feel for a very broad field within the geosciences. Structural geology is concerned with the deformation of geological materials at scales that range from the tiniest mineral grains to plate-scale displacements. We are going to focus on a few aspects of rock deformation that will help us interpret some simple geologic maps and leave the rest of structural geology for another time.

To understand deformed rock, we begin by recalling what we know about undeformed rock. In particular, we are going to work primarily with layered sedimentary rock in this lab, although any kind of rock or sediment can be subject to deformation. For example, the foliation observed in metamorphic rock is the product of deformation.

The **principle of superposition** tells us that in an undeformed sequence of sedimentary strata, the oldest layers are on the bottom and the youngest layers on top. The **principle of original horizontality** indicates that sedimentary layers are typically deposited in horizontal layers. This is what we expect to find in an undeformed sequence of sedimentary rock: near-horizontal layers that get progressively younger upward.

Deformation and Stress

Rock deforms when differential stress is applied to it in excess of the strength of the rock. The simplest way of defining stress is that it is force applied over a unit area, recalling that Isaac Newton defined force as mass times acceleration.

ACTIVITY 10.3

Fault Analysis Using Orthoimages, (p. 286)

Think About It What does a fault look like from above?

Objective Analyze and interpret selected faults using aerial photographs.

Before You Begin Read the section: How Does Rock Deform? Pay particular attention to the subsections Brittle Deformation and Types of Faults.

Plan Ahead You will need a pencil for initial sketching of fault traces and a pen for your final answer.

of the stress states that generated these big structures without much more information than is supplied in a geologic map. We can say, in a general and qualitative way, that some structures probably involve shortening of the crust and that others probably involve stretching or extension of the crust. For now, we will just try to describe the geometry of the deformed rock and note some of the different types of structures produced during deformation.

Brittle Deformation

Fracturing and faulting are two of the most important mechanisms of **brittle deformation**, which is quite common in at least the upper part of Earth's crust. Rock that deforms without apparent fracturing or faulting is often said to have been subject to **ductile deformation**.

Fractures or **cracks** develop in many solid materials when sufficient differential stress is applied to them. Imagine that you have a piece of classroom chalk and you hold tightly onto one end while another person holds the other. Now, the two of you move apart, pulling the ends of the chalk until it breaks. The chalk will break along a fracture plane that is perpendicular to the direction that the chalk was pulled. The two sides of the fracture moved apart from each other. You and your co-investigator put the chalk into a state of tension and applied enough **tensile stress** to open a fracture that broke through the piece of chalk.

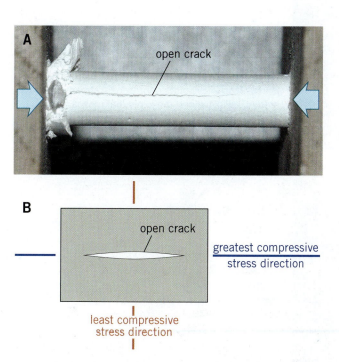

Figure 10.7 Experimental generation of an extension crack. **A.** A stick of chalk is shortened in a vise as indicated by the arrows, and an extension crack develops in the chalk. The two sides of the crack move away from each other, opening the crack. **B.** Sketch of an idealized extension crack, showing the greatest and least compressive stresses that resulted in the crack.

What would happen if we try to shorten the chalk by pushing on its two ends, applying a **compressive stress**? It is more difficult to deform chalk when we shorten or compress it, so we will need to use a vise. When the chalk is shortened, it develops the same general kind of crack it developed before—one in which the two sides or faces of the crack move away from each other (**Fig. 10.7**). The crack propagates or grows parallel to the direction in which we are shortening the chalk and parallel to the direction of greatest compressive stress. We call this an **extension crack**. The essential characteristics of an extension crack are that the crack *opens* perpendicular to the crack faces, and the crack propagates in the direction of the greatest compressive stress.

An extension crack is one of the most common structures in Earth's crust, and it occurs on scales ranging from smaller than a mineral grain to cracks that affect just a single layer of rock (**Fig. 10.8**) all the way up to regional fractures that can be traced for many kilometers across the countryside. Mud cracks are extension cracks that develop when mud dries and contracts. The cooling fractures that develop in lava are also extension cracks that develop as the rock contracts. The bigger extension fractures that can be seen cutting across all layers of rock and that persist for many meters are called **joints**. Some of the sets of joints that help shape the topography of Yosemite, Grand Canyon, and Zion National Parks are so extensive that they can be seen from Earth orbit (**Fig. 10.9**).

Faults are surfaces or zones along which there is shear displacement—displacement parallel to the surface (**Fig. 10.10**). As with extension fractures, faults can be observed crossing individual grains (microfaults) or can form systems of faults that are hundreds of kilometers long, such as the Alpine (New Zealand), Atacama (Chile), Great Glen (Scotland), North Anatolian (Turkey), and San Andreas Faults.

Types of Faults

A fault can be classified on the basis of the direction that one side moves relative to the other along the fault. We are going to need some descriptive terms to help us classify faults, so let's begin by imagining a fault surface that is inclined perhaps 45° relative to a horizontal plane. The block on one side of the fault that is above the fault is called the **hanging-wall block**, and the other block—the one that is below the fault—is called the **footwall block**. The story goes that as near-horizontal mine tunnels (adits) are excavated along a fault, the miners stand on the footwall while the hanging wall is hanging over their heads.

A fault along which the hanging-wall block slides down the fault, parallel to the dip vector, is called a **normal fault** (**Fig. 10.11A**). If you put a box near the top of a slick inclined surface, it is *normal* to expect that box to slip *down* the surface. Normal faults develop in areas where the crust is stretching or extending, as it does along divergent plate boundaries, in continental rift zones, and in a number of

Figure 10.8 Extension cracks in an outcrop. Extension cracks in brittle layers of a sedimentary rock. The lower layer with cracks is about 2 cm thick.

Approximate orientation of prevalent joint set

North

0   5 km

Figure 10.9 Regional joint set at Zion National Park, Utah. The joints control local stream flow, eroding valleys along the joints that makes them easy to see from above. (NASA Landsat 8 Operational Land Imager scene, **http://earthobservatory.nasa.gov/IOTD/view.php?id=88228)**

fault

fault

Figure 10.10 Small fault in sedimentary rock. The displacement along this fault exposed on a near-vertical seacliff is probably on the order of a couple of meters. The camera was pointed approximately parallel to the strike of the fault surface.

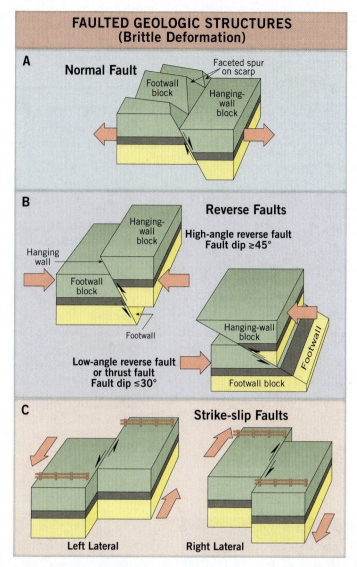

FAULTED GEOLOGIC STRUCTURES
(Brittle Deformation)

A

Normal Fault

Faceted spur on scarp

Footwall block

Hanging-wall block

B

Reverse Faults

Hanging-wall block

Hanging wall

Footwall block

Footwall

High-angle reverse fault
Fault dip ≥45°

Hanging-wall block

Footwall

Low-angle reverse fault
or thrust fault
Fault dip ≤30°

Footwall block

C

Strike-slip Faults

Left Lateral

Right Lateral

Figure 10.11 Three types of faults. A. Normal faults develop when brittle rock is extended or stretched. **B.** Reverse faults develop when brittle rock is shortened. **C.** Strike-slip faults.

more local geological situations involving extension of the crust. Landslides slip downslope on small normal faults.

If you put a box on a slick inclined surface, let go, and it slipped *up* the surface, that would be the absolute *reverse* of what you expected to happen. And yet that does in fact happen with another type of fault: a **reverse fault**. The hanging-wall block slips up the fault surface parallel to the dip direction in a reverse fault. A reverse fault with a dip angle of greater than ~45° is often called a **high-angle reverse fault**, and a reverse fault with dip angles of less than ~30° is commonly called a **thrust fault** (**Fig. 10.11B**). Reverse faults develop as a result of the shortening of the crust as occurs along some convergent plate boundaries among other areas affected by compressional stress.

A near-vertical fault typically has a horizontal or near-horizontal direction of slip between the two sides. This is the same as saying that the direction of slip is parallel or nearly parallel to the strike of the fault surface. We refer

ACTIVITY 10.4

Appalachian Mountains Geologic Map, (p. 288)

Think About It How do geologists visualize geologic structures using geologic maps and cross-sections?

Objective Construct and interpret a geologic cross-section of folded rocks.

Before You Begin Read the following sections about folds. Refer back to the section Geologic Mapping and Map Symbols as needed.

Plan Ahead You will need a ruler and a pencil for initial sketching of contacts on a cross-section.

We use the same logic to define a **right-lateral strike-slip fault** like the Hayward Fault in California. The photographs in **Fig. 10.12** were taken at a place where the Hayward Fault crosses a road producing an obvious deflection of the curb with **Fig. 10.12A** taken looking in one direction and **Fig. 10.12B** taken at the same curb looking in the opposite direction. Two buried utility boxes in the planting strip next to the curb are labeled X and Y, so you can relate the two photographs together by the points they

to this kind of fault as a **strike-slip fault**. A **left-lateral strike-slip fault** is one in which the other side has moved to the left over time. It doesn't matter which side of a left-lateral fault you are standing on when you are making this observation—the other side will have moved to the left (**Fig. 10.11C**).

Figure 10.12 Right-lateral fault. The curb in these photographs was originally straight, and its current shape is due to strike-slip along the Hayward Fault since the curb was laid. **A.** Curb distorted by faulting with the more distant part of the curb having moved to the right. Two buried utility boxes are marked X and Y as reference points. **B.** The same curb viewed from the opposite direction with reference points X and Y.

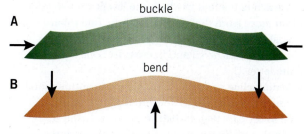

Figure 10.13 Some ways of making a fold. A. Shortening parallel to a strong layer (green) produces a fold by buckling. **B.** Applying unbalanced loads oblique to a strong layer (brown) produces a fold by bending.

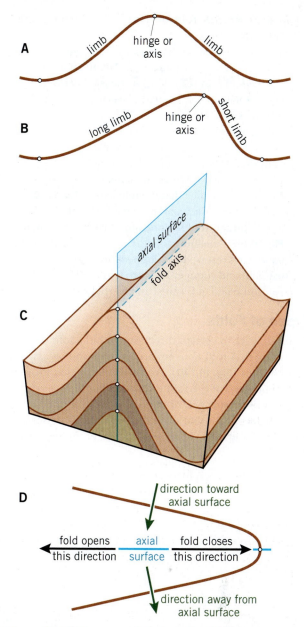

Figure 10.14 Terms used to describe folds. A. The fold hinge (or axis) is where the fold is most tightly curved. This symmetric fold has limbs that are the same length. **B.** This asymmetric fold has limbs that are different lengths. **C.** The axial surface is formed by combining all of the fold axes. **D.** Explanation of the opening and closing directions of a fold and the meaning of "toward" or "away from" the axial surface of a fold.

share in common. As you can see, it doesn't matter which side of the Hayward Fault you are standing on—the other side moves to the right. That makes the Hayward Fault a right-lateral strike-slip fault.

Normal and reverse faults in which the motion along the fault is parallel to the dip of the fault are called **dip-slip faults**. As previously explained, a fault in which the relative motion is entirely horizontal is a **strike-slip fault**. Slip along many faults involves both dip-slip and strike-slip components, and these types of faults are called **oblique faults**. We will work primarily with dip-slip and strike-slip faults in this lab.

Folds

We are all at least a little bit familiar with folding because many things in our daily experience can be folded: fabric, paper, metal, plastic, and thin sheets of wood. Some layers fold because the layers are strong, exert a mechanical influence on the folding, and can shear over the layers next to them. We can fold a thick paper-back phone book because the pages are free to slip over adjacent pages. When the layers are relatively strong, folding might occur because of shortening parallel to the layers (**Fig. 10.13A**) or because of a system of forces applied at a high angle to the layer (**Fig. 10.13B**).

In other cases, folding can result from ductile flow in which the layers exert no significant influence on the process except to mark the changes caused by flow. You can experiment with ductile folding by forming a layer of vanilla ice cream on some aluminum foil or wax paper and topping it with a layer of chocolate ice cream. Manipulating the foil or paper can create a variety of fold forms along the contact between chocolate and vanilla. The layers are simply flowing in response to the differential stress.

Describing the Geometry of Folds

There are many, many terms that geoscientists use to describe folds, but we will use just a few in this lab.

- The **hinge** of a fold is where the layer is curved most tightly (**Fig. 10.14A**). The term arises by analogy with a door hinge, and the **hinge line** or **hinge curve** is

analogous to the hinge pin. Many geologists use the word **axis** as a synonym for hinge, and for the sake of simplicity, we will adopt that practice here. (Structural geologists generally restrict the word *axis* for use with a particular type of fold called a *cylindrical fold*.)

- The **limbs** of a fold are the relatively flat or planar sides of the fold on either side of the hinge, analogous to the plates on a door hinge that attach to the door and doorframe (**Fig. 10.14A, B**).
- In a **symmetric fold**, the limbs are about the same length (**Fig. 10.14A**).

- In an **asymmetric fold**, the limbs are different lengths (**Fig. 10.14B**).
- The **axial surface** can be thought of as connecting the axes of all of the layers in a fold (**Fig. 10.14C**).
- A **horizontal fold** is a fold whose axis is horizontal (**Figs. 10.15A** and **B**). Similarly, a **vertical fold** has a vertical axis.
- A **plunging fold** has an axis that is not horizontal or vertical but plunges or is inclined (**Figs. 10.15C, D**).
- An **upright fold** has a vertical axial surface. Similarly, an **inclined fold** has an inclined axial surface (**Fig. 10.15E**). Notice that the fold axis in **Fig. 10.15E** is horizontal, but the axial surface is inclined.
- An **overturned fold** has one limb that is overturned (**Fig. 10.15E**), meaning that the normal superposed sequence is reversed—older beds are higher along that limb and younger beds are lower. If both limbs are overturned, it is called an **inverted fold**.

Types of Folds

We often refer to certain directions within a fold when describing it. A fold *closes* toward the axis and *opens* away from the axis (**Fig. 10.14D**). We describe layers as located in the *inner part* of the fold toward the fold hinge surface or on the *outer part* of the fold away from the fold hinge surface. Now that we know this little bit of descriptive terminology, let's learn about some of the basic types of folds.

- An **anticline** is a fold in which the layers are *older* as you move from the outer part of the fold toward the axial surface (**Fig. 10.15B**). A horizontal anticline opens down and closes up and broadly resembles a book placed open on a table with the cover facing up.

- A **syncline** is a fold in which the layers are *younger* as you move from the outer part of the fold toward the axial surface (**Fig. 10.15A**). A horizontal syncline opens up and closes down and in cross-section can resemble the letter U.
- A **monocline** generally has a planar, near-horizontal upper part that flexes down to a planar, near-horizontal lower part (**Fig. 10.15F**). It is like a curved step or a 1-limb anticline connected to a 1-limb syncline.
- A **structural basin** is like a bowl set on the table ready to be filled with candy, so a basin opens up and closes down (**Fig. 10.15G**). Geologic maps of basins generally show a concentric arrangement of formations with the youngest formation in the middle.
- A **dome** is like a bowl set upside-down on a table, so a dome opens down and closes up (**Fig. 10.15H**). Geologic maps of domes generally show a concentric arrangement of formations with the oldest formation in the middle.

FOLDED GEOLOGIC STRUCTURES (Ductile Deformation)

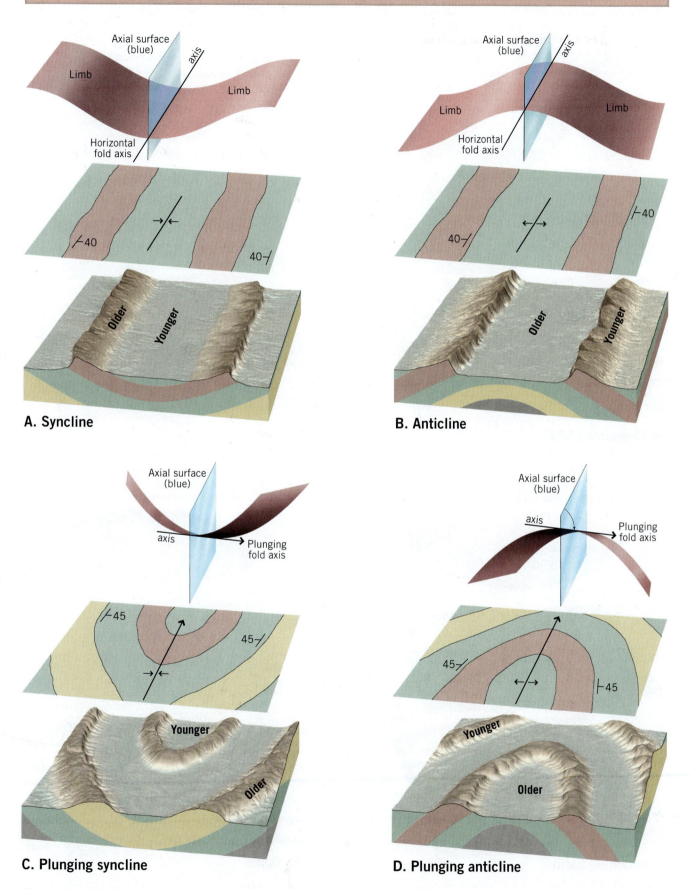

Axial surface (blue)
axis
Limb
Limb
Horizontal fold axis

40
40

Older
Younger

A. Syncline

Axial surface (blue)
axis
Limb
Limb
Horizontal fold axis

40
40

Older
Younger

B. Anticline

Axial surface (blue)
axis
Plunging fold axis

45
45

Younger
Older

C. Plunging syncline

Axial surface (blue)
axis
Plunging fold axis

45
45

Younger
Older

D. Plunging anticline

Figure 10.15 Folds in block diagrams with geologic maps. The symbols used in these figures are defined in **Fig. 10.4**.

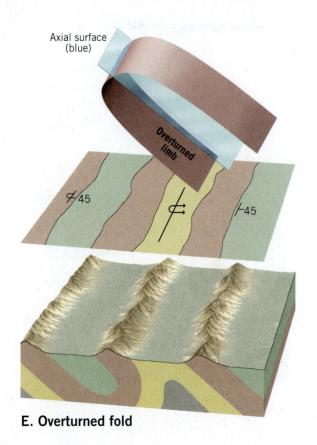

E. Overturned fold

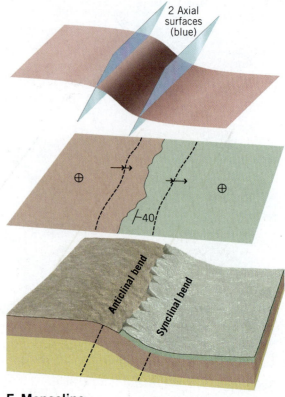

F. Monocline

Basins are somewhat circular or oval, with the youngest strata in the middle. Think of a bowl.

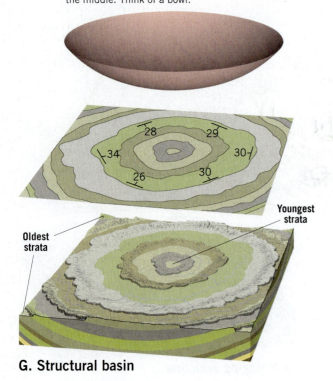

G. Structural basin

Domes are somewhat circular or oval, with the oldest strata in the middle. Think of an upside-down bowl.

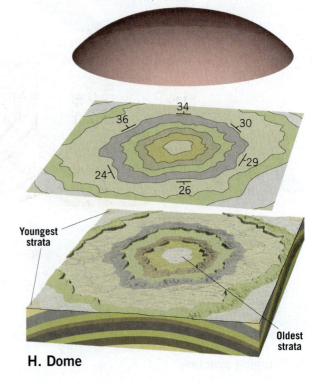

H. Dome

Figure 10.15 (continued)

Name: _____ **Course/Section:** _____ **Date:** _____

A Fig. A10.1.1 is an aerial photograph of part of the Grand Canyon centered on latitude 36.334°N, longitude 112.725°W. The geology of this area is presented on a map by George Billingsley published in 2000 by the USGS (**http://pubs.usgs.gov/imap/i-2688/i-2688.pdf** and **http://pubs.usgs.gov/imap/i-2688/i-2688_pamphlet.pdf**). Part of several contacts are shown as dashed lines at the bottom and top of the photograph. The mapped rock units are explained in the table below the photograph.

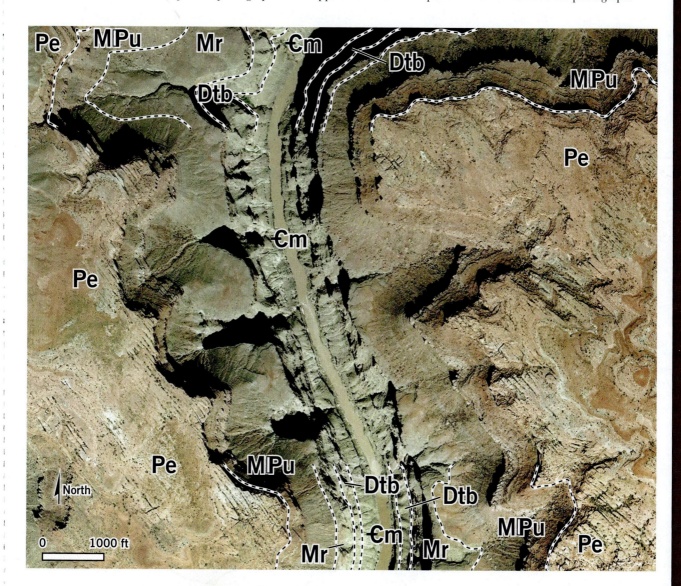

EXPLANATION

Pe	Esplanade Sandstone (Lower Permian)
MPu	Watahomigi Formation (L Pennsylvanian and U Mississippian)
Mr	Redwall Limestone (Upper and Lower Mississippian)
Dtb	Temple Butte Formation (Upper and Lower Devonian)
Єm	Muav Limestone (Middle Cambrian)

Figure A10.1.1

1. Examine the photograph carefully. In a few words, describe what features on the slopes or visible in the rocks seem to define the place where Billingsley put each of the contacts.
   (a) Between the Esplanade Sandstone and the Watahomigi Formation

   (b) Between the Watahomigi Formation and the Redwall Limestone

   (c) Between the Redwall Limestone and the Temple Butte Formation

   (d) Between the Temple Butte Formation and the Muav Limestone

2. Complete the geologic map by tracing each of the contacts from the bottom of the map to join with the same contact at the top of the map. Make sure your teacher can see where the contacts are.

B Figure A10.1.2 is part of the USGS 7.5-minute topographic map of Havasu Falls Quadrangle, Arizona (2014). As with the aerial photograph in **Fig. A10.1.1**, George Billingsley's contacts are partially mapped.

1. Complete the geologic map in **Fig. A10.1.2** by carefully drawing the contacts while carefully noting how they might relate to the topographic contours.
2. Draw the profile along the line between X and X′ in the profile box at the bottom of **Fig. A10.1.2** using just the index contours. Four points are already plotted in the profile box for you.
3. Use the contacts that you have drawn to create a geological cross-section in the profile box. Use a pencil to work on the cross-section, and don't be shy about erasing. Refer to **Figs. 10.1B** and **10.5** for help making a cross-section.

C **REFLECT & DISCUSS**   Examine your geologic map, and imagine how the Watahomigi Formation once extended across the canyon before the river eroded it. Do you think the Watahomigi Formation is deformed, or is it relatively undeformed? Why?

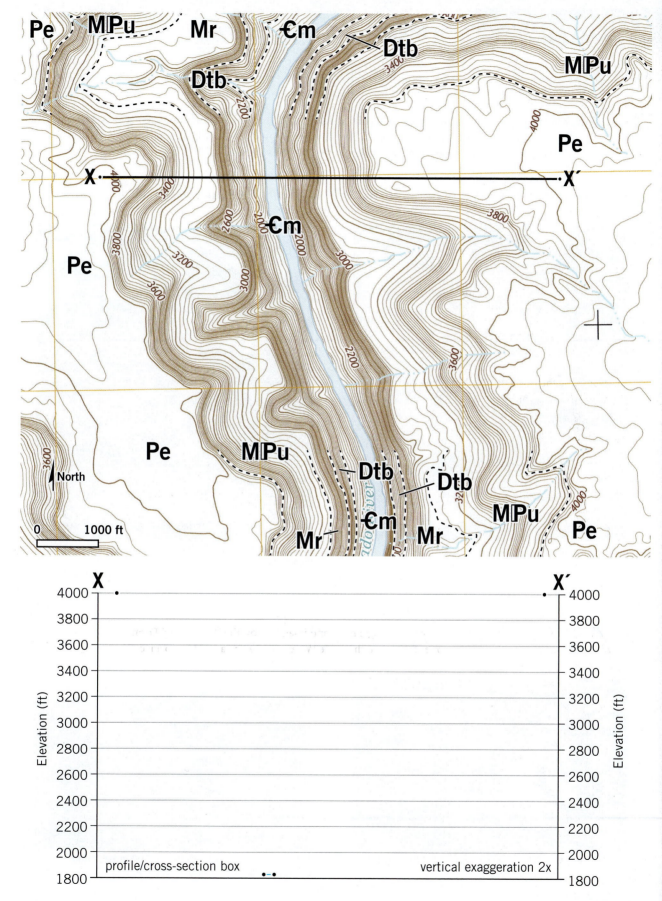

Figure A10.1.2

Name: _____   Course/Section: _____   Date: _____

A The principles of superposition and original horizontality remind us of what an undeformed sequence of sedimentary rock layers should be like. Information presented in this lab describes several types of deformation that can be seen in rock. Examine each image in **Fig. A10.2.1**. Beneath each image, answer these two questions:

1. Is the rock in the photograph deformed or not deformed?
2. If you interpret that the rock is deformed, describe the specific features of the rock that led you to that interpretation.

1. Grand Canyon, Arizona rock layers.

2. Cliff face about 400 m tall, south-central Alaska (USGS photograph by N.J. Silberling).

3. Quartzite, Maria Mts, Riverside County, California (USGS photograph by W.B. Hamilton).

4. Sandstone on a steep wall about 100 m tall, Little Colorado River Gorge, Navajo Nation, Arizona

Figure A10.2.1

REFLECT & DISCUSS   Based on your analysis of deformed rocks in part **A**, classify rock deformation into two categories, and note what images in **Fig. A10.2.1** would be in each category. Be prepared to explain your classification to your classmates and teacher.

Figure A10.2.2 is an oblique view of a small outcrop of fine-grained layered rock that contains fossils. The layers probably represent bedding planes. The layers are inclined, and the sense of inclination can be determined by noticing the water in the lower-left corner of the photograph.

1. Using the water line as a guide, carefully draw a strike line near the top of one of the flat layers using the black-and-white image for your drawing. (Make sure your teacher can see the strike line you draw.) Remember that a strike line is a horizontal line on the inclined surface and that the waterline is also a horizontal line on that surface.
2. Examine the photograph carefully, and imagine a ball rolling down the inclined surface. Now draw what you infer to be a dip vector extending from the strike line using the black-and-white image for your drawing.
3. Identify which end of the strike line points toward the Right-Hand-Rule reference strike. Label that end of the strike line with "RHR." Refer to **Fig. 10.2** for help visualizing the RHR reference strike.
4. Let's assume that the geoscientist was pointing the camera due north when this photograph was taken. In what general direction do these beds dip? _____
5. You cannot directly measure the dip angle in this view, but you can estimate the dip angle within perhaps 10° or so. About what is the dip angle of these beds? _____°

REFLECT & DISCUSS   We have been looking at a small outcrop with inclined beds in **Fig. A10.2.2**.

1. What is the strike of a horizontal surface?
2. What is the dip angle of a vertical surface?
3. The dip direction is a general compass direction like north or southwest. What is the dip direction of a vertical surface, or does it even have a dip direction?

White glint off a film of ice on water

Figure A10.2.2

A Figure A10.3.1 is part of a USGS orthophotomap of the Frenchman Mountain Quadrangle, near Las Vegas, Nevada. The center of this map area is 36.1596°N, 114.9477°W. The Nevada Geological Survey has determined that faulting occurred here between 11 and 6 Myr ago.

North

0   500 m

Figure A10.3.1

1. Use a pencil to trace the faults that cross this area. When you are content with your interpretation in pencil, trace over it in pen and add half-arrow symbols (**Fig. 10.4**) to show relative motion across each fault.

2. Interpret the type of fault or faults you have you mapped. What is your interpretation based on?

B Figure A10.3.2 shows an area just west of **Fig. A10.3.1** centered on 36.1387°N, 114.9783°W.

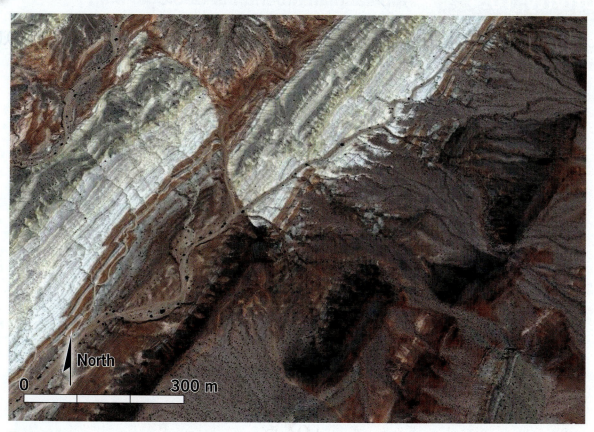

Figure A10.3.2

1. Use a pencil to trace all of the faults that you can detect in the image. When you are content with your interpretation in pencil, trace over it in pen and add half-arrow symbols to show the relative motion along the faults.

2. Interpret the type(s) of fault(s) you have mapped.

C **REFLECT & DISCUSS** Aerial photographs like these primarily give us information about possible horizontal separation of formations along faults. Make a case for the hypothesis that these inclined beds might have been separated by a vertical component of slip along a fault in addition to or instead of horizontal slip along the fault. (*Hint*: Use your hands to represent tilted beds on opposite sides of the fault to help you visualize the situation.)

Activity 10.4   Appalachian Mountains Geologic Map

Name: Journie Dyehouse Ferguson   Course/Section: _____   Date: 11/5/2020

A Complete the geologic cross-section in **Fig. A10.4.1** using the steps described in **Fig. 10.5**.

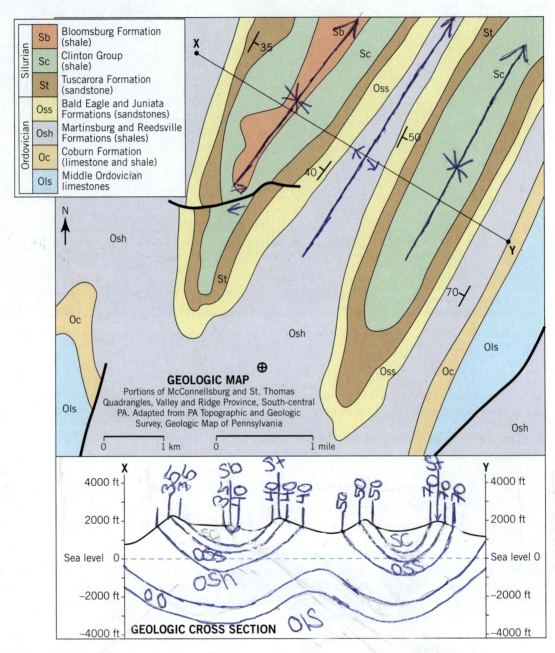

Silurian	Sb	Bloomsburg Formation (shale)
	Sc	Clinton Group (shale)
	St	Tuscarora Formation (sandstone)
Ordovician	Oss	Bald Eagle and Juniata Formations (sandstones)
	Osh	Martinsburg and Reedsville Formations (shales)
	Oc	Coburn Formation (limestone and shale)
	Ols	Middle Ordovician limestones

GEOLOGIC MAP
Portions of McConnellsburg and St. Thomas Quadrangles, Valley and Ridge Province, South-central PA. Adapted from PA Topographic and Geologic Survey, Geologic Map of Pennsylvania

GEOLOGIC CROSS SECTION

Figure A10.4.1

B Label the kind(s) of geologic structure(s) revealed by your work. Then add the appropriate symbols from **Fig. 10.4** to the geologic map to show the approximate traces of the axial surfaces for the most prominent folds on the map. Refer to **Figs. 10.14** and **10.15** for help.

C Add half-arrows to the fault near the center of the geologic map to show the relative motions of its two sides. What kind of fault do you interpret this to be? Right Lateral strike-slip fault

288

Name: _____ Course/Section: _____ Date: _____

Tear Cardboard Models 1–6 from the back of your lab manual. Cut and fold them as noted in red on each model.

A **Cardboard Model 1** This model shows Ordovician (O, green), Silurian (S, tan), Devonian (D, gray), Mississippian (M, light brown), Pennsylvanian (yellow), and Permian (P, peach) formations striking due north and dipping 24° to the west. Provided are a complete geologic map on the top of the diagram and three of the four vertical cross-sections (south, east, and west sides of the block diagram).

1. Finalize Cardboard Model 1 as follows. First construct the vertical cross-section on the north side of the block so it shows the formations and their orientations (strike and dip). On the map, draw a strike and dip symbol on the Mississippian sandstone that dips 24° to the west (see **Figs. 10.2** and **10.4** for the strike-and-dip symbol).

2. Explain the sequence of events that led to the existence of the formations and the relationships that now exist among them in this block diagram.

B **Cardboard Model 2** This model is slightly more complicated than the previous one. The geologic map is complete, but only two of the cross-sections are available. Letters **A–G** indicate ages from oldest (**A**) to youngest (**G**).

1. Finalize Cardboard Model 2 as follows. First complete the north and east sides of the block. Notice that the rock units define a fold. This fold is an anticline because the fold closes upward and the oldest formation (**A**) is toward the axial surface of the fold (**Figs. 10.14** and **10.15**). It is an upright horizontal fold because its axial surface is vertical and its axis is horizontal. On the geologic map, draw strike-and-dip symbols to indicate the orientations of formation **E** (gray formation) at points **I**, **II**, **III**, and **IV**. On the map at the top of the model, draw the proper symbol along the axis of the fold (**Fig. 10.4**).

2. How do the strikes at all four locations compare with each other?

3. How does the dip direction at points **I** and **II** compare with the dip direction at points **III** and **IV**? *In your answer, include the dip direction at all four points.*

C **Cardboard Model 3** This cardboard model has a complete geologic map. However, only one side and part of another are complete. Letters **A–E** are ages from oldest (**A**) to youngest (**E**).

1. Finalize Cardboard Model 3 as follows. Complete the remaining two-and-one-half sides of this model using as guides the geologic map on top of the block and the one-and-one-half completed sides. On the map, draw strike-and-dip symbols showing the orientation of formation **C** at points **I**, **II**, **III**, and **IV**. On the map at the top of model, draw the proper symbol along the axis of the fold (**Fig. 10.4**).

2. How do the strikes of all four locations compare with each other?

3. How does the dip direction (of formation **C**) at points **I** and **II** compare with the dip direction at points **III** and **IV**? *Include the dip direction at all four points in your answer.*

4. Is this fold plunging or horizontal? _____

5. Is it an anticline or a syncline? _____

6. On the basis of this example, how much variation is there in the strike at all points in a horizontal fold?

D **Cardboard Model 4**   Letters **A–H** indicate rock ages from oldest (**A**) to youngest (**H**). This model shows an anticline that plunges to the north.

1. Finalize Cardboard Model 4 as follows. Complete the north and east sides of the block. Draw strike-and-dip symbols on the map at points **I**, **II**, **III**, **IV**, and **V**. On the map at the top of model, draw the proper symbol along the axis of the fold including its direction of plunge (**Fig. 10.4**). Also draw the proper symbol on the geologic map to indicate the orientation of beds in formation **E**.

2. How do the directions of strike and dip differ from those in Cardboard Model 3?

E **Cardboard Model 5**   Letters **A–H** indicate rock ages from oldest (**A**) to youngest (**H**). This model shows a plunging syncline. Two of the sides are complete and two remain incomplete.

1. Finalize Cardboard Model 5 as follows. Complete the north and east sides of the diagram. Draw strike-and-dip symbols on the map at points **I**, **II**, **III**, **IV**, and **V** to show the orientation of layer **G**. On the map at the top of model, draw the proper symbol along the axis of the fold including its direction of plunge (**Fig. 10.4**).

2. In which direction does this syncline plunge, and what is the plunge angle (the angle from horizontal to the fold axis, measured in a vertical plane)? (*Hint:* Use the cross-sections on the sides of the model to measure the plunge angle.)

F **Cardboard Model 6**   This model shows a fault that strikes due west and dips 45° to the north. Three sides of the diagram are complete, but the east side is incomplete.

1. Finalize Cardboard Model 6 as follows. At point **I**, draw a symbol from **Fig. 10.4** to show the strike and dip of the fault. On the cross-section along the west edge of the block, draw arrows parallel to the fault, indicating relative motion across the fault. Label the hanging wall and the footwall. Complete the east side of the block. Draw half-arrows parallel to the fault to indicate relative motion across the fault. Now look at the geologic map and at points **II** and **III**. Write **U** on the side that went up and **D** on the side that went down. At points **IV** and **V**, draw strike-and-dip symbols for formation **B**.

2. Is the fault in this model a normal fault or a reverse fault (**Fig. 10.11**)? On what evidence do you base your interpretation?

3. On the geologic map, what happens to the contact between units **A** and **B** where it crosses the fault?

Could the same offset of dipping contacts along this fault have been produced by strike-slip motion?

G **REFLECT & DISCUSS**   There is a general principle that, as erosion of the land proceeds, *contacts migrate in the down-dip direction*. Is this true in this example? Explain.

Block Diagram Analysis and Interpretation

Activity 10.6

Name: _____ Course/Section: _____ Date: _____

For each block diagram in **Fig. A10.6.1**:

1. Complete the diagram so that contact lines between rock formations are drawn on all sides.

2. Add symbols (**Fig. 10.4**) to indicate the orientations of all structures.

3. On the lines provided below each of the block diagrams, write the name of the geologic structure represented in the diagram.

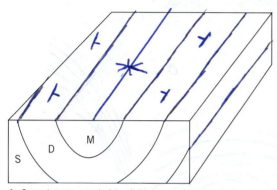

A. Complete top and side. Add appropriate symbols from Fig. 10.4. What type of fold is this?

Syncline

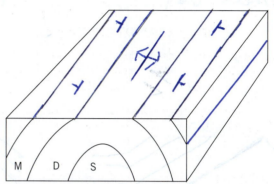

B. Complete top and side. Add appropriate symbols from Fig. 10.4. What type of fold is this?

anticline

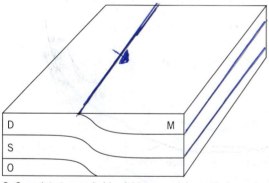

C. Complete top and side. Add appropriate symbols from Fig. 10.4. What type of fold is this?

Monocline

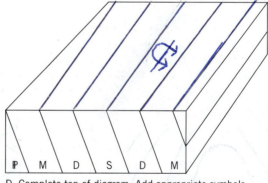

D. Complete top of diagram. Add appropriate symbols from Fig. 10.4. What geologic structure is this?

Overturned fold

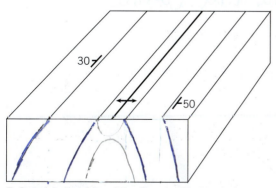

E. Complete the sides of the diagram. What geologic structure is this?

anticline

F. Complete the sides of the diagram. What geologic structure is this?

Plunging Syncline

Figure A10.6.1

291

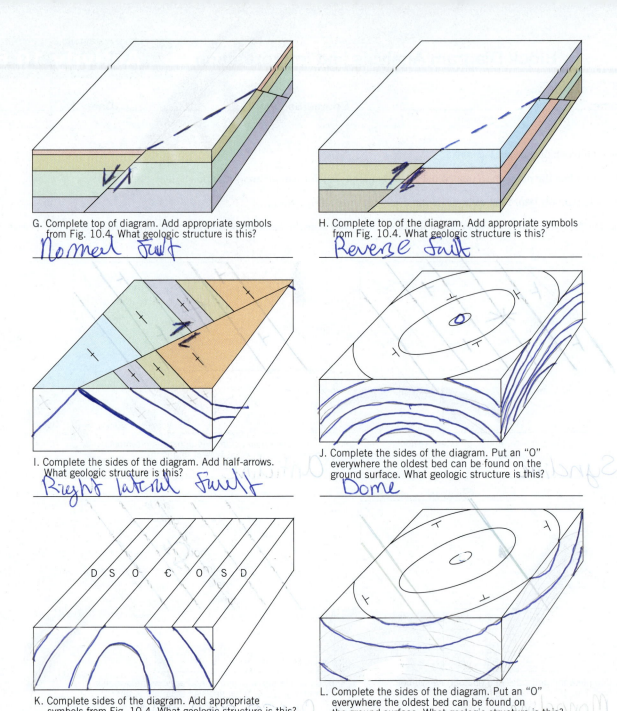

G. Complete top of diagram. Add appropriate symbols from Fig. 10.4. What geologic structure is this?
Normal Fault

H. Complete top of the diagram. Add appropriate symbols from Fig. 10.4. What geologic structure is this?
Reverse fault

I. Complete the sides of the diagram. Add half-arrows. What geologic structure is this?
Right lateral fault

J. Complete the sides of the diagram. Put an "O" everywhere the oldest bed can be found on the ground surface. What geologic structure is this?
Dome

K. Complete sides of the diagram. Add appropriate symbols from Fig. 10.4. What geologic structure is this?
Anticline

L. Complete the sides of the diagram. Put an "O" everywhere the oldest bed can be found on the ground surface. What geologic structure is this?
Structural Basin

Figure A10.4.1   *(continued)*

LABORATORY 11

Pre-Lab Video 11

https://goo.gl/225WpT

Stream Processes, Geomorphology, and Hazards

Contributing Authors

Pamela J.W. Gore • *Georgia Perimeter College*

Richard W. Macomber • *Long Island University–Brooklyn*

Cherukupalli E. Nehru • *Brooklyn College (CUNY)*

▲ The meandering Middle Fork of the Koyukuk River near Wiseman, Alaska, flowing south from the mountains of the Brooks Range in the background. Sandy point bars and steep cut banks can be seen along the channel. To the right of the river is the Dalton Highway (Route 11) and the trans-Alaska pipeline, through which crude oil from Prudhoe Bay along the Arctic Ocean flows south to Valdez. The center of the photo is ~67.369°N, 150.145°W.

BIG IDEAS

Streams shape the landscape and provide water for communities and agriculture. Flood and landslide hazards are also associated with streams. Streams erode the landscape, move and deposit sediment, and provide habitat for life in many forms. And the landscape has its effect on streams in return. Studying the shape of the land surface as given on topographic maps provides us with many helpful clues as we try to understand the streams in an area. In the United States, streamflow data and flood modeling supported by federal, state, and local governments provide us with basic information needed to make good public-policy decisions about water resources and avoiding flood hazards.

FOCUS YOUR INQUIRY

Think About It How are you affected by streams?

ACTIVITY 11.1 Streamer Inquiry (p. 294, 307)

Think About It How do stream erosion and deposition shape the landscape?

ACTIVITY 11.2 Introduction to Stream Processes and Landscapes (p. 295, 309)

ACTIVITY 11.3 A Mountain Stream (p. 295, 313)

ACTIVITY 11.4 Escarpments and Stream Terraces (p. 301, 315)

ACTIVITY 11.5 Meander Evolution on the Rio Grande (p. 301, 317)

ACTIVITY 11.6 Retreat of Niagara Falls (p. 304, 319)

Think About It How do geologists determine the risk of flooding along rivers and streams?

ACTIVITY 11.7 Flood Hazard Mapping, Assessment, and Risk (p. 305, 321)

Introduction

Earth's surface is constantly changing, and streams are an important part of the process. Streams cover almost all parts of Earth's surface above sea level—all parts except those covered by ice or that are simply too arid to sustain surface water. Even in some of the most arid deserts, there are hints that streams have coursed the land surface in the past. And streams flow on, in, under, and around glaciers.

The earliest human communities were located along streams as are most of our towns and cities today. Yes, many great cities grew from towns located along the ocean, but because we can't survive by drinking seawater, those coastal towns were invariably located near fresh water supplied by a stream or, less frequently, a lake. Streams have been an important source of fresh water, food, transportation, and recreation. Even the occasional floods along streams are ultimately beneficial as they replenish the nutrients in our agricultural fields along river floodplains.

A stream is also the focus of ecosystems that can have a profound effect on the chemical and physical characteristics and evolution of the stream. Streams and the wetlands associated with them replenish and are replenished by the vast groundwater system that lies just below the surface. Even among *geomorphologists*—geoscientists who study the processes that shape the land surface—there is a clear recognition that we are only beginning to understand the basics of streams.

Throughout this lab, the approximate locations of some of the rivers and streams you will study are given, with positions expressed in latitude and longitude. This should allow you to explore the examples for yourself at whatever scale you choose using Google Earth (http://earth.google.com). Just type the coordinates into the search box in Google Earth (e.g., 0.3N, 50.2W), and GE will find the spot for you. Once GE gets you to the spot, zoom back out a little bit and get a feel for the surroundings. You will encounter a world of intriguing streams and stream-dominated landscapes as you skim above Earth, without needing an airplane or spacecraft.

ACTIVITY 11.1

Streamer Inquiry, (p. 307)

> **Think About It** How are you affected by streams?

Objective Analyze where a community's stream water comes from and where it goes, and then infer how a community may benefit from such knowledge.

Before You Begin Read the Introduction and the following section: What Are the Components of Stream Systems?

Plan Ahead This activity requires the use of a device with a web browser and internet access.

What Are the Components of Stream Systems?

A **stream** is a natural system that transports water and sediment from a higher elevation to a lower elevation and modifies the terrain that it crosses. Large streams are called rivers, and small streams have a number of different names including creek, brook, and rill. **Perennial streams** flow continuously throughout the year and are represented on topographic maps as blue curves. Many perennial streams exist in areas where groundwater is so close to the ground surface that the water table (the surface between water-saturated material below and undersaturated material above) intersects the ground surface along the stream channel. Hence, the perennial stream is the surface expression of the groundwater system. **Intermittent streams** flow only at certain times of the year, such as rainy seasons or when snow melts in the spring. They are represented on topographic maps as blue dashed curves with three dots between each line segment.

Geoscientists think of a **stream system** as having three basic components: **tributary streams** that collect the water and sediment from throughout a drainage basin, the main **trunk stream** that transports the collected water and sediment, and the **distributary streams** that discharge the water and sediment at the stream's base level. A **drainage basin** or **watershed** is the ground-surface area that drains into a particular stream. A stream's **base level** is the lowest level that the stream flows to, where it might discharge its water and sediment load into another stream, a lake or dry basin, or the ocean.

One example of a complete stream system was captured by NASA in an image of the ~16,000 km² drainage basin of the Ocoña River in Peru (**Fig. 11.1A**). The small tributary streams at the headwaters of this drainage basin collect rain and the meltwater from snow and glaciers along the great ridge of the Andes Mountains on the northwest edge of the *Altiplano* (**Fig. 11.1B**), a vast high plane that is exceeded in size only by the Tibetan Plateau of central Asia. Water and eroded sediment are transported downstream through increasingly large tributaries. One of these tributaries, the Cotahuasi River, flows through the deepest canyon on Earth, passing more than 3354 m below the ~5–6000 m high volcanoes on either side of it (**Fig. 11.1B**). The main trunk stream of this system is the Ocoña River, which flows south to where it discharges into the Pacific Ocean near the coastal town of Ocoña.

Two different types of distributary system are shown in **Fig. 11.2**: a delta formed by river sediment deposited at the edge of a body of water and an alluvial fan formed by sediment deposited at the mouth of a mountain canyon as the creek flows onto a more nearly horizontal surface. Deltas occur in many forms, including the classic rounded triangular shape of the Nile Delta (~31.0°N, 31.1°E), the complex birds-foot delta of the Mississippi River (~29.2°N, 89.3°W), and the braided form of the Ganges (~22.3°N, 89.6°E), Amazon (~0.3°N, 50.2°W), and Madison River Deltas (**Fig. 11.2**).

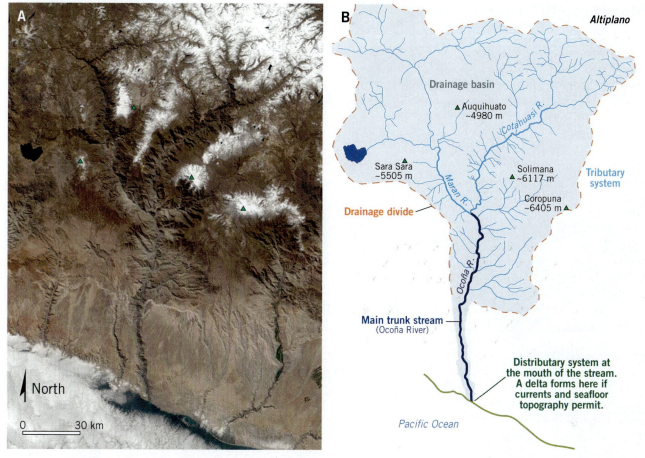

Figure 11.1 **River system from source to base.** The drainage basin of the Ocoña River in Peru from its upland source area in the Andes Mountains to its mouth at the Pacific Ocean. Data were collected by the Moderate Resolution Imaging Spectroradiometer (MODIS) instrument on NASA's Terra satellite in 2004 and were used by Jacques Descloitres of the MODIS Rapid Response Team at NASA-Goddard Space Flight Center to create the image. This image is centered near 15.73°S, 73.09°W. (Source http://visibleearth .nasa.gov/view.php?id=71761)

Alluvial fans have a flat, half-cone shape that looks like a semicircle in map view (**Fig. 11.2**). Sometimes, multiple alluvial fans coalesce into an alluvial slope extending out from a mountain front. Classic alluvial fans and alluvial slopes occur in Death Valley, California (~36.1°N, 116.8°W), where streams discharge into the lowest basin in North America—as much as 86 m (282 ft.) below sea level.

ACTIVITY 11.2

Introduction to Stream Processes and Landscapes, (p. 309)

> **Think About It** How do stream erosion and deposition shape the landscape?

Objective Analyze and interpret stream valley features using maps and an orthoimage, stream profile, and graph.

Before You Begin Read the section: How Does Water Flow Across a Surface?

Plan Ahead You will need ~40 cm (15 inches) of thin string or thread to measure the channel length along a meandering stream. This might be provided by your teacher.

How Does Water Flow Across a Surface?

If you set a ball on a concrete driveway that is gently inclined downward toward the street, the ball is going to start rolling toward the street. If the driveway is more steeply inclined, the ball will probably be moving at a

ACTIVITY 11.3

A Mountain Stream, (p. 313)

Objective Investigate the changes in a stream as it flows from a relatively steep bed-rock channel in the mountains onto the flat floodplain of a much bigger river.

Before You Begin Read the section: How Does Water Flow Across a Surface?

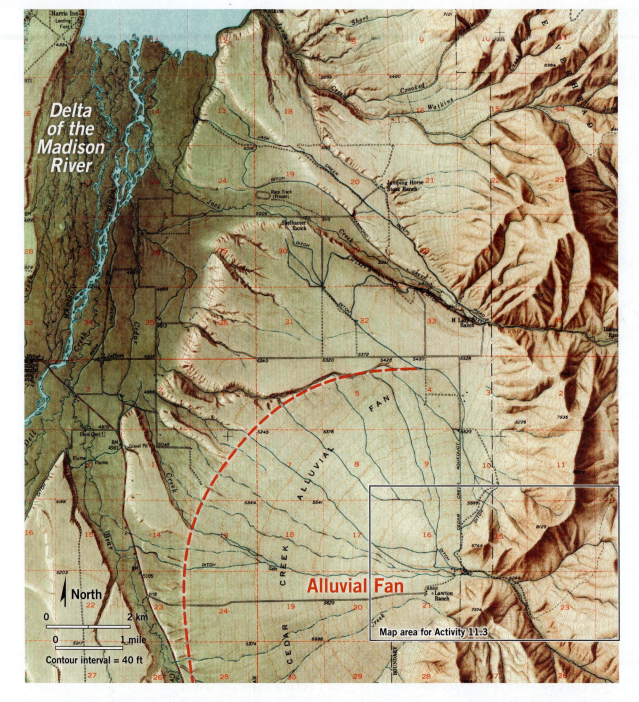

Figure 11.2 A delta and an alluvial fan. This topographic map includes the delta of the Madison River at Ennis Lake as well as the alluvial fan of Cedar Creek. Part of the outer edge of the Cedar Creek alluvial fan is outlined with the red dashed curve. This portion of the USGS 15-minute quadrangle map of Ennis, Montana (1969), is a shaded relief map with illumination from the west. Map centered near 45.35°N, 111.66°W. (Courtesy of USGS)

higher velocity when it reaches the street. Apparently, the inclination of the driveway matters in determining how the ball moves.

Gradient

A characteristic or variable that changes from one point to another point is said to have a **gradient** between those two points. A gradient describes the rate of change in that variable. We are going to think about stream slope and channel gradients in this lab, but we could also think about

temperature gradients, pressure gradients, economic gradients, color gradients, and many other changes that occur over a distance. A gradient is simply a ratio—one number divided by another number.

A **slope gradient** between any point A and any other point B is simply the difference in the elevation of the two points divided by the horizontal distance between the two points (**Fig. 11.3**). The units we use to express the gradient are largely chosen for convenience and can be units like centimeters per meter (cm/m), m/km, in./ft.,

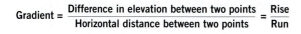

$$\text{Gradient} = \frac{\text{Difference in elevation between two points}}{\text{Horizontal distance between two points}} = \frac{\text{Rise}}{\text{Run}}$$

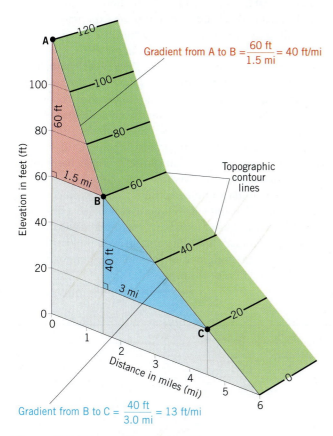

Gradient from A to B $= \dfrac{60 \text{ ft}}{1.5 \text{ mi}} = 40$ ft/mi

Topographic contour lines

Gradient from B to C $= \dfrac{40 \text{ ft}}{3.0 \text{ mi}} = 13$ ft/mi

Figure 11.3 The gradient of a surface. The gradient of a surface is a measure of the steepness of its slope, measured along the fall line—the steepest part of the slope. Several methods for expressing stream gradient are used, all of which divide the difference in elevation measured between two points on the surface by the horizontal distance between those two points.

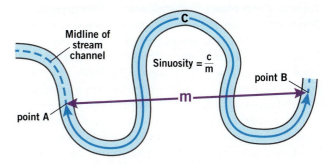

Sinuosity $= \dfrac{c}{m}$

Figure 11.4 Channel gradient and sinuosity. The *channel gradient* is the difference in elevation between two points A and B divided by the channel length measured along the midline between the two points. Streams are classified as linear, sinuous, or meandering based on their *sinuosity*, which is the channel length between two points A and B along the midline of a channel (c), divided by the horizontal distance between the two points (m).

or feet/mile. Gradients can also be dimensionless if we use the same unit to measure change in elevation as we use to measure change in distance, such as m/m. In that case, the units cancel each other, and the result is called a *dimensionless number*. Sometimes these types of gradients are expressed as a percentage, so a change of 1 m in elevation over a horizontal distance of 1 m would have a gradient of 1 or 100%. Because most slopes are not flat planar surfaces, the slope gradient is considered an average over the distance between A and B.

A **channel gradient** between two points A and B that are located along the midline of a stream channel is the difference in elevation between the two points divided by the midline channel length (c) between the two points (**Fig. 11.4**).

Sinuosity

Virtually all stream channels have some curviness, or *sinuosity*, to them, and some are quite convoluted, as if the river could not decide which way to go. Of course, rivers don't decide anything. Geoscientists measure a stream's

sinuosity as a way of describing whether to classify it as a linear, sinuous, or meandering stream. Given two points A and B along the midline of a stream channel, spaced far enough apart so that the stream segment between them fairly represents the overall shape of the stream, the sinuosity is the ratio of the midline channel length (c) divided by the straight-line distance (m) between the two points (**Fig. 11.4**). **Meandering streams** have a sinuosity greater than 1.5, **linear streams** have a sinuosity less than 1.3, and **sinuous streams** are between the two.

Potential Energy, Topographic Contours, and Flow

A spherical ball at rest on a horizontal surface in still air will have no tendency to roll. We can pick up the ball and move it to any other spot on the horizontal surface, set it down so that it is at rest, and the ball will still have no tendency to roll. We science nerds call the horizontal surface an **equipotential surface**—a surface along which every point has equal gravitational potential energy because it has the same elevation as every other point on the surface. A ball at a higher elevation has greater gravitational potential energy than a ball at a lower elevation.

We can think of topographic contours as curves on the ground surface along which every point has the same elevation. All of the points along a given topographic contour have the same elevation, so they are all part of the same horizontal plane that intersects the ground surface along that particular topographic contour. All of the points along a given topographic contour have the same gravitational potential energy.

When you have high potential energy in one place and low potential energy in another place, flow from high to low will occur unless there is some barrier to flow. Flow of what, you ask? Flow of balls, air, water, electrons, heat, and other forms of energy, for example. And the flow will occur along a path that is perpendicular to lines of equal potential energy.

Surface Water Flows Perpendicular to Topographic Contours. To visualize how gravitational potential energy affects the surface flow of water, imagine you and two friends have spread out across a hillside and are sitting where the three circles are located in **Fig. 11.5A.** (For the purpose of this example, suppose the hillside is made of material that water will not soak into and that has topographic contours painted on it.) Next, both of your friends and you pour some water out of your water bottles to see how it flows over the surface. The water from each of you will flow from upslope to downslope. At every point along its trajectory, the water will flow down the steepest path in front of it as shown by the blue curves extending from the three circles. A topographic contour is horizontal, so the line of steepest descent is always perpendicular to (i.e., at a right angle or 90° to) the topographic contour. You and your friends would observe that the water flows perpendicular to the topographic contours. Eventually, all of the water that you have poured will flow together or coalesce along the drainage course at the bottom of the little valley.

Surface Water Flow in Drainage Basins. Now, imagine that you bring *all* of your friends and family to that same special hillside, position them where the circles are located in **Fig. 11.5B,** and repeat the experiment. You carefully map out the path along which the water flowed from each of your co-investigators. In the central part of the area that is bounded on three sides (west, north, and east) by ridges, you find that water flows together into the valley and then down the valley to the river below. The blue flow lines are all within the same **drainage basin** because they ultimately coalesce into the same drainage channel. On the other side of those bounding ridges, the water flows *away from* that central drainage basin into other drainage basins. You draw a red dashed curve to mark the boundary of the central drainage basin containing the blue flow lines. This boundary is called a **drainage boundary** or *watershed boundary.* Everything inside the drainage boundary flows into and through the drainage basin, and everything outside the drainage boundary flows into a different basin.

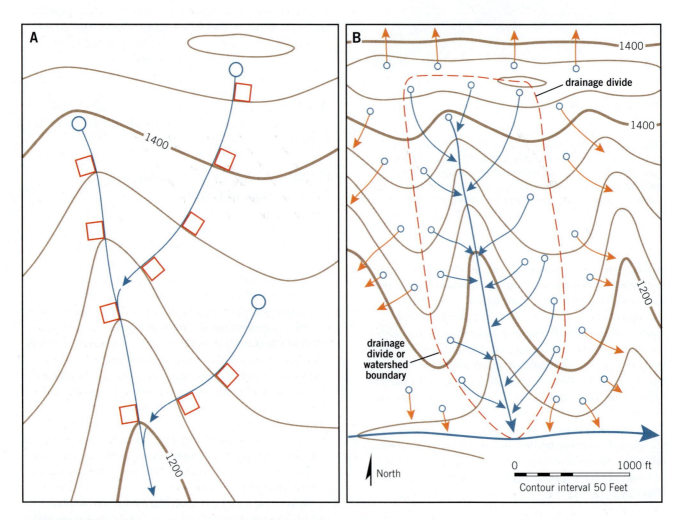

Figure 11.5 Flow lines across a topographic surface. The steepest path on a topographic surface is always perpendicular to topographic contour at the point where the flowline crosses the contour. **A.** Using the fact that flow lines are perpendicular to topographic contours, a flowline can be approximated from any point on a topographic surface. The blue curves are flow lines that begin at the circles. **B.** A set of flow lines for a topographic surface indicates that surface water flow lines converge within a drainage basin (blue flow lines), and diverge outside of that basin (orange flow lines). The boundary between the blue and orange flow lines is the drainage boundary.

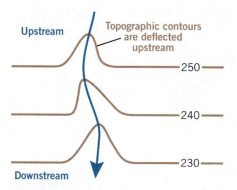

Figure 11.6 Rule of Vs for topographic contours. As a result of the erosion of a stream channel on an inclined surface, each topographic contour forms a "V" that points like an arrowhead in the upstream direction. Stream flow is downstream.

Rule of Vs

In a valley with a stream flowing through it, topographic contours form a "V" that points in the upstream direction (**Fig. 11.6**). This important observation is called the "rule of Vs", which is very useful when interpreting maps on which stream courses are not all included with some form of blue curve. But what does it mean to say that the "V" points upstream? Think of the V-shape of the contours along a stream channel as having the shape of an arrowhead. The arrowhead points upstream. (Water, of course, flows downstream toward lower elevations.) By learning the rule of Vs, you can use an accurate topographic map marked with a stream course to interpret the direction in which water flows in that stream even if the contours are not labeled with elevations.

Stream Order

Geoscientists who study rivers and their effects on landscapes—fluvial geomorphologists—often focus on a particular segment of a larger river system. In order to describe the relationship of a particular segment to the total system, a way of classifying stream segments was developed (**Fig. 11.7**). A **first-order stream** is the smallest stream in the system and has no tributaries. A **second-order stream** starts where two first-order streams flow together and continues until it flows into another second-order stream. The result of two second-order streams converging is a **third-order stream**, and so on. First-order streams are the most abundant streams in a stream system and also the shortest. As stream order increases, the number of those streams decreases and their average length tends to increase. There are also predictable changes in average stream gradient, stream width, and stream discharge with increasing stream order.

Stream Discharge

The volume of water flowing by a particular point along a stream during a specified unit of time is the stream's **discharge**, which is measured in cubic feet per second

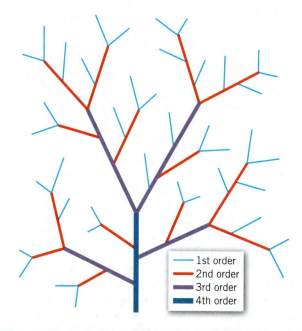

Figure 11.7 Stream order. First-order streams flow into second-order streams, which flow into third-order streams, and so on. Any description of stream order depends on the scale of observation and the specific definition of first-order stream used in the analysis.

($ft.^3/s$). Let's start with a simpler model of discharge to get the basic idea. Imagine that you know that the cross-sectional area of a water pipe is A square feet ($ft.^2$) and you know that the velocity at which the water moves through the pipe is V feet per second ($ft./s$). Assuming the pipe is entirely full of water that is moving at that velocity, the discharge from the pipe is the product of the cross-sectional area times the velocity:

$$\text{discharge} = A \, ft.^2 \times V \, ft./s = D \, ft.^3/s$$

So if $A = 2 \, ft.^2$ and $V = 30 \, ft./s$, the discharge D is $60 \, ft.^3/s$.

The situation is a little bit more complicated for streams because stream channels are not enclosed pipes with a constant cross-sectional area. A device called a **stream gage** measures the elevation of a stream's surface (the **stage**) at a given point along the stream. The stage is then converted to a discharge (**streamflow**) using a **rating function** developed for that specific stream gage that relates stage to discharge expressed in cubic feet per second. The U.S. Geological Survey in cooperation with other agencies operates more than 8000 automatic stream gages throughout the United States. These typically record stage information every 15 minutes and transmit those data via satellite, cellular network, or radio every 1–4 hours (or more frequently during flood events) to be processed by the National Water Information System (NWIS), which calculates the discharge. Those data are made available to the public via the *NWISWeb* at http://waterdata.usgs.gov/nwis and through the *WaterWatch* web site at http://water.usgs.gov/waterwatch/.

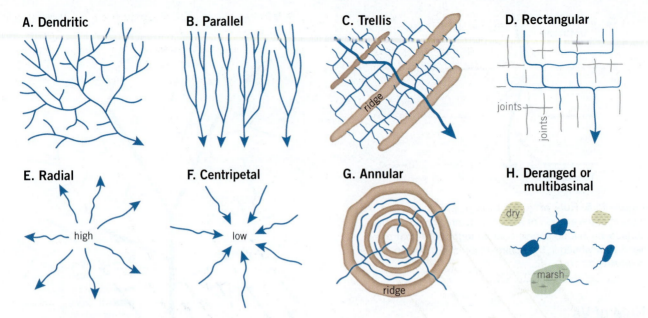

A. Dendritic

B. Parallel

C. Trellis
ridge

D. Rectangular
joints
joints

E. Radial
high

F. Centripetal
low

G. Annular
ridge

H. Deranged or multibasinal
dry
marsh

Figure 11.8 **Some stream drainage patterns.** Different types of drainage systems develop in response to many factors, including the shape of the ground surface, the type and uniformity of the material at the ground surface, the presence of fractures and faults, and climate conditions.

Pattern of a Stream System

The geometry in which various orders of tributary streams flow together in an area can provide clues about the underlying geology and geologic history of an area. Of course, the detailed pattern in each stream system is unique, but certain general **stream drainage patterns** have been identified and named (**Fig. 11.8**).

- **Dendritic pattern**—resembles the branching of a tree (**Figs. 11.8A** and **11.9**). Water flow is from the branch-like tributaries to the trunk-like main stream or river. This pattern is common where a stream cuts into flat, gently sloping, uniform layers of rock or sediment that are relatively free of fractures, faults, or other characteristics that might concentrate drainage courses in a particular direction.
- **Parallel pattern**—a network of channels on a more steeply inclined slope developed in uniform rock or sediment (**Figs. 11.8B** and **11.10**).
- **Trellis pattern**—resembles a vine growing on a trellis where the main stream is long and intersected at nearly right angles by its tributaries (**Fig. 11.8C**). This pattern commonly develops where alternating layers of hard and soft rock have been tilted and eroded to form a series of parallel or near-parallel ridges and valleys. The hard rock forms the ridges because it is more resistant to erosion than the soft rock that forms the valleys. The main stream channel cuts through the ridges. The main tributaries flow parallel to each other along the valleys while the first-order streams flow off the ridges nearly perpendicular to the valleys.

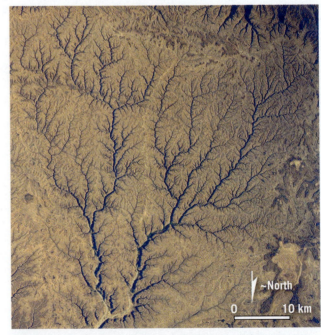

~North

0   10 km

Figure 11.9 **Dendritic drainage system.** Drainage systems that develop on very gently sloping, flat surfaces made of very uniform material tend to form dendritic drainage patterns. This dendritic drainage system is in the Hadramaut Mountains of Yemen and developed on a surface with a ~1.2% gradient that slopes just ~0.7° over ~50 km. Yemen is one of the driest places on Earth, so this drainage network probably formed during the wetter times of the last ice age. This astronaut-taken photograph (NASA image STS41G-36.36) is centered at 15.03°N, 48.33°E.

Figure 11.10 Parallel drainage system. Drainage systems developed on moderately or steeply sloping, flat surfaces made of very uniform material tend to form parallel drainage patterns. This parallel drainage system developed on the planar slope of a waste pile at the Bingham Canyon open-pit mine, Utah, near 40.53°N, 112.11°W. The base of the slope is approximately 120 m wide in this photograph.

- **Rectangular pattern**—a network of channels with right-angle bends that form a pattern of interconnected rectangles and squares (**Fig. 11.8D**). This pattern often develops in areas that have fractures called **joints** that extend for many meters (sometimes many kilometers), faults, or tilted beds that intersect each other at right angles. Streams erode channels along the joints, faults, or weak beds.
- **Radial pattern**—channel flow outward from a central upland area, resembling the spokes of a wheel (**Fig. 11.8E**). This pattern develops on round or conical hills, such as volcanoes and some structural domes.
- **Centripetal pattern**—channels converge on a low area at the center of a closed basin that might not have an outlet stream (**Fig. 11.8F**). Many of the dry playa lake beds in the western United States have some form of centripetal drainage around them.
- **Annular pattern**—a set of incomplete, concentric rings of streams connected by short radial channels (**Fig. 11.8G**). This pattern commonly develops on eroding structural domes or basins that contain alternating folded layers of resistant and nonresistant rock types.

- **Deranged or multibasinal pattern**—a random pattern of stream channels that seem to have no relationship to underlying rock types or geologic structures (**Fig. 11.8H**). This is common in areas where older drainage patterns were obliterated by glacial erosion, and new drainage patterns have yet to fully develop in the glaciated terrain.

ACTIVITY 11.4

Escarpments and Stream Terraces, (p. 315)

Think About It How do stream erosion and deposition shape the landscape?

Objective Analyze and interpret escarpments and terraces along the Souris River using topographic contours and a river valley profile.

Before You Begin Read the section below: Stream Channels and Floodplains.

ACTIVITY 11.5

Meander Evolution on the Rio Grande, (p. 317)

Objective Analyze a map of changes in the course of the Rio Grande to interpret the evolution of meanders.

Before You Begin Read the section below: Stream Channels and Floodplains

Stream Channels and Floodplains

Streams flow from higher elevations to lower elevations, perpendicular to topographic contours. But that is not the end of the story. Streams continuously change the shape of the ground surface through erosion and deposition, and so they change the shape of the topographic contours. The shape of the river channel can also evolve over time. Streams are systems that evolve in ways that minimize the amount of energy expended to move water and sediment from higher elevations to lower elevations.

Types of Stream Channels

Individual stream channels are observed in a continuous spectrum of different forms that result from the interplay of many different factors, including gradient, discharge, sediment load, vegetation, erodability of the channel or floodplain, climate, and urbanization. We will consider three examples along this spectrum of channel types.

Figure 11.11 **A linear-sinuous river.** The Hudson River has a sinuous channel between Schuylerville and Mechanicville, New York. Saratoga Lake is on the west side of the image. Image data are from the Operational Land Imager on NASA's Landsat 8 satellite. The middle of this image is around 43.01°N, 73.64°W.

Figure 11.12 **A braided river.** The Dart River of the South Island, New Zealand, drains part of the Southern Alps that includes several glaciers and tributary streams. It has an abundant supply of sediment and (seasonally) of water. As a result, the Dart River has many gravel bars that separate several active stream channels. The center of this photograph is near 44.75°S, 168.33°E.

- **Straight channel**—Stream segments that are very young or whose shape is controlled either by a narrow valley between parallel valley slopes or ridges, by joints (fractures) or faults, or by an unusually steep gradient tend to have channels that are classified as either linear or sinuous (**Fig. 11.11**). **Linear channels** have a sinuosity of less than 1.3, and **sinuous channels** have a sinuosity of 1.3 to 1.5 (**Figs. 11.4** and **11.11**).

- **Braided channel**—Stream segments that have more than one active channel in which the channels seem to weave into and out of each other in a pattern that resembles braided hair are called **braided streams** (**Fig. 11.12**). In a braided stream, the channels branch and join, separated by sand and gravel bars. Braided streams develop as a result of several factors acting alone or together, including abundant supply of water, abundant supply of sediment, or moderately steep gradient. In arid regions, the multiple channels in a braided stream might have flowing water only during flash floods.

- **Meandering channel**—The word "meander" in English means to wander aimlessly or to follow a winding course. Its origins are traced to a river in western Turkey that is currently known as the Büyük Menderes River but that was known during classical Greek times as the Maeander or Meander River (**Fig. 11.13**).

A meandering stream has a sinuosity of more than 1.5 so its channel is very curvy, winding its way along a floodplain that has a low gradient. Each of the major curves of the channel is called a **meander bend**. If you imagine walking out across the land inside of a meander bend, you could see the river flowing on both sides of you as you walk toward a point of land, somewhat like walking out onto a peninsula jutting out into the ocean. As you arrive at the end of that rounded point of land, you are likely to encounter a sandy beach or bar at the point that slopes gently into the stream—a **point bar**. Looking across the stream, the bank on the other side is likely to be much steeper and will lack a sand bar. In fact, the river is cutting into that opposite bank little by little each day and does so quite vigorously during flooding. That opposite bank is called a **cutbank**. Sediment is deposited on the inside of a meander loop at the point bar and is eroded from the outside of a meander loop along the cutbank. The effect of deposition on one side and erosion on the other is that the position and shape of the channel changes over time.

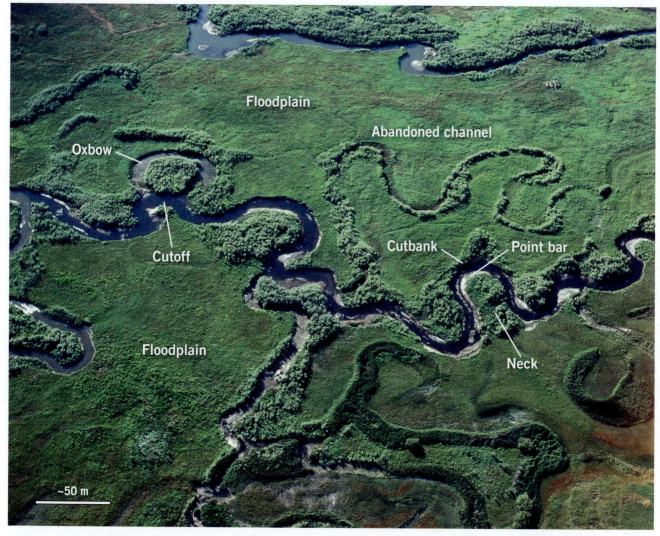

Figure 11.13 A meandering river and its floodplain features. The meandering channel of a small stream near Elko, Nevada, is typical of streams across a very broad range of sizes, including the Mississippi River.

The Floodplain

The dynamic nature of meandering streams is easy to recognize in **Fig. 11.13**, which shows a modern stream on a floodplain that has many fragments of older abandoned channels. If we could hover above this floodplain and watch it change over hundreds or thousands of years of flood cycles, we would see a channel that appears to wriggle back and forth across the floodplain like a snake. Along the sides of the floodplain, parallel to the overall trend of the stream, we generally find slopes or escarpments that mark the boundaries of the active stream system. An **escarpment** is a relatively steep face that connects a more gently sloping surface at a higher level to a gently sloping surface at a lower level. Escarpments are usually caused by stream or coastal erosion or by faulting.

Streams not only change because of the dynamics of eroding, depositing, and moving water and sediment, but they also adjust in response to broader regional changes. Global sea level rises and falls over time, and this changes the base level to which rivers flow. Within continents,

areas can also rise or subside slowly due to deformation of the crust, and this triggers a variety of responses throughout the stream system. And the climate changes over time, affecting the amount of water supplied to a stream system.

When there is an abundant supply of sediment and enough water to move it or when the base level rises, streams are likely to be in a depositional phase in which older floodplain and stream channel sediments are buried by newer stream sediments. On the other hand, lowering of a stream's base level or increase in gradient due to uplift in the stream's source area is likely to have the opposite effect: erosion of all or part of older floodplain deposits. Remnants of these older floodplains are sometimes observed along the edge of the current floodplain, seen as flat surfaces that are parallel to but a bit higher than the active floodplain. These remnants are called **stream terraces**. The escarpment between the active floodplain and the older stream terraces commonly marks the edge of the current floodplain.

Retreat of Niagara Falls, (p. 319)

Think About It How do stream erosion and deposition shape the landscape?

Objective Describe erosional processes at Niagara Falls and calculate the rate at which the falls is retreating upstream.

Before You Begin Read the section: Stream Profile.

Plan Ahead This activity involves measuring the length of a curving channel, which can be done with a piece of thin string or thread. You will need perhaps 15 cm (6 inches) of string.

Stream Profile

Imagine that we project the shape of a stream channel onto a vertical plane that is parallel to the average direction the stream is flowing in. We will call the resulting curve the **stream profile** and notice that it extends from the upland area in which the stream originates down to the stream's base level (**Fig. 11.14A**). The stream gradient is steeper in the upland part of the profile than it is downstream. The physics of a stream system favors a stream profile that uses the least work to move water and sediment and a stream shape that distributes that work almost evenly across all parts of the stream system.

If the shape of a stream profile is changed, as might happen due to faulting (**Fig. 11.14B**), natural processes of erosion and deposition will adjust the profile until it returns to a minimum work shape. Displacement along the fault caused by an earthquake generates an

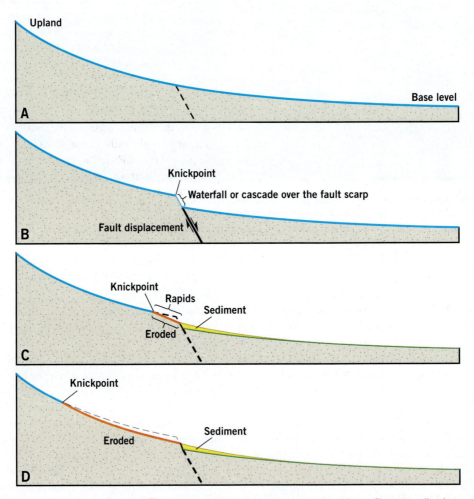

Figure 11.14 Streams adjust their own profiles. A. A stream's shape when viewed in profile generally shows a smooth change from steeper gradient in the uplands to much less steep gradient near base level. **B.** If that profile is disrupted by faulting, the stream adjusts through erosion and deposition. **C.** The knickpoint migrates from the fault upstream. **D.** Eventually, the stream re-establishes a smooth profile.

Flood Hazard Mapping, Assessment, and Risk, (p. 321)

Think About It How do geologists determine the risk of flooding along rivers and streams?

Objective Construct a flood magnitude/frequency graph, map floods, and flood hazard zones and assess flood hazards along the Flint River in Georgia.

Before You Begin Read the section below: Flood Hazard Assessment and Risks.

---

escarpment or **fault scarp** that crosses the stream channel. The top of that new step in the channel is called a **knickpoint**, and a waterfall or cascade is likely to develop at the knickpoint right after the earthquake. The stream will erode the top of the waterfall, causing the knickpoint to slowly migrate upstream and the waterfall to evolve into a set of rapids (**Fig. 11.14C**). The stream profile will continue to change until it returns to a smooth shape (**Fig. 11.14D**).

Flood Hazard Assessment and Risks

Humans have been living along rivers and their flood plains from the earliest times, but that coexistence has not been without difficulty. Rivers pose challenges for us through sedimentation, erosion, changing course, drying up, and flooding among other lesser hazards. With some notable exceptions, most catastrophic floods in human history have been along rivers. Even flooding from the tsunami associated with the great Tōhoku earthquake of 2011 in Japan was focused along river valleys as the flood waters extended beyond the coastline.

Flood Hazard Assessment and Mitigation

The least expensive and most successful strategy for protecting lives and property from the dangers of flooding is simple avoidance. Geoscientists and surface-water hydrologists have developed a networked system for collecting and modeling streamflow data that are intended to identify the areas that are most susceptible to flooding. Areas known to be especially susceptible to flooding are called **Special Flood Hazard Areas** (**SFHA**). An SFHA is an area that would be inundated by a flood that has a 1% chance of being equaled or exceeded in any given year,

which is also known as a **100-year flood** or **base flood**. It is safe to say that you should avoid buying or renting a residence located within the 100-year floodplain, or SFHA. If your home or business is located in a 100-year floodplain, you should expect to have your property damaged or destroyed and your loved ones or coworkers harmed (or worse) by floodwaters. It is far better to avoid (most) flood hazards by choosing to live and work *above* the 500-year floodplain.

Given that our life expectancy is less than 100 years, people tend to interpret the term "100-year flood" as a once-in-a-lifetime event. But that is not what hydrologists mean when they use the term "100-year flood." The meaning of the term "100-year flood" is that there is a one-in-100 chance that a flood of that magnitude or greater will occur *in any given year*. It is possible that a flood of that magnitude or greater will occur twice in the same year or not at all for more than a century.

Our concerns do not end with 100-year floods because larger floods occur. Texas reportedly experienced 500-year floods in 2015 and again in 2016 (**Fig. 11.15**). Flooding associated with a slow-moving weather front in mid-April, 2016, caused at least 8 deaths in Texas. Just over a month later in late May–early June, another 24 people in Texas and at least 7 in neighboring Oklahoma died as a result of flooding. Flooding killed people in rural areas as well as in cities as large as Houston. Over the half century from 1959 to 2008, an average of 17 people died each year due to flooding in Texas. Worldwide, flooding is thought to be the most deadly of all natural hazards.

Many flood deaths in the United States are associated with people in vehicles who try to drive through flooded roads, not understanding that it takes just 18 to 24 inches of moving water to sweep most cars off the road. On June 2, 2016, nine soldiers drowned when their Light Medium Tactical Vehicle (LMTV) weighing more than 11.5 tons was overturned by flooding at a low-water crossing on Owl Creek at Fort Hood, Texas. Your family car or pickup truck is certainly not safer than a LMTV in a flood.

Tropical Storm Alberto entered Georgia early in July 1994 and remained in a fixed position for several days. More than 20 inches of rain fell in west-central Georgia over those three days and caused severe flooding along the Flint River. Montezuma was one of the towns along the Flint River that was flooded, and is the subject of Activity 11.7. Some of the flood damage experienced by Montezuma in 1994 can be prevented in the future through better planning.

Flood Safety

If you should ever find your path obstructed by a flooded area, common sense would dictate that you should not attempt to cross it on foot or in a car or truck. However, common sense sometimes eludes even the most rational of us. Perhaps the following brief consideration of the different velocities of streamflow across a floodplain and in a

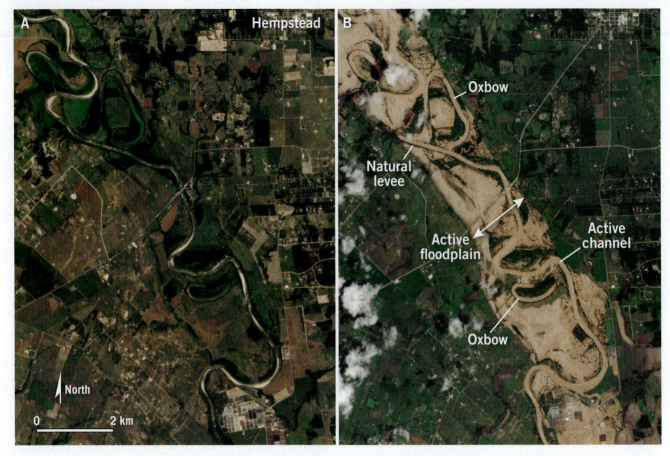

Figure 11.15 Spring 2016 flood along Brazos River, Texas. A week of heavy rains on soil that was already saturated from prior rainfall led to flooding along the Brazos River in May–June 2016, shown here in images of the same area downstream of Hempstead, Texas. Image data are from the Operational Land Imager on NASA's Landsat 8 satellite. **A.** Brazos River on May 4, 2013, at normal levels. **B.** Brazos River on May 28, 2016, during a flood that rose as much as 54.81 feet above normal (gage) level near Richmond, Texas, downstream from this area.

channel might indicate that crossing a flooded area is foolish and might be fatal.

The main channel is difficult or impossible to see when the whole floodplain is covered in muddy water. New channels might have formed during the flood that you can't see. The water in all of the deeper channels is flowing faster below the surface than the shallow water across the rest of the floodplain. The road that you think is still there under the muddy water might have collapsed into the torrent. You should also understand that flood waters are generally a witch's brew of fallen trees and branches, dead animals, raw sewage, and chemicals from spilled or submerged tanks and holding ponds. There might be submerged live electrical wires or fire from oil or gasoline floating on the water's surface to make the hellish scene complete. In short, avoid floods and cross them only when it can be done safely in an appropriate watercraft while wearing an adequate life preserver.

Name: _____ Course/Section: _____ Date: _____

Have you ever stood beside a stream and wondered where the water comes from or where it goes? *Streamer* is a map-based database of stream maps and information that allows you to find out. It is a component of the U.S. National Atlas project, managed by the U.S. Geological Survey, that allows you to trace streams upstream to their sources or downstream to where they empty into larger streams or the ocean. To use *Streamer*, go to http://water.usgs.gov/streamer/web/ and click on the "Go to Map" button. Then follow the directions below.

A Where does the water come from?

1. Pick a community in the United States that you are interested in. What is the name of your community?

2. Locate your community on the *Streamer* map. You can use the zoom slider on the map to zoom in, use your mouse wheel to scroll in or out, or hold the shift key and drag a box around the area where your community is located. You can also type the location of the community (city, state) in the "Location Search" panel and then press "Enter/Return" on your keyboard to locate your community.

3. Choose the largest stream located in or near your community to study. Click on the "Trace Upstream" button and then click on a point on the stream to display in red all of the streams that supply water to that point on the stream.

4. Now click on the "Trace Report" button, and select "Detailed Report" to get a Stream Trace Detailed Report.

   (a) What stream did you study? Trace Origin Stream Name: _____
   (b) What are the elevation and coordinates of the point on the stream that you selected?
       Trace Origin Elevation (ft.): _____ feet above sea level
       Trace Origin (latitude, longitude): _____ _____
   (c) Through how many communities does the stream flow before it gets to this point? Cities (count): _____
   (d) In how many named streams does the water flow to this point in the stream? Stream Names (count): _____
   (e) What is the total length of the stream(s) named in part **4d**? Total Length of Traced U.S. Streams (miles): _____
   (f) Close the Stream Trace Detailed Report (but *not* the map) by clicking on the small gray "x" of the right-hand tab labeled "Streamer Report" near the top of the browser window and proceed to part **B** below.

B Where does the water go?

1. Click on the "Trace Downstream" button. Then click on approximately the same point of the same stream that you studied in part **A** to display in red where the water goes after that point on the stream.

2. Click on the "Identify" button. Now when you click on any part of the red downstream trace of the water, it will identify the name of the stream at that point of the trace. List the names of all of the streams in the downstream trace of the water from upstream (the starting red point in your map) to where it enters the "Outlet Waterbody" at the downstream end of the red line on your map.

3. Click on the "Trace Report" button, and select "Detailed Report" to get a new Stream Trace Detailed Report.

   (a) In how many named streams does the water flow downstream from this point? Stream Names (count): _____
   (b) Through how many communities does the water flow downstream from this point? Cities (count): _____
   (c) What is the total length of the stream(s) named in part **3b**? Total Length of Traced U.S. Streams (mi.): _____
   (d) Into what "Outlet Waterbody" does the stream's water eventually empty? _____
   (e) What is the name of the last community or feature that the stream passes through before it enters the "Outlet Waterbody"? _____
   (f) Close the Stream Trace Detailed Report (but *not* the map), and proceed to part **C**.

C Your stream system.

1. Click on the "Trace Upstream" button again. Then click on the downstream end of the downstream trace that you identified in part **B**. It will be located near the place you identified in part **B3e**. This will display an entire stream drainage network from the smallest upland tributaries to the largest river (main stream or main river).

2. Comparing a tributary stream near the upstream headwaters of this river system with the main river near the mouth of the system,

   (a) Infer whether the upstream or downstream segment handles the most water in a given unit of time. _____

   (b) Infer whether the upstream or downstream segment has a greater channel width. _____

   (c) Infer which part of a stream system flows down a steeper slope: the upstream or downstream segment. _____

3. Click on the "Trace Report" button, and select "Detailed Report" to get a new *Streamer* Report. How many USGS stream gages are used to monitor this river system from source to its mouth, where it discharges to the outlet waterbody? _____

   This is a small indication of the continuing investment made by the people of the United States to understand and manage its surface-water resources and to protect people from flood hazards through the efforts of the U.S. Geological Survey, U.S. Army Corps of Engineers, U.S. Bureau of Reclamation, and other federal, state, and local agencies.

D **REFLECT & DISCUSS** Why would a community located on or near a stream want to know where its stream water comes from, and what else might it want to know about the water?

E **REFLECT & DISCUSS** Why would a community located on or near a stream want to know where its stream water goes after passing the community?

Name: _____   Course/Section: _____   Date: _____

A Trout Run Drainage Basin:

1. Complete parts **a** through **h** in **Fig. A11.2.1**.

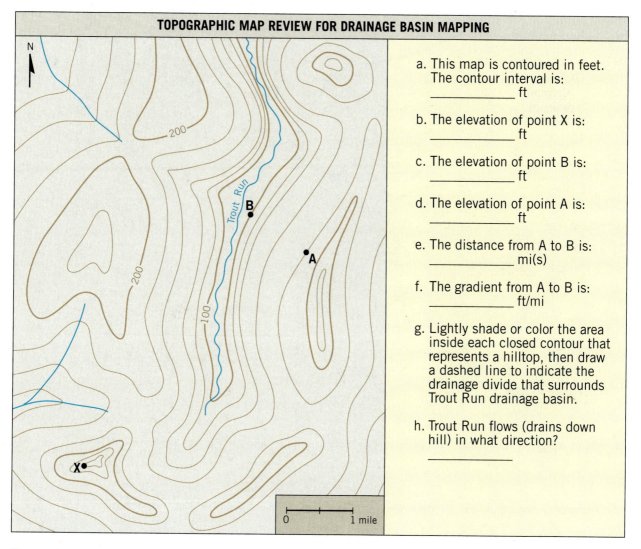

TOPOGRAPHIC MAP REVIEW FOR DRAINAGE BASIN MAPPING

a. This map is contoured in feet. The contour interval is: _____ ft

b. The elevation of point X is: _____ ft

c. The elevation of point B is: _____ ft

d. The elevation of point A is: _____ ft

e. The distance from A to B is: _____ mi(s)

f. The gradient from A to B is: _____ ft/mi

g. Lightly shade or color the area inside each closed contour that represents a hilltop, then draw a dashed line to indicate the drainage divide that surrounds Trout Run drainage basin.

h. Trout Run flows (drains down hill) in what direction? _____

Figure A11.2.1

2. Imagine that heavy rain fell on a saturated landscape at location X in **Fig. A11.2.1**. Is it likely that runoff water would flow from X downhill into Trout Run? Explain your reasoning.

B **Figure A11.2.2** is a portion of the Lake Scott 7.5-minute quadrangle, Kansas, that is centered around latitude 38.7°N, longitude 100.95°W. This area is an ancient upland surface that slopes gently eastward from an elevation of about 5500 feet along the Rocky Mountains to about 2000 feet above sea level in western Kansas. Streams in western Kansas drain eastward and cut channels into the ancient upland surface. Small tributary streams merge to form larger streams that eventually flow into the Mississippi River.

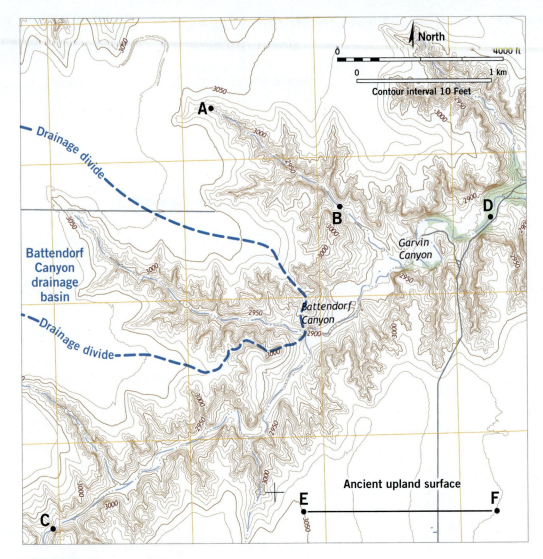

Figure A11.2.2

1. What is the gradient of the ancient upland surface between points E and F in **Fig. A11.2.2**? Show your work.

2. What type of drainage pattern are the modern streams in this area developing (see **Figs. 11.8–11.10**)?

3. Notice in **Fig. A11.2.2** that small tributaries merge to form larger streams. We will consider the intermittent stream in Garvin Canyon to be a first-order stream because it has no significant tributaries (refer to **Fig. 11.7**). What is the gradient and sinuosity, from A to B on **Fig. A11.2.2**, of the first order stream in Garvin Canyon? Refer to **Figs. 11.3** and **11.4** for help measuring gradient and sinuosity. Show your calculations. You will graph these data later in the activity.

   Gradient: _____ ft./mi. Sinuosity: _____

4. Notice how the drainage divide of the stream in Battendorf Canyon is defined by a blue dashed curve in **Fig. A11.2.2**. Draw a similar curve, as exactly as you can, to show the boundary of the Garvin Canyon drainage basin. Points A and B are in Garvin Canyon. Refer to **Fig. 11.5** for help in deciding how to draw the drainage divide. Draw the divide.

5. What is the gradient and sinuosity of the stream that flows between points C and D (**Fig. A11.2.2**), and what is the stream order? (Refer to **Figs. 11.3** and **11.4** for help measuring gradient and sinuosity.) Show your calculations. You will graph these data later in the activity.

Gradient: _____ft./mi.   Sinuosity: _____   Stream order: _____

6. REFLECT & DISCUSS By some accounts, the Mississippi River is a 10th-order stream. Based on your answers to the parts **3** and **5** above, what do you think happens to the *gradient* of streams as they increase in order?

7. REFLECT & DISCUSS What do you think happens to the *discharge* of streams as they increase in order?

C Examine the enlarged part of the Strasburg, Virginia, quadrangle map in **Fig. A11.2.3**, which is centered near latitude 38.92°N, longitude 78.34°W.

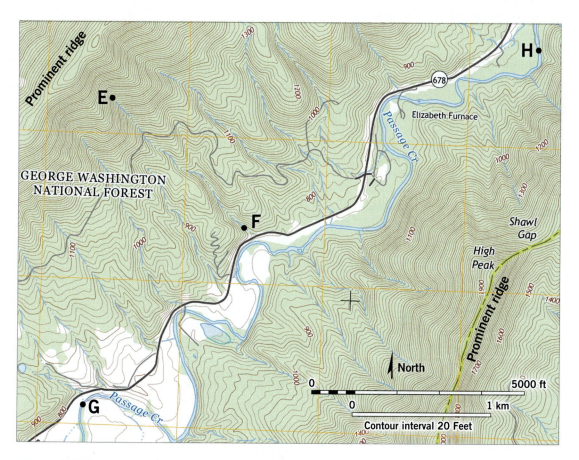

Figure A11.2.3

1. What drainage pattern is developed in this area, taking note that this map area is between two prominent, nearly parallel ridges? Refer to **Fig. 11.8** for help. Explain your reasoning.

2. In **Fig. A11.2.3**, what is the gradient and sinuosity of the small stream, from E to F? (Refer to **Figs. 11.3** and **11.4** for help measuring gradient and sinuosity.) Show your calculations. You will graph this data later in the activity.

Gradient: _____ft./mi.   Sinuosity: _____

3. In **Fig. A11.2.3**, what is the gradient and sinuosity of Passage Creek from G to H? (Refer to **Figs. 11.3** and **11.4** for help measuring gradient and sinuosity.) Show your calculations. You will graph this data later in the activity.

Gradient: _____ft./mi.

Sinuosity: _____

D We will use semi-logarithmic graph paper (**Fig. A11.2.4**) to investigate whether there might be a relationship between gradient and sinuosity.

1. Plot points for the following streams, and draw a best-fit line through the points:

   - Stream segment A–B (Garvin Canyon stream) from part **B4** of this activity.
   - Stream segment C–D from part **B5** of this activity.
   - Stream segment E–F (tributary of Passage Creek) from part **C2**.
   - Stream segment G–H (Passage Creek) from part **C3**.

2. Based on the summary graph that you just completed, do you think there is a relationship between a stream's gradient and whether its channel is straight, sinuous, or meandering? If you do, what do you think that relationship is?

E **REFLECT & DISCUSS** Compare the landscapes that you studied in this activity. What do you think are some of the factors that might determine the kind of drainage pattern that develops on a landscape and whether a stream is eroding its channel or depositing sediment?

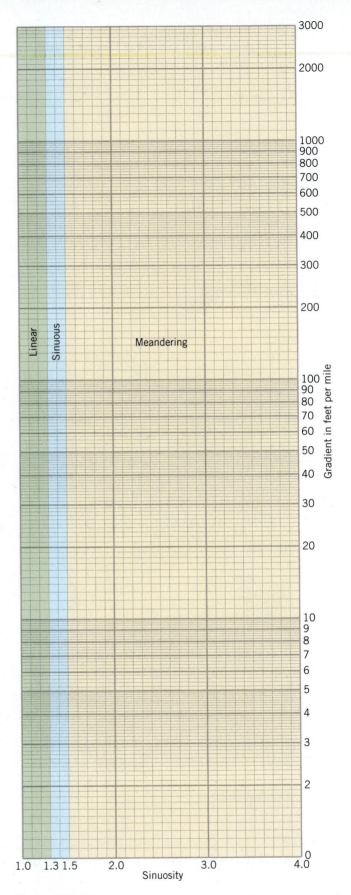

Figure A11.2.4

Activity 11.3

Name: _____ **Course/Section:** _____ **Date:** _____

Refer to the portion of the Ennis, Montana, 15-minute quadrangle in **Fig. 11.2**.

A What types of stream channel (shown in **Figs. 11.11–11.13**) are present on:

1. The streams in the mountains along the east side of this map?

2. The Cedar Creek Alluvial Fan?

3. The valley of the Madison River (northwestern portion of **Fig. 11.2**)?

B Figure A11.3.1 is an expanded portion of **Fig. 11.2**, centered near latitude 45.31°N, longitude 111.59°W. **Fig. A11.3.1** includes a gridded rectangular area called a *profile box* that we will use to plot the profile of Cedar Creek as it flows from the

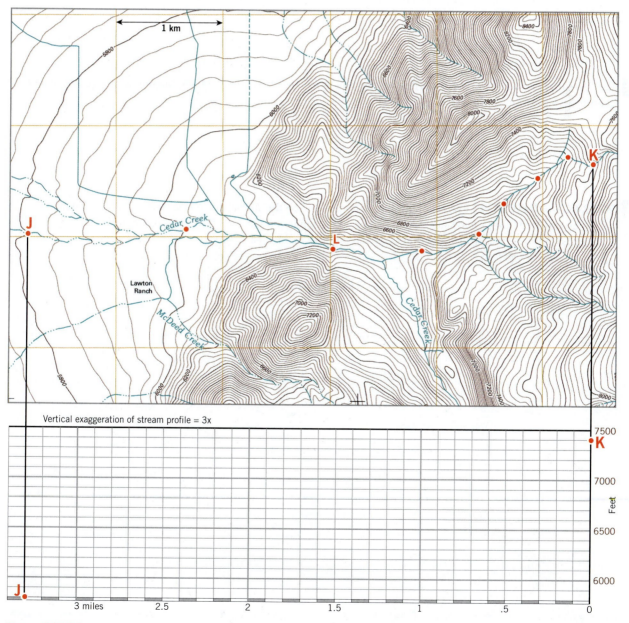

Figure A11.3.1

mountains onto the alluvial fan. The horizontal axis is east–west distance along Cedar Creek, in miles, and the vertical axis is elevation, in feet. Complete the profile from point J to point K along Cedar Creek by plotting and connecting the nine red elevation points. Points J and K have already been plotted in the profile box as examples.

C On the profile you constructed in part **B**, identify and label the parts of the profile that represent (1) where the channel is eroded into bedrock and (2) where the channel is established on sediment that was deposited by Cedar Creek.

D What is the average gradient of Cedar Creek

1. from point K to point L? Gradient: _____ft./mi.

2. from point L to point J? Gradient: _____ft./mi.

E How does that stream's gradient change downstream as it enters the alluvial fan, and how might this change in gradient contribute to the formation of the alluvial fan?

F **REFLECT & DISCUSS** Can you find any streams mapped on the alluvial fan in **Figs. 11.2** or **A11.3.1** that do not follow the expected flow path for water moving across topographic contour lines (**Fig. 11.5**)? What might have caused this unexpected flow direction? *Hint:* Notice that there are ranches in the area.

Escarpments and Stream Terraces

Activity 11.4

Name: _____ **Course/Section:** _____ **Date:** _____

Escarpments and terraces are found along many stream valleys. Escarpments are long cliffs or steep narrow slopes that separate one relatively level part of the landscape from another. Terraces are long, narrow, or broad almost-level surfaces bounded on one or both sides by an escarpment. Stream terraces parallel the stream. The difference in elevation between two terraces can range from centimeters to tens of meters.

The 7.5-minute quadrangle map of Voltaire, North Dakota, includes part of the course of the Souris River (**Fig. A11.4.1**). Part of a large continental ice sheet extended south and west from the Hudson Bay area of eastern Canada during the Pleistocene Ice Age, covering this area. As the ice sheet melted and became smaller, meltwater streams flowed where the ice had been, transporting and depositing vast amounts of sediment. Streams have been forming and modifying this landscape since the ice retreated, about 11,000–12,000 years ago.

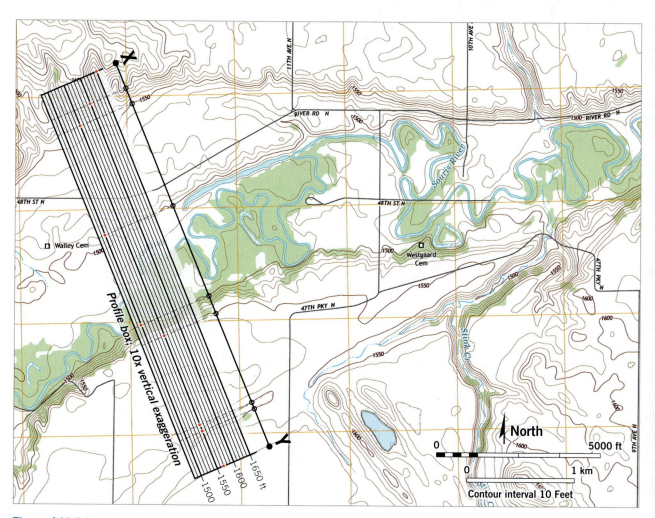

Figure A11.4.1

A The modern floodplain of the Souris River near latitude 48.11°N, longitude 100.81°W can be identified by the flat area on both sides of the active stream channel. What other meandering stream features named in **Fig. 11.13** do you recognize in this image?

B On the basis of the image and topographic contours, make a profile of the landscape from X to Y in the profile box provided in **Fig. A11.4.1**. Red circles mark several points along that profile that have already been plotted for you. Label terraces with a T and escarpments with an E.

C Describe how the escarpments might have formed along the Souris River.

D On your profile, label the modern floodplain of the Souris River and record its width along line X–Y.

E What was the maximum width of the Souris River floodplain in the past (measured along line X–Y), and how can you tell?

F Give one possible reason why the Souris River floodplain was wider in the past.

G **REFLECT & DISCUSS**   Notice along line X–Y that the terrace on the south side of the Souris River is 30–40 feet higher than the terrace on the north side of the river. Suggest how these two different levels of terraces may have formed and which one is older based on your hypothesis.

Name: _____ Course/Section: _____ Date: _____

The Rio Grande River forms part of the national border between Mexico and the United States (**Fig. A11.5.1**). Notice that the position of the river changed in many places between 1936 (red line and leaders by lettered features) and 1992 (blue water bodies and leaders by lettered features). Study the meander terms provided in **Figs. 11.13** and **A11.5.1**, and then proceed to the questions on the next page.

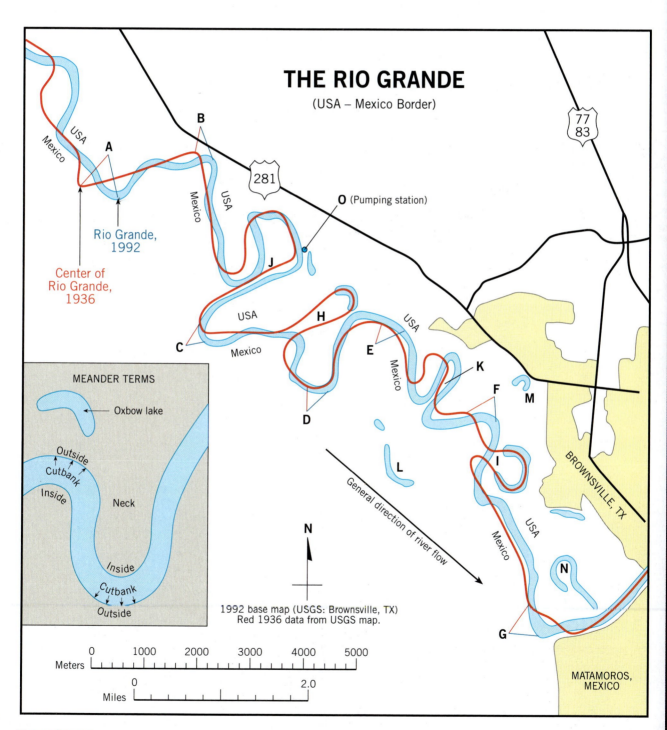

Figure A11.5.1

A Study the meander cutbanks labeled A through G. The red leader from each letter points to the cutbank's location in 1936. The blue leader from each letter points to the cutbank's location in 1992. In what two general directions (relative to the meander, relative to the direction of river flow) have these cutbanks moved?

B Study locations H and I.

1. In what country were H and I located in 1936?

2. In what country were H and I located in 1992?

3. Explain a process that probably caused locations H and I to change from meanders to oxbow lakes.

C Based on your answer in part **B3**, predict how the river might change in the future at locations J and K.

D What are features L, M, and N, and what do they indicate about the historical path of the Rio Grande?

E What is the average rate at which meanders like A through G migrated here (in meters per year) from 1936 to 1992? Explain your reasoning and calculations.

F **REFLECT & DISCUSS** Explain in steps how a meander evolves from the earliest stage of its history as a broad, slightly sinuous meander to the stage when an oxbow lake forms.

Name: _____   **Course/Section:** _____   **Date:** _____

The Niagara River flows north from Lake Erie to Lake Ontario and forms part of the border between the United States and Canada. Over this straight-line distance of approximately 43 km, the river drops approximately 100 m in elevation to Lake Ontario. Just over half of that total drop in elevation occurs at Niagara Falls, where the river flows over a hard dolostone caprock called the Lockport Formation (**Fig. A11.6.1**). A series of soft shales and harder dolostone beds are below the hard caprock. As the water tumbles over the edge, its turbulence is a potent agent of erosion that removes rock from the weaker formations and undercuts the stronger Lockport dolostone caprock. Blocks of the caprock break off, and the edge that marks the top of Niagara Falls migrates upstream (south), eroded block by block.

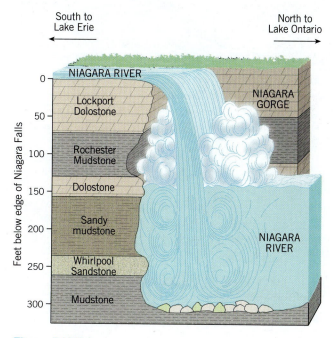

Figure A11.6.1

Today, Niagara Falls is located about halfway between the two lakes at the head of Niagara Gorge (**Fig. A11.6.2**). The gorge begins at Niagara Falls, and ends at the base of the escarpment along the east-northeast-trending line of contact between the Whirlpool Sandstone and the Queenston Shale. Geologic evidence indicates that the Niagara River became the primary outlet river for the upper Great Lakes into the Ontario Basin about 11,000 years ago, as an enormous continental ice sheet retreated north out of the area. At that time, the Niagara River cascaded over the edge of the Lockport Formation at the Niagara Escarpment near Lewistown, New York, and Queenston, Canada, near 43.16°N, 79.05W. Since that time, retreat of the waterfall has resulted in the excavation of Niagara Gorge downstream (north) of the falls.

A About how long is the Niagara Gorge today? To measure the approximate length of the Niagara Gorge, manipulate a piece of string so that it follows the centerline of the Niagara Gorge as mapped in **Fig. A11.6.2**. Mark the beginning and end of the gorge on the string. Then pull the string straight and measure the distance between the two marked points to give you the map distance along the channel in the gorge. Then compare that map distance with the bar scale on the map, and calculate the true length of the gorge using proportions. Approximate length of Niagara Gorge: ~ _____ km.

B Based on this geochronology and the length of Niagara Gorge as shown in **Fig. A11.6.2**, calculate the average rate at which Niagara Falls has migrated southward along the Niagara River course in cm/year. Show your calculations. Rate of retreat: ~ _____ cm/year.

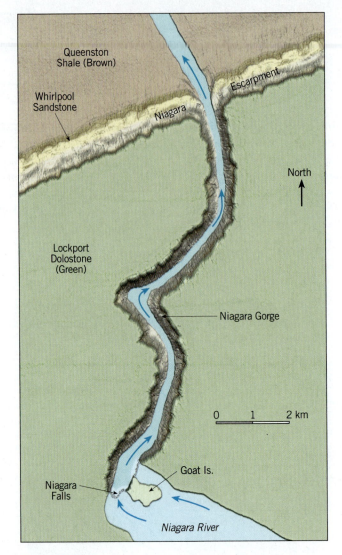

Figure A11.6.2

C Name as many factors as you can that could cause Niagara Falls to retreat at a faster rate.

D Name as many factors as you can that could cause Niagara Falls to retreat more slowly.

E Niagara Falls is about 35 km north of Lake Erie, and it is retreating southward. If the falls were to continue its retreat at the average rate calculated in part **A**, then how many years from now would the falls reach Lake Erie? _____ years.

F **REFLECT & DISCUSS** Look at the cross-section of Niagara Falls in **Fig. A11.6.1**. Describe how the process that formed the falls could have begun. (*Hint*: Use your knowledge of stream erosion and the effects of stream gradient.)

Name: _____ **Course/Section:** _____ **Date:** _____

A On the portion of the 7.5-minute topographic map of Montezuma, Georgia (**Fig. A11.7.1**), locate the gaging station on Flint River in the middle of the map. The center of the map is near latitude 32.30°N, longitude 84.05°W. The elevation from which the gage measures changes in river level—the **gage datum**—is at an elevation of 255.83 feet above sea level, and the height of the river relative to this datum is called the **gage height** or **stage**. The river is considered to be at flood stage when the gage height is 20 feet above the gage datum, or an elevation of 275.83 feet. A flood in July 1994 established a record gage height of 34.11 feet, or 289.94 feet above sea level. This corresponds to the 290-foot contour line on the map. Trace the 290-foot contour line on both sides of the Flint River and label the area within these contours where the land is lower than 290 feet, "1994 Flood Hazard Zone."

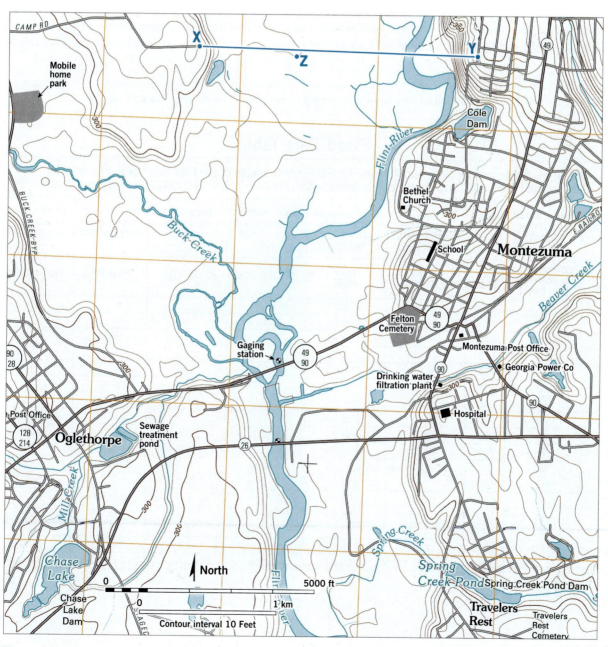

Figure A11.7.1

B Name two essential public utilities or facilities that were submerged by the flood, and infer the possible impact that flooding those facilities might have had on the environment and on quality of life in Montezuma, Oglethorpe, or downstream communities after the flood.

C Notice line X–Y near the top center part of **Fig. A11.7.1**.

1. The map shows the Flint River at its normal stage. What is the width of the Flint River at its normal stage along line X–Y? _____ ft.

2. What was the width of the river along this line when it was at maximum flood stage (290 feet) during the July 1994 flood? _____ ft.

D Notice the floodplain of the Flint River along line X–Y on the map in **Fig. A11.7.1**. The floodplain is the relatively flat, marshy land between the river and the steep escarpments that develop by erosion during floods.

1. What is the elevation of the floodplain at point Z near line X–Y, expressed in feet above sea level? _____ ft.

2. How deep was the water that covered that floodplain at point Z during the 1994 flood? _____ ft.

3. How did you derive your answer?

4. Did the 1994 flood stay within the floodplain and its bounding valley slopes? Does this suggest that the 1994 flood was of normal or abnormal magnitude for this river? Explain your reasoning.

E The USGS has compiled a list of the highest water levels in the Flint River at the Montezuma gaging station for 112 years (1897 and 1905–2015). Parts of the data have been summarized in the Flood Data Table in **Fig. A11.7.2**.

Flood Data Table

Recurrence Intervals for Selected, Ranked, Annual Highest Stages of the Flint River over 112 Years of Observation (1897 and 1905–2015) at Montezuma, Georgia (USGS Station 02349605, data from USGS)								
Number of years of data (n)	Rank of annual highest river stage (S)	Year	River elevation above gage (gage height), in feet (GH)	Gage elevation (datum) above sea level, in feet (GD)	River elevation above sea level, in feet (RE)	Recurrence interval (RI), in years	Probability of occurring in any given year	Percent chance of occurring in any given year
From USGS	Result of sorting by maximum gage height	From USGS	From USGS	From USGS	RE = GH + GD	RI = (n+1)/S	"1 in RI"	S/(n+1) or 1/RI
112	1	1994	34.11	255.83	289.94	_____	1 in _____	_____%
112	2	1929	27.40	255.83	283.23	_____	1 in _____	_____%
112	3	1990	26.05	255.83	281.88	_____	1 in _____	_____%
112	4	1897	26.00	255.83	281.83	_____	1 in _____	_____%
112	5	1949	25.20	255.83	281.03	_____	1 in _____	_____%
112	10	1961	24.00	255.83	279.83	11.3	1 in 11.3	9%
112	15	1913	22.30	255.83	278.13	7.5	1 in 7.5	13%
112	20	1978	21.56	255.83	277.39	5.7	1 in 5.7	18%
112	30	1952	20.70	255.83	276.53	3.8	1 in 3.8	27%
112	40	2009	19.37	255.83	275.20	2.8	1 in 2.8	36%
112	50	1962	18.70	255.83	274.53	2.3	1 in 2.3	45%
112	60	1926	17.90	255.83	273.73	1.9	1 in 1.9	54%
112	70	1977	17.39	255.83	273.22	1.6	1 in 1.6	63%
112	80	1991	15.84	255.83	271.67	1.4	1 in 1.4	71%
112	90	1986	14.13	255.83	269.96	1.3	1 in 1.3	80%
112	100	2011	11.93	255.83	267.76	1.1	1 in 1.1	89%
112	112	2012	7.39	255.83	263.22	1.0	1 in 1.0	100%

Figure A11.7.2

1. The annual maximum elevations of the Flint River were ranked in magnitude from S = 1 (the highest annual river level ever recorded—the 1994 flood) to S = 112 (the lowest annual maximum elevation). Data for 17 of these ranked years are provided in the Flood Data Table and can be used to calculate recurrence interval for each magnitude (rank, S). The **recurrence interval** (or **return period**) is the average number of years between occurrences of a flood of a given rank (S) or greater. The recurrence interval for a flood of a given rank can be calculated as: RI = (n + 1)/S. Calculate the recurrence interval, probability of occurring in any given year ("1 in RI"), and percentage chance of occurring in any given year (1/RI) for ranks 1–5, and write them in the table.

2. Notice that a recurrence interval of 5.0 means that there is a one-in-five probability (or 20% chance) that an event of that magnitude will occur in any given year. This is known as a *5-year flood.* What is the probability that a *100-year flood* will occur in any given year?

3. Plot (as carefully as you can) points on the flood magnitude/frequency graph (**Fig. A11.7.3**) for all 17 ranks of annual high river stage in the Flood Data Table. Use the vertical axis (blue) to plot the "river elevation above sea level" listed in the blue column of the Flood Data Table (**Fig. A11.7.2**) and the horizontal axis (pink) to plot the "recurrence interval" listed in the pink column. Then use a pencil and ruler to draw a line through the set of points so that approximately as many points are above the line as are below the line. Extend the line to the left and right edges of the graph.

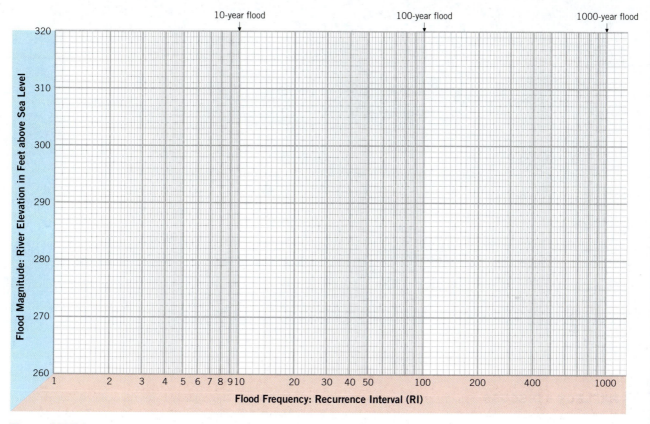

Figure A11.7.3

4. Your completed flood magnitude/frequency graph can now be used to estimate the probability of future floods of a given magnitude and frequency. A 10-year flood on the Flint River is the point where the line in your graph crosses the vertical line that corresponds to a flood frequency of 10 years. What is the probability that a future 10-year flood will occur in any given year, and what will be its magnitude expressed in river elevation in feet above sea level?
Probability: 1 in _____   Magnitude of 10-year flood: _____ ft.

5. What is the probability for any given year that a flood on the Flint River at Montezuma, Georgia, will reach an elevation of 275 feet above sea level? Probability: 1 in _____

F The hazard areas on a Flood Insurance Rate Map are defined using a *base flood elevation* (BFE)—the computed elevation to which floodwater is estimated to rise during a *base flood.* The regulatory-standard base flood elevation is the 100-year flood elevation. Based on your graph, what is your interpretation of the BFE for Montezuma, Georgia? _____ ft. above sea level.

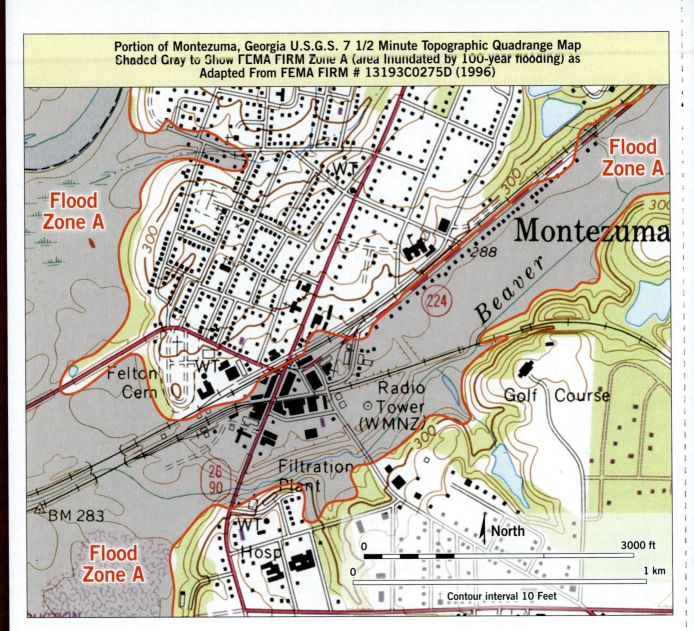

Portion of Montezuma, Georgia U.S.G.S. 7 1/2 Minute Topographic Quadrange Map Shaded Gray to Show FEMA FIRM Zone A (area inundated by 100-year flooding) as Adapted From FEMA FIRM # 13193C0275D (1996)

Flood Zone A

Flood Zone A

Flood Zone A

Flood Zone A

WT

Montezuma

Beaver

288

224

300

300

300

300

Felton Cem

Radio Tower (WMNZ)

Golf Course

Filtration Plant

26 90

BM 283

WT

WT

Hosp

North

0   3000 ft

0   1 km

Contour interval 10 Feet

Figure A11.7.4

G **REFLECT & DISCUSS**   The 1996 FEMA FIRM for Montezuma, Georgia, shows hazard areas designated *Zone A*. Zone A is the official designation for areas expected to be inundated by 100-year flooding. The location of Zone A is the gray area bounded by the red curve on **Fig. A11.7.4**. The analysis you completed in part **F** of this activity allows you to estimate the BFE corresponding to 100-year flooding at Montezuma, Georgia, based on approximately two decades more data than was available when the 1996 FIRM map was developed. Using the topographic contours as a guide, trace your estimate of the 100-year base flood elevation on **Fig. A11.7.4**, using a pen or pencil that will make your tracing legible. How does your work compare with the red line on the 1996 FEMA FIRM map?

Groundwater Processes, Resources, and Risks

Pre-Lab Video 12

https://goo.gl/TQWzrF

Contributing Authors

Garry D. McKenzie • *Ohio State University*

James R. Wilson • *Weber State University*

Richard N. Strom • *University of South Florida, Tampa*

▲ Travertine terraces developed by precipitation of calcium carbonate from hot groundwater springs at Pamukkale, Turkey, below the ancient city of Hierapolis (~ 37.923°N, 29.122°E).

BIG IDEAS

Groundwater is an essential resource for society because it is the source of most fresh liquid water on Earth. Aquifers are rock bodies from which useful or economic quantities of fresh, drinkable water can be obtained from springs or wells. We can create maps and cross-sections of the water table that help us to predict how groundwater flows. Acidic rainwater and groundwater can slowly dissolve carbonate and evaporite bedrock, leading to the formation of karst topography and features such as sinkholes, caves, and disappearing streams. Excessive pumping of groundwater can cause ground subsidence that can increase a community's susceptibility to flooding. Hydrogeology is a subfield of the geosciences that directly impacts society by addressing some of its most pressing resource issues.

FOCUS YOUR INQUIRY

Think About It How does groundwater move underground?

ACTIVITY 12.1 Groundwater Inquiry (p. 326, 337)

- -

Think About It How can we estimate the flow of groundwater along or below the water table?

ACTIVITY 12.2 Where Is the Nasty Stuff Going? (p. 327, 339)

ACTIVITY 12.3 Using Data to Map the Flow of Groundwater (p. 329, 340)

- -

Think About It What is karst topography, and how does water flow beneath it?

ACTIVITY 12.4 Karst Processes and Topography (p. 331, 341)

ACTIVITY 12.5 Floridan Aquifer System (p. 333, 343)

- -

Think About It What can happen if groundwater is withdrawn faster than it is replenished?

ACTIVITY 12.6 Land Subsidence from Groundwater Withdrawal (p. 335, 345)

- -

Introduction

Water that exists in fractures and in the tiny spaces between grains below Earth's surface is called **ground water**. It is part of an interconnected global hydrologic system that includes soil moisture, streams, lakes, oceans, glaciers, permafrost, water vapor in the atmosphere, and the water stored in biomass.

Most of the fresh liquid water on Earth is groundwater. UNESCO estimates that Earth's total groundwater volume is about 23.4 million cubic kilometers, of which less than half is *potable*, or suitable for drinking. The rest includes natural salt water, water that contains microbial life that is harmful to humans, and water that has become polluted with various chemical compounds. Potable groundwater is an essential resource.

With the passage of environmental legislation in the United States in the mid-1970s that recognized the need to control and mitigate the pollution of groundwater resources, the field of hydrogeology underwent a dramatic expansion both within universities and as a profession serving the public. **Hydrogeology** is the study of groundwater, the transport of energy and chemicals by groundwater, and the laws that govern its flow and interactions with the materials it flows through. It is said that the vast majority of all hydrogeologists who have ever lived are still in practice today, and employment opportunities continue to be good because of the steady need for reliable sources of fresh water worldwide.

Porosity and Permeability

The **porosity** of a volume of subsurface material—soft sediment to hard rock—is the percentage of that volume that is not composed of solid material like mineral grains and cement. The larger spaces between solid grains are called **pores**. Pores provide the storage space for groundwater and are an important part of the path along which groundwater flows. Fluids can pass through a rock only if pores are connected with one another to form a flow path. The small spaces between pores are called **pore throats**, and the width of pore throats helps to determine how easily fluids can pass through. Pore throats are like the neck in an hourglass, which controls how fast sand moves from the top bulb to the bottom.

Only the pores that are interconnected with one another are of interest in studying groundwater. Water-filled vesicles that are effectively isolated from one another within an igneous rock contribute to the porosity of that rock but do not contribute to the groundwater flow system. The **effective porosity** is the percentage of the total volume that is composed of interconnected pores.

Open fractures within and between grains are an important form of porosity. Fluids can often flow more rapidly along the relatively straight paths provided by open fractures than through the more complicated paths moving around grains through pores and pore throats in an intergranular pore system. Water can move through cracks as narrow as 10 millionths of a meter ($10 \mu m$), although it moves very slowly through these tight spaces.

Permeability is the property that expresses how easily a gas or liquid can pass through rock or sediment. A material that is more permeable to water will allow water to pass through it at a higher rate than will a less permeable material. For example, a rock that is crossed by many open fractures will be more permeable than a similar rock without the fractures. An unfractured material that has larger pore throats will be more permeable than a material with smaller pore throats. A sedimentary stratum that is mostly sand or gravel with very little clay or silt will have larger pores and pore throats than a stratum that is mostly composed of clay or silt. As a consequence, coarser-grained sedimentary strata have a higher permeability than finer-grained strata. Sandstone is more permeable than shale because the larger pores and pore throats in sandstone allow sandstone to store and transmit water more rapidly than shale.

Although the ideas here involve new words, you probably already know the basic principles from your personal experience. In the kitchen, you might use a colander (a kind of bowl with holes or slots in the bottom) to drain cooked pasta. You might use a wire

---

ACTIVITY 12.1

Groundwater Inquiry, (p. 337)

Think About It   How does water move underground?

Objective   Experiment with water to determine its behavior in confined and unconfined spaces and in relation to shale and sandstone.

Before You Begin   Read the following section: Porosity and Permeability.

Plan Ahead   For Experiment 2, you will need (or your teacher will need to supply you with) two pieces of 1.25 cm (½ in.) inside-diameter clear-plastic flexible tubing, each about 1 m long; at least 4 cotton balls; some loose sand; some sediment that is about half clay-size particles—a dry soil will do nicely; and a funnel or a plastic laboratory wash bottle to help you to get water into the tubing.

If your teacher wants you to re-create Experiments 1 or 3 in this activity, you will need (or your teacher will supply you with) the following items:

- Experiment 1: Specimens of shale, sandstone, and perhaps other common rock types and a dropper bottle containing water.

- Experiment 3: An empty 2-L plastic bottle, some tape, a ruler, some way of making a small hole in the plastic bottle safely, a tray or a square aluminum foil cake pan to contain the water jet, and some paper towels or a cloth towel to clean up spilled water.

mesh strainer to drain vegetables or to strain tea. Those kitchen implements allow water to pass through rapidly and are very permeable to water. A coffee filter, on the other hand, has holes that are small enough to trap the coffee grounds while allowing the precious liquid to pass through slowly. Coffee filters are much less permeable than colanders or strainers. Finally, you might put the coffee in a styrofoam cup. Styrofoam is made of small plastic balls that are molded together, but there are still tiny spaces between and within the plastic balls. If you let your coffee sit for hours in a typical styrofoam cup, you will see tiny beads of brown coffee on the outside of the cup. The cup is also permeable, but the rate at which water (coffee) can pass through is very slow. It is not an absolute barrier to water, but it is enough of a barrier to serve its purpose.

Saturated and Unsaturated Zones

For most of the distance between the ground surface and the depth at which interconnected pores in rock cease to exist because of the great pressure (~5 km), the pores are filled with air, water, natural gas, or petroleum fluids. The part of the subsurface in which the pores are completely filled with water is called the **saturated zone**. In most places, there is a transitional zone immediately below the ground surface that is not entirely saturated with water and through which water passes on its way down to the saturated zone. This transitional zone goes by several names, including the **vadose zone**, the **zone of aeration**, and the **unsaturated zone**.

The boundary between the unsaturated and saturated zones is a surface along which the fluid pressure in the pores is the same as the atmospheric pressure. That boundary is called the **water table** (**Fig. 12.1**). The 3-dimensional shape of the water table in an area where it has not been altered by pumping wells is a subdued replica of the shape of the ground surface. There are usually high spots in the water table underneath high spots in the ground surface, and depressions in the water table below depressions in the ground surface. In some places, the water table intersects the ground surface, resulting in springs or surface-water systems such as lakes and flowing streams. A topographic map of the shape of the water table can be used to determine the general direction of groundwater flow at the top of the saturated zone.

Aquifers

A saturated, permeable rock unit from which water can be produced in economic quantities is called an **aquifer**. This definition is based on one published by the National Ground Water Association and includes economics as part of the definition. In that sense, this definition is similar to the definition of an *ore body* as a rock unit from which some economic mineral can be produced at a profit. Some hydrogeologists would leave economics to economists and replace the word "economic" with the word "useful." But this does not solve all of our definitional problems because

oil companies pump briny or polluted wastewater into an oil field to increase pressure and improve oil production. That water is plainly useful, but the formation that contains the wastewater would not be considered an aquifer by most hydrogeologists because the water is not potable. And should we use the word *aquifer* for a formation that has potable water but whose permeability is so small that it costs much more to produce the water than the water is worth?

With these issues in mind, we will adopt a working definition. An *aquifer* is a volume of rock (that might include more than one formation or rock type) from which useful or economic quantities of potable or fresh water can be produced, either in springs or through wells. An aquifer can be a rock like sandstone that stores and transmits water through interconnected pores or even a fractured igneous or metamorphic rock in which the water is stored and transmitted through an interconnected network of open fractures.

Unconfined aquifers (including the recharge area of confined aquifers) have a water table that is in contact with the atmosphere either directly or through interconnected pores or fractures in an unsaturated zone (**Fig. 12.2**). The direct connection between an unconfined aquifer and the atmosphere and ground surface is why unconfined aquifers are particularly susceptible to damage by pollutants that infiltrate downward from the ground surface.

Confined aquifers are separated from the atmosphere by a layer of material that has a significantly smaller permeability—the **confining layer** (**Fig. 12.2**). A rock layer that inhibits the flow of water so that it flows very slowly is sometimes called an **aquitard**. The confining layer or aquitard can be mudstone, shale, unfractured igneous or metamorphic rock, cemented sandstone, rock crushed or altered to clay along a fault, or some types of unfractured "tight" limestones or dolostones. Water typically enters the confined aquifer through a recharge area (**Fig. 12.2**).

ACTIVITY 12.2

Where Is the Nasty Stuff Going?
(p. 339)

> **Think About It** How can we estimate the flow of groundwater along or below the water table?

Objective Construct a water table contour map and determine the direction of groundwater movement along that surface.

Before You Begin Read the following sections: Hydraulic Gradient of the Water Table and Mapping Water Table Topography. Refer back to previous sections as needed.

Plan Ahead You will need a pencil for initial sketching of contours and a pen.

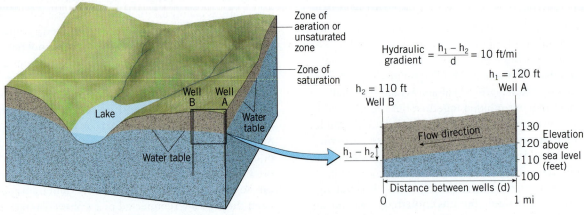

A. Groundwater zones and the water table

Zone of aeration or unsaturated zone

Zone of saturation

Well B   Well A

Lake

Water table

Water table

$$\text{Hydraulic gradient} = \frac{h_1 - h_2}{d} = 10 \text{ ft/mi}$$

$h_1 = 120$ ft
Well A

$h_2 = 110$ ft
Well B

Flow direction

$h_1 - h_2$

130
120   Elevation above
110   sea level
100   (feet)

Distance between wells (d)

0   1 mi

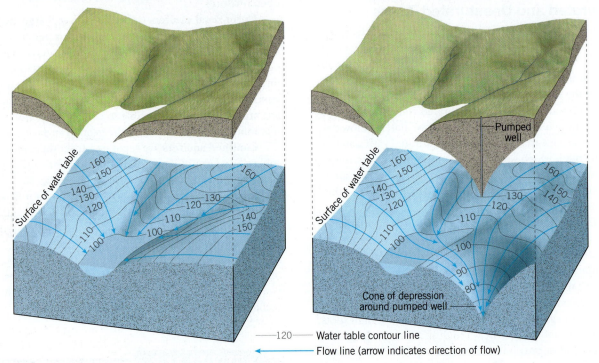

B. Normal water table contours and flow lines: note that flow direction is downhill to streams and the lake

Surface of water table

160 150 140 130 120 110 100

120 130 140 150 160

110 100

C. Water table contours and flow lines changed by a cone of depression developed around a pumped well

Pumped well

Surface of water table

160 150 140 130 120 110 100

160 150 140

130 120

110 100 90 80

Cone of depression around pumped well

—120— Water table contour line

⟵ Flow line (arrow indicates direction of flow)

Figure 12.1 Groundwater movement through an unconfined aquifer. A. Rainwater seeps into the *zone of aeration* or *unsaturated zone*, where the pores are filled with air and water. Some of the water seeps down to the *saturated zone* where all of the interconnected pores are filled with water. Its upper surface is the water table. The shape of the water table in an area without pumping wells is similar to that of the ground surface. Groundwater flows down the hydraulic gradient in unconfined aquifers. **B.** Contours map the topography of the water table and allow us to estimate flow lines that indicate the path taken by flowing groundwater. Flow lines—the blue curves with arrowheads—are perpendicular to the contours. Flow lines converge or diverge but never cross. **C.** A pumped well that withdraws water faster than it can be replenished, causing development of a cone of depression in the water table and a change in the groundwater flow lines.

Hydraulic Gradient of the Water Table

The water table in many unconfined aquifers is inclined. Just as the slope gradient can be used to describe the steepness of a hillside, the **hydraulic gradient** of the water table can be used to describe its steepness over some horizontal distance (**Fig. 12.1A**). The hydraulic gradient between a pair of contours of different elevations on the water table is the elevation difference between the contours divided by the horizontal distance between the contours measured perpendicular to the average orientation of the contours.

Mapping Water-Table Topography

Imagine that you take a break from your studies and go to the beach to build a sand castle. It doesn't matter whether it is a point bar on a river, a lakeshore, or a beach

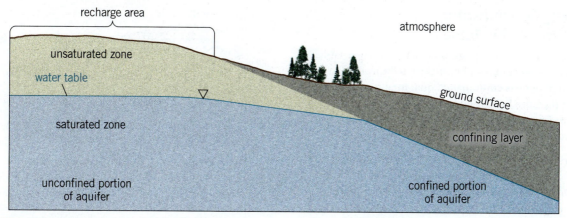

Figure 12.2 **Unconfined and confined aquifers.** An unconfined aquifer is in direct connection with the atmosphere across the water table and unsaturated zone. A confined aquifer has a confining layer between the aquifer and the atmosphere. The triangle is used in cross-sections to indicate the water table.

ACTIVITY 12.3

Using Data to Map the Flow of Groundwater, (p. 340)

> **Think About It** How can we estimate the flow of groundwater along or below the water table?

Objective Determine the direction of ground-water movement in a vertical plane through an unconfined aquifer using information from wells.

Before You Begin Read the following sections: Hydraulic Gradient of the Water Table, Mapping Water Table Topography, and Mapping Groundwater Below the Water Table. Refer back to previous sections as needed.

Plan Ahead You will need a pencil for initial sketching of contours and a pen.

by the ocean as long as it is a sandy beach. You take your bucket and little shovel and go right down near the water but stay up on the dry sand. You start digging to build the moat around your sand castle, and a little way down, you encounter water that seeps into the hole from below. Being a novice scientist, you notice that the elevation of the water in your moat is about the same as the water in the river, lake, or ocean next to you. In fact, you have dug down to the water table, which is at the level of the water in the moat. You can measure the distance from the ground surface to the water table at that point on the beach to find the water table elevation.

Geoscientists measure the depth to the water table in essentially the same way but usually dig the hole with a drill rig or auger. We then determine the elevation to which groundwater rises in a hole dug just below the water

table. The location of the well and elevation of the ground surface is determined by careful surveying, and the depth to the water is measured using a measuring tape or laser rangefinder. When a sufficient number of water-table elevations are determined, a topographic contour map of the water table can be constructed from those elevation data as in Activity 12.2.

It is not too difficult to see *how* we can make a topographic map of the water table, but *why* might it be useful to make a water-table map? Determining the shape of the water table is an important part of the process of estimating the volume of water in an unconfined aquifer. We can use a water-table map superimposed on a topographic map of the ground surface to predict where springs will occur—in places where the water-table surface is above the ground surface. We can see the effect of pumping groundwater as the water table is pulled down into a **cone of depression** around the pumping well (**Fig. 12.1C**). This effect is similar to the depression you can see in the top surface of a thick milkshake when you use a straw to drink the shake quickly. And a liquid pollutant that does not mix with water and is less dense than water—known as a **LNAPL** or a *light non-aqueous phase liquid*—will flow along the water table perpendicular to the water-table contours along a path called a **flow line** (e.g., **Fig. 12.3**). Hence, constructing a topographic map of a water table allows us to predict the spread of some types of groundwater pollutants.

Figure 12.3 **Groundwater flow and water-table contours.** Groundwater flows perpendicular to the topographic contours on the water-table surface.

Mapping Groundwater Below the Water Table

Just as we can create a topographic map of a water table, a vertical cross-section can be constructed through an aquifer that depicts contours of equal total head. "Head" is a fluid pressure term used by hydrogeologists. The **total head** is the sum of three other quantities: the *velocity head*, the *elevation head*, and the *pressure head* (**Fig. 12.4**).

- The **velocity head** is related to the energy of moving water. In most groundwater situations, the velocity head is usually so small that we can safely ignore it (as we will in this lab), simplifying the total head calculation to elevation head plus pressure head.
- The **elevation head** is simply the elevation of the bottom of a monitoring well or piezometer, which is the point for which these various heads are determined. A typical **piezometer** is a plastic or steel pipe with a screen on the bottom end to keep sediment from rising into the well.
- The **pressure head** is the distance from the bottom of the piezometer to the top of the water in the pipe. Water rises in the piezometer pipe in response to the fluid pressure at the bottom of the pipe. The greater the distance that the water has risen inside of the piezometer, the higher is the fluid pressure at that point in the aquifer.

If we can compile a set of total-head values at different depths and locations along a line of section, we can plot the data on the section and draw contours of equal total head (black dashed curves in **Fig. 12.5**). Groundwater flow lines can be drawn along that vertical plane so that they intersect the **equal total-head contours** at right angles as in **Fig. 12.3**. (Equal total-head contours are also called **potentiometric** or **piezometric contours** by hydrogeologists.) Water within the aquifer flows from high total head to low total head. You have an opportunity to create a groundwater flow map in a vertical section in Activity 12.3.

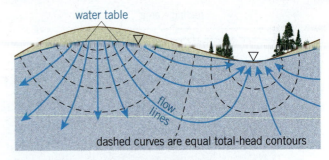

Figure 12.5 Flow map in an unconfined aquifer. Groundwater flow depicted in a vertical cross-section through an unconfined aquifer. The dashed black curves are surfaces of equal total head, and the blue curves with arrowheads are groundwater flow lines. A flow line at any given point is perpendicular to the surface of equal total head and is directed from higher to lower head. Flow lines point down under hills in the ground surface. Under the low spots (lakes and valleys), flow lines point up because water flows from high head to low head.

Although we will think about it simply in this lab and its activities, groundwater flow is actually a rather challenging problem. A particle of water that starts on an inclined water table will probably move beneath the water table rather than along it, and the rest of its flow path will involve some sort of 3-dimensional curve. So along any vertical section through an aquifer, at least some of the water particles will be moving into and out of the vertical plane of any flow map we might draw for that vertical plane.

Each of the equal total-head contours is part of a 3-dimensional **potentiometric surface**—a surface on which all points have the same potential energy or total head—and the contours simply mark where the potentiometric surface intersects either the water table or the vertical plane in our cross-section. The hydraulic gradient at any place within an aquifer is the change in total head in whatever direction that change is the greatest divided by the distance over which that change occurs. A particle of water flows in the direction of greatest change in total head, perpendicular to the potentiometric surface, from higher total head to lower total head.

Mapping the flow of groundwater enables hydrogeologists to investigate the sources of groundwater pollution as well as provide knowledge necessary to avoid such contamination. If you relied on a surface stream for your drinking water, you would probably not choose to live downstream from a feedlot where farm animals wallow (and do other things) in or next to that stream. Using the same logic, would you place your well in an unconfined aquifer down-gradient from the same feedlot if that well supplied your drinking water?

Caves and Karst Topography

The term **karst** describes a distinctive landscape or topography that is dominated by the effects of the dissolution of limestone, dolostone, or evaporites (**Figs. 12.6** and **12.7**).

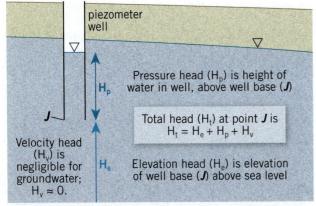

Figure 12.4 Definition of total, pressure, and elevation heads. The total head is the sum of the elevation head, pressure head, and velocity head. The velocity head can be neglected in most cases for groundwater because the velocities are so small.

Pressure head (H_p) is height of water in well, above well base (J)

Total head (H_t) at point J is
$$H_t = H_e + H_p + H_v$$

Velocity head (H_v) is negligible for groundwater; $H_v \approx 0$.

Elevation head (H_e) is elevation of well base (J) above sea level

A. Early stage of karst development

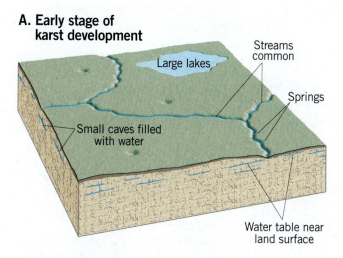

Large lakes

Streams common

Springs

Small caves filled with water

Water table near land surface

B. Mid-stage of karst development

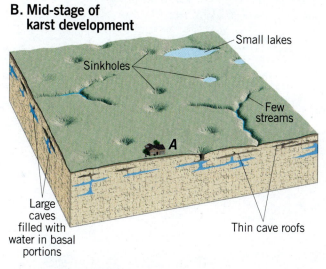

Small lakes

Sinkholes

Few streams

A

Large caves filled with water in basal portions

Thin cave roofs

C. Late (advanced) stage of karst development

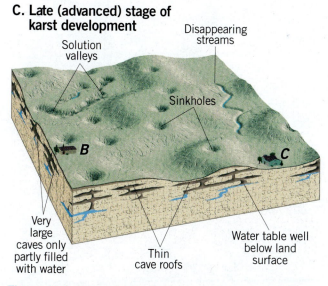

Disappearing streams

Solution valleys

Sinkholes

B

C

Very large caves only partly filled with water

Thin cave roofs

Water table well below land surface

Figure 12.6 Stages in the evolution of karst topography. Karst topography is the result of dissolution of water-soluble bedrock, which is usually limestone or dolostone but can be evaporite (e.g., gypsum or halite).

ACTIVITY 12.4

Karst Processes and Topography, (p. 341)

Think About It What is karst topography, and how does water flow beneath it?

Objective Explore and evaluate the topographic features, groundwater movements, and hazards associated with karst topography.

Before You Begin Read the following section: Caves and Karst Topography.

Plan Ahead You will need a colored pencil for this activity.

Caves, sinkholes, and underground streams are all common features of karst areas. Groundwater typically flows through networks of open joints (**Fig. 12.8**) or caverns (**Fig. 12.9**) that are enlarged by dissolution of the soluble bedrock.

Rainwater may contain several acids, but the most common is carbonic acid (H_2CO_3). It forms when water (H_2O) and carbon dioxide (CO_2) combine in the atmosphere ($H_2O + CO_2 = H_2CO_3$). All natural rainwater is mildly acidic with an average pH of around 5.4. As it infiltrates the soil, the water chemistry continues to evolve. Bacteria, other organisms, and various decay processes in soil generate CO_2 that makes soil moisture even more acidic as it eventually migrates downward toward the water table. This natural acid is able to dissolve the limestone through a complex series of processes that can be summarized in a deceptively simple chemical equation:

$$CaCO_3 + H_2CO_3 \leftrightarrow Ca^{2+} + 2HCO_3^-$$

| calcite | carbonic acid | calcium ions dissolved in groundwater | bicarbonate ions dissolved in groundwater |

Much of the dissolution of limestone seems to occur near or just below the water table, and the resulting ions are carried away in solution by the flowing groundwater. In parts of a cavern system that are persistently above the water table, groundwater percolating downward from the ground surface with a dissolved load of calcium and carbonate ions can evaporate inside the cavern, causing precipitation of the ions to form calcite grains. The tiny calcite grains accumulate to form stalactites (on the ceiling), stalagmites (on the floor), and an almost limitless assortment of other cave features (e.g., **Fig. 12.9**).

Figure 12.7 Orthophoto map of an area affected by karst. Portion of the Park City, Kentucky, 7.5-minute orthophoto quadrangle map by the USGS (2010). The grid squares are 1 km².

Figure 12.8 Water flow through fractures. Looking east toward the Arkansas River from Vap's Pass, Oklahoma, 15 miles northeast of Ponca City. The Fort Riley Limestone is exposed at the surface here. There is no soil, but plants have grown naturally along linear joints or fractures in the bedrock.

Figure 12.9 Stalactites. These stalactites formed on part of the ceiling of Cave of the Winds, which has formed in Paleozoic limestones near Manitou Springs, Colorado. Notice how they cluster along lines in the cave ceiling, probably indicating that water is passing through joints or fractures in the overlying rock and dripping into the cave.

Typical karst topography has the following features, which are illustrated in **Fig. 12.6** and visible on the USGS orthoimage of the Park City Quadrangle in Kentucky in **Fig. 12.7**.

- **Sinkholes**—surface depressions formed through several different processes including the collapse of caves or other large underground void spaces.
- **Solution valleys**—valleylike depressions formed by a linear series of sinkholes or collapse of the roof of a linear cave.
- **Springs**—places where water flows naturally from the ground because the water table is above the ground surface.
- **Disappearing streams**—streams that drop below the ground surface, flowing into open joints or caverns.

ACTIVITY 12.5

Floridan Aquifer System, (p. 343)

Think About It What is karst topography, and how does water flow beneath it?

Objective Construct a water-table contour map and determine the rate and direction of groundwater movement.

Before You Begin Read the following section: The Floridan Aquifer System. Refer back to the Introduction and Caves and Karst Topography section as needed.

Plan Ahead You will need a calculator, ruler, pencil for initial sketching of contours, and a pen.

The Floridan Aquifer System

Thick limestone beds that are susceptible to dissolution by acidic groundwater underlie most of the state of Florida. **Fig. 12.10** shows karst features developed in the Floridan aquifer system in the northern part of Tampa, Florida. The many lakes and ponds in this area occupy old sinkholes, and many other sinkholes can be recognized on the map as dry closed depressions indicated by the hachured contours—contours with small tick marks that point inward, indicating a closed depression.

By determining and mapping the elevations of water surfaces in the lakes, you can determine the slope of the water table and the direction of groundwater flow. This is because the surface water in all of these lakes is directly connected to the groundwater aquifer system through a network of open joints and caverns. It is somewhat like having a set of swimming pools throughout a neighborhood in which all of the pools are interconnected through meterwide pipes so that water can freely flow from pool to pool. More information about sinkholes in west-central Florida and the Floridian aquifer system is available in an article by Anne Tihansky in USGS Circular 1182, which is accessible via **http://fl.water.usgs.gov/PDF_files/cir1182 _tihansky.pdf.**

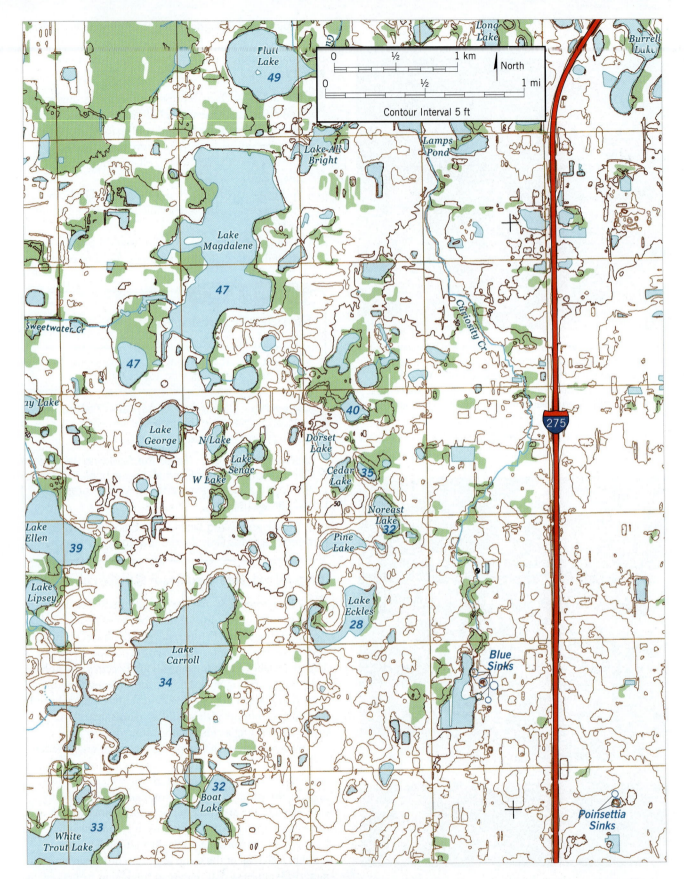

Figure 12.10 Topographic map in a karst area. Portion of the Sulphur Springs, Florida, 7.5-minute topographic quadrangle map by the USGS (2015). The center of this map is located approximately at 28.068°N, 82.472°W. Water-surface elevations for several of the lakes are indicated in feet above sea level. The grid squares are 1 km².

ACTIVITY 12.6

Land Subsidence from Groundwater Withdrawal, (p. 345)

> **Think About It** What can happen if groundwater is withdrawn faster than it is replenished?

Objective Evaluate the way that groundwater withdrawal can cause subsidence (sinking) of the land.

Before You Begin Read the following section: Land Subsidence Hazards Caused by Groundwater Withdrawal.

Plan Ahead You will need a ruler, a calculator, a pen, and a pencil for this activity.

Land Subsidence Hazards Caused by Groundwater Withdrawal

Land subsidence caused by human withdrawal of groundwater is a serious problem in many places throughout the world. For example, in the heart of Mexico City, the land surface has gradually subsided more than 13.5 m (>44 ft.). At the northern end of California's Santa Clara Valley, about 17 square miles (mi.²) of land have subsided below the highest tide level in San Francisco Bay and now must be protected by earthworks. Other centers of subsidence include New Orleans, Houston, Tokyo, Venice, and Las Vegas. With increasing withdrawal of groundwater and more intensive use of the land surface, we can expect the problem of subsidence to become more widespread.

Subsidence in the Santa Clara Valley

During the first half of the 20th century, the Santa Clara Valley of California was an important agricultural area that depended on groundwater for irrigation (**Fig. 12.11**). It was

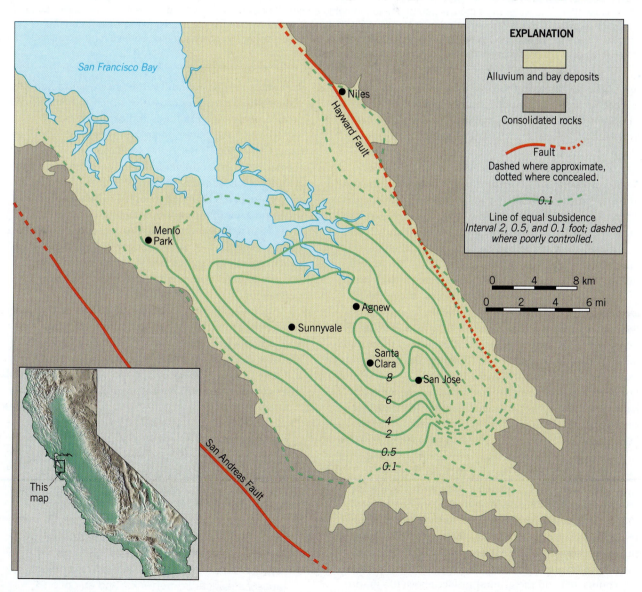

Figure 12.11 Land subsidence, 1934–1967, in the Santa Clara Valley, California. (Map data from J.F. Poland and R.L. Ireland, 1988, USGS Professional Paper 497-F, Fig. 21.)

Laboratory 12 Groundwater Processes, Resources, and Risks ■ **335**

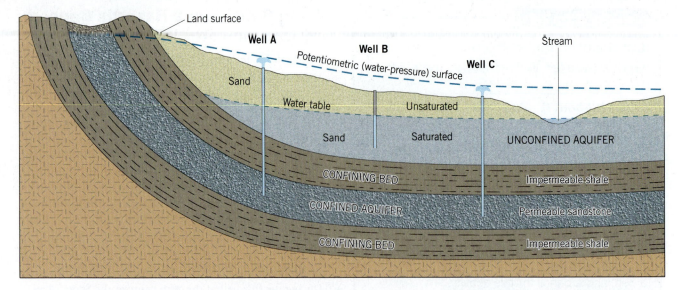

Figure 12.12 Geologic cross-section of unconfined and confined aquifers. The upper unconfined aquifer has a water table that is separate and independent of the water table for the confined aquifer below it. Well B in the unconfined aquifer has a water level that does not reach the ground surface. Water enters the confined aquifer from a recharge area in the highlands on the left side of the illustration. The water table there is part of a potentiometric surface that extends away from the recharge area. The slope of that surface away from the recharge area is caused by wells that produce water from the confined aquifer. Wells A and C are flowing artesian wells that do not need to be pumped because the top of the well is below the potentiometric surface for the confined aquifer from which they are producing water. Vertical scale is exaggerated.

one of the first areas in the United States where land subsidence due to withdrawal of groundwater was recognized. The Santa Clara Valley is a large structural trough filled with sand and gravel transported primarily by streams, along with muddy beds that are rich in montmorillonite clay, totaling more than 460 m (1500 ft.) in thickness. Montmorillonite is a clay that expands dramatically when wet and contracts when dry. Sand-and-gravel aquifers predominate near the valley margins, but the major part of the sediment in the middle of the valley is silt and clay.

Initially, at least some of the wells that reached the confined aquifer 150–365 m (500–1200 ft.) below the ground surface in the Santa Clara Valley were flowing artesian wells similar to wells **A** and **C** in **Fig. 12.12** because the recharge area along the valley margins was above the land surface in the middle of the valley. However, drilling of as many as 1700 wells per year for irrigation in the 1920s lowered the groundwater level dramatically. Whereas flowing artesian wells existed in the San Jose area in the late 1880s, the USGS monitoring well in downtown San Jose indicated the water level had fallen to 235 ft. below ground level by 1964. In 85 years since 1910, the elevation of the ground surface at the P7 benchmark in downtown San Jose had subsided by 14 ft. due to groundwater withdrawal and compaction of sediments in both the aquifer and the clay-rich aquitard layers.

Land subsidence is related to the compressibility of water-saturated sediments. Withdrawing water from wells that tap into confined and unconfined aquifers not only removes water from the aquifer system, but it also reduces the water pressure in the aquifers and confining beds. As the water pressure is reduced, the aquifer system is gradually compacted and the ground surface above it is gradually lowered. The water pressure can be restored in the aquifer system by replenishing (or **recharging**) the aquifer

with water. Permanent land subsidence results when the reduced fluid pressures cause the largely irreversible compaction of clay-rich interbeds and confining beds.

Through the extraordinary and effective efforts of the Santa Clara Valley Water District and cooperating agencies, the decline in groundwater levels was reversed starting in ~1969 to such an extent that by 1995, groundwater levels in San Jose had rebounded to just 35 ft. below ground level, and a few of the once-flowing artesian wells in the area began to flow at the ground surface again. Subsidence of the ground surface was also largely halted by 1969, but the changes in ground-surface elevation caused by excessive groundwater pumping prior to 1969 are likely to be permanent.

Additional information about the successful efforts to restore the groundwater system in the Santa Clara Valley and to stop additional subsidence of the ground surface is available in an article by S.E. Ingebritsen and David Jones in USGS Circular 1182, which is accessible via **http://pubs .usgs.gov/circ/circ1182/pdf/05SantaClaraValley.pdf.**

Name: _____ **Course/Section:** _____ **Date:** _____

The rate of groundwater flow depends on properties of the rock and on the pressure gradient that drives the flow.

A **EXPERIMENT 1** Analyze what happened when water was dropped on the rock specimens **Fig. A12.1.1**.

1. In general, what effect do you think layers of fine-grained sedimentary rock like mudstone or shale are likely to have on the speed of groundwater movement or the volume of groundwater that can be stored? Why?

2. In general, what effect do you think layers of sandstone are likely to have on the speed of groundwater movement or the volume of groundwater that can be stored? Why?

EXPERIMENT 1: What happens when a drop of water is applied to shale and sandstone?
Procedure: A drop of water was placed on four different rocks. This is what happened after 5 seconds.

| Shale | Sandstone | Shale | Sandstone |

Figure A12.1.1

B **EXPERIMENT 2** Obtain two pieces of 1.25 cm (½ in.) inside-diameter clear-plastic flexible tubing, each about 1 m long, from the plumbing section of a hardware store. In one tube, put enough sand to fill perhaps 10 cm of the tube's length, and in the other tube place a similar amount of finer-grained dry soil that is at least half clay-sized particles. Put a ball of cotton loosely on each side of the sediment to hold the sediment in place (**Fig. A12.1.2A**). Hold each tube so that it is in the shape of a U with the sediment at the bottom, and gently add water to both sides until the water reaches about a third of the way up each side (**Fig. A12.1.2B**).

1. Starting with the tube that contains sand, lift one side of the tube ~10–20 cm above the other so that water from the high side flows into the lower side (**Fig. A12.1.2C**) and measure the time needed for the water levels in the two sides to return to the same elevation. _____ sec

2. Do the same experiment with the tube that contains the finer sediment. _____ sec

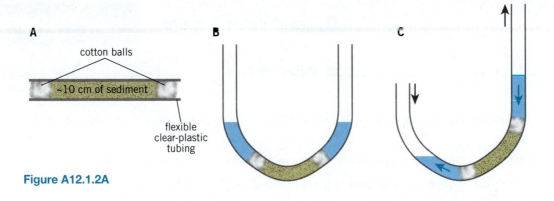

Figure A12.1.2A

3. How does the rate at which water flows through sand differ, if at all, from the rate at which water flows through the finer-grained sediment? Why?

C **EXPERIMENT 3** Analyze the procedures in Fig. A12.1.3 and then use a pencil to plot the data on the graph. When you are confident in the accuracy of your plot, use a pen to make your data points more visible. How is the distance of the water jet related to the height of water in the bottle? Why?

EXPERIMENT 3: What happens when water drains from a hole in the side of a bottle?

Procedure:
1. A small hole was punched in the side of a 2-liter plastic bottle, 6 cm above its bottom.
2. Tape was placed over the hole.
3. The bottle was filled with water to a height of 22 cm.
4. The tape was removed, and a jet of water shot out of the hole. The distance that the water jet shot from the bottle to the table top was recorded for specific water heights in the bottle.
5. Plot the data on the graph paper to see if there is a trend.

Water height in bottle (cm)	Water jet distance (cm)
22.0	10.0
18.5	9.5
16.5	9.0
14.0	8.0
13.0	7.5
11.0	6.0
10.0	5.5
9.0	4.5
8.0	3.5
7.5	2.5
7.0	2.0
6.0 (hole height)	0

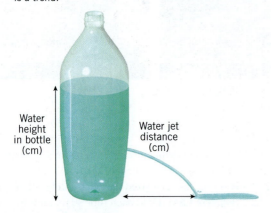

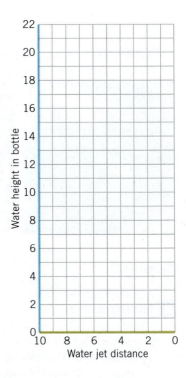

Figure A12.1.3

D **REFLECT & DISCUSS** In the initial moments of Experiment 3 after the hole was first opened, the jet of water seemed to have a lot of energy as it shot through the air. Imagine changing the experiment so that instead of shooting into the air, the water entered a clear plastic tube that curved back up toward the top of the bottle. About how high do you think the water would rise in the tube as compared with the water level in the bottle? (*Hint:* Think about the water levels in the two sides of the plastic tube in Experiment 2.)

Where Is the Nasty Stuff Going?

Activity 12.2

Name: _____ Course/Section: _____ Date: _____

Figure A12.2.1 is a map showing a grid of well-located sites where the elevation of the water table has been measured. The water-table elevations are listed (in ft.) next to the piezometer locations. The elevations were measured relative to a local permanent benchmark whose location and elevation were determined by a professional surveyor. The aquifer is a permeable sandstone with no confining layer in the map area.

A Using a pencil, make a topographic contour map of the water table at a contour interval of 1 foot. When you think your contours are accurate, label and trace over your contours with a pen. Parts of the 4-foot and 8-foot contours are done for you.

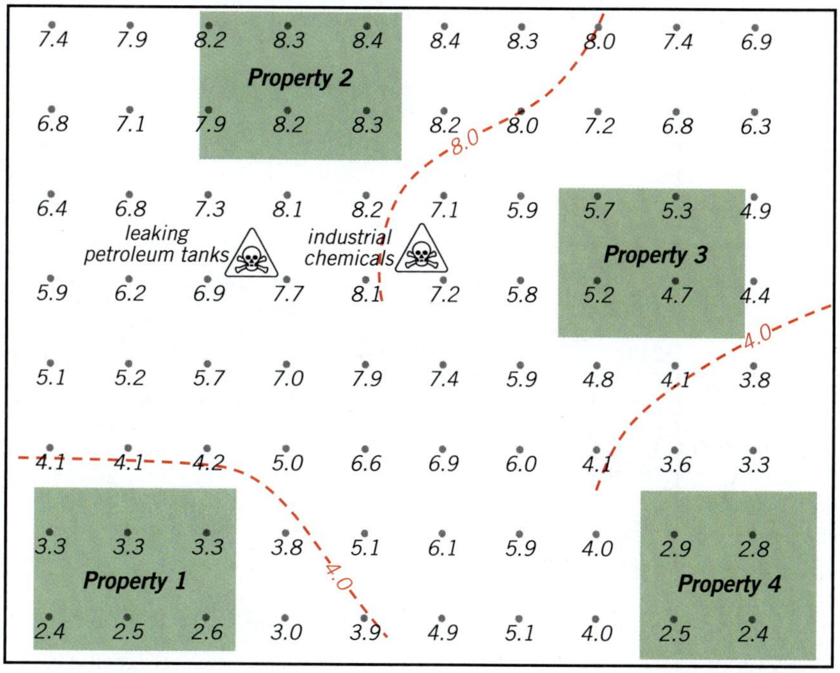

Figure A12.2.1

B Four properties are for sale in the map area, and you need to choose one for new athletic fields that will be used by thousands of children over many years. All properties have private wells, so well water would be used to irrigate the playing fields and for food services at the facility.

   Unfortunately, there are two sources of groundwater pollution in the area: one that has leaking underground petroleum tanks and another where industrial chemicals were improperly stored and disposed. Pollutants from these sites are toxic and can be expected to move with the groundwater. The location of the pollution sources are indicated by symbols on the map. Groundwater along the water table flows perpendicular to the water-table contours from high to low elevation (**Fig. 12.3**). Use the water-table contours as a guide, and draw a flow line from each of the pollution sources that extends to the edge of the map.

C Using the flow lines you just constructed, choose the property where children will be least likely to be endangered by the toxic chemicals.

Activity 12.3 — Using Data to Map the Flow of Groundwater

Name: _____   **Course/Section:** _____   **Date:** _____

We want to understand how groundwater flows below the water table as viewed in a vertical plane through an unconfined aquifer. At several locations along the line of section, a set of piezometers was installed that extended to different depths. This set is called a nest of piezometers. The elevation of the open (screened) base of each of the piezometers was determined—the elevation head. The height to which the water rose in each piezometer was also measured—the pressure head. The total head can be determined by adding the elevation head and the pressure head.

The profile of the ground surface, the water table (marked by the triangles), and the total head at dozens of points along the line of section are given in **Fig. A12.3.1**. Elevations and heads are given in feet.

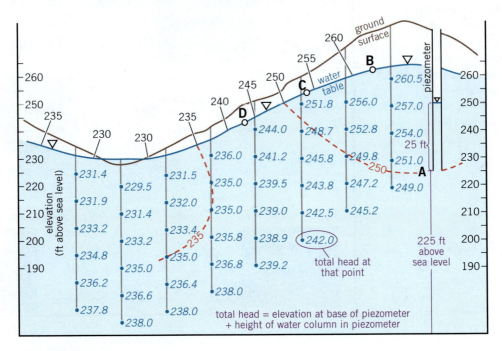

Figure A12.3.1

A Examine the piezometer on the right side of the section at whose base is point **A**.

1. What is the *pressure head* for point **A**? Refer to **Fig. 12.4** if you need help. _____ ft.

2. What is the *elevation head* for point **A**? _____ ft.

3. What is the *total head* for point **A**? _____ ft.

B Using the total head for point **A** and all of the other total head values noted on the section, construct contours of equal total head at a contour interval of 5 ft., similar to the contours shown as dashed red curves in **Fig. 12.5**. It is best to start contouring in pencil and then go over your contours in pen after you are confident about your work. Parts of the 235 ft. and 250 ft. contours are provided in **Fig. A12.3.1** to help you get started.

C The contours you just constructed are part of 3-dimensional equal total-head (or *potentiometric*) surfaces. Groundwater flows perpendicular to these potentiometric surfaces from high total head to low total head (**Fig. 12.3**). Draw a flow line from each of the labeled points—**A**, **B**, **C**, and **D**—and continue the flow lines until they end at the water-table surface or reach the edge of the cross-section.

Name: _____ Course/Section: _____ Date: _____

A Analyze **Figs. 12.8** and **12.9**.

1. In the area photographed in **Fig. 12.8**, there is little or no soil developed on the limestone bedrock surface, yet abundant plants are growing along linear features in the bedrock. What does this indicate about how water travels through bedrock under this part of Oklahoma?

2. If you had to drill a water well in the area pictured in **Fig. 12.8**, where would you drill (relative to the pattern of plant growth) to find a good supply of water? Why?

3. **REFLECT & DISCUSS** How is **Fig. 12.9** related to **Fig. 12.8**?

B It is common for buildings to sink into newly formed sinkholes as they develop in karst regions. Consider the three new home construction sites (labeled **A**, **B**, and **C**) in **Fig. 12.6**, relative to sinkhole hazards.

1. Which new home construction site (**A**, **B**, or **C**) might be the *most* hazardous? Why?

2. Which new home construction site (**A**, **B**, or **C**) might be the *least* hazardous? Why?

3. **REFLECT & DISCUSS** Imagine that you are planning to buy a vacant lot on which to build a new home in the region portrayed in **Fig. 12.6**. What could you do to find out if there is a sinkhole hazard in the location where you are thinking of building your home?

C Study the orthophoto map of part of the Park City (Kentucky) 7.5-minute quadrangle in **Fig. 12.7**. The center of this map area is located at 37.073°N, 86.090°W. This entire area is underlain by limestone, although the limestone is overlain by sandstone in the small northern part of this image (Bald Knob, Opossum Hollow) that is covered by dense dark green trees.

1. How can you tell in the area on this orthoimage where limestone crops out at Earth's surface?

2. Recall that on a topographic map, a depression is shown by a contour line with hachures (tic marks) that form a closed loop. Describe the pattern of depressions on the topo map of Park City (**Fig. A12.4.1**). Why do some of the depressions contain ponds, but others do not?

3. **REFLECT & DISCUSS** Notice that there are many naturally formed circular ponds in the northwest half of the image. (The triangular ponds are surface water impounded behind dams constructed by people.) How could you use the elevations of the surfaces of the ponds to determine how groundwater flows through this region?

D Refer to **Fig. A12.4.1**, which is a topographic map of the northernmost ~75% of the map area in **Fig. 12.7**.

1. Compare the map and orthoimage, and then use a pencil to trace on **Fig. A12.4.1** the contact that separates limestone with karst topography from forested, more resistant sandstone. Color the sandstone bedrock with a colored pencil on **Fig. A12.4.1**.

2. Gardner Creek is a *disappearing stream*. On **Fig. A12.4.1**, place arrows along Gardner Creek to show its direction of flow, and then circle the location where it disappears underground. Circle the disappearing end of two other disappearing streams in the southeast quarter of the map.

3. Find and label a solution valley anywhere on **Fig. A12.4.1**.

341

Figure A12.4.1

4. **REFLECT & DISCUSS**   Notice that a pond has been constructed on the sandstone bedrock on top of Bald Knob and filled with water from a well. If the well is located on the edge of the pond, how deep below that surface location was the well drilled just to reach the water table? Show your work.

Name: _____ **Course/Section:** _____ **Date:** _____

Refer to **Figs. 12.10** (Sulphur Springs Quadrangle) and **A12.5.1**.

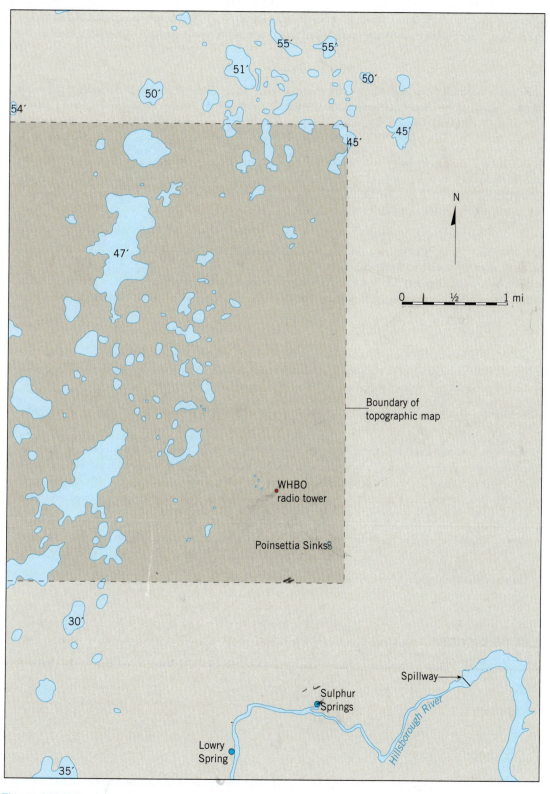

Figure A12.5.1

A On **Fig. A12.5.1**, mark the elevations of water levels in the lakes (obtain this information from **Fig. 12.10**). The elevations of Lake Magdalene and some lakes beyond the boundaries of the topographic map are marked already for you.

B Contour the water-table surface (use a 5-foot contour interval) on **Fig. A12.5.1**. Draw only contour lines representing whole fives (40, 45, and so on). Do this in the same manner that you contoured land surfaces in the topographic maps lab.

C The flow of shallow groundwater in **Fig. A12.5.1** is at right angles to the contour lines. The groundwater flows from high elevations of the water table to lower elevations, just like a stream. Draw three or four flow lines with arrows on **Fig. A12.5.1** to indicate the direction of shallow groundwater flow in this part of Tampa. The southeastern part of **Fig. 12.10** shows numerous closed depressions but very few lakes. What does this indicate about the level of the water table in this region?

D Note the Poinsettia Sinks, a pair of sinkholes with water in them in the southeast corner of the topographic map (**Fig. 12.10**). Next, find the cluster of five similar sinkholes, called Blue Sinks, about 1 mile northwest of Poinsettia Sinks (just west of the WHBO radio tower). Use asterisks (*) to mark their locations on **Fig. A12.5.1**, and label them Blue Sinks.

E On the sketch map, draw an arrow (vector) along the shortest path between Blue Sinks and Poinsettia Sinks. The water level in Blue Sinks is 15 ft. above sea level, and the water level in Poinsettia Sinks is 10 ft. above sea level. Calculate the hydraulic gradient (show your work below in ft./mi.) along this arrow and write it next to the arrow on the sketch map. (Refer to the hydraulic gradient in **Fig. 12.1** if needed.)

F On **Fig. 12.10**, note the stream and valley north of Blue Sinks. This is a fairly typical disappearing stream. Draw its approximate course onto **Fig. A12.5.1**. Make an arrowhead on one end of your drawing of the stream to indicate the direction that water flows in this stream. How does this direction compare to the general slope of the water table?

G In March 1958, fluorescent dye was injected into the northernmost of the Blue Sinks. It was detected 28 hr. later in Sulphur Springs on the Hillsborough River to the south (see sketch map). Use these data to calculate the approximate velocity of flow in this portion of the Floridan aquifer system (show your work):

1. in feet/hour: _____   **2.** in miles/hour: _____   **3.** in meters/hour: _____

H The velocities you just calculated are quite high, even for the Floridan aquifer system. But this portion of Tampa seems to be riddled with solution channels and caves in the underlying limestone. Sulphur Springs has an average discharge of approximately 44 cubic feet per second (cfs), and its maximum recorded discharge was 165 cfs (it once was a famous spa). During recent years, the discharge at Sulphur Springs has decreased. Water quality has also worsened substantially.

1. The area shown in **Fig. 12.10** is a fully developed suburban residential community located about 8 mi. north of downtown Tampa. Why do you think the discharge of Sulphur Springs might have decreased in recent years?

2. Why do you think the water quality has decreased in recent years?

I **REFLECT & DISCUSS** Name two potential groundwater-related hazards that might have a negative impact on homeowners in this area.

Name: _____ **Course/Section:** _____ **Date:** _____

A Santa Clara Valley, California.

1. Based on **Fig. 12.11**, where are the areas of greatest subsidence in the Santa Clara Valley?

2. What was the total subsidence at San Jose (**Fig. A12.6.1**) from 1934 to 1967 in feet?

Year	Total Subsidence (feet) from 1912 level
1912	0.0
1920	0.3
1934	4.6
1935	5.0
1936	5.0
1937	5.2
1940	5.5
1948	5.8
1955	8.0
1960	9.0
1963	11.0
1967	12.7

Figure A12.6.1 Subsidence at benchmark P7 in San Jose, California.

3. What was the average annual rate of subsidence for the period of 1934 to 1967 in feet/year?

4. Analyze **Fig. 12.11**. At what places in the Santa Clara Valley would subsidence cause the most problems? Explain your reasoning.

5. Would you expect much subsidence to occur in the darker shaded (tan) areas of **Fig. 12.11**? Explain.

6. By 1960, the total subsidence at San Jose had reached 9.0 ft. (**Fig. A12.6.1**). What was the average annual rate of subsidence (in feet/year) for the seven-year period from 1960 through 1967? Show your work.

7. Refer to **Fig. A12.6.2**, which shows the variation in several factors relevant to the subsidence of the Santa Clara Valley since ~1915.

   (a) Using **Fig. A12.6.2A**, about how much has the elevation of benchmark P7 changed since 1970? _____ ft.

   (b) The water table in San Jose was just about at the ground surface in 1915 but its depth steadily increased to a maximum in the mid-1960s. Using **Fig. A12.6.2B**, estimate the maximum water-table depth below the ground surface. _____ ft.

   (c) Since the mid-1960s, the water-table depth has steadily decreased. What effect has the rising water table had on the elevation of the ground surface in San Jose?

   (d) Think about the trend in precipitation since ~1916 shown in **Fig. A12.6.2C**. Precipitation and snowmelt recharge the aquifer. What effect might the observed trend in precipitation have had on the water-table depth?

   (e) Groundwater has been pumped in support of agriculture throughout the 20th century. Examine **Fig. A12.6.2D** and describe any trend that you can see in the rate of pumping between 1915 and the mid-1960s. What effect did that have on the water-table depth?

(f) Changes in water policy in the mid-1960s resulted in projects to artificially recharge the aquifer, using water brought to the area from the Sierra Nevada Mountains by aqueducts. Examine **Fig. A12.6.2D** to determine how the rate of pumping changed since the mid-1960s. What effect did artificial recharge and changes in pumping seem to have had in the Santa Clara Valley?

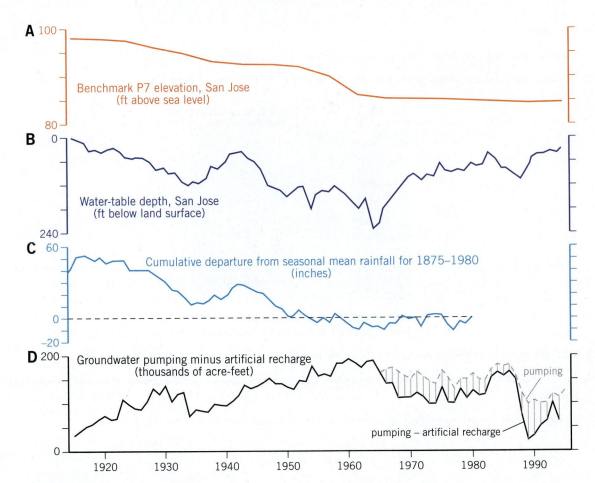

Figure A12.6.2 **Variation in several factors affecting groundwater and subsidence in the Santa Clara Valley, California.**

B **REFLECT & DISCUSS**   The Santa Clara Valley and surrounding areas have changed from being an agricultural area into a global center for digital technology that is commonly known as Silicon Valley. How might water usage change in an area like this as it transitions from agricultural to urban?

Glaciers and the Dynamic Cryosphere

Pre-Lab Video 13

https://goo.gl/U610dh

Contributing Authors

Sharon Laska • *Acadia University*

Kenton E. Strickland • *Wright State University–Lake Campus*

Nancy A. Van Wagoner • *Acadia University*

▲ Agassiz Glacier (left) and Malaspina Glacier (right) in Alaska are the largest of several glaciers in this NASA Landsat 8 image. Both begin as valley glaciers near the top of the image but then leave the mountains to flow across a flat coastal plain, becoming piedmont glaciers in which the ice forms a wild assortment of folds. The image center is near 60.05°N, 140.70°W.

BIG IDEAS

The cryosphere—made up of ice, snow, and frozen ground—is the part of the solid-Earth system that is at or below 0°C. It is both an important source of historical information about environmental change over the past ~800,000 years and a sensitive indicator of recent climate change resulting from human activities. Sea level varies by hundreds of meters between times when Earth is heavily glaciated compared with times when there is little or no persistent ice on the surface. During the past century, mountain glaciers in the Cascades, Sierra Nevada, Rocky Mountains, and elsewhere have receded measurably. An important effect of increased melting of continental glaciers, ice caps, and ice sheets is an increase in sea level that will endanger low-lying coastal communities worldwide.

FOCUS YOUR INQUIRY

Think About It What is the cryosphere, and how does the extent of sea ice vary over time?

ACTIVITY 13.1 The Cryosphere and Sea Ice (p. 348, 367)

Think About It How do mountain glaciers affect landscapes?

ACTIVITY 13.2 Mountain Glaciers and Glacial Landforms (p. 351, 369)

Think About It How might glaciers be affected by climate change?

ACTIVITY 13.3 Nisqually Glacier Response to Climate Change (p. 361, 371)

ACTIVITY 13.4 Glacier National Park Investigation (p. 361, 373)

Think About It How do continental glaciers affect landscapes?

ACTIVITY 13.5 Some Effects of Continental Glaciation (p. 363, 374)

Introduction

On February 14, 1990, the camera on the Voyager 1 spacecraft was directed back toward Earth, 6 billion km away, for one last photograph before the imaging system was shut down to save power. The resulting image has 640,000 pixels of which Earth occupied less than 1. Reflected light modified by water vapor, liquid water, and ice in the Earth environment provides the ultimate image of our world as the water planet—a pale blue dot in the vastness of space.

Our global environment is dominated by water, and as water transitions among its three phases (solid, liquid, gas), it is a sensitive indicator of temperature. The **cryosphere** is the part of Earth's interconnected systems that involves frozen water as well as rock that is persistently below the freezing point of water (**Fig. 13.1**). We monitor variations in the cryosphere with the same interest that a physician monitors a patient so that we can use our improved knowledge of environmental change to help society make good decisions about living sustainably within a global climate system.

It has become essential for us to understand better how Earth's climate has changed over time and how human activities affect climate. The cryosphere not only is affected by a changing climate but also provides us with a detailed historical record of climate change during most of the past million years through the study of ice cores. Ice near the bottom of a deep core collected from Dome C in Antarctica (75.01°S, 123.35°E) by the *European Project*

for Ice Coring in Antarctica is more than 800,000 years old and contains a wealth of information about the last eight glacial cycles. Combined with other ice cores collected in Antarctica and Greenland, and information from marine sediment cores and continental glacial deposits, we are compiling a detailed picture of how Earth's cryosphere and climate have varied over time.

ACTIVITY 13.1

The Cryosphere and Sea Ice, (p. 367)

Think About It   What is the cryosphere, and how does the extent of sea ice vary over time?

Objective   Analyze global and regional components of the cryosphere and then infer how they may change and ways that such change could affect other parts of the Earth system.

Before You Begin   Read the following sections: The Cryosphere and The Cryosphere and Climate Change.

Plan Ahead   You will need a pencil, ruler, and a calculator.

Map of Regional Variations in the Cryosphere

ICE SHELF: A sheet of ice attached to the land on one side but afloat on the ocean on the other side.

SEA ICE: A sheet of ice that originates from the freezing of seawater.

SEASONAL SNOW: Snow and ice may accumulate here in winter, but it melts over the following summer.

PERMAFROST CONTINUOUS: The ground is permanently frozen over this entire area.

PERMAFROST DISCONTINUOUS: The ground is permanently frozen in isolated patches within this area.

MOUNTAIN GLACIERS AND ICE CAPS: This area contains permanent patches of ice on mountain sides (cirques), river-like bodies of ice that flow down and away from mountains (valley and piedmont glaciers), and dome-shaped masses of ice and snow that cover the summits of mountains so that no peaks emerge (ice cap).

ICE SHEET: A pancake-like mound of ice covering a large part of a continent (more than 50,000 km²).

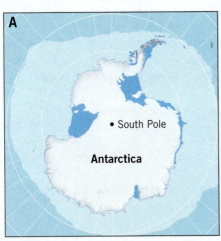

Figure 13.1 Cryosphere components. These maps show the general distribution of the most important parts of the cryosphere: the parts of Earth where there is frozen water. **A.** Southern hemisphere: Antarctica. **B.** Northern hemisphere: Arctic and adjacent areas. (Courtesy of UNEP/GRID-Arendal.)

The Cryosphere

Parts of the active cryosphere can be found on most continents although it is most evident in the polar regions. The cryosphere has several major components.

Seasonal Snow

Frozen precipitation falls throughout the temperate and polar regions and at high altitudes anywhere that there is sufficient atmospheric moisture. Most seasonal snow melts within days or weeks, but some snow can accumulate to form ice that can last for years.

Mountain Glaciers, Ice Caps, and Ice Fields

A **glacier** is a volume of ice that persists for many years, that has a surface area of at least 0.1 km², and that is thick enough (usually >50 m thick) so that it is able to flow. **Mountain glaciers** develop in cooler, higher-altitude source areas where there is sufficient snow to sustain them. There are small glaciers on Mount Kilimanjaro (5895 m) near the equator in Tanzania (3.065°S, 37.358°E). The elevation necessary to support development of a mountain glacier decreases with increasing distance from the equator so that glaciers at 60°N or S latitude can originate very near to sea level. An **ice cap** has a surface area of <50,000 km² and covers the summit area of a mountainous area so that no peaks emerge from the ice surface. An **ice field** is similar to an ice cap, but at least some peaks are above the ice surface. Both ice caps and ice fields can be source areas for glaciers that flow along established valleys extending from the summit area.

Ice Sheets and Ice Shelves

An **ice sheet**, also known as a **continental glacier**, is a very large mass of ice with a surface area of more than 50,000 km² that can be more than 3 km thick and that flows under its own weight. Ice sheets can be so massive that their weight depresses Earth's crust, forming a bowl-shaped depression. They flow from the thickest part in the center of the ice sheet outward toward their thinner margins. Ice sheets are currently found on Greenland and Antarctica, but during the height of Pleistocene glaciation, there were also ice sheets on northern Europe, North America, and South America.

An **ice shelf** originates on a continent but flows out onto the sea surface. While it is connected to the land-based glacier, an essential characteristic of an ice shelf is that it is afloat. Blocks that break away from an ice shelf form icebergs. Ice sheets and ice shelves are composed of fresh water because they were originally formed of accumulated snow.

Permafrost and Seasonal Frozen Ground

A volume of rock or sediment whose temperature persists at 0°C or less for many years, regardless of whether it has much or any frozen water in it, is called **permafrost**. As much as 20% of Earth's land mass is underlain by permafrost. Rock or sediment that is frozen for at least part of the year but that rises above freezing during the summer is called **seasonal frozen ground**. It is analogous to permafrost and underlies as much as 50% of the land mass in the northern hemisphere at the height of winter but thaws each year.

River Ice and Lake Ice

In areas where winter temperatures dip below freezing for extended periods of time, water on the surface of rivers and lakes can freeze. This ice thaws during the following spring or summer.

Sea Ice

As its name implies, sea ice forms on the sea surface and is predominantly composed of frozen seawater. Snow can also contribute to the thickness of existing sea ice. The largest masses of sea ice occur in the Arctic Ocean and around the continent of Antarctica (**Fig. 13.2**). In both locations, sea ice reaches its maximum thickness and extent during the winter months and then melts back to a minimum extent and thickness during the summer months. Arctic sea ice reaches its minimum thickness and extent by September whereas the sea-ice minimum around Antarctica occurs in February. Sea ice helps moderate Earth's climate because its bright white surface reflects sunlight back into space. Without sea ice, the ocean absorbs the sunlight and warms up. Sea ice in the Arctic also provides the ideal environment for animals like polar bears, seals, and walruses to hunt, breed, and migrate as survival dictates. Some Arctic human populations rely on subsistence hunting of such species to survive.

The Cryosphere and Climate Change

Climate is the characteristic weather of a region as experienced over a significant time interval—decades, centuries—with particular emphasis on temperature and precipitation. The long time frame of climate is the factor that differentiates it from weather, which reflects precipitation, wind, atmospheric pressure, and temperature at a given place and time. Major cycles of cooling and warming that are about 100,000 years long have dominated the climate during the past ~2.6 million years—that is, during the Quaternary period and primarily during the Pleistocene epoch of geologic time. The long cold weather parts of these cycles are called **glacial periods** or **glacial stages**, and the warmer period between two glacial periods is called an **interglacial period**. The current major interglacial period during the Holocene epoch has lasted for the last ~11,700 years. Through study of ice cores, we are developing a good understanding of the last eight glacial stages although we know the most about the recent Wisconsin (or Wisconsinan) stage.

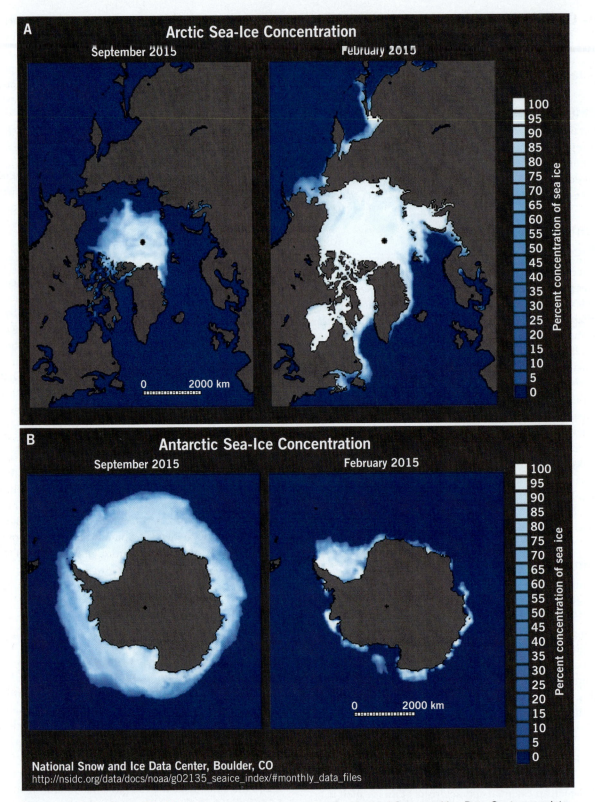

Figure 13.2 Sea ice at Earth's poles. These maps made by the National Snow and Ice Data Center use data from orbiting satellites that measure the concentration of sea ice. A 15% concentration for a given area means that 15% of the sea surface is covered in ice. **A.** Sea ice in the northern hemisphere, summer and winter 2015. **B.** Sea ice in the southern hemisphere, around Antarctica, winter and summer 2015.

Milankovitch Cycles

The periodic timing of the Pleistocene glaciations in ~100,000-year cycles is interpreted to be controlled by the variation in certain characteristics of Earth's orbit around the Sun and the variation in tilt of its axis. The contribution to these astronomical causes to variation in Earth's surface temperature was worked out in detail by Milutin Milanković between 1912 and the early 1940s, so we call these Milanković cycles or, more commonly, **Milankovitch cycles**. The cooling part of Milankovitch cycles results in shorter and cooler summers in the northern hemisphere. That increases the chance that winter accumulations of ice formed from snow on the continents of the northern hemisphere will persist through the following summer. Accumulation of this persistent ice over the course of many years results in formation of continental ice sheets and glaciers.

Changes in the abundance of greenhouse gases in the atmosphere amplify the Mikankovitch effect on climate. Such changes seem to occur in tandem with the cooling and warming of continental landmasses in the northern hemisphere.

Geoscience is not aware of global-scale "ice ages" during the ~260 Myr before the Quaternary period. Hence, Milankovitch cycles are not the only factor that controls the occurrence of glacial periods. Changes in ocean circulation and the position of the continents relative to the poles and equator—that is, changes related to plate tectonics—are interpreted to be able to turn these ice ages on and off over longer time intervals.

The Cryosphere and Climate Systems

The various elements of the cryosphere are very sensitive to changes in the temperature of shallow ocean water and to air temperature at Earth's surface. At the same time, changes in the cryosphere exert feedback effects on Earth's climate that can be quite complex. For example, an increase in snow, ice on continents, sea ice, and white clouds of water vapor in the atmosphere makes Earth's surface whiter and more reflective (i.e., increases Earth's **albedo**), causing more of the Sun's radiation to reflect back into space. This tends to have a cooling effect. Deposition of soot or "black carbon" on ice, snow, and sea ice can lower the albedo, promote warming, and contribute to more rapid melting. Such carbon deposition is a byproduct of burning fossil fuels or forests.

Increasing temperature causes changes to the cryosphere that can trigger additional warming in a process scientists call a positive feedback—a small change in one direction causes a larger change in that same direction. For example, melting of continental glaciers and ice sheets increases the flow of fresh water into the oceans. Melting of the Greenland Ice Sheet in particular can introduce cold fresh water into the northern end of a worldwide ocean circulation pattern that seems to be driven primarily by variations in water density. Normally, cold salty water sinks in the North Atlantic Ocean near Iceland; however, the influx of cold fresh water from Greenland might interrupt this pattern, affecting the way the oceans distribute heat energy across the globe. It has been proposed that interruption of this **thermohaline ocean circulation** pattern might have led to rapid and significant changes in climate in the past.

Another example of a positive feedback involves methane (CH_4)—a **greenhouse gas** whose presence in the atmosphere causes some of the long wavelength radiation that reflects off Earth's surface to be absorbed in the atmosphere and reflected back to the surface. The result is additional warming. Methane can be trapped in a water-ice structure called a **clathrate** (also known as **methane hydrate**). Methane gas is released when methane clathrate melts. An increase in surface temperature on land and in the global oceans tends to reduce the area of permafrost and ground-covering ice, which can release trapped stores of methane when frozen biomaterials thaw and decay, leading to further warming. There are also stores of methane clathrate in marine sediments surrounding continents. Geoscientists are concerned that an increase in ocean temperature will melt the clathrate and release methane gas into the atmosphere, resulting in additional surface warming.

The Cryosphere and Effects of Anthropogenic Climate Change

Improving our knowledge and monitoring the cryosphere are essential to gaining a fuller understanding of how Earth's climate changes over time and of the effects of human activities on climate change.

In discussions of climate change, a frequently asked question is "Who cares if Earth gets a little warmer? Some people who live in the Snow Belt spend a lot of money each year to travel someplace warmer during the winter!" The problem is that a warming climate accelerates the rate of melting of glaciers, ice caps, and ice sheets on land, including the Greenland and Antarctic Ice Sheets. Water that is not now in the ocean basins will flow into the ocean basins, causing sea level to rise. Water will expand as it becomes warmer, causing an additional rise in sea level. When there were no continental glaciers or ice sheets (for example, ~65 to 100 Myr ago), sea level is estimated to have been about 250 m higher than it is today. During the most recent glacial maximum ~20,000 years ago, sea level was about 130 m lower than it is today because so much water was converted to ice on the continents. A rise in sea level of even a few meters would have disastrous effects for low-lying communities worldwide.

ACTIVITY 13.2

Mountain Glaciers and Glacial Landforms, (p. 369)

> **Think About It**   How do mountain glaciers affect landscapes?

Objective   Analyze features of landscapes affected by mountain glaciation and infer how they formed.

Before You Begin   Read the following sections: Glaciers and Mountain Glaciers.

Plan Ahead   You will need a pencil, ruler, and a calculator.

Glaciers

Recall that a glacier is a large mass of ice that persists for many years on land and that is thick enough and heavy enough to be able to flow under its own weight.

How Glaciers Form

Glaciers can form wherever the winter accumulation of snow and ice exceeds the loss of ice volume during the summer. Layers of ice accumulate from snowfall that occurs in **snowfields** or **zones of accumulation**—areas of persistent snow and ice that remain from year to year (**Fig. 13.3**). The ice becomes a glacier when it reaches at least 50 m in thickness. The loss of snow and ice is called **ablation** or **wastage** and includes the processes of melting and **sublimation**—direct change from ice to water vapor without melting.

How Glaciers Move

Glacial ice is a rock composed of grains of solid (frozen) water. Like other rocks, it deforms if it is subjected to enough differential stress. (Recall that differential stress is a system of forces in which the magnitude of the force varies systematically in different directions.) Ice that is subjected to stress equal to the weight of about 50 m of ice is able to flow like Silly Putty, so one way that glaciers move is by flow. Ice that is thinner than about 50 m is unable to move on its own and is just a block of **stagnant ice**.

The upper 50 m of a glacier is a brittle layer that breaks when stressed beyond its strength. The big cracks in the surface of a glacier are called **crevasses**. At the upper end of a mountain glacier is a large crevasse called the **bergschrund** (German for "mountain crack") that separates the flowing ice from the ice that is frozen to the bedrock of the headwall or cirque. Crevasses form in the upper layer as a glacier flows and changes shape.

Some glaciers are frozen to the rock or sediment below them (the **substrate**) in which case the motion observed at the top of the glacier is entirely due to the flow of ice below the brittle layer. Other glaciers slide over the substrate in addition to flowing within the glacier. As the temperature of the glacial system warms, a glacier can change from being frozen to the substrate to sliding. This might result in a glacier accelerating or

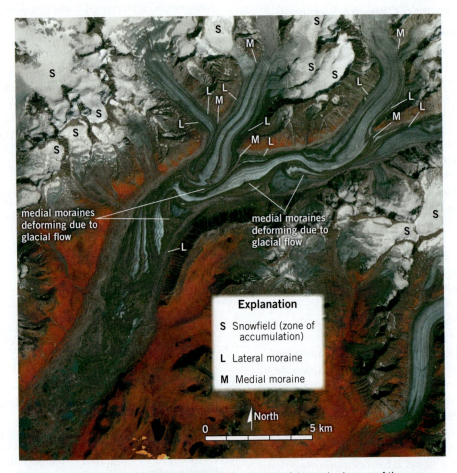

Figure 13.3 Mountain glaciers in Alaska. This is a false-color image of the area around the Susitna Glacier in the Alaska range. The red areas are covered in vegetation. White is snow, light blue-gray is ice, darker blue is water, and where rock and sediment cover the ice in moraines, the glacier appears brown. Some of the moraines are deformed by surging tributary glaciers. (The data for this image were collected by the ASTER instrument on NASA's Terra satellite and were processed by the NASA/METI/AIST/Japan Space Systems and United States/Japan ASTER Science Team.)

surging forward when it is no longer frozen to the substrate. A surging glacier can move as much as ~100 times its normal rate for days or months. Less extreme surges can persist for years but are still relatively short-lived events in the history of a glacier. The presence of melt water at the base of the glacier can also facilitate surging.

Glaciers flow from where they are thickest, in the zone of accumulation, outward to their thinner margins. For example, the **Laurentide Ice Sheet** that covered most of Canada and parts of the northern and northeastern states in the United States during the Pleistocene glaciations had its thickest part in the area now occupied by Hudson Bay in Canada where the ice was more than 3 km thick. If you were standing on the summit of the Laurentide Ice Sheet, the flow of the surface layer would have been downslope and away from you in all directions. Mountain glaciers have a zone of accumulation that is at a higher altitude than the thin outer margin of glacial ice, so flow in mountain glaciers is generally downslope.

Glacial Retreat

The leading edge of a shrinking glacier melts back closer to the glacier's source, and the surface area of the glacier decreases. This is called **glacial retreat**. Although we use the word *retreat* to describe how glaciers leave an area behind as the climate warms, glaciers have no reverse gear. The ice in a glacier is always moving in the same direction away from the zone of accumulation and on toward the far margin of the glacier. A retreating glacier is simply getting smaller because it is losing mass to melting and sublimation.

Deposition of Glacial Sediment

Deposits of rocky gravel, sand, silt, and clay accumulate where there once was ice. These deposits collectively are called **drift**. Drift that accumulates directly from the melting ice is said to be **unstratified drift** or **till**. Drift that is transported by meltwater becomes more rounded, sorted by size, and deposited in layers. This material is called **stratified drift**. Wind also can transport the sand, silt, and clay particles from drift. This wind-transported sediment can form dunes or **loess** deposits—wind-deposited, unstratified accumulations of clay and silt.

As with other topics in geoscience, many specialized words are used in the study of the cryosphere. Terms related to glaciers and glacial landforms that are frequently used in this chapter are explained in Figs. 13.4–13.6.

EROSIONAL FEATURES OF GLACIATED REGIONS		MOUNTAIN GLACIATION	CONTINENTAL GLACIATION
Cirque	Bowl-shaped depression on a high mountain slope, formed by a cirque glacier	X	
Arête	Sharp, jagged, knife-edge ridge between two cirques or glaciated valleys	X	
Col	Mountain pass formed by the headward erosion of cirques	X	
Horn	Steep-sided, pyramid-shaped peak produced by headward erosion of several cirques	X	
Headwall	Steep slope or rock cliff at the upslope end of a glaciated valley or cirque	X	
Glacial trough	U-shaped, steep-walled, glaciated valley formed by the scouring action of a valley glacier	X	
Hanging valley	Glacial trough of a tributary glacier, elevated above the main trough	X	
Roche moutonnée	Asymmetrical knoll or small hill of bedrock, formed by glacial abrasion on the smooth stoss side (side from which the glacier came) and by plucking (prying and pulling by glacial ice) on the less-smooth lee side (down-glacier side)	X	X
Glacial striations and grooves	Parallel linear scratches and grooves in bedrock surfaces, resulting from glacial scouring	X	X
Glacial polish	Smooth bedrock surfaces caused by glacial abrasion (sanding action of glaciers analogous to sanding of wood with sandpaper)	X	X

Figure 13.4 Erosional features of mountain or continental glaciation.

DEPOSITIONAL FEATURES OF GLACIATED REGIONS		MOUNTAIN GLACIATION	CONTINENTAL GLACIATION
Ground moraine	Sheet-like layer (blanket) of till left on the landscape by a receding (wasting) glacier.	X	X
Terminal moraine	Ridge of till that formed along the leading edge of the farthest advance of a glacier.	X	X
Recessional moraine	Ridge of till that forms at terminus of a glacier, behind (up-glacier) and generally parallel to the terminal moraine; formed during a temporary halt (stand) in recession of a wasting glacier.	X	X
Lateral moraine	A body of rock fragments at or within the side of a valley glacier where it touches bedrock and scours the rock fragments from the side of the valley. It is visible along the sides of the glacier and on its surface in its ablation zone. When the glacier melts, the lateral moraine will remain as a narrow ridge of till or boulder train on the side of the valley.	X	X
Medial moraine	A long narrow body of rock fragments carried in or upon the middle of a valley glacier and parallel to its sides, usually formed by the merging of lateral moraines from two or more merging valley glaciers. It is visible on the surface of the glacier in its ablation zone. When the glaciers melt, the medial moraine will remain as a narrow ridge of till or boulder train in the middle of the valley.	X	
Drumlin	An elongated mound or ridge of glacial till (unstratified drift) that accumulated under a glacier and was elongated and streamlined by movement (flow) of the glacier. Its long axis is parallel to ice flow. It normally has a blunt end in the direction from which the ice came and long narrow tail in the direction that the ice was flowing.		X
Kame	A low mound, knob, or short irregular ridge of stratified drift (sand and gravel) sorted by and deposited from meltwater flowing a short distance beneath, within, or on top of a glacier. When the ice melted, the kame remained.		X
Esker	Long, narrow, sinuous ridge of stratified drift deposited by meltwater streams flowing under glacial ice or in tunnels within the glacial ice.		X
Erratic	Boulder or smaller fragment of rock resting far from its source on bedrock of a different type.	X	X
Boulder train	A line or band of boulders and smaller rock clasts (cobbles, gravel, sand) transported by a glacier (often for many kilometers) and extending from the bedrock source where they originated to the place where the glacier carried them. When deposited on different bedrock, the rocks are called erratics.	X	X
Outwash	Stratified drift (mud, sand, and gravel) transported, sorted, and deposited by meltwater streams (usually muddy braided streams) flowing in front of (down-slope from) the terminus of the melting glacier.	X	X
Outwash plain	Plain formed by blanket-like deposition of outwash; usually an outwash braid plain, formed by the coalescence of many braided streams having their origins along a common glacial terminus.	X	X
Valley train	Long, narrow sheet of outwash (outwash braid plain of one braided stream, or floodplain of a meandering stream) that extends far beyond the terminus of a glacier.	X	
Beach line	Landward edge of a shoreline of a lake formed from damming of glacial meltwater, or temporary ponding of glacial meltwater in a topographic depression.		X
Glacial-lake deposits	Layers of sediment in the lake bed, deltas, or beaches of a glacial lake.	X	X
Loess	Unstratified sheets of clayey silt and silty clay transported beyond the margins of a glacier by wind and/or braided streams; it is compact and able to resist significant erosion when exposed in steep slopes or cliffs.		X

Figure 13.5 Depositional features of mountain or continental glaciation.

WATER BODIES OF GLACIATED REGIONS		MOUNTAIN GLACIATION	CONTINENTAL GLACIATION
Tarn	Small lake in a cirque (bowl-shaped depression formed by a cirque glacier). A melting cirque glacier may also fill part of the cirque and may be in direct contact with or slightly up-slope from the tarn.	X	
Ice-dammed lake	Lake formed behind a mass of ice sheets and blocks that have wedged together and blocked the flow of water from a melting glacier and or river. Such natural dams may burst and produce a catastropic flood of water, ice blocks, and sediment.	X	X
Paternoster lakes	Chain of small lakes in a glacial trough.	X	
Finger lake	Long narrow lake in a glacial trough that was cut into bedrock by the scouring action of glacial ice (containing rock particles and acting like sand paper as it flows downhill) and usually dammed by a deposit of glacial gravel (end or recessional moraine).	X	X
Kettle lake or kettle hole	Small lake or water-saturated depression (10s to 1000s of meters wide) in glacial drift, formed by melting of an isolated, detached block of ice left behind by a glacier in retreat (melting back) or buried in outwash from a flood caused by the collapse of an ice-dammed lake.	X	X
Swale	Narrow, shallow depression between two moraines.	X	X
Marginal glacial lake	Lake formed at the margin (edge) of a glacier as a result of accumulating meltwater; the upslope edge of the lake is the melting glacier itself.	X	X
Meltwater stream	Stream of water derived from melting glacial ice, that flows under the ice, on the ice, along the margins of the ice, or beyond the margins of the ice.	X	X
Misfit stream	Stream that is not large enough and powerful enough to have cut the valley it occupies. The valley must have been cut at a time when the stream was larger and had more cutting power or else it was cut by another process such as scouring by glacial ice.	X	X
Marsh or swamp	Saturated, poorly drained areas that are permanently or intermittently covered with water and have grassy vegetation (marsh) or shrubs and trees (swamp).	X	X

Figure 13.6 Water bodies resulting from mountain or continental glaciation.

Mountain Glaciers

Mountain glaciers usually flow down valleys that were established initially by rivers (**Figs. 13.3, 13.7,** and **13.8**). Mountain glaciers can originate in a **cirque** (a hollow or depression shaped like a half bowl on the side or near the top of a mountain), along a steep headwall on the side of a ridge or peak, or in an ice cap or ice field. Sediment eroded by mountain glaciers is carried by the glacier and by meltwater streams. We can see the eroded debris on top of the glacier in **Fig. 13.3**, but there is also eroded debris within the glacier and along its base.

How Mountain Glaciers Cause Erosion

Mountain glaciers exert significant vertical and horizontal stress on valley sides as the glaciers flow down the valley. Glaciers can erode sediment as well as hard rock as they move, so mountain glaciers modify the shape of the pre-existing valleys that they flow down. The downslope movement and extreme weight of glaciers cause them to scrape and erode rock materials that they encounter. In the process called **plucking**, glacial ice freezes around rock material and rips it from bedrock. The rock debris is then incorporated into the glacial ice and transported many kilometers by the glacier. The debris gives glacial ice extra power to erode through **abrasion**. This process occurs as the heavy rock-filled ice moves over the land, scraping surfaces like a giant sheet of sandpaper. Rock debris falling from valley walls commonly accumulates on the surface of a moving glacier and is transported downslope. Thus, glaciers transport huge quantities of sediment, not only *in* but also *on* the ice.

Erosional and Depositional Landforms of Mountain Glaciers

The various parts of an active mountain glacier system are illustrated in **Fig. 13.7**, and the resulting glacial landforms and deposits are illustrated in **Fig. 13.8**.

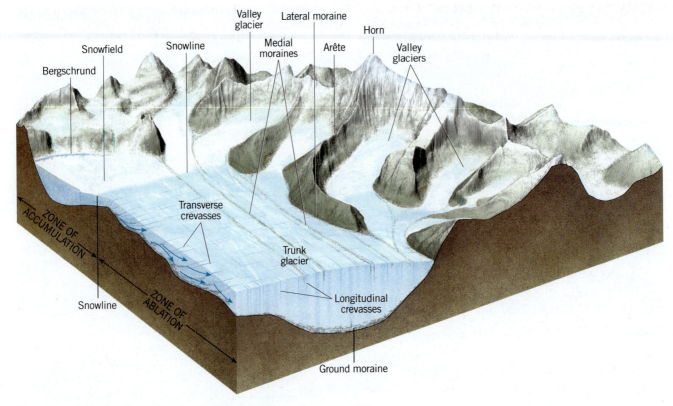

Figure 13.7 Typical features of an active mountain glacier system.

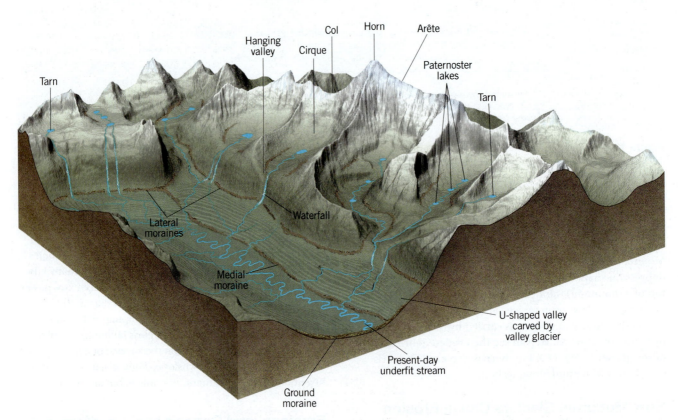

Figure 13.8 Typical erosional and depositional effects of mountain glaciation. Most of the terms used in this figure are defined in **Figs. 13.4–13.6**.

Mountain glaciers leave evidence of their past location. In Fig. 13.9A, we can see the Byron Glacier in Alaska flowing from its source area in a cirque on the side of Byron Peak. The glacier flows down toward a U-shaped valley that it once filled when the climate was colder. The linear valley just south (left) of Mitre Peak in the Milford Sound area of New Zealand has the same U shape but no obvious glacier (**Fig. 13.9B**). We can infer from its shape and its location in a mountainous area that it might have been modified by a glacier, and indeed, there is abundant evidence in the area that this is a valid inference. The Pasu Glacier in the Karakoram Range of the Himalayan Mountains is smaller than it once was but still extends along most of the Pasu Valley (**Fig. 13.10A**). Lateral moraines and a recessional end moraine are clearly evident. The Minapin Glacier that flows north from Rakaposhi (7788 m) has left behind several lateral moraines and recessional moraines just beyond its active terminus, or end (**Fig. 13.10B**). Moraines help to establish where a glacier's edge was at different times in the past. Most moraines are preserved only when a glacier is in retreat. Glacial advances in the same area destroy most if not all of the old moraines.

In an area that has recently been covered by an active glacier, the bedrock that was once directly under the sliding ice typically shows evidence of having been polished and scratched (or striated) by the moving glacier (Fig. 13.11). Careful study of these striations can indicate which direction the glacier flowed.

Yosemite Valley, California

There are many excellent examples of glacially modified valleys in the western United States and Alaska, but the most familiar is undoubtedly the Yosemite Valley in the central Sierra Nevada Mountains of California (Fig. 13.12). There, you can see moraines, glacial polish and striations, spectacular hanging valleys with some of the highest waterfalls in the world, glacial erratics (boulders carried great distances from where they were plucked by a glacier), and classic U-shaped and V-shaped valleys eroded into granitic bedrock. The shape of the main part of Yosemite is a bit misleading because the ground surface is underlain by as much as 2000 feet of glacial sediment and lake beds. The actual bedrock bottom of Yosemite is hidden from view under sediment except at the far east and west ends of the valley.

The source area for most of the glaciers that sculpted Yosemite Valley is in the Tuolumne Meadows area to the east of the valley where an ice field sent glaciers down through the valleys of the Merced and Hetch Hetchy Rivers. Two small cirque glaciers remain in the Tuolumne

Figure 13.9 Typical U-shaped glacially modified valleys. Mountain glaciers usually follow valleys that were originally established by stream erosion. Subsequent glaciation widens and deepens the existing valley. **A.** Valley of the Byron Glacier, Alaska (60.754°N, 148.855°W). **B.** Tributary valley adjacent to Milford Sound, a glacially carved fiord in Fiordland National Park, South Island, New Zealand (44.659°S, 167.894°E).

Figure 13.10 Moraine development by valley glaciers in the Karakoram-Himalaya Mountains. A. Lateral and recessional (end) moraines of Pasu Glacier (36.457°N, 74.870°E), which flows from a great ridge that contains Shispare (7611 m) in the upper left corner of the photo and Pasu (6478 m) in the right background. B. Moraines in the area were recently occupied by the Minapin Glacier that flows from the north side of Rakaposhi (7788 m). This view is from just north of the Hunza River (36.227°N, 74.548°E).

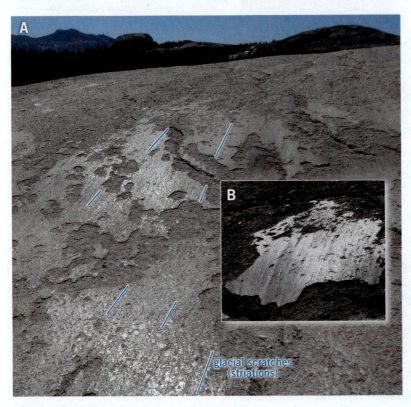

Figure 13.11 Glacial striations and polish. A. This rock was polished and scratched or striated as glacial ice from the Tuolumne ice field flowed over Pothole Dome (37.879°N, 119.393°W) on its way toward Yosemite Valley or the Hetch Hetchy Valley. The trends of several glacial striations are indicated with blue line segments. B. The inset photo shows another example of striated glacial polish, which is about 0.5 m across.

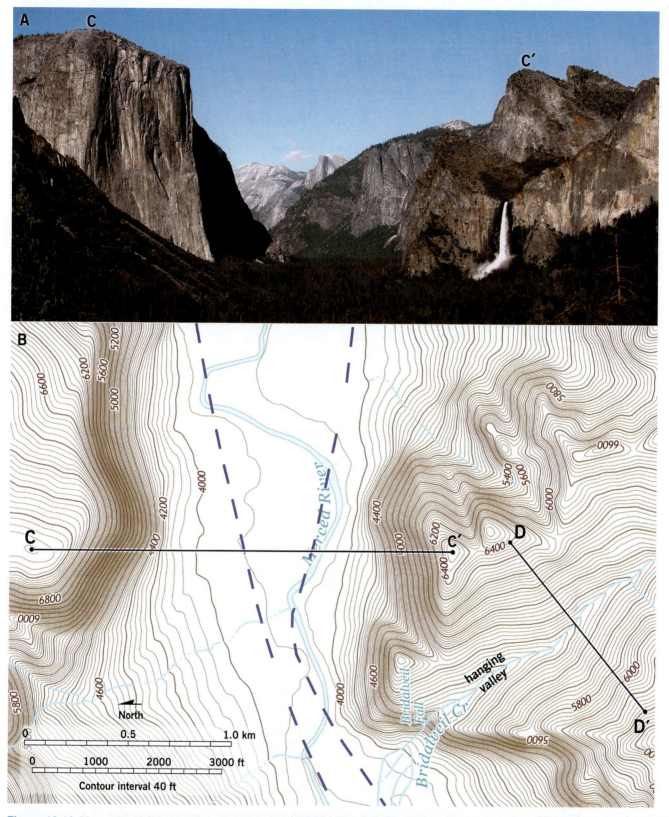

Figure 13.12 Yosemite Valley. A. View of glacially modified Yosemite Valley, looking east from Inspiration Point (37.7156°N, 119.6769°W). Points C and C′ are the endpoints of a profile constructed in Activity 13.2. Point C is on El Capitan, and C′ is on Middle Cathedral Rock. Bridal Veil Falls is to the right, and Half Dome is in the background at the center of the photo. **B.** Portion of the USGS 7.5-minute topographic quadrangle map of El Capitan, California. The purple dashed curve is a topographic contour on the top of the granitic basement below all of the glacial sediment in the valley based on geophysical investigations by Gutenberg, Buwalda, and Sharp in the mid-1950s. Sections C–C′ and D–D′ are used in Activity 13.2.

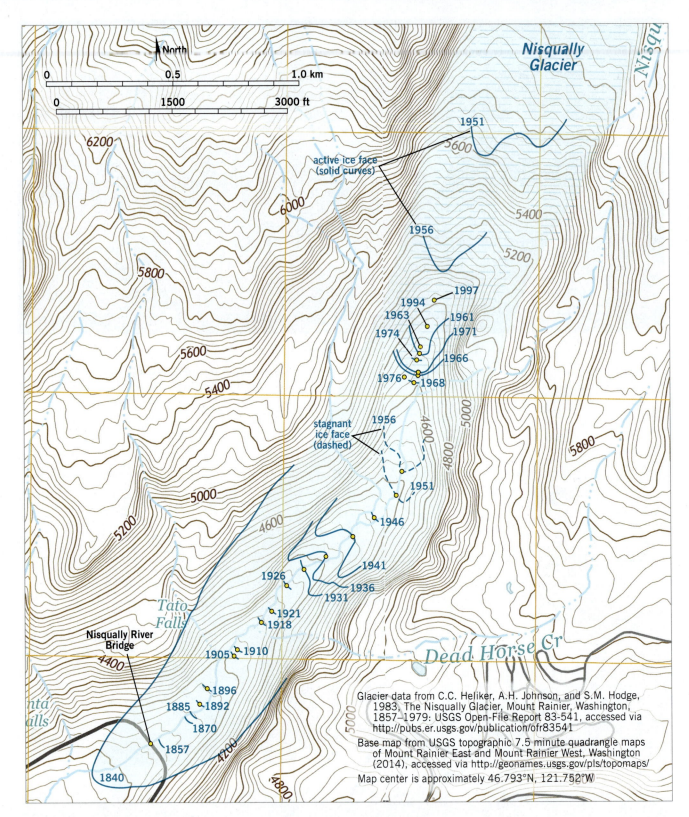

Figure 13.13 Variation in length of Nisqually Glacier on Mt. Rainier, Washington. This map shows where the end (terminus) of Nisqually Glacier was located between 1840 and 1997. Yellow dots are points used in Activity 13.3.

area—the Lyell and McClure Glaciers, which were first studied by John Muir around 1869–1872. Muir measured the rate of motion of McClure Glacier at about an inch per day, and nearly 150 years later, geoscientists from the National Park Service measured about the same rate even though the glacier is now much smaller.

Nisqually Glacier, Washington

Nisqually Glacier is one of many active valley glaciers that occupy the radial drainage of Mt. Rainier—an active volcano in the Cascade Range located near Seattle, Washington. Nisqually Glacier is located on the southern side of Mt. Rainier and flows south toward the Nisqually River Bridge (**Fig. 13.13**). The position of the glacier's

---

ACTIVITY 13.3

Nisqually Glacier Response to Climate Change, (p. 371)

Think About It   How might glaciers be affected by climate change?

Objective   Evaluate the use of Nisqually Glacier as an indicator of climate change.

Before You Begin   Read the following section: Nisqually Glacier, Washington. Refer to previous sections as needed.

Plan Ahead   You will need a pencil, ruler, and a calculator.

ACTIVITY 13.4

Glacier National Park Investigation, (p. 373)

Objective   Analyze glacial features in Glacier National Park and infer how glaciers there might change in the future.

Before You Begin   Read the following section: Glacier National Park, Montana. Refer to previous sections as needed.

Plan Ahead   You will need a pencil, ruler, and a calculator.

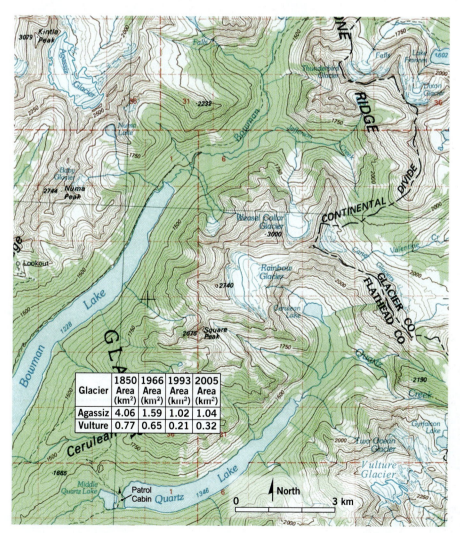

Glacier	1850 Area (km²)	1966 Area (km²)	1993 Area (km²)	2005 Area (km²)
Agassiz	4.06	1.59	1.02	1.04
Vulture	0.77	0.65	0.21	0.32

Figure 13.14 Portion of Glacier National Park, Montana. This small map area includes nine named glaciers, including the Agassiz and Vulture Glaciers that are considered in Activity 13.4. Glacier data in the inset table are from the USGS and its research partners at Portland State University. The center of this map is 48.8845°N, 114.0921°W, and it is a portion of USGS topographic quadrangle map of the Whitefish Range (1981).

terminus or downflow end was first recorded in 1840, and it has been measured and mapped by many geologists since that time. The map in **Fig. 13.13** contains data published by the U.S. Geological Survey (USGS) in 1983 as well as more recent data acquired through aerial image analysis. Notice how the glacier has retreated with a couple of advances since 1840.

Glacier National Park, Montana

Glacier National Park is located on the northern edge of Montana across the border from Alberta and British Columbia, Canada. Many if not all of the erosional features formed by glaciation in the park developed during the Wisconsinan glaciation or a shorter cold period in the Holocene, known as the "little ice age" between ~1300 to 1850 AD. It has been estimated that there were approximately 150 small cirque glaciers in the park in 1850, but the USGS counts only 25 today that meet its criteria for an active glacier. Nine of those can be observed on the topographic map in **Fig. 13.14**. The USGS reports that a

climate model intended to predict the rate of glacial retreat among the Glacier National Park's largest glaciers suggests that they might all vanish by the year 2030.

Continental Glaciation

During the Pleistocene glaciation, thick ice sheets covered most of Alaska and Canada, extending south into the continental United States. Continental ice sheets also formed in northern Europe and South America, and they persist to this day in Greenland and Antarctica. These continental glaciers produced a variety of characteristic landforms that are illustrated in **Figs. 13.15** and **13.16**.

Kames are low hills of stratified sand and gravel formed by one of several possible mechanisms (**Fig. 13.17A**). Some kames form when sediment-laden water flowed into a hole in stagnant ice. Other kames are remnants of meltwater deltas formed near the ice margin. **Drumlins** are composed of till and are interpreted to have been remolded under the pressure exerted by the weight of glacial ice on

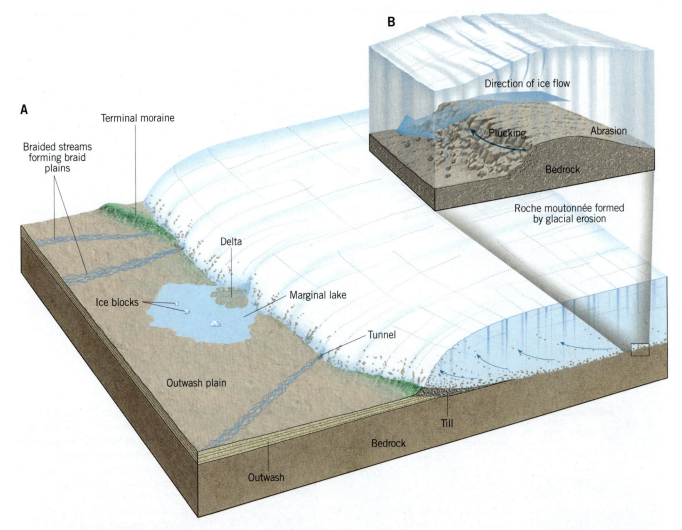

Figure 13.15 Typical features of an active continental glacier system. A. Continental glaciers are highly effective agents of erosion, sediment transport, and deposition. **B.** As the ice advances, it erodes bedrock to form a roche moutonnée, plucking blocks from the down-flow side.

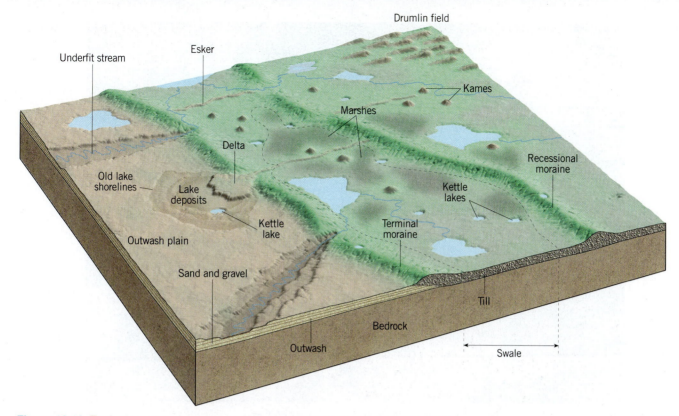

Figure 13.16 Typical erosional and depositional effects of continental glaciation. Continental glaciation leaves behind many characteristic landforms after the ice melts. Most of the terms used in this figure are defined in **Figs. 13.4–13.6**.

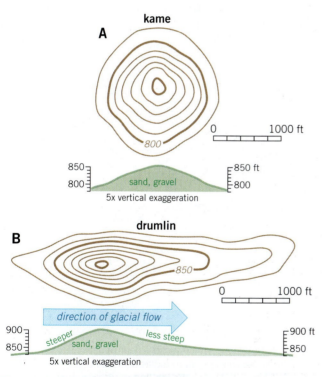

Figure 13.17 Kames and drumlins. A. Topographic map and profile of a kame—a mound or short ridge of stratified (layered) sediment originally deposited on or below stagnant ice and that was left behind when that ice melted. **B.** Topographic map and profile of a typical drumlin—a streamlined ridge composed of compacted till formed under a glacier by glacial flow. The long axis of the drumlin is parallel to glacial flow, and the long tail of the less steep side of the drumlin points in the direction of glacial flow.

ACTIVITY 13.5

Some Effects of Continental Glaciation, (p. 374)

Think About It How do continental glaciers affect landscapes?

Objective Analyze features of landscapes affected by continental glaciation and infer how they formed.

Before You Begin Read the section: Continental Glaciation.

Plan Ahead You will need a pencil, ruler, and a calculator.

wet sediment that can be molded under the glacier near its terminus (**Fig. 13.17B**). Another suggestion is that drumlins form when the edge of a glacier advances over an older recessional end moraine, remolding or carving parts of the moraine into drumlins. They have a streamlined shape that we can use to interpret the direction of glacier movement (**Fig. 13.17B**). **Eskers** are sinuous deposits of sand and gravel that reflect transport and deposition in a stream that flowed on, in, or under the margin of a retreating glacier or stagnant block of ice (**Fig. 13.18**). The principal streams associated with a glacier are the **outwash streams** that carry meltwater and sediment away from the end of the glacier across an **outwash plain** (**Fig. 13.15**).

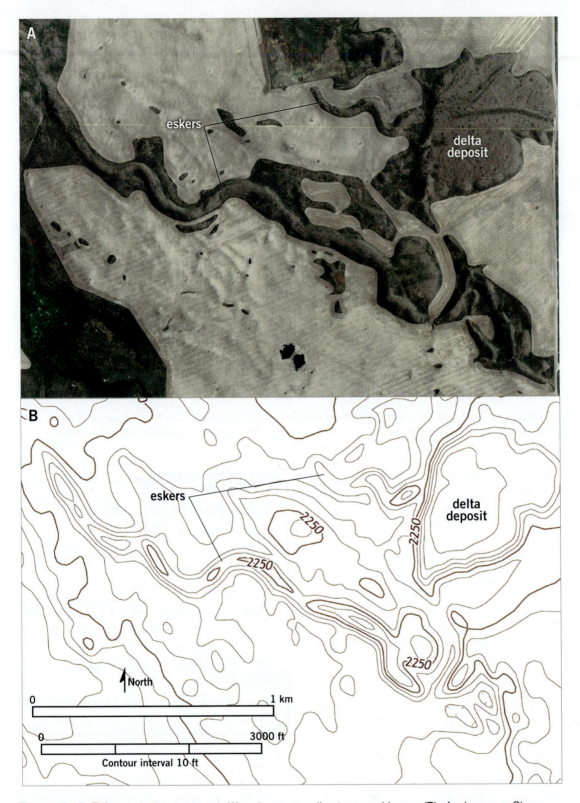

Figure 13.18 Eskers. Aerial photograph (**A**) and corresponding topographic map (**B**) of eskers near Sims Corner, Washington. Eskers are composed of sediment carried on, in, or below a glacier in a glacial stream. Eskers preserve the shape of the stream channel or ice tunnel and are deposited when the ice that contains them melts.

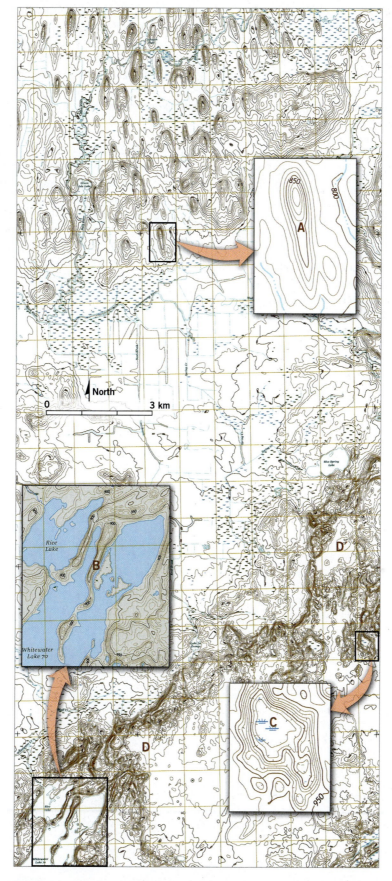

Figure 13.19 Continental glacier landforms near Whitewater, Wisconsin. Composite map processed to accentuate glacial landforms, derived from USGS 7.5-minute topographic maps of Rome, Palmyra, Little Prairie, and Whitewater, Wisconsin. The inset maps expand the areas indicated by the smaller black rectangles so that key features can be recognized in Activity 13.5.

Kettles are depressions in moraines, till, or outwash deposits that form when an isolated stagnant block of ice melts. They are recognizable on topographic maps by noting the hachures that indicate a closed depression. The middle of a kettle is often marshy and might even contain a kettle lake if the groundwater table is higher than the bottom of the kettle. **End moraines** form along the edge of an ice sheet when ice is supplied to the edge at the same rate at which the edge is melting, resulting in a terminus that remains in about the same place for many years. In that case, the rock and sediment debris carried by the glacier is transported to the edge and dropped in the same place, causing the development of a ridge of this transported material. The end moraine that is farthest from the middle of the ice sheet is called the **terminal moraine**, and all end moraines that are created inside of the terminal moraine during glacial retreat are called **recessional moraines**. These features can be seen in many states in the upper Midwest of the United States, including the Whitewater area of Wisconsin (Fig. 13.19).

Recognizing and interpreting these landforms is important in conducting work such as regional soil analyses, studies of surface drainage and water supply, and exploration for sources of sand, gravel, and minerals. The thousands of lakes in the Precambrian Shield area of Canada also are legacies of this continental glaciation, as are the fertile soils of the north-central United States and south-central Canada.

MasteringGeology™

Looking for additional review and test prep materials? Visit the Study Area in MasteringGeology to enhance your understanding of this chapter's content by accessing a variety of resources, including Pre-Lab Videos, Self-Study Quizzes, Geoscience Animations, Mobile Field Trips, *Project Condor* Quadcopter videos, *In the News* articles, glossary flashcards, web links, and an optional Pearson eText.

Name: _____ **Course/Section:** _____ **Date:** _____

A The cryosphere is composed of all parts of Earth where water is frozen, whether on the surface (snow, ice, sea ice) or below the surface (permafrost).

  1. Notice that Mexico is a beige- to yellow-colored region with no snow or ice near the bottom of **Fig. 13.1B**. What is the order in which you would encounter different parts of the cryosphere if you traveled overland from Mexico to the North Pole?

  2. While continental glaciation occurs only in Greenland and Antarctica today, mountain glaciers and ice caps occur in many places, including Canada, Russia, Alaska, the Rocky Mountains, the Andes, the Alps, and the Himalayas. Some mountain glaciers exist very close to the equator. How do you think it is possible for glaciers to exist at the equator?

B Scientists at the National Snow and Ice Data Center (NSIDC) have measured the extent of Arctic and Antarctic sea ice every month since at least 1978 using satellite data. A table of data from NSIDC for September in the Arctic and February in the Antarctic is provided in **Fig. A13.1.1**.

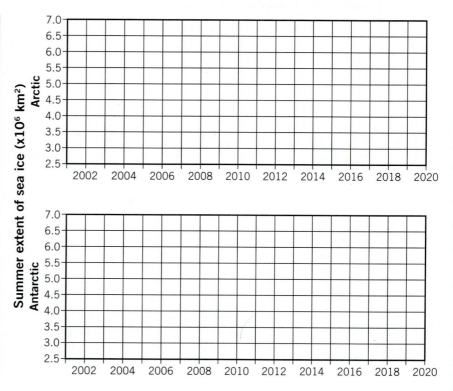

Summer Extent of Sea Ice

Year	September in Arctic, in millions of square km	February in Antarctic, in millions of square km
2015	4.68	3.77
2014	5.29	3.90
2013	5.35	3.91
2012	3.63	3.61
2011	4.63	2.52
2010	4.93	3.20
2009	5.39	2.98
2008	4.73	3.95
2007	4.32	2.95
2006	5.95	2.69
2005	5.59	2.99
2004	6.08	3.66
2003	6.18	3.92
2002	5.98	2.98
2001	6.78	3.81

Figure A13.1.1

  1. What was the average extent of Arctic sea ice in September from 2001 to 2015 in millions of km^2? Show your work.

  2. Plot all of the data for the extent of Arctic sea ice from 2001 to 2015 on the graph labeled "Arctic" in **Fig. A13.1.1**, and then use a ruler to estimate a best-fit line through the points so that the number of points above the line is about the same as the number below the line.

3. Based on your plot and calculations in part **2**, would you say that the amount of Arctic sea ice as measured in September of each year is decreasing, increasing, or staying about the same? Explain.

4. What do you predict the extent of Arctic sea ice will be in 2020?

5. What was the average annual extent of Antarctic sea ice from 2001 to 2015 in millions of km^2? Show your work.

6. Plot all of the data for the extent of Antarctic sea ice from 2001 to 2015 on the graph labeled Antarctic in **Fig. A13.1.1**, and then use a ruler to estimate a best-fit line through the points.

7. What do you predict the extent of Antarctic sea ice will be in 2020?

8. Based on your work in response to parts B2 and B5–B7, would you say that the annual amount of Antarctic sea ice as measured in February of each year is decreasing, increasing, or staying about the same? Explain.

C **REFLECT & DISCUSS**   How do the changes in Arctic sea ice extent over time compare with the Antarctic changes? If they are different, why do you think that might be? Consider as many possibilities as you can.

Name: _____ **Course/Section:** _____ **Date:** _____

A **Figure 13.12** includes a topographic map of the area between El Capitan to the north (left) and the Cathedral Rocks and Bridalveil Valley to the south (right). This was once the stream-cut valley of the Merced River, but it has repeatedly been reshaped by glaciers during the past ~2.6 million years. The most recent glacier retreated from Yosemite Valley around 10,000 years ago.

1. Use the profile box in **Fig. A13.2.1** to construct a profile across Yosemite Valley from El Capitan to Middle Cathedral Rock (C and C′, respectively, in **Fig. 13.12**). The vertical lines in the profile box are where index contours (the thicker contours with labeled elevations) cross the line of section C–C′ in **Fig. 13.12**, and the profile is started for you on both sides of the profile box. Continue the profile down all the way to the two points connected by a dashed line between elevations 2800 and 3000 feet. This is the depth of the top of the granitic bedrock that was excavated by glaciers, and everything above that curve to the current ground surface is glacial or postglacial sediment.

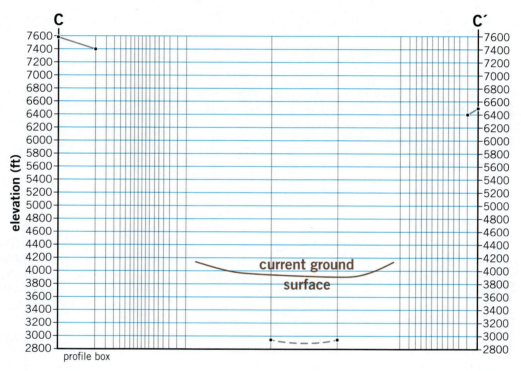

Figure A13.2.1

2. Use the profile box in **Fig. A13.2.2** to construct a profile across Bridalveil Valley from D to D′ in **Fig. 13.12**. The points at D and D′ are provided for you, as is the point along Bridalveil Creek at the bottom middle of the profile box. The vertical lines are where the index contours cross the line of section D–D′ in **Fig. 13.12**.

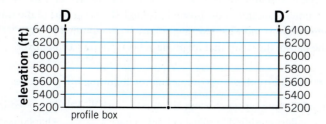

Figure A13.2.2

3. **Figure A13.2.3** includes a topographic map and profile box across the Merced River downstream from Yosemite Valley. Carefully construct the profile of section E–E′ in the profile box. There aren't any vertical lines this time to help you, so learn from your previous experience in parts **A1** and **A2** of this activity, or ask your teacher for help.

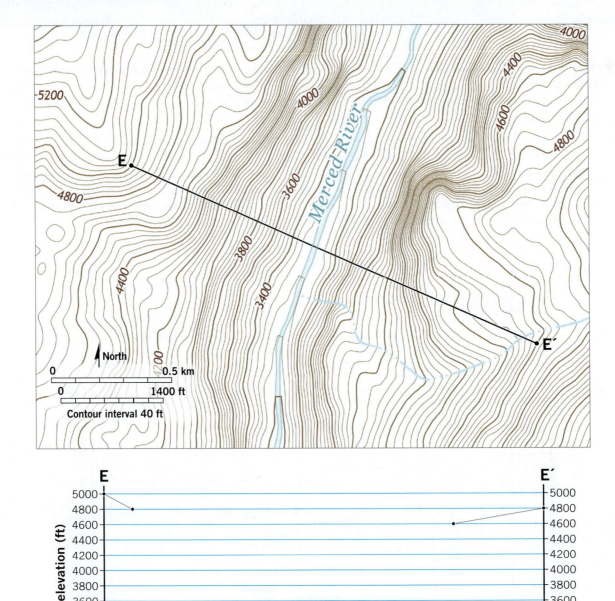

Figure A13.2.3

4. Geoscientists think that the upper part of this valley was modified by a very large glacier early in the Pleistocene Ice Age, but that the lower part of the valley was primarily or exclusively cut by the Merced River. Of course, the water and sediment from every glacial episode in Yosemite Valley was carried away by the Merced River, so it had a significant amount of erosive power at those times. Examine your finished profile E–E′. What part of the valley might you interpret as having been carved by the Merced River, and what part might have been modified by a glacier? That is, below what elevation along profile E–E′ is the part of the valley that might have been cut only by the river?

B Generalize your observations into an initial hypothesis by completing the following two problems.

1. Based on your work, complete the following sentence using either a "U" or "V." Valleys eroded by rivers tend to have a _____-shaped profile, whereas valleys modified by glaciers tend to have a _____-shaped profile.

2. Interpret whether Bridalveil Valley (section D–D′) is more likely to have been shaped by a river or modified by a glacier or some contribution from both. Explain your reasoning.

Name: _____ Course/Section: _____ Date: _____

Nisqually Glacier is a mountain glacier located on the south side of Mt. Rainier, Washington. Mt. Rainier is considered by the USGS to be one of the most threatening volcanoes in the Cascade Mountains. It has not erupted for more than a century, perhaps not for the past ~500 years.

A Complete the following tasks to compile a data table showing the location of the end of Nisqually Glacier relative to the Nisqually River Bridge at different times.

1. Measure the horizontal distance from Nisqually River Bridge to each of the small yellow dots on **Fig. 13.13** that show where the end (terminus) of Nisqually Glacier was in the past. Record your map measurements (in mm) on the data table in **Fig. A13.3.1**.

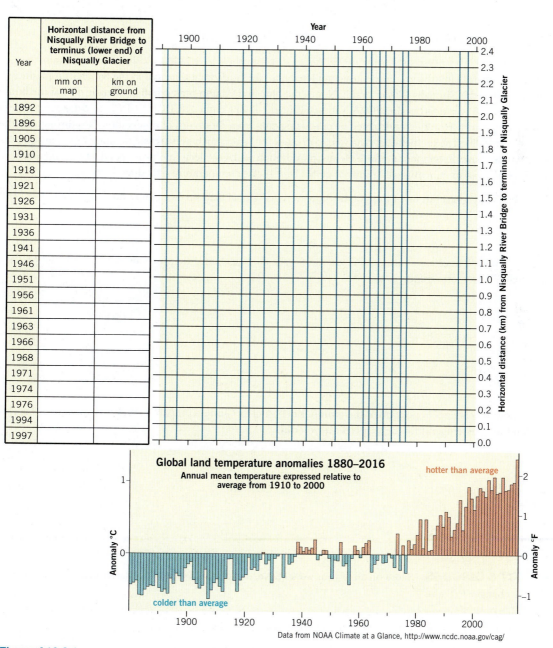

Year	Horizontal distance from Nisqually River Bridge to terminus (lower end) of Nisqually Glacier	
	mm on map	km on ground
1892		
1896		
1905		
1910		
1918		
1921		
1926		
1931		
1936		
1941		
1946		
1951		
1956		
1961		
1963		
1966		
1968		
1971		
1974		
1976		
1994		
1997		

Global land temperature anomalies 1880–2016
Annual mean temperature expressed relative to average from 1910 to 2000

hotter than average

colder than average

Data from NOAA Climate at a Glance, http://www.ncdc.noaa.gov/cag/

Figure A13.3.1

2. Use the bar scale on **Fig. 13.13** and your knowledge of proportions to convert your map measurements (in mm) to distance measurements at full scale (in km). Record your calculated distances, rounded to the hundredth of a km, on the data table in **Fig. A13.3.1**.

B Plot your data.

1. The vertical lines on the graph in **Fig. A13.3.1** represent the years for which data have been compiled. The horizontal lines correspond to horizontal distances between the bridge and the glacier terminus in tenths of km. Carefully plot the data you have compiled on the graph as a set of points.

2. Lightly draw a smooth pencil line through each of the data points in the correct sequence.

3. Notice that the glacier terminus retreated up the valley at some times but advanced back down the valley at other times. Summarize these changes in a brief written description that includes the specific time intervals when the glacier retreated or advanced.

C There is a bar chart below the graph you just completed, and the horizontal time axis in the bar chart is identical to the horizontal axis in the Nisqually Glacier graph. Notice the blue and salmon pink graph of climatic data at the bottom of your graph (part **B**) provided by the NOAA National Climatic Data Center (NCDC). NCDC's global mean temperatures are mean temperatures for Earth calculated by processing data from thousands of observation sites throughout the world (from 1880 to 2009). The temperature data were corrected for factors such as increases in temperature around urban centers and decreases in temperature with elevation. Although NCDC collects and processes data on land and sea, this graph shows the variation in annually averaged global land surface temperature only since 1880.

1. Describe the long-term trend in this graph—how averaged global land surface temperature changed from 1880 to 2015.

2. Lightly in pencil, trace any shorter-term pattern of cyclic climate change that you can identify in the graph. Describe this cyclic shorter-term trend.

D Describe how the changes in position of the terminus of Nisqually Glacier compare to variations in annually averaged global land surface temperature. Be as specific as you can.

E **REFLECT & DISCUSS** Based on all of your work above, do you think Nisqually Glacier can be used as an indicator of climate change? Explain.

Glacier National Park Investigation — Activity 13.4

Name: _____ **Course/Section:** _____ **Date:** _____

Refer to the map of Glacier National Park in **Fig. 13.14**.

A List the features of glaciation from **Figs. 13.3–13.10** that you can observe in **Fig. 13.14**.

B Locate Quartz Lake and Middle Quartz Lake in the southwest part of the map. Notice the patrol cabin located between these lakes. Infer the chain of geologic/glacial events (steps) that led to formation of Quartz Lake, the valley of Quartz Lake, the small piece of land on which the patrol cabin is located, and the cirque in which Rainbow Glacier is located today.

C Based on your answers above, what kind of glaciation (mountain versus continental) has shaped this landscape?

D Locate the continental divide on **Fig. 13.14**, and recall that it divides surface water that flows west into the Pacific Ocean from water that flows east into the Atlantic Ocean or Gulf of Mexico. Think of ways that the continental divide may be related to weather and climate in the region. Recall that weather systems generally move across the United States from west to east.

1. Describe how modern glaciers of this region are distributed in relation to the Continental Divide.

2. Based on the distribution you observed, describe the weather/climate conditions that may exist on opposite sides of the Continental Divide in this region.

3. Look at the location of glaciers in this map area in relation to ridges and mountain peaks. Do the glaciers tend to occur on the north, south, east, or west sides of these landforms? If so, on what side do most tend to be located? Suggest at least one explanation for this observation.

E Using the data table in **Fig. 13.14**, describe how the area of Agassiz Glacier changed from 1850 to 2005. Agassiz Glacier is in the northwest part of the map.

F Describe how the area of Vulture Glacier changed from 1850 to 2005. Vulture Glacier is in the southeast part of the map.

G REFLECT & DISCUSS What do you expect the area (km^2) of Agassiz and Vulture Glaciers to be in 2020? Explain.

A A topographic map of an area near Whitewater, Wisconsin, shows many landscape features that are characteristic of continental glaciation (**Fig. 13.19**). Many of those features are listed in **Figs. 13.4–13.6** and **13.15–13.18**.

1. Study the size and shape of the short, oblong rounded hills in the northwestern part of **Fig. 13.19**. Detail map A shows one of these hills. Fieldwork has revealed that they are made of till. What type of feature are they, and how did they form?

2. Toward what direction did the glacial ice flow in this area, and how can you tell?

3. Find the long, narrow, sinuous ridge that extends into a lake, shown in detail map B. What do you interpret this feature to be, and how do you think it formed?

4. In the southeast part of **Fig. 13.19** are many enclosed depressions marked by hachures on topographic contours like the one shown in detail map C. What do you interpret the depression in detail map C to be?

5. The features we just looked at in part **4** are part of an area that is a bit higher than the land to the north and has many small hills and depressions within a topography that seems chaotic. That area starts parallel to a line from point D to D′, and extends to the southeast corner of the map. What glacial landform do you interpret this area to be?

6. Note the marshy area running from the west-central edge of **Fig. 13.19** to the northeastern corner, separating the features shown in detail map A from those labeled B, C, and D–D′. Describe the probable origin of this flat marshy area. (More than one answer is possible.)

7. List the features of glaciated regions that you can recognize in **Fig. 13.19**.

B **REFLECT & DISCUSS** How is the glaciated area of **Fig. 13.19** different from areas affected by mountain glaciation, and how are they the same?

Desert Landforms, Hazards, and Risks

Pre-Lab Video 14

https://goo.gl/4LfeLC

Contributing Authors

Charles G. Oviatt • *Kansas State University*

James B. Swinehart • *Institute of Agriculture & Natural Resources, University of Nebraska*

James R. Wilson • *Weber State University*

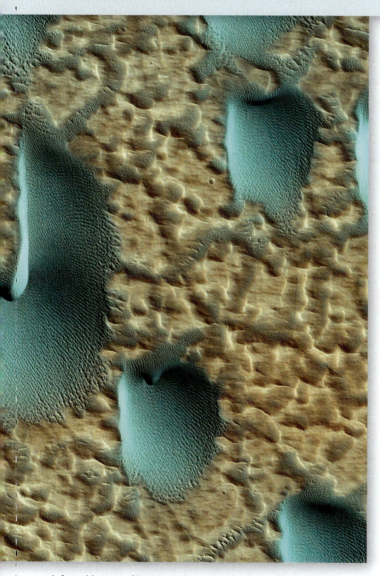

▲ Infrared image of barchan dunes on the floor of a crater in the Noachis Terra region of Mars (39.519°S, 4.598°W). The dunes are interpreted to be composed of sand derived from weathered basalt. (Image created by NASA/JPL and the University of Arizona using data from the HiRISE [High Resolution Imaging Science Experiment] camera on the Mars Reconnaissance Orbiter.)

BIG IDEAS

Deserts and dry lands are areas with arid-to-dry climates that generally have sparse vegetation and receive precipitation just a few days or one season of the year. Even so, water is one of the primary agents that produces characteristic desert landforms and flood hazards. Wind is also a factor in the erosion and transportation of sediment, especially silt and the sand that forms dunes. Although many people live in dry lands, true deserts do not support any agriculture without imported water or pumped groundwater.

FOCUS YOUR INQUIRY

Think About It   What are some characteristic processes, landforms, and hazards of dry lands?

Activity 14.1   Dryland Inquiry (p. 376, 387)

- -

Think About It   What can we learn from topographic maps and satellite images about dryland processes and landforms?

Activity 14.2   Sand Seas of Nebraska and the Arabian Peninsula (p. 379, 389)

- -

Think About It   How can topographic maps and aerial photographs of dry lands be used to interpret how their environments have changed?

Activity 14.3   Dry Land Lakes of Utah (p. 380, 391)

- -

Think About It   What desert landforms can be found in Death Valley, and how did the valley form?

Activity 14.4   Death Valley, California (p. 380, 393)

- -

Introduction

Deserts are areas where the annual precipitation averages less than 10 inches of rain or equivalent snow, and the vegetation is insufficient to sustain a significant human population. Availability of water usually limits the sustainable development of large towns or cities. Much of the precipitation that occurs in a desert is lost through evaporation or infiltration to aquifers below the surface. Deserts can be hot or cold, but their defining characteristics are dryness and scarcity of precipitation.

Types of deserts include polar and high-latitude deserts like Antarctica that are very cold and dry, coastal deserts like the Atacama Desert of South America, mid-latitude deserts in basins within a continental interior like the Gobi and Sonoran deserts, rain-shadow deserts like the ones east of the mountain ranges of western North America (e.g., Cascades, Sierra Nevadas, and Rocky Mountains), and trade-wind deserts like the Sahara that form a global belt in subtropical latitudes, where climatic factors produce dry conditions. Western North America has many areas that are sufficiently arid to be considered deserts totaling an estimated 500,000 km². They include the Mojave Desert of California and Nevada, the Sonoran and Chihuahuan Deserts of northern Mexico and the southwestern United States, and the Great Basin between the Rocky Mountains on the east and the Sierra Nevada and Cascade Mountains on the west.

A broader category of landscape includes **dry lands**—lands in arid and semiarid climates. Estimates vary, but dry lands make up between about 30–40% of Earth's land surface and support one-third of the world's human population. The current distribution of dry lands is concentrated in the subtropics and middle latitudes—from about latitude 10° to 45° north and south. Within those bands, a variety of local conditions such as topography and elevation, proximity to an ocean, and prevailing winds and weather patterns affect whether a particular area is humid or dry. Sixteen percent of dry lands are true deserts and cannot support any agriculture without imported water or pumped groundwater. Groundwater pumping that does not balance usage with recharge will deplete the aquifer, making agriculture unsustainable.

Dryland ranching or farming is possible if sufficient soil remains to support plants, but the number of animals and the types of crops must be adapted to the conditions. We must recognize the potential for **land degradation**—a state of declining agricultural productivity due to natural and/or human causes. Lands in humid climates might undergo degradation from factors such as soil erosion, farming without crop rotation or fertilization, overgrazing, or dramatic increases or decreases in soil moisture. However, degraded humid lands always retain the capability of some level of agricultural production. This is not true in dry lands where degradation may cause the land to be transformed into a desert in a process called **desertification**. The United Nations Environmental Program estimates that 70% of all existing dry lands are now in danger of desertification from factors related to human population growth, climate change, poor groundwater-use policies, overgrazing, and other poor land management practices.

Deserts are an integral part of the global ecosystem. Their importance is not dependent on whether we can find ways to use deserts. We should be wise enough to protect desert ecosystems so that we can continue to learn from the remarkable organisms that have adapted to a desert environment. Preservation of biological diversity in extreme environments is likely to provide us with essential knowledge we need for our own survival in an ever-changing world. As adaptable as humans are, we should be aware of the imperative for our land use to be sustainable and compatible with natural conditions, if only because of the abundant historical evidence of what can go wrong when we establish agriculture, industries, or communities in environments that cannot support them.

ACTIVITY 14.1

Dryland Inquiry, (p. 387)

Think About It   What are some characteristic processes, landforms, and hazards of dry lands?

Objective   Analyze satellite images and photographs of American dry lands and infer processes and hazards that occur there.

Before You Begin   Read the following section: Typical Parts of a Desert Landscape.

Plan Ahead   You will need a pencil, ruler, and a calculator.

Typical Parts of a Desert Landscape

About half of the land surface in desert areas is exposed bedrock, which is subject to erosion by water and wind. A quarter of the surface is a relatively thin layer of poorly sorted sediment with grains ranging in size from clay to small boulders on top of bedrock. Most of the rest is covered in loose sand, which can form dunes and move around the surface if there is enough wind and if the sand is persistently dry.

Most sediment in a typical desert environment is moved by water even though flowing water might occur only during infrequent rainstorms. Sediment of all sizes that is deposited by streams is called **alluvium**. Geologists use the term *alluvium* for loose sediment that has not been consolidated or lithified into sedimentary rock and that has been deposited during recent geologic time—usually during the most recent part of the Quaternary Period.

Deserts have characteristic geologic or geomorphic features, including those in the following list.

- **Bedrock**: hard rock of any type (igneous, metamorphic, sedimentary) exposed at the surface and subjected to both chemical and physical weathering

- **Pediments**: flat surfaces that slope gently away from highland areas, that were eroded onto bedrock, and that are typically covered by alluvium
- **Alluvial fans, bajadas, alluvial slopes and alluvial valley fill**: composed of clastic or detrital sediment—boulders, sand, silt, clay—deposited by streams
- **Intermittent streams**: flow when there is runoff water available but whose channels are frequently dry during the summer if not most of the year
- **Sand dunes**: form and move in response to the wind; can be stabilized by vegetation if there is sufficient moisture
- **Sabkhas**: flat areas in an arid environment where sediment has been removed by wind or moving water down to very near the water table, and evaporite minerals crystallize at or near the ground surface
- **Playas**: dry, flat, unvegetated areas underlain by layers of clay, silt, or sand at the bottom of desert basins where evaporite minerals can form. Playas might be occupied by lake water during relatively wet seasons but are typically dry most if not all of the time
- **Hardpan**: a hard layer that can develop at or just below the surface where clays and other minerals are cemented together by silica, iron oxides, calcite, or other minerals that are not very soluble in water.

Sand Dunes

Alluvium deposited by rivers is composed of whatever materials and grain sizes that were available in the source area. In many desert environments that include exposed bedrock, freshly deposited alluvium contains a full range of grain sizes from clay through small boulders. As the landscape dries after infrequent rains, wind goes to work on the unconsolidated alluvium, moving sand-sized grains up to ~0.5 mm in diameter and transporting them to where they might accumulate as dunes (**Fig. 14.1**). Some silt might also be carried by the wind, but very fine-grained materials are cohesive and can be difficult to erode. Wind-borne silt-sized sediment is called **loess** and can be transported far beyond the sandy dune fields (**Fig. 14.1A**). Loess can form extensive deposits adjacent to glaciated areas or deserts. Loess from the extensive glaciation of North America during the Quaternary Period contributed to the highly fertile soils of the Great Plains and California's Central Valley.

In a dune field, sand-sized particles that are typically ~0.1–0.3 mm in diameter are moved by the wind, bouncing their way up the gently inclined **stoss slope** of a dune (**Fig. 14.1B**). This bouncing movement is called **saltation**. When sand reaches the top of the stoss slope, it is blown over the edge onto a more steeply inclined **lee slope**, which is inclined about 30° in dry sand. Sand deposited on the lee slope is buried by the next volley of sand grains. Over time, dunes can move in the direction of the wind

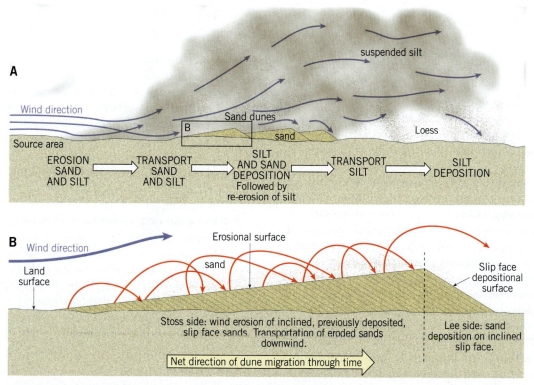

Figure 14.1 Eolian (wind-related) erosion, transportation, and deposition. A. Strong winds erode sand and silt from a source area and transport the sediment to new areas. As the wind velocity decreases, sand accumulates first (closest to the source) whereas silt and clay are carried further downwind. **B.** Sketch of a cross-section through a sand dune. Wind transports sand up the gently inclined *stoss side* (upwind side) of the dune. Sand is deposited on the slip face, forming the *lee side* (downwind side) of the dune. This continuing process results in net downwind migration of the dune.

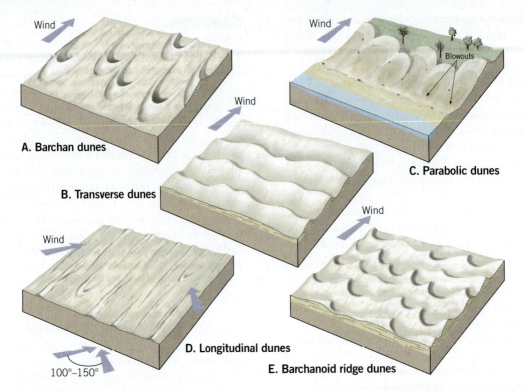

Figure 14.2 Common types of sand dunes. Some of the most common dune forms are illustrated with the prevailing wind direction that leads to their formation.

as grains are eroded from the stoss side and deposited on the lee side.

Some common types of dunes are illustrated in **Fig. 14.2** and described below:

- **Barchan dunes** are crescent-shaped dunes in which the horns or tips point downwind. Barchan dunes have a gently sloping side (stoss side) facing into the wind and a more steeply inclined face (lee side) on the downwind side of the dune (**Fig. 14.2A**). Barchan dunes form on planar, relatively hard surfaces where there is a limited sand supply, and wind direction is fairly constant.
- **Transverse dunes** have crests that are linear to sinuous and oriented perpendicular to the prevailing wind direction (**Fig. 14.2B**). Some transverse dunes move across relatively hard surfaces where there is a limited sand supply, like barchan dunes. Where sand is more abundant, a transverse dune might move over the stoss side of the next dune in the downwind direction, eventually creating stacked sequences of cross-bedded dune deposits.
- **Parabolic dunes** are crescent- or parabolic-shaped dunes in which the horns or tips point upwind (**Fig. 14.2C**). As the name implies, parabolic dunes have the shape of a parabola. They are often found along coastlines where the prevailing wind blows toward land and there is an abundant supply of sand from the beach. Parabolic dunes are frequently stabilized in part by grass or other sparse vegetation and

are commonly localized around a blowout where vegetation has been removed by the wind. Once the vegetation is gone, the sand is exposed to wind erosion and moves from the blowout area onto the dune. The blowout occupies the interior of the parabolic dune between the horns.
- **Longitudinal dunes** are long, relatively narrow dunes that are aligned parallel to the average prevailing wind direction (**Fig. 14.2D**). Most longitudinal dunes develop where wind alternates from two different directions, blowing with approximately equal force and frequency. Longitudinal dunes can also be localized behind an obstruction that cannot be moved by the wind. Longitudinal dunes can be more than 100 km long.
- **Barchanoid ridge dunes** form when barchan dunes are numerous and the crests of adjacent barchan dunes merge into transverse ridges, giving the resulting composite dune crest a scalloped shape (**Fig. 14.2E**). They can easily be distinguished from true transverse dunes that have long sinuous to very sinuous crests.

Blowouts

The most common wind-eroded landform is probably a **blowout**—a shallow depression developed where wind has eroded vegetation and soil (**Fig. 14.2C**). Wind erosion might leave behind fragmented rock or vegetated bits of

Figure 14.3 Algodones Dune Field, Sonoran Desert, near the California–Mexico border. Migrating dune sand presents many maintenance challenges for the All-American Canal and Interstate Highway 8. The image is centered at 32.738°N, 114.891°W. (NASA photograph taken from the International Space Station on January 31, 2009 [ISS018-E-24949].)

the previous ground surface in small pillars or patches. Blowouts might have an adjacent sand dune or dunes that formed where sand-sized grains were deposited after being removed from the blowout. Blowouts range in size from a few meters to a few kilometers in diameter.

Hazards of Moving Sand

Dunes tend to migrate slowly in the direction of the prevailing wind and are composed of a well-sorted but poorly packed mass of sand grains. Well-sorted deposits are mostly composed of the same-size grains, and dunes are mostly formed by sand grains from ~0.1 to 0.3 mm in diameter. Poorly packed grains are in a haphazard arrangement rather than being packed together as tightly as possible. As a result of being well sorted and poorly packed, sand dunes are very porous and permeable to water, and can easily be eroded by wind or water. Structures built on sand dunes or sand sheets (flat continuous layers of unconsolidated sand) are susceptible to a number of foundation problems. Motorists on roads that cross dune fields can be subjected to diminished visibility and loss of traction as sand is blown across the roadway. And mechanical systems with moving parts that are exposed to airborne quartz sand or silt are susceptible to damage by abrasion. (You might recall that quartz is harder than steel.)

The construction of the All-American Canal across the Algodones Dunes in southeastern California illustrates some of the hazards associated with dunes (Fig. 14.3). The All-American Canal is reportedly the largest irrigation canal in the world, carrying 740.6 m³ per second (26,155 ft.³/s) of water from the Colorado River to the Imperial Valley, the city of San Diego, and several smaller cities in southern California. The part of the canal that crosses the Algodones Dune field has suffered significant loss through leakage in the past because of the great permeability of the dune sand, requiring that the canal be rebuilt with a specialized lining. Blowing sand can enter the open top of the canal, which generally carries water

that has little or no sediment. The sand increases the erosive power of the flowing water and can degrade the walls of the canal downstream. It can also reduce the effective size of the canal. Sand needs to be removed before the water can be put into pipes; otherwise, the effectiveness and life expectancy of the pipes, valves, and pumps might be degraded. Without constant maintenance, the moving dunes might even fill the canal.

ACTIVITY 14.2

Sand Seas of Nebraska and the Arabian Peninsula, (p. 389)

Think About It What can we learn from topographic maps and satellite images about dryland processes and landforms?

Objective Identify landforms, including types of sand dunes, in dry lands and analyze dry lands to determine their risk of desertification.

Before You Begin Read the following section: Sand Seas of Nebraska and the Arabian Peninsula.

Plan Ahead You will need a yellow and a blue pencil.

Sand Seas of Nebraska and the Arabian Peninsula

Deserts can be rocky or sandy. An extensive sandy desert is called a sand sea, or **erg**. The largest erg on Earth is the Rub' al Khali of the Arabian Peninsula. Rub' al Khali is pronounced *ROOB al KHAL ee* and is Arabic for "the Empty Quarter." It covers an area of about 250,000 km²—nearly the size of Oregon. Many kinds of

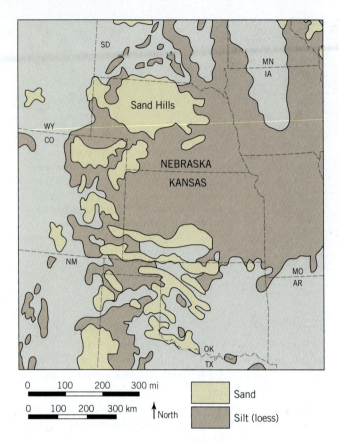

Figure 14.4 **The Sand Hills of Nebraska.** Map of part of the midwestern United States showing the location of Nebraska's Sand Hills and associated sand and silt (loess) deposits.

0 100 200 300 mi	Sand
0 100 200 300 km ↑North	Silt (loess)

This sand sea was active in Late Pleistocene and early Holocene time. It was supplied with sand and silt largely derived from the glaciated areas to the west and north, but the dunes have not been actively forming or moving for about the past 8000 years. This was determined by dating the radioactive carbon of organic materials that have been covered up by the large dunes. The large dunes are now covered with short-grass prairie that is suitable for limited ranching. About 17,000 people, mostly ranchers, now live in the Sand Hills.

Dryland Lakes

The amount of rain that falls on a particular dry land normally fluctuates over time periods ranging from months to millennia. Therefore, a dry land may actually switch back and forth between arid, semiarid, and slightly wetter conditions. Where lakes persist in the midst of dry lands, their water levels fluctuate up and down in relation to such periodic changes in precipitation and climate. Lakes might develop in an area during periods of higher precipitation and reduced aridity and evaporation, and those lakes might dry up during intervening periods of less precipitation and greater aridity and evaporation. Great Salt Lake, Utah, is an example of a lake in an arid environment with a history of changing lake levels.

Great Salt Lake is located in an enclosed basin, so water can escape from the lake only by evaporation. When it rains or when snow melts in the surrounding hills, the water level of the lake rises. Therefore, the level of Great Salt Lake has varied significantly in historic times over periods of months, years, and decades. During one dry period of many years, people ignored the dryland hazard of fluctuating lake levels and constructed homes, roads, farms, and even a 2.5-million-dollar resort, the Saltair, near the shores of Great Salt Lake. Many of these structures, including the resort, were submerged when a wet period

active dunes occur there, and some reach heights of more than 200 m. Rub' al Khali is a true desert with rainfall less than 35 mm (1.4 in.) per year.

The Sand Hills of Nebraska (**Fig. 14.4**) are only one-fifth the size of the Rub' al Khali, or about 50,000 km² of land, but form the largest erg in the Western Hemisphere.

ACTIVITY 14.3

Dryland Lakes of Utah, (p. 391)

Think About It How can topographic maps and aerial photographs of dry lands be used to interpret how their environments have changed?

Objective Analyze an aerial photo and topographic map of the Utah desert to evaluate the history of Lake Bonneville.

Before You Begin Read the following section: Dryland Lakes.

Plan Ahead You will need a blue pencil.

ACTIVITY 14.4

Death Valley, California, (p. 393)

Think About It What desert landforms can be found in Death Valley, and how did the valley form?

Objective Identify desert landforms of Death Valley, California, and infer how the valley is forming.

Before You Begin Read the following section: Stretching the Crust to Form Faults, Basins, and Ranges.

Plan Ahead You will need some colored pencils: brown, yellow, blue, and orange.

Figure 14.5 Southern part of Wah Wah Valley, Utah. This desert valley was half filled with part of Lake Bonneville during the Pleistocene when the climate there was cooler and wetter. Ancient lake levels can be seen like bathtub rings around the dry lakebed of the Wah Wah Valley Hardpan. The image is centered at 38.593°N, 113.316°W. (Aerial photograph from the USGS National High Altitude Photograph collection accessed via Earth Explorer; photo NC1NHAP830453173.tif.)

occurred from 1982–1987. The State of Utah installed huge pumps in 1987 to pump lake water into another valley, but the pumps were left high and dry during a brief dry period that lasted for 2 years (1988–89) after they were installed.

Geologic studies now suggest that the historic fluctuations of Great Salt Lake are minor in comparison to those that have occurred over millennia. Great Salt Lake is actually all that remains of a much larger lake that covered 20,000 mi.² of Utah—Lake Bonneville. Lake Bonneville reached its maximum depth and geographic extent about 15,500 years ago as glaciers were melting near the end of the last Ice Age. One arm of the lake at that time extended into Wah Wah Valley, Utah, which is now a dry land (**Fig. 14.5**).

Stretching the Crust to Form Faults, Basins, and Ranges

The Basin and Range Province of North America includes all or part of the Chihuahuan, Sonoran, Mojave, and Great Basin Deserts. The Basin and Range includes part of the North American Plate that has been stretched during the past ~17 million years. In response to that stretching, normal faults have developed in the upper continental crust. The lack of vegetation and soil covering rock outcrops makes this area a favorite for geologists to study active as well as ancient structures.

At the western edge of the Basin and Range Province in California and Nevada is the Sierra Nevada Mountains block, which is being dragged slowly toward the northwest because of interactions between the Pacific and North American Plates (**Fig. 14.6**). A transitional zone called the Walker Lane is between the Basin and Range and the Sierra block. The Walker Lane is a zone dominated by right-lateral strike-slip displacement, but that also includes extensional features and even some rotating crustal blocks. As much as about 20% of the motion of the Pacific Plate relative to the North American Plate occurs across the Walker Lane with more occurring along the San Andreas Fault System farther to the west. Death Valley is located within the Walker Lane.

While deserts are not dependent on these structural features, the close association of deserts and active faulting in North America makes them interesting to study together. Faults are surfaces or narrow zones along which one block slides relative to another block, a bit like the surface between two book covers when you pull a book down from a shelf. A **strike-slip fault** has a near-vertical surface along which one block moves horizontally past the other. A **normal fault** is an inclined surface along which the block above the fault surface (the **hanging-wall block**) slides down the surface relative to the block below the fault (the **footwall block**). Normal faults are very common in areas where the crust is being stretched or extended (**Fig. 14.7**).

The result of crustal stretching in western North America is a series of parallel valleys or basins that are separated from one another by parallel ranges (**Figs. 14.6 and 14.7**). The ranges are mostly formed by **horsts** that constitute the footwall block for normal faults on both sides of the range (**Fig. 14.7D**). The valleys are mostly **graben** (pronounced *GRAH ben*) that have subsided along normal faults for which they are the hanging wall block. A graben is bound by normal faults that are inclined in opposite directions and toward the center of the graben, giving it the look of a keystone at the top of an arch.

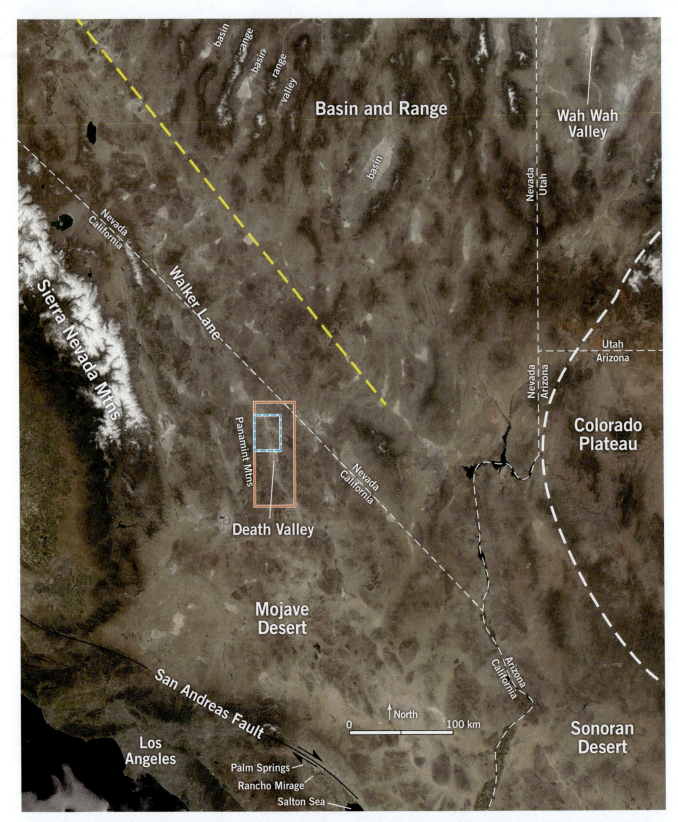

Figure 14.6 Structural setting, deserts of the southwestern US. Death Valley is shown in its structural context within the southern Walker Lane north of Los Angeles and the San Andreas Fault, east of the southern Sierra Nevada Mountains, and south and west of the active Basin and Range Province. The dashed yellow line is the approximate location of the boundary between the Walker Lane and the Basin and Range. The rectangle with the dashed blue line is the area in **Fig. 14.9** and the orange rectangle is the area in **Fig. 14.10**. The image is centered at 36.510°N, 116.113°W. (Image from the MODIS sensor on NASA's Terra satellite.)

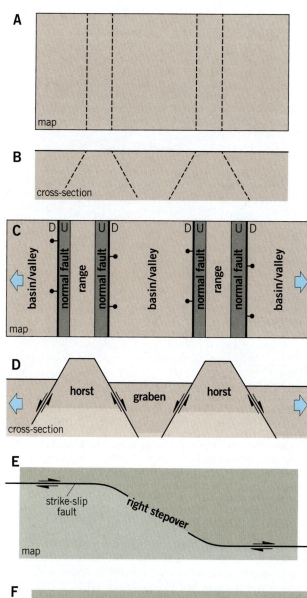

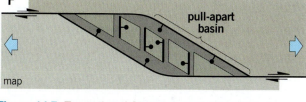

Figure 14.7 Extensional Structures. A. Map view of a hypothetical area of continental crust before extension. Faults will develop along dashed lines. **B.** Cross-section of the same area before extension. **C.** As the crust is stretched, normal faults develop that divide the crust into blocks. The black circle and bar symbols indicate normal faulting with the black circle on the block that moved down along the fault. Fault-bounded ridges and basins or valleys form as the crust is extended. **D.** Cross-section after extension showing typical horsts and graben. **E.** Map view of a strike-slip fault system with a stepover. **F.** Displacement along the two strike-slip fault segments results in extension within the stepover area between the faults, generating normal faults (marked by circle and bar symbols) and subsidence within the pull-apart basin that results. The faults and other structures within an actual pull-apart basin are more complex than depicted in this schematic diagram.

Stretching can also occur between two segments of a strike-slip fault that are parallel but not aligned with each other (**Fig. 14.7E–F**). This seemingly odd geometric arrangement can happen for a variety of reasons as a strike-slip fault zone evolves in a particular area. The area between the ends of the two faults shown in **Fig. 14.7E** form what structural geologists call a **right stepover** (or **releasing bend**). If you are walking along one of the segments, you would have to step over to the right in order to continue along the other segment. Stepovers have been observed at scales ranging from millimeters to kilometers. **Fig. 14.7E** shows two right-lateral strike-slip fault segments that are parallel but not in line with each other. These segments are shown at an early phase of their development before they have joined along the stepover. Continued movement along the strike-slip fault system causes the area between the fault segments to stretch or extend, which typically results in development of normal faults (marked by the circle-and-bar symbols in **Fig. 14.7F**). As stretching continues, the area between the strike-slip faults sinks or subsides to form a **pull-apart basin**. The bottom of the Dead Sea is one of the lowest points in continental crust on Earth at 733 m (2405 ft.) below sea level and is a pull-apart basin along a strike-slip fault system—the Dead Sea Transform Fault.

Typical Landforms in Mountainous Deserts

Two characteristics of dryland precipitation combine to create some of the most characteristic desert landforms other than blowouts and dunes. First, rainfall in dry lands is minimal, so there are few plants to trap and bind sand or aid in the development of soil that would absorb rainwater. Second, when rainfall does occur, it generally is in the form of violent thunderstorms. The high volume of water falling from such storms causes flash floods over dry ground. These floods develop suddenly, have high discharge, and last briefly.

Alluvial Fans, Bajadas, and Inselbergs. Relatively hard, impermeable bedrock forms the ground surface in many desert highland areas, so most of the rainwater that falls in the highland areas flows off rather than seeps in. Runoff water takes eroded bits of the bedrock with it on its downslope journey. Steep-walled bedrock canyons that channel the sediment-laden storm water can be eroded into these highland areas. At the base of the canyon, the channel becomes less steep, the water slows, and the coarser part of the sediment load begins to settle out to form a cone- or fan-shaped deposit called an **alluvial fan** (**Figs. 14.8** and **14.9**). Alluvial fans function as the distributary system of mountain streams in arid climates. Where alluvial fans from two or more adjacent canyons coalesce, a **bajada** is formed. As alluvium accumulates at the base of the mountains, it can bury all but the highest points along the mountain front. Isolated knobs of bedrock that are surrounded by alluvium are called **inselbergs** (**Fig. 14.8**).

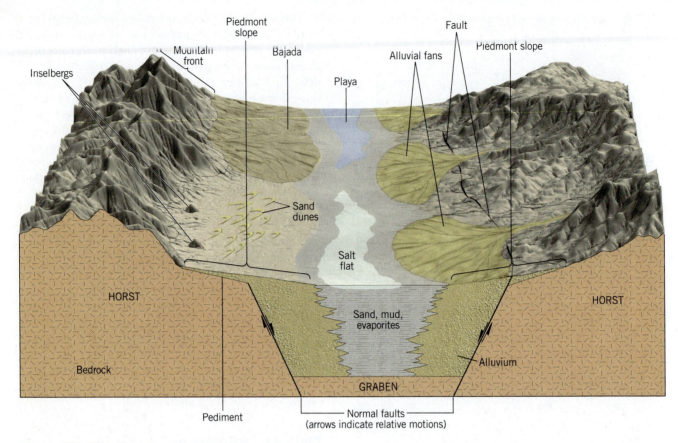

Figure 14.8 **Typical graben in an arid environment.** As the crust is extended, the graben subsides along normal faults. Sediment is shed off adjacent highlands and collects in the basin, resulting in landforms that are typical of a desert landscape in the Basin and Range Province of the western United States.

Stream channels on alluvial fans tend to be braided and switch back and forth across the fan surface over time (**Figs. 14.9** and **14.10**). Many if not all of the channels on an alluvial fan in a desert environment are dry most of the time except during infrequent rainfall events. Channels on alluvial fan surfaces are often filled by debris flows, diverting future runoff water into other channels. As the rapidly flowing water makes it down the mountain front and onto the alluvial slopes and flats, it can carve steep-walled ravines. The ravines are often floored with gravel deposited as the flow decreases and ends. Such steep-walled ravines with gravel floors are called **arroyos**, **wadis**, or **dry washes**.

Flood and Debris Flow Hazards. Water simply runs off of the rocks when it rains in mountainous dry lands because there is no soil to absorb it. This leads to the development of severe flash floods, which have the cutting power to erode rock and transport sediment. Heavy rains in dry or poorly vegetated highland areas often trigger **debris flows** in which a tremendous amount of sediment is carried downslope by water. In undeveloped areas, debris flows contribute sediment to alluvial fans and bajadas and to valleys and basins. Debris flows can move very rapidly—as fast as you can drive away from them in a car—and have the

consistency of freshly poured concrete. Unlike **mudflows** that are composed of clay, silt, and sand-size particles, debris flows can contain mud as well as boulders, vegetation, cars, houses, and the proverbial kitchen sink. When debris flows occur in populated areas, trees and buildings are knocked down, roadways demolished, people are buried, and the destruction can be devastating. Flash floods and debris flows do millions of dollars of property damage and claim many lives every year.

Playas and Playa Lakes. The lowest areas on the ground surface of a desert basin are the sites of **playas** that can fill with water during wet periods to form **playa lakes**. The water ultimately evaporates, leaving behind a dry lakebed or hardpan. The playa can be very close to the water table in which case it will function as a **sabkha**. Groundwater containing salts in solution will rise to the surface of the playa, and the water will evaporate, leaving behind the newly crystallized salts at or near the ground surface. In other cases, the runoff water from countless isolated rain events makes its way to the playa and evaporates, leaving behind whatever soluble material it picked up as it moved across the bedrock and over and through the alluvium. The result is a **salt flat** that typically contains a variety of evaporite minerals.

Within the image, the following labels appear:

0 1 2 3 miles
0 1 2 3 4 km

↑ North

Amargosa
Mountains

Outline of Figure A14.4.1

Furnace
Creek

Panamint
Mountains

Amargosa
Mountains

Panamint
Mountains

Badwater
Basin

Figure 14.9 Death Valley, California. Aerial image of the central part of Death Valley, California. Badwater Basin, at the bottom of the image, is the lowest area in North America at 86 m (282 ft.) below sea level. The image is centered at 36.384°N, 116.875°W.

(NASA Earth Observatory image processed by Jesse Allen and Robert Simmon using NASA EO-1 data.)

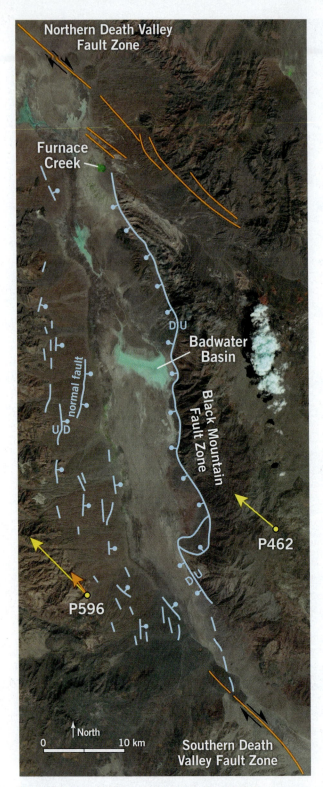

Figure 14.10 **Quaternary faults in Death Valley, California.** Blue curves are normal faults with the circle and bar symbols on the down side of the fault. Orange faults are (primarily) right-lateral strike-slip faults. The two yellow circles identify Plate Boundary Observatory GPS sites that allow us to directly observe relative motion across the valley. Yellow arrows indicate velocities relative to the stable interior part of the North American Plate (NAM08), and the orange arrow indicates the velocity of GPS site P596 relative to P462, discussed in Activity 14.4. Image centered at 36.215°N, 116.780°W. (From the USGS *Quaternary Fault and Fold Database of the United States* [http://earthquake.usgs.gov/hazards/qfaults/]. Base photo is a composite of NASA EO-1 images.)

Death Valley, California

The desert scenery of southeastern California has been the backdrop for so many films during the past century that it is broadly familiar to many people. The features of desert landscapes seem particularly concentrated in Death Valley, California, which has the distinctions of having the lowest spot on the North American continent at Badwater Basin (86 m or 262 ft. below sea level) and the hottest reliably measured air temperature ever recorded on Earth (56.7°C or 134°F on July 10, 1913) according to the World Meteorological Organization. It is also the largest national park in the lower 48 states. For hydrogeologists and sedimentologists, Death Valley has travertine springs, fresh water, brackish water, playas and playa lakes, sabkha conditions, modern sand dunes, talus cones, alluvial fans and bajadas, old sedimentary rock, and fresh sediment. Like the Wah Wah Valley of Utah, Death Valley was once filled with a great inland lake. Lake Manley was the ultimate base level for all of the drainages in the Great Basin during wetter times in the Quaternary Period. That is one of the reasons why Death Valley is so rich in evaporite minerals. It also has igneous, metamorphic, and sedimentary rock exposed at the surface. And for the structural geologist, it has ancient folds and both ancient and active faults.

Right-lateral strike-slip faults are apparent at the north and south ends of Death Valley (orange fault traces in **Fig. 14.10**). Along the valley sides are fresh and older scarps from normal faults that are mostly inclined down toward the middle of the valley (light blue fault traces with circle-and-bar symbols in **Fig. 14.10**). Scarps along the Black Mountain Fault on the east side of the valley are particularly clear to visitors to Death Valley, and the smaller scarps along the bajadas of the western side of the valley are more subtle and discontinuous. When viewed looking either east or west across the middle of Death Valley near Badwater Basin, it is easy to visualize this as a typical Basin-and-Range-style valley (e.g., **Figs. 14.7A–D**) with normal faults along both sides. But given its setting in the Walker Lane (**Fig. 14.6**) and the presence of the Northern and Southern Death Valley Fault Zones at either end of the valley (**Fig. 14.10**), other interpretations are certainly possible (e.g., **Figs. 14.7E–F**) as we will investigate in Activity 14.4.

MasteringGeology™

Looking for additional review and test prep materials? Visit the Study Area in MasteringGeology to enhance your understanding of this chapter's content by accessing a variety of resources, including Pre-Lab Videos, Self-Study Quizzes, Geoscience Animations, Mobile Field Trips, *Project Condor* Quadcopter videos, *In the News* articles, glossary flashcards, web links, and an optional Pearson eText.

Name: _____   Course/Section: _____   Date: _____

A When most people think of dry lands and deserts, they imagine hot sandy landscapes. Most of the southwestern United States is desert, including the Sonoran Desert of southern California and Arizona. However, rocky landscapes are typical of most of the Sonoran Desert. Sandy areas like the Algodones Dune Field in **Fig. 14.3** are present but limited. This is the location where *Star Wars* producers filmed desert scenes of the film's planet Tatooine.

1. Notice the sand dunes of the Algodones Dune Field in **Fig. 14.3**. Why do you think there are so few plants growing on the dunes?

2. Winds in the Algodones Dune Field can reach velocities up to 60 mph. This can create hazardous conditions and the need for maintenance on the canal and Interstate Highway 8. What would be the hazard and what maintenance would be needed periodically on the canal and Interstate Highway 8 as a result of the hazard?

3. This region is managed by the U.S. Bureau of Land Management as the Imperial Sand Dunes Recreation Area. Portions of the dunes are available for operating off-road vehicles. What effect do you think the operation of off-road vehicles here would have on plant growth and the hazards you described above?

B Death Valley is along the border between the Mojave Desert and the Great Basin Desert. Analyze the images of the Death Valley region in **Figs. 14.8–14.10**.

1. Steep mountainous slopes occur on both the east and west sides of Death Valley. There is almost no soil or vegetation on the slopes. Describe what you think conditions would be like in the bedrock canyons on these mountain slopes when a heavy rain falls on them.

2. Notice the alluvial fans and bajadas that form at the base of the bedrock canyons along the mountain fronts in **Figs. 14.9** and **14.10**. Explain how you think these landforms were formed.

3. Notice that there is little, if any, standing water in Death Valley on a typical day, even though you can see evidence that water sometimes flows into the valley from the mountains. Death Valley is a closed basin, meaning that water has no way to drain from it. It is also the hottest and driest place in North America. When there is water on the floor of the valley, it is alkaline to salty and not potable (drinkable). How do you think the water gets so alkaline and salty?

4. Suppose you could walk down to the white patches on the floor of Death Valley in **Fig. 14.9** and examine them. Predict what materials and conditions you would find there.

5. Residents of Furnace Creek have grassy lawns, trees, and potable water to drink. Why do you think their water is potable?

C Open Google Earth. Type coordinates 33.69°N, 116.25°W into the search box and press Search to go to the location. View the area between the Salton Sea and Palm Springs from an elevation of around 70 km.

1. This area stands out amid the desert mountains to the northeast and southwest because of the irrigated, cultivated land between Palm Springs and the Salton Sea. Where do you think the water that is used to make this valley green comes from?

2. The northeastern edge of the valley is bounded by the southern San Andreas Fault Zone (e.g., 33.767°N, 116.220°W through 33.838°N, 116.310°W), and the southwestern edge of the valley is roughly parallel to the San Andreas. The San Andreas fault zone is primarily a zone of right-lateral displacement along strike-slip faults, but there are other faults within this zone. What kind(s) of faults are probably involved in the continuing development of this fault-bounded valley? *Hint:* Refer to **Fig. 14.7**.

3. Now use the Google Earth search box to navigate to the triangular community of Rancho Mirage (33.738°N, 116.417°W), viewing it from an altitude of around 3 km. Using your mouse, hover over the icons at the top of the Google Earth screen to find and select the "Show historical imagery" feature. Use the slider to go back in time and view how this desert valley has changed, and describe some of the changes you see in the historical imagery.

4. In July of 1979, a violent storm developed here, and nearly 6 in. of rain fell on the San Jacinto Mountains, uphill and to the southwest of Rancho Mirage. At that time, flood control at Rancho Mirage included a system of earthen channels and concrete walls that were overwhelmed by a flash flood. Many homes were damaged, and two lives were lost in the event. Analyze recent images of Rancho Mirage and describe any evidence of flood control measures.

5. Would you feel safe living at Rancho Mirage? Explain.

D REFLECT & DISCUSS The United Nations Convention to Combat Desertification (UNCCD; http://www.unccd.int/en/) points out that many people live in dry lands and that more than half of the world's productive land is dry land. Make a list of hazards that might affect people living a desert community or on a farm or ranch in a dry land along with measures that can be taken to manage the risks associated with those hazards.

Activity 14.2

A Analyze **Fig. A14.2.1**, which shows part of the Rub' al Khali of the Arabian Desert, centered at 21.794°N, 54.802°E. The Rub' al Khali is reportedly the largest contiguous sand desert or erg in the world. The sand is mostly quartz with a reddish hematite coating transported south by strong winds from Jordan, Syria, and Iraq. These winds, called *shamals*, can reach speeds of nearly 50 mph that last for days. The dunes move across a flat surface formed by clay beds that were deposited in ancient lakes. White patches between the dunes are mostly gypsum. This NASA image was created by Robert Simmon from data collected by the Enhanced Thematic Mapper on Landsat 7 on August 26, 2001, provided by the USGS.

Figure A14.2.1

1. What kind(s) of dunes are visible in this image? Refer to **Fig. 14.2**.

2. What is the predominant wind direction here, and how can you tell?

B Analyze the USGS orthoimage of part of Nebraska's semiarid Sand Hills region in **Fig. A14.2.2** (centered at 42.148°N, 102.321°W) on the next page. Rainwater quickly drains through the porous sand, so the hilltops are dry and support only sparse grass. There is a shallow water table, so there are lakes, marshes, and moist fields between the hills.

1. Several ponds are evident in **Fig. A14.2.2A**, and their upper surfaces coincide with the local water table. Referring to the topographic map in **Fig. A14.2.2B**, what is the elevation of the water table in this area?

2. The top of one of the dunes is marked with a star. Is the slope on the northwest side of that dune more or less steep than the slope on the southeast side of the dune?

3. Which side of the dune marked with a star seems to have a crescent or concave shape?

4. Compare the shape of the dune marked with a star with dunes shown in **Fig. 14.2**, and interpret the type of dune it is.

5. Based on your answers to the previous three questions, interpret the direction the wind blew while these dunes were forming. The wind blew from _____ to _____. Explain your reasoning.

6. Some of the sand hills in **Fig. A14.2.2** are mostly isolated from the rest like the one marked with a star. Other dunes line up to form a linear or sinuous ridge. Referring to **Fig. 14.2**, what do we call a ridge of these kinds of dunes?

7. Use a yellow pencil to color the more isolated dunes, and use a blue pencil to color the ridges of dunes.

8. List some of the ways that the Sand Hills dunes are similar to the sand dunes of the Rub' al Khali.

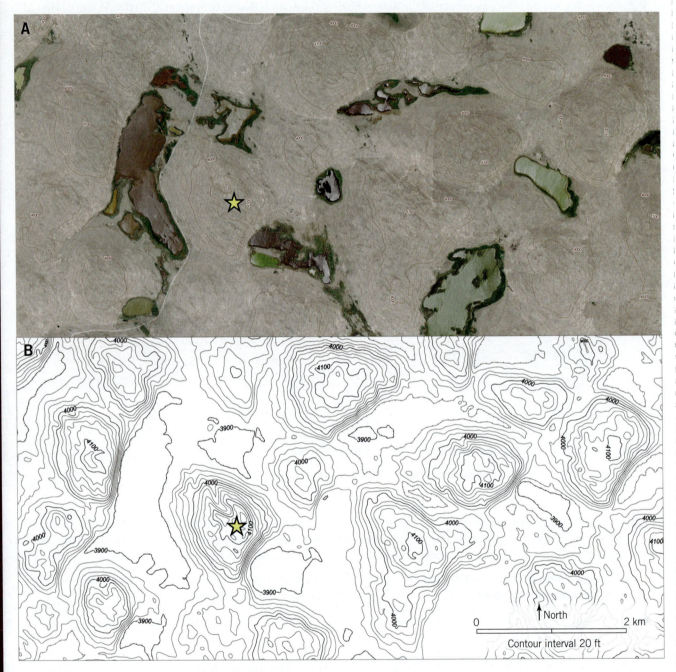

Figure A14.2.2

C REFLECT & DISCUSS Many cities in central and eastern Nebraska rely on groundwater for consumption, industry, and pleasure. As these cities continue to grow and their use of groundwater increases, what effect might this have on the environments and people of the Sand Hills?

Name: _____   Course/Section: _____   Date: _____

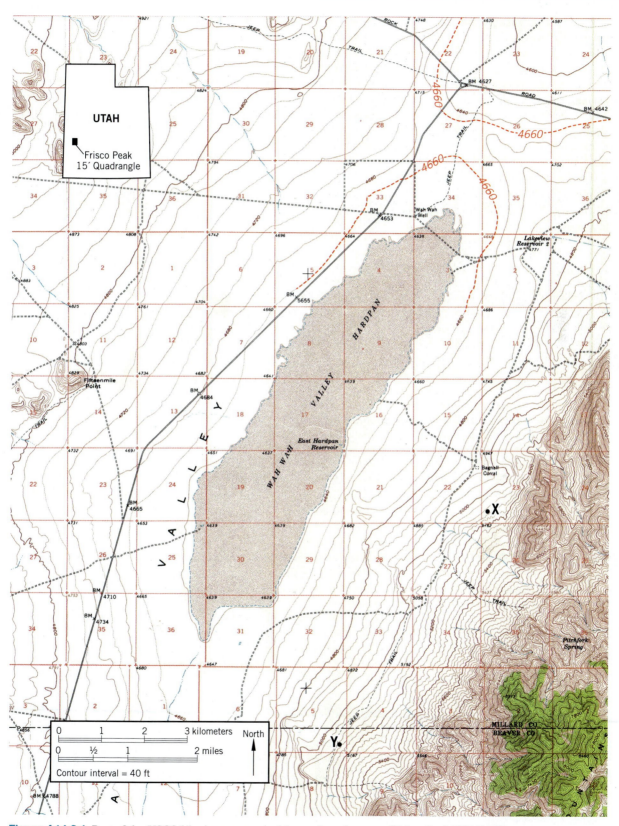

Figure A14.3.1 Part of the USGS 15-minute topographic quadrangle map of Frisco Peak, UT (1960).

A What specific type of feature is the Wah Wah Valley Hardpan?

B If the Wah Wah Valley Hardpan were to fill with water, how deep could the lake become before it overflows to the northeast along the jeep trail, near the red section number 27 located just above the northeast end of the hardpan in **Fig. A14.3.1**? Show your work.

C On **Fig. 14.5**, notice the curves that resemble topographic contours such as the one along which the points labeled X and Y are located. These curves are low, step-like terraces that go all around the valley, like bathtub rings. How do you think these terraces formed?

D On the aerial photo (**Fig. 14.5**) and topographic map (**Fig. A14.3.1**), what evidence can you identify for a former deeper lake—an arm of Lake Bonneville—in Wah Wah Valley at locations X and Y? What was the elevation of the upper surface of that ancient lake?

E On the topographic map, use a blue colored pencil to draw where the shoreline of ancient Lake Bonneville was when the lake filled to its highest elevation using your answer for part **D**. Then use the blue pencil to color in the area that used to be under water.

F Studies by geologists of the Utah Geologic Survey and USGS indicate that ancient Lake Bonneville stabilized in elevation at least three times before present: 5100 ft. about 15,500 years ago, 4740 ft. about 14,500 years ago, and 4250 ft. about 10,500 years ago.

1. What is the age of the lake level that you identified in part **D**?

2. Modern Great Salt Lake has an elevation of about 4200 ft. and is 30 ft. deep. It occupies part of the large basin that was once filled by Lake Bonneville. How deep was Lake Bonneville at the current location of the Great Salt Lake when Lake Bonneville was at its highest level?

G **REFLECT & DISCUSS**   Explain how the climate must have changed in Utah over the past 15,500 years to cause the fluctuations in levels of Lake Bonneville investigated above. In your answer, consider the times identified in part **F** and conditions in Utah today.

Name: _____ **Course/Section:** _____ **Date:** _____

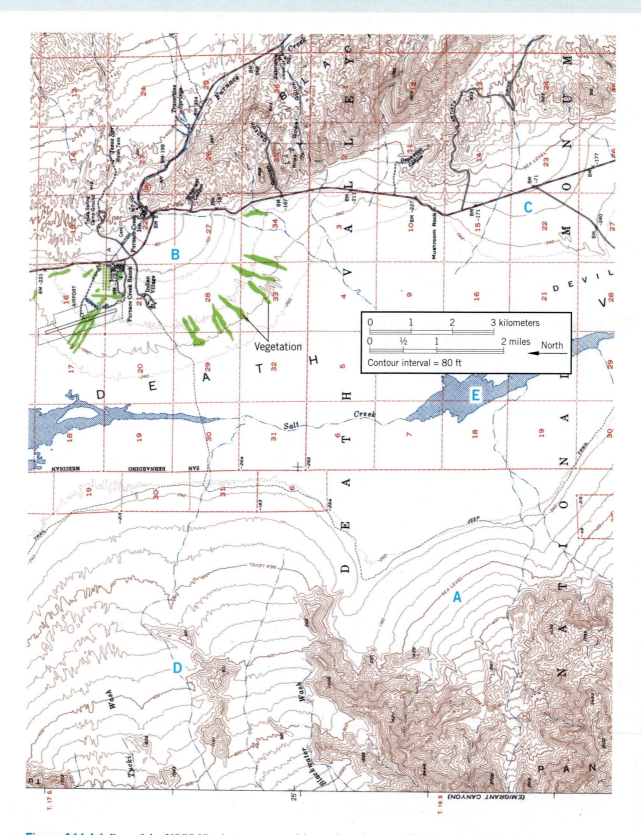

Figure A14.4.1 Part of the USGS 15-minute topographic quadrangle map of Furnace Creek, CA (1952).

A Analyze the topographic map in **Fig. A14.4.1** and complete the following tasks. Refer to **Fig. 14.8** for guidance.

1. Carefully and neatly color alluvial fan A yellow, including the places at the top (upslope end) of the fan where it extends up into two canyons. Do the same for alluvial fans B and C.

2. Color the inselbergs brown in the vicinity of location D.

3. Color the 00 (sea level) topographic contours blue on both sides of the valley.

4. Make an orange line along the downhill edge of the *mountain front* (**Fig. 14.7**) on both sides of the valley.

5. What is the elevation of the lowest point on the map? _____ ft.

B Notice the intermittent stream that drains from the upstream end of the alluvial fan system A (that you have already colored yellow) to the playa at E. How would the grain size of the sediments along this stream change as you walk downslope from the highest part of the fan, up in the canyons, to the playa (E)? Why?

C Carefully examine **Fig. A14.4.1** for evidence of normal faults on either or both sides of Death Valley. Refer to **Fig. 14.10** for ideas. Draw a dark dashed line with a regular pencil or a black colored pencil wherever you think a normal fault might reach the ground surface on either the east or west sides of the valley.

D Almost a decade of measurements indicate that relative to the stable interior of the North American Plate (NAD08), GPS site P462 moves toward azimuth ~309.7° at a rate of 3.90 mm/yr., and site P596 moves toward 315.5° at 5.79 mm/yr. (yellow arrows in **Fig. 14.10**). Both GPS sites are part of the Plate Boundary Observatory developed as part of the EarthScope Project. Relative to P462, GPS site P596 moves toward 327.1° at 1.95 mm/yr. (orange arrow in **Fig. 14.10**).

1. Block faulting typical of the Basin and Range tends to involve stretching roughly perpendicular to normal faults as shown in **Figs. 14.7A–D**. Extension in a pull-apart basin tends to be roughly parallel to strike-slip faults as shown in **Figs. 14.7E–F**. Based on the orange arrow originating at GPS site P596 in **Fig. 14.10** that shows the motion of the west side of Death Valley relative to the east side, is the current widening of Death Valley more likely related to Basin-and-Range block faulting or to development of a pull-apart basin? Explain your reasoning.

2. The strike-slip faults to the north and south of Death Valley have been active for around 12 Myr (**Fig. 14.10**). If P596 moved in the same direction at the same rate relative to P462 for 12 Myr, how far would P596 move? _____ km About how far is it from the bedrock on the east side of the valley to the top of the alluvial slope on the west side of the valley measured parallel to the orange arrow and the orange strike-slip faults? _____ km

E **REFLECT & DISCUSS** Notice that people chose to build a ranch on alluvial fan B even though this entire region is dry land. What do you think was the single most important reason why those people chose alluvial fan B for their ranch instead of one of the other fans?

Coastal Processes, Landforms, Hazards, and Risks

Contributing Authors

James G. Titus • *U.S. Environmental Protection Agency*

Donald W. Watson • *Slippery Rock University*

Pre-Lab Video 15

https://goo.gl/z8Fy4c

▲ Tilted marine sedimentary rock layers of the Monterrey Formation (Miocene) emerge from beach sand on the east side of Point Dume along the Malibu coastline, southern California (34.005°N, 118.803°W).

BIG IDEAS

A coastline is the boundary between land, atmosphere, and an ocean or lake. It is also an important transition zone between the marine and nonmarine environments and their interdependent ecosystems. Specific factors like waves, erosion, sediment supply, biological activity (mangroves, coral reefs), storms, and sea-level changes combine to shape coastal landforms. The dynamic nature of the coastal environment presents a host of potential hazards and risks for humans who choose to live there. While a variety of engineered structures have been designed to modify coastlines and mitigate some of the risks, the most sustainable solution might be to leave the most dynamic coastal environments in their natural state, thereby avoiding many of the risks to humans.

FOCUS YOUR INQUIRY

Think About It What factors affect the shape and position of coastlines?

ACTIVITY 15.1 Coastline Inquiry (p. 396, 405)

ACTIVITY 15.2 Introduction to Coastlines (p. 398, 406)

- -

Think About It How successful are efforts to protect coastlines from erosion by building artificial structures?

ACTIVITY 15.3 Coastline Modification at Ocean City, Maryland (p. 400, 407)

- -

Think About It How will rising sea level affect coastal communities?

ACTIVITY 15.4 The Threat of Rising Seas (p. 402, 408)

- -

Introduction

When viewed from an airplane or satellite, coasts appear to be very simple—the strips of land that are next to the salt water of the ocean. But at ground level, one cannot help but notice that coastlines are dynamic systems characterized by constant change. Wind is blowing, water is flowing, rocks are eroding, sediment is moving about, and landscapes are being shaped. Organisms are an important part of this transitional environment, both as anchors of ecosystems and as suppliers of sediment. The United Nations Environment Program (UNEP) reports that more than half of the world's population lives within 60 km of an ocean, and three-quarters of all large cities are located on a coast. Humans find many resources along coasts, but they also face hazards associated with living in such a dynamic environment. Rising global sea level will add to these challenges in the decades to come.

ACTIVITY 15.1

Coastline Inquiry, (p. 405)

Think About It What factors affect the shape and position of coastlines?

Objective Compare and contrast photographs of coastlines and determine what factors primarily affect them.

Before You Begin Read the following section: Dynamic Natural Coastlines.

Dynamic Natural Coastlines

A **coastline** is the boundary where atmosphere, ocean, and land meet. It can look like a simple line on the map, but it is constantly changing as waves, tides, storms, and climate exert their influences on it. Biological activity also exerts a strong influence on coastlines from the stabilizing effects of grasses and mangroves to the countless incremental changes wrought by all of the organisms that live and die along a coast.

Some examples of coastlines are pictured in **Fig. 15.1**. In each case, the land is acted upon by water, wind, organisms, and sometimes ice in ways that vary in both intensity and time. For example, there is constant water and air motion, but their intensities vary throughout the day in response to tides and the weather. At some times, **erosional processes** (those that remove sediment and cut into rock, reefs, and marshes) might be dominant over **depositional processes** (those that cause sediment to accumulate and marshes or reefs to grow). At other times, the depositional processes might be dominant over the erosional ones. Over longer periods of time, erosion or deposition is generally dominant, so most coastlines are either receding (moving landward, eroding back) or advancing (building seaward).

Factors That Affect Coastlines

There are many factors that affect the shapes of coastal landforms and the overall positions of coastlines, but here are some of the most important:

- **What the land is made of** determines how much the agents of change must work on the land to shape it. The land may be hard rock, clay, sand, large loose rocks, or a combination of these. Humans might modify the coastline by "armoring" it with large boulders (called *rip rap*) or rigid concrete structures, including seawalls, breakwaters, jetties, and groins.
- **Supply of sediment** carried to a specific location along a coastline by rivers, coastal currents, or people often determines whether the coastline is made of sand or of bare rock and whether the position of the coastline is advancing or receding. Marine organisms also supply sediment, and along beaches dominated by carbonate sands, these organisms supply almost all of the sediment. Sediment supply can be seasonal as alternating wet and dry seasons lead to the rise and fall of local rivers supplying sediment to the beach. Some sediment may be eroded locally along the coast as when waves undercut a cliff and rocks collapse into the water.
- **Waves** carry sediment onto beaches when they are gentle (**low-energy waves**), but they remove sediment from beaches and erode the land when they are large and forceful (**high-energy waves**). Particles moved by waves and blasted against rocky surfaces will cause abrasion—smoothing and wearing down of the rocky surfaces. The direction of the waves is a factor in what direction sediment is moved and what parts of a coastline are eroded the most.
- **Wind** interacts with the surface of the water to generate the waves and blows beach sand into dune forms on the adjacent land.
- **Currents** running along the coastline (**longshore currents**), in streams reaching the coastline (stream currents on deltas), and back-and-forth through coastal environments (**tidal currents**) move sediment about and redeposit it on beaches, sand bars and spits, and tidal flats.
- **Storms** are highly energized systems, so they are one of the main factors that determines the shape of coastal landforms. A single storm, like a hurricane, can significantly erode one part of a coastline and deposit a large volume of sediment on another part of the coastline.
- **Tsunamis** are pulses of water generated by earthquakes or large submarine landslides that can cause massive flooding, erosion, and the transportation and deposition of sediment. Tsunamis can dramatically alter coastal environments and ecosystems and pose a significant hazard to people and structures within their reach.
- **Changing sea level** is generally a gradual process that leads to the migration of the coastline either toward the land (a transgression) or toward the ocean (a regression). As we will see in a later section, sea level changes in a particular area can be due to local

A. Maryland coastline with saltmarsh (NOAA)

B. San Francisco, California coastline (NOAA)

C. Oregon coastline (NOAA)

D. North Carolina coastline (NOAA)

E. Destin, Florida urbanized coastline

F. Florida Keys coastline with mangrove plants

G. Maine coastline (NOAA photo by Albert E. Theberge)

H. Caribbean island coastline with reefs (NOAA)

Figure 15.1 **Photographs of eight different coastlines.**

causes like uplift or subsidence of the coast, and sea level can change globally due to the effects of climate change. Global sea level is currently rising at a rate of about 3 mm/year. On longer timespans, tectonic movements cause ocean basins to open and close and affect their size, shape, and depth.

- **Humans and other organisms** modify their environment. Corals construct reefs that armor the coastline against erosion. Marsh plants and mangroves trap and bind sediment with their roots and absorb the energy of storms. Humans use a variety of methods to preserve and build up coastlines, but they also destroy marshes and reefs and otherwise degrade elements of the coastline.

Submergent and Emergent Coasts

Over decades of time, geologists characterize coastal areas as submergent or emergent.

- A **submergent coast** is one that is subject to an increase in sea level relative to the local land surface, so the coastline is moving landward (**Fig. 15.2**). This can be due to a variety of causes, including local or regional subsidence—sinking—of the land surface or global sea level rise. The result is that coastal land that was once above sea level is now flooded. Geologists refer to this as a **transgression** of the ocean and its sediments onto the land. There is a characteristic sequence of transgressive sedimentary deposits that can be recognized in the rock record of ancient coastal environments.

- An **emergent coast** is one that is subject to a decrease in sea level relative to the local land surface, so the coastline is moving seaward (**Fig. 15.3**). Local or regional uplift or a decrease in global sea level can cause emergence, resulting in a larger area of land exposed above sea level. This is called a **regression** of the ocean and its sediments, which leads to a set of characteristic features that can be recognized in the sedimentary record of ancient coasts. A regression typically involves more erosion than deposition compared with a transgression.

Coasts can also be affected by processes that are considered **constructional processes**, such as the growth of a delta that extends the land surface seaward or the development of a fringing reef that serves as a breakwater for a coastline or island. These features can exist along either emergent or submergent coastlines. For example, the Louisiana coastline throughout the Mississippi Delta is submergent except where sediment is being added to the top of the delta, so dikes and levees have been built in an attempt to keep the ocean from flooding New Orleans. Some of those flood-control structures failed in Hurricane Katrina. However, the leading edge of the Mississippi Delta continues to extend out into the Gulf of Mexico because of the vast supply of sediment being carried there and deposited from the Mississippi River.

Figs. 15.2 and **15.3** illustrate some coastal features that you will need to identify in **Figs. 15.4**, **15.5**, and **15.6**. Study these features and their definitions below.

- **Delta**—a sediment deposit at the mouth of a river where it enters an ocean or lake (**Fig. 15.4**).
- **Barrier island**—a long, narrow island that parallels the mainland coastline and is separated from the mainland by a lagoon, tidal flat, or salt marsh (**Figs. 15.2** and **15.5**).
- **Beach**—a gently sloping deposit of mobile sediment along a coastline, usually extending to just above the highest-tide line. Beach sediment generally includes sand through gravel-sized particles.
- **Washover fan**—a fan-shaped deposit of sand or gravel transported and deposited landward of a beach or barrier island during a "washover" event that occurs during a storm or very high tide.
- **Berm crest**—the highest part of a beach; it separates the **foreshore** (seaward part of the beach) from the **backshore** (landward part of the beach) and often represents the highest tide level.
- **Estuary**—a river valley flooded by a rise in the level of an ocean or lake (**Fig. 15.2**). A flooded glacial valley is called a **fjord**.
- **Headland with cliffs**—projection of land that extends into an ocean or lake and generally has cliffs along its water boundary (**Fig. 15.2**).
- **Spit**—a sand bar extending from the end of a beach into the mouth of an adjacent bay as a result of sand transported by longshore currents (**Fig. 15.2** and **Fig. 15.3**).
- **Tidal flat**—muddy or sandy area that is covered with water at high tide and exposed at low tide.
- **Salt marsh**—a marsh that is flooded by ocean water at high tide (**Fig. 15.2** and **Fig. 15.3**).
- **Wave-cut cliff** (or *sea cliff*)—a seaward-facing cliff along a steep coastline produced by wave erosion (**Fig. 15.2**).
- **Wave-cut platform**—a bench or shelf at sea level (or lake level) along a steep shore and formed by wave erosion. Wave-cut platforms are best developed along emergent coastlines (**Fig. 15.3**).
- **Marine terrace**—an elevated wave-cut platform that is bounded on its seaward side by a cliff or steep slope and formed when a wave-cut platform is elevated by uplift or regression along an emergent coastline (**Fig. 15.3**).

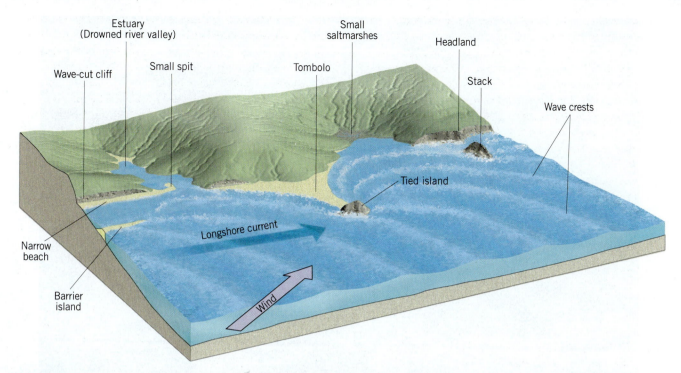

Figure 15.2 Submergent (transgressive) coastline features. A transgressive coastline is caused by a rise in sea level, subsidence of the land, or both. As the ocean transgresses onto the land and the land is submerged, waves erode cliffs to create wave-cut platforms, valleys are flooded to form estuaries, old wetlands are submerged and new ones form at higher levels, bays deepen, beaches narrow, some existing islands are submerged, and new islands form from highlands that become separated from the mainland as sea level rises.

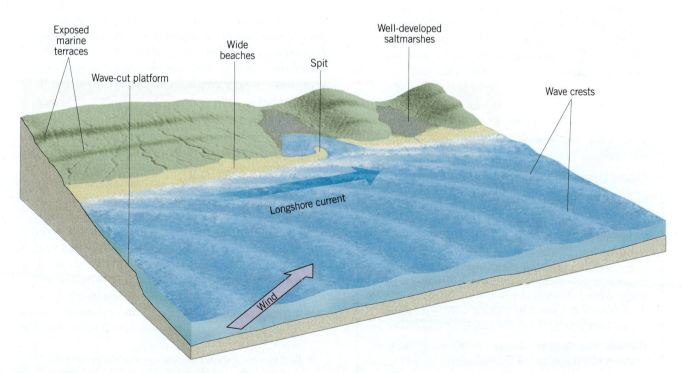

Figure 15.3 Emergent (regressive) coastline features. A regressive or emergent coastline is caused by a drop in sea level, uplift of the land, or both. Emergence can result in expansion of tidal flats and coastal wetlands, exposure of wave-cut terraces, and the development of wide stable beaches.

Figure 15.4 Photograph of the Po Delta region, northern Italy. Plumes of sediment from the southern Alps, northern Apennine Mountains, and Po Valley enter the Adriatic Sea through the Po River Delta. Photo centered around 44.974°N, 12.205°E. (NASA astronaut aboard Space Shuttle *Challenger* on October 9, 1984 [STS41G-41-19].)

- **Stack**—an isolated rocky island near a headland cliff (**Fig. 15.2**).
- **Tombolo**—a sand bar that connects an island with the mainland or another island. Tombolos are best developed along submergent coastlines (**Fig. 15.2**).
- **Tied island**—an island connected to the mainland or another island by a tombolo (**Fig. 15.2**).

ACTIVITY 15.3

Coastline Modification at Ocean City, Maryland, (p. 407)

Think About It How successful are efforts to protect coastlines from erosion by building artificial structures?

Objective Identify the common types of artificial structures used to modify coastlines and explain their effects on coastal environments.

Before You Begin Read the following section: Human Modification of Coastlines.

Plan Ahead You will need a pencil, ruler, and calculator.

Human Modification of Coastlines

Humans create several types of coastal structures using durable materials like reinforced concrete or boulders in order to protect harbors, build up sandy beaches, or extend the coastline. Study the following four kinds of structures and their effects on the motion of water and sediment in **Fig. 15.6**.

- **Seawall**—a wall or embankment constructed along a coastline to prevent erosion by waves and currents.
- **Breakwater**—an offshore wall constructed parallel to a coastline to block waves. The longshore current is halted behind such walls, so the sand accumulates there, and the beach widens. Where the breakwater is used to protect a harbor from currents and waves, sand often collects behind the breakwater and might have to be dredged.
- **Groin**—a short wall constructed perpendicular to a coastline in order to trap sand along a beach. Sand carried parallel to the beach by the longshore current accumulates behind the groin, on the up-current side of the groin.
- **Jetties**—long walls extending from shore at the mouths of harbors and used to protect the harbor entrance from filling with sand or being eroded by waves and currents.

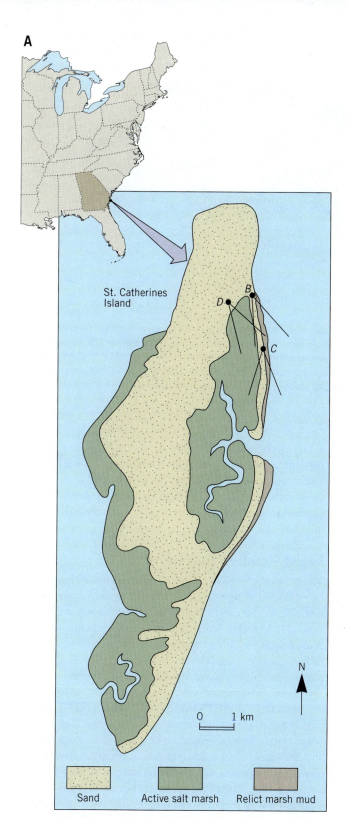

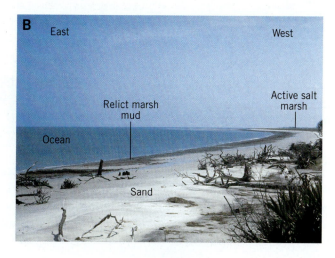

Figure 15.5 St. Catherines Island, Georgia. St. Catherines Island is part of a barrier island complex that has been actively evolving for thousands of years. **A.** Geological sketch map showing distribution of sand, active salt marsh with its living community of plants and animals, and relict marsh mud on St. Catherines Island, centered around 31.634°N, 81.157°W. Some of the sand in the interior of the island is ancient (Pleistocene), but the sand along the eastern edge of the island is thinner and more recently deposited. The locations where the three photographs were taken are indicated at points *B*, *C*, and *D*, along with lines indicating the field of view of the photos. **B.** View south–southeast from point *B* on map at low tide. The dark-brown "ribbon" adjacent to ocean is relict marsh mud. Light-colored area is sand. **C.** View south from point *C* on map at low tide. **D.** View southeast from point *D* (Aaron's Hill) on map.

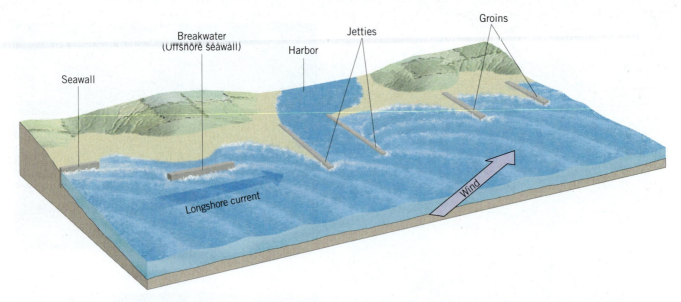

Figure 15.6 Coastal structures. Seawalls are constructed along the shore to stop erosion of the shore or extend the coastline (as sediment is used to fill in behind them). A breakwater is a type of offshore seawall constructed parallel to coastline. The breakwaters stop waves from reaching the beach, so the longshore drift is broken and sand accumulates behind them (Instead of being carried downshore with the longshore current). Groins are short walls constructed perpendicular to shore. They trap sand that is transported by longshore currents. Jetties are long walls constructed at entrances to harbors to keep waves from entering the harbors. However, they also trap sand just like groins.

ACTIVITY 15.4

The Threat of Rising Seas, (p. 408)

Think About It   How will rising sea level affect coastal communities?

Objective   Describe the likely prospect of global sea-level rise and analyze the coastal hazards and increased risks it may cause.

Before You Begin   Read the following section: Changing Sea Level.

Plan Ahead   You will need a pencil and calculator.

Changing Sea Level

Coastal areas are uniquely sensitive to changes in sea level. The average or **mean sea level** is the average height of the world's ocean surface once the effects of wind waves, tidal variations, and other short-duration processes are factored out. The **geoid** is a measured or modeled surface that passes along the mean sea level of the world's oceans and is a surface of equal gravitational potential energy. Knowing where the geoid is in a given place away from the world oceans allows us to determine the elevation of the ground surface. Since the early 1990s, data from orbiting satellites that carry instruments designed to measure Earth's gravitational field and surface shape with great accuracy have been used by geoscientists to map the geoid. Differences between the geoid and the actual level of the sea surface

at any given location and time arise due to a variety of factors, including ocean currents, variations in salinity and temperature of sea water that result in differing water densities, variation in air pressure, tides, wind-driven waves, and so on. Analysis of all the relevant data allows us to measure and monitor change in the global sea level.

Local Relative Sea Level

The mean sea level at a given place along a coastline can be estimated using tide gauge records to yield a **relative sea level**. There are two high tides and two low tides every day due to the rotation of Earth and the gravitational pull of the Sun and Moon. Tide gauge records are affected by a variety of factors other than tides, including storm surges. A **storm surge** is an abnormal rise of water pushed landward by high winds and/or low atmospheric pressure associated with storms. The National Oceanic and Atmospheric Administration (NOAA) expresses the storm surge as the height of the sea surface above the expected tide level, reflecting the idea that the storm surge is added to the normal tide. NOAA also measures **storm tide**—which it defines as the height of the sea surface caused by a combination of the normal tide level and storm surge. Storm surges can cause the ocean to rise by as much as about 4 m above the normal astronomical tide, depending on the magnitude of the storm and other factors. However, except for hurricanes, most storm surges are in the range of 1 m or less. The effects of high and low atmospheric pressure, wind, and storm surges must be removed in order to obtain a reasonable estimate of mean sea level at a given location. NOAA tide gauge records indicate that sea level at Ocean City, Maryland (**Fig. 15.7**), has been rising at about 5.5 mm/yr. since 1975, or nearly twice as fast as the increase in global sea level.

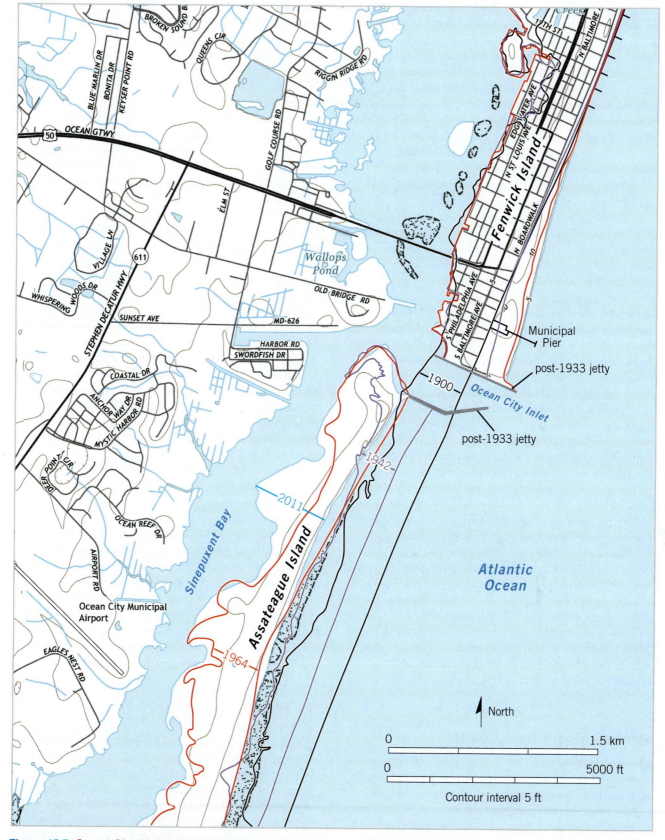

Figure 15.7 Ocean City, Maryland. Variation in shape and position of the barrier island at Ocean City between 1901 and 2011, shown on a USGS topographic map published in 2014.

Long-term changes in local sea level over tens to thousands of years can result from changes in the elevation of the land surface and from changes in global sea level. Areas that were covered by continental ice sheets within the past 20,000 years, such as Canada and the northern part of the continental United States, are rising as a result of **glacial isostatic adjustment** (GIA), also known as **glacial rebound**. Continental lithosphere is bowed down by the weight of a glacier and slowly rebounds elastically when that weight is removed through melting. Near Hudson's Bay, Canada, the rate of uplift related to GIA is as great as ~1 cm/yr., but this effect drops to near zero along most of the Atlantic coastline in the United States.

In other places, the land surface is sinking or subsiding. For example, the land surface in the Mississippi Delta area is sinking due to sediment loading, compaction, and normal faulting, so the City of New Orleans continues to subside below sea level. Other coastal cities such as Houston experience subsidence due to withdrawal of groundwater, petroleum, and other hydrocarbons from subsurface reservoirs. Normal faulting toward the Gulf of Mexico is another cause of subsidence in the Houston area. Gulf Coast cities are already subject to flooding during heavy rains and storm surges, so they are particularly vulnerable to the risks associated with rising global sea level.

Tectonic processes involving lithospheric plates also cause the land surface to rise and fall. Divergent motion along continental rifts causes subsidence, as is seen in the East African Rift and in the Basin and Range Province and Rio Grande Rift of western North America. By analyzing very accurate location data collected over decades using GPS, we have learned that points along the coastline from northern California through Washington into British Columbia have been rising and moving eastward relative to points along the Cascade Mountains. The shallow part of the Cascadia Subduction Zone has not slipped in more than 300 years, so three centuries of convergent motion has accumulated through the elastic deformation of the leading edge of the upper (North American) plate. During a future major earthquake along the Cascadia Subduction Zone, points along the current coastline are expected to move west by several meters and down as part of the accumulated elastic deformation is relaxed.

Global Mean Sea Level

Changes in **global mean sea level** over periods of tens to thousands of years occur due to changes in climate, and all indications are that our climate is warming. Increasing the temperature of seawater results in an increase in volume because warm seawater is less dense than cold seawater. Increasing ocean temperature causes sea level to rise. Melting of ice in continental glaciers and ice sheets adds water to the oceans. Melting of the Greenland and West Antarctic ice sheets will be particularly important in

causing sea level to rise during the coming centuries. The complete melting of the Greenland and West Antarctic ice sheets would cause sea level to rise by about 10 m above the present level. The U.S. Geological Survey (USGS) has estimated that a sea-level increase of 10 m would flood the area where about 25% of the U.S. population currently resides.

Sea level was 6–9 m higher during the last major interglacial period (between 129,000 and 116,000 years ago) than it is today. Perhaps this provides a reasonable estimate of the increase in sea level in the foreseeable future beyond 2100. The impact of a 6–9 m rise in global sea level will be severe in many places beyond the coastlines of North America. Most of the Netherlands is below an elevation of 10 m. Major population centers in China, Vietnam, Thailand, Indonesia, Egypt, and Bangladesh are below 10 m. Some entire countries such as the Maldives and other island nations will be submerged when the Greenland and West Antarctic ice sheets are dramatically reduced or completely melted by warming temperatures.

The National Aeronautics and Space Administration (NASA) reports that sea level has risen at a rate of about 3 mm/yr. since 1993 and that ice loss by glaciers in Greenland doubled between 1996 and 2005. It will likely take centuries to melt these ice sheets, but that is no justification for complacency or inaction now. Continuing to produce greenhouse gases at our current rate commits us to temperatures that will make the melting of the Greenland and West Antarctic ice sheets inevitable.

There is an ethical dimension to the choices that we make, individually and as a society. While climate is the result of many natural processes, the anthropogenic contribution to climate change is the only part of this system that we can control. We need to consider the safety and needs of people who are at the most immediate risk of sea-level rise. We need to think beyond our own self-interest to consider the effects of our choices and decisions on the world that we will leave to our grandchildren's children. The geosciences will continue to provide reliable information to inform those decisions, but society must be wise enough to use that information to avoid the most severe potential consequences of a warming climate.

MasteringGeology™

Looking for additional review and test prep materials? Visit the Study Area in MasteringGeology to enhance your understanding of this chapter's content by accessing a variety of resources, including Pre-Lab Videos, Self-Study Quizzes, Geoscience Animations, Mobile Field Trips, *Project Condor* Quadcopter videos, *In the News* articles, glossary flashcards, web links, and an optional Pearson eText.

Name: _____ Course/Section: _____ Date: _____

A Refer to the photographs of coastlines in **Fig. 15.1** and the list of Factors Affecting Coastlines on pages 396 and 398.

1. For each of the following coastal areas (see **Fig. 15.1**), describe the type of geologic material(s) that are along the coastline if visible. Then name the two or three most important factors that determine the characteristics of the coastline.

   (a) Maryland coastline with salt marsh grasses rooted in clay

   (b) San Francisco, California, coastline

   (c) Oregon coastline

   (d) North Carolina coastline

   (e) Destin, Florida, urbanized coastline

   (f) Florida Keys coastline with mangrove plants

   (g) Maine coastline (note person for scale)

   (h) Caribbean island coastline with fringing reefs (i.e., reefs attached to the island) and a barrier reef

B **REFLECT & DISCUSS**

1. Which of the eight coastlines shown in **Fig. 15.1** is building out or expanding into the water, and what is causing that to happen?

2. Which coastline gives you the impression that the people who live there are not very concerned about rising sea level?

3. Along which coastline(s) will the average position of the coastline on a map change the most when sea level rises by a few meters?

Name: _____ Course/Section: _____ Date: _____

A Refer to the photograph of the Po Delta, Italy (**Fig. 15.4**). During Etruscan times around 600 BC, the city of Adria was a busy seaport on the coastline at the mouth of the Po River. Adria was so important that the Adriatic Sea was named after it. The Po River has continued to deposit sediment at its mouth and has extended its delta far beyond Adria, which is no longer located on the coastline.

1. What has been the average annual rate (in cm/yr.) at which the Po Delta has extended toward the Adriatic Sea since Adria was a thriving seaport on the coast? Show your work.

2. Based on the average annual rate that you just calculated, how many centimeters does the leading edge of the Po Delta move seaward during the lifetime of someone who lives to be 60 years old? Show your work.

3. **REFLECT & DISCUSS** Sea level is rising and submerging coastlines adjacent to the Po Delta, and yet the delta is still extending out into the Adriatic Sea. Why?

B Refer to the map and photographs of St. Catherines Island, Georgia (**Fig. 15.5**). Note that on the southwestern and east-central parts of the island there are large areas of salt marsh (**Fig. 15.5A**). Living salt marsh plants are present there as shown on the right (west) in **Figs. 15.5B** and **15.5C**. A sandy beach that occurs on the east side of the island can be seen in **Figs. 15.5B** and **15.5C** bounded on its seaward side (left) by another strip of marsh mud. This is a relict salt marsh consisting of mud from an ancient salt marsh that is eroding along the coastline.

1. What type of sediment is probably present beneath the beach sands in **Figs. 15.5B** and **15.5C**?

2. Explain how you think the beach sands were deposited landward of the relict marsh mud.

3. Portions of the living salt marsh wetland in **Fig. 15.5C** recently have been buried by white sand that was deposited from storm waves that crashed over the beach and sand dunes. What is the name given to this type of sand body?

4. **Fig. 15.5D** was taken from a landform called Aaron's Hill. It is the headland of this part of the island. What do you think will eventually happen to Aaron's Hill as sea level rises? Why?

5. Based upon your answer in part 4, would Aaron's Hill be a good location for a resort hotel? Explain your answer.

6. **REFLECT & DISCUSS** Based upon your inferences, observations, and explanations above, what will eventually happen to the living salt marsh in **Figs. 15.5B** and **15.5C**?

Name: _____ **Course/Section:** _____ **Date:** _____

Ocean City is located on Fenwick Island, a long, narrow barrier island (**Fig. 15.7**). During a severe hurricane in 1933, the island was breached by tidal currents that formed Ocean City Inlet and split the barrier island in two. Ocean City is still located on what remains of Fenwick Island. The city is a popular vacation resort that has undergone much property development over the past 50 years. South of Ocean City Inlet, Assateague Island has remained undeveloped as a state and national seashore.

Rising sea level at Ocean City has increased the risk of beach erosion there, so barriers have been constructed there to trap sand. Examine the portion of the Ocean City, Maryland, topographic quadrangle map provided in **Fig. 15.7**. Outlines of the coastline as mapped in 1900 (black curves), 1942 (purple), 1964 (orange), and 2011 (blue) are shown on a portion of the USGS topographic map published in 2014, illustrating how this coastline has evolved over time.

A After the 1933 hurricane carved out a tidal channel through the barrier island just south of Ocean City, the Army Corps of Engineers constructed a pair of jetties on each side of the inlet to keep it open. These are labeled *post-1933 jetty* on **Fig. 15.7**. Sand filled in behind the northern jetty, so it is now a seawall forming the straight southern edge of Ocean City on Fenwick Island. Based on this information, would you say that the longshore current along this coastline is traveling north to south or south to north? Explain your reasoning.

B Notice that since 1933 Assateague Island has migrated landward (west) relative to its 1900 position (**Fig. 15.7**).

1. Why did Assateague Island migrate landward?

2. Field inspection of the west side of Assateague Island reveals that muds of the lagoon (Sinepuxent Bay) are being covered up by the westward-advancing island. What was the rate of Assateague Island's westward migration from 1942–1964 in m/year? Show your work.

3. Based on your last answer in part **B2** and extrapolating from 1964, in what approximate year would you predict that Sinepuxent Bay should cease to exist as the west side of Assateague Island moves westward? Show your work.

4. Notice from the position of Assateague Island in 2011 that it has not merged with salt marshes of the mainland. What natural processes and human activities might have prevented this?

C Notice the short black lines that represent groins that have been constructed on the east side of Fenwick Island (Ocean City) in the northeast corner of **Fig. 15.7**, about 2 km north of the inlet.

1. Why do you think these groins have been constructed there?

2. What effect could these groins have on the beaches around Ocean City's Municipal Pier at the southern end of Fenwick Island? Why?

D Hurricanes normally approach Ocean City from the south–southeast. In 1995, Hurricane Felix approached Ocean City but turned back out into the Atlantic Ocean before making landfall. How does the westward migration of Assateague Island increase the risk of hurricane damage to Ocean City?

E **REFLECT & DISCUSS** The westward migration of Assateague Island might be halted or even reversed if all of the groins, jetties, and seawalls around Ocean City were removed. How might removal of all of these structures affect the risk of environmental damage to properties in Ocean City?

F **REFLECT & DISCUSS** There has been a community of Ocean City on Fenwick Island since before 1900—for more than a century. The highest elevation in Ocean City is about 3 m (10 ft.), and most of the city is below 1.5 m (5 ft.). Do you think Ocean City is sustainable for the next century? Explain your answer.

Name: _____   **Course/Section:** _____   **Date:** _____

In planning for coastal management and safe and economical coastal development, responsible planning commissions and real estate developers should "play it safe" and assume that sea level will continue to rise. There are many predictions of future rises in global mean sea level, but regional trends should also be considered as in these examples.

A Imagine that you are planning to buy a shorefront property in Ocean City, Maryland, this year. You plan to use the property for family vacation getaways over the next 50 years and then sell the property. The ground floor of the property was 1.2 m above mean sea level in 2010.

1. According to NOAA, the historic rate of sea level rise here since 1975 has been 5.48 +/− 1.67 mm/year. Using the "plus or minus" error, what has been the minimum rate and the maximum rate of mean sea level rise here in mm/year?

   (a) _____ mm/yr. minimum rate
   (b) _____ mm/yr. maximum rate

2. Using the minimum and maximum rates above and recalling that 1 in. − 25.4 mm, calculate how much sea level will rise in mm and in. at Ocean City over the next 50 years.

   (a) _____ mm minimum
   (b) _____ in. minimum
   (c) _____ mm maximum
   (d) _____ in. maximum

3. Local mean sea level is the average position of sea level between low and high tides. High tides occasionally reach 0.88 m (2.9 ft.) above mean sea level here, and storm surges often raise sea level an additional 0.3 m (1 ft.). When Hurricane Sandy passed offshore of Ocean City in 2012, the storm surge caused a total storm tide of 3.59 feet. (Further north at Bergen Point tide gauge, Staten Island, New York, the National Weather Service reports a 2.87 m storm surge on top of a 1.57 m high tide, bringing the water level to 4.44 m above local mean sea level during Hurricane Sandy.) Given these natural day-to-day variations in sea level and the prospect of sea level rise calculated above, would it be a wise decision to purchase the shorefront property that you planned to buy? Explain your reasoning.

4. The city of Ocean City expects the following temporary increases in sea level due to storm surges in hurricanes. How would this affect your purchasing decision? Why?

   Category 1 hurricane: 74–95 mph winds, storm surge 1.22–1.52 m (4–5 ft.)
   Category 2 hurricane: 96–110 mph winds, storm surge 1.83–2.44 m (6–8 ft.)
   Category 3 hurricane: 111–130 mph winds, storm surge 2.74–3.66 m (9–12 ft.)
   Category 4 hurricane: 131–155 mph winds, storm surge 3.96–5.49 m (13–18 ft.)
   Category 5 hurricane: >156 mph winds, storm surge >5.49 m (>18 ft.)

B **REFLECT & DISCUSS**   Based on your analysis of data in part A, what would you suggest as the minimum elevation of the floor of any new house or commercial building constructed along the Ocean City coast? Explain.

Earthquake Hazards and Human Risks

Pre-Lab Video 16

https://goo.gl/gMj34o

Contributing Authors

Thomas H. Anderson • *University of Pittsburgh*
David N. Lumsden • *University of Memphis*

Pamela J.W. Gore • *Georgia Perimeter College*

▲ The Gorkha earthquake in Nepal on April 25, 2015 caused ~9,000 deaths and ~22,000 injuries, and left hundreds of thousands homeless. Casualties were primarily due to building collapse, landslides, and avalanches. The magnitude 7.8 earthquake occurred on a thrust fault that contributes to the uplift of the Himalaya Mountains.

BIG IDEAS

Earthquakes are natural vibrations that originate below Earth's surface. An earthquake in the upper crust occurs when energy that had been stored in elastically deforming rock is quickly released and moves outward from the earthquake source area in the form of seismic waves. Most earthquakes recorded by seismographs are too small to be felt or cause damage, but damage associated with the largest earthquakes is often catastrophic when they occur at shallow depths near populated areas. Vertical displacement of a fault under water during an earthquake can generate a tsunami that can cause great destruction far beyond the area shaken by the earthquake. Information collected by networks of seismographs not only tells us about the location of the earthquake source but also provides a wealth of data about Earth's interior. Geoscientists provide information to engineers, architects, builders, public planners, politicians, and society in general that can help us to avoid the most dangerous effects of earthquakes.

FOCUS YOUR INQUIRY

Think About It How do bedrock and sediment behave during earthquakes, and how does this affect human-made structures?

ACTIVITY 16.1 Earthquake Hazards Inquiry (p. 410, 417)

Think About It How can information about seismic waves be used to locate the epicenter of an earthquake?

ACTIVITY 16.2 How Seismic Waves Travel through Earth (p. 411, 419)

ACTIVITY 16.3 Locate the Epicenter of an Earthquake (p. 412, 421)

Think About It How do geologists use landscape features and focal mechanism studies to analyze fault motions?

ACTIVITY 16.4 San Andreas Fault Analysis at Wallace Creek (p. 413, 423)

ACTIVITY 16.5 New Madrid Seismic Zone (p. 413, 425)

Introduction

A few times every year, we hear of some country or region that is trying to cope with the devastating effects of a large earthquake. Journalists tell us about the magnitude of the earthquake, the estimated numbers of deaths and injuries, and the number of people left homeless. Security camera video footage of goods shaking off of store shelves, of people trying to remain standing, and of windows vibrating until they break out of their frames are common. Buildings collapse, landslides bury communities, and critical access roads are destroyed during earthquakes. Fires caused by broken gas lines can sweep through an area with a diminished capacity to respond. The video record of the deadly tsunamis associated with the Indian Ocean earthquake in 2004 and the Tohoku earthquake in Japan in 2011 has added another dimension to our perception of earthquakes as disasters.

Seismology is the scientific study of earthquakes, the way seismic waves move through the Earth, and the information about Earth's structure that we can glean from analysis of seismic waves. Geoscientists who specialize in seismology are called, cleverly enough, **seismologists**. Seismology is used in a wide variety of ways in addition to the study of natural earthquakes. Seismologists work to discover the structure of the crust below Earth's surface and are involved in the exploration for oil, gas, and economic minerals. Seismology is used to map the sea floor, to detect submarines and surface vessels, and to determine whether a seismic event is an earthquake or the detonation of a nuclear device. In this chapter, we will concentrate on earthquake seismology and the information it provides about active faults.

Seismologists and their networks of seismographs detect more than a million earthquakes every year, most of which are too small for anyone to feel. Earthquakes occur below Earth's surface, beginning at the earthquake **focus** and radiating out from that point (**Fig. 16.1**).

The **epicenter** of an earthquake is on Earth's surface directly above the focus. Most earthquakes in the upper crust occur because of frictional slip along faults, but some result from processes related to the movement of magma. Earthquakes that originate at depths of ~70 to 700 km are associated with subduction processes.

Large-magnitude earthquakes that occur within the upper crust are typically associated with a significant displacement along a fault. The great San Francisco earthquake of 1906 was associated with slip along ~476 km of the San Andreas Fault from central California to Cape Mendocino north of San Francisco. The maximum ground displacement along the fault was about 6 m just north of San Francisco. An estimated 3000 people died in that earthquake, whose magnitude is estimated to have been around 7.8. About 80% of San Francisco was destroyed by the 1906 earthquake and the subsequent fire.

The **San Andreas Fault** is the most well-known fault in the United States, but it is far from the only fault that can generate damaging earthquakes. The **Cascadia Subduction Zone** extending from northern California to British Columbia, Canada, is thought capable of producing a magnitude 9 earthquake. The **New Madrid Seismic Zone** (NMSZ) has produced magnitude 7 and perhaps 8 earthquakes in the past few thousand years. More earthquakes occur in Alaska than in any other state, and the magnitude 9.2 Alaskan earthquake of 1964 is one of the largest earthquakes ever recorded anywhere. Many smaller fault systems along the Appalachian Mountains, St. Lawrence Valley, and between the Rocky Mountains and the Pacific Coast can generate magnitude 6 to 7 earthquakes. The **Quaternary Fault and Fold Database of the United States** catalogs the faults that geoscientists have identified as having been active within the past 1.6 million years (http://earthquake.usgs.gov/hazards/qfaults/).

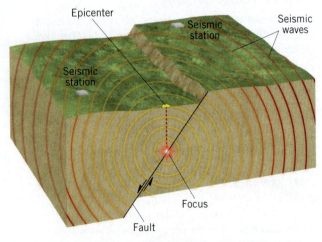

Figure 16.1 The focus and epicenter of an earthquake. Earthquakes originate at the focus, and seismic waves move out from the focus like waves in a pond moving from where a pebble hit the water. The epicenter is on Earth's surface directly above the focus.

ACTIVITY 16.1

Earthquake Hazards Inquiry, (p. 417)

> **Think About It**   How do bedrock and sediment behave during earthquakes, and how does this affect human-made structures?

Objective   Experiment with models to determine how earthquake damage to buildings is related to the Earth materials on which they are constructed. Apply your experimental results to evaluate earthquake hazards and human risks in San Francisco.

Before You Begin   Read the following section: Earthquakes and Seismic Waves.

Plan Ahead   You will need a paper or plastic cup and enough sand to fill it about ⅔ full, enough water to fill it about ¼ full, some coins, and a towel to deal with any spills.

How Seismic Waves Travel through Earth, (p. 419)

Think About It   How can information about seismic waves be used to locate the epicenter of an earthquake?

Objective   Graph seismic data to construct and evaluate travel-time curves for P-waves, S-waves, and surface waves, then use seismograms and travel-time curves to locate the epicenter of an earthquake.

Before You Begin   Read the following section: Earthquakes and Seismic Waves.

Plan Ahead   You will need a pencil and ruler.

Earthquakes and Seismic Waves

Earthquakes involve rapid releases of energy from rock that has been strained, like the snap of a rubber band that has been stretched and released or broken. That energy moves out from the source or focus of an earthquake in the form of **body waves** that move through Earth's interior. There are two forms of body waves. A primary wave or **P-wave** is a compressional wave that is best known to us as a sound wave. As a P-wave moves through a liquid, solid, or gas, particles move very slightly back and forth in the direction the wave is moving, eventually returning to their original position. The other body wave is a secondary wave or **S-wave**, which is a shear wave that causes particles to move back and forth perpendicular to the direction the wave is moving. S-waves can travel only through solids and move at a slower velocity than P-waves.

When body waves reach the ground surface of Earth, they generate another type of wave called a **surface wave** that is confined to the material along Earth's surface in much the same way that wind-driven water waves are confined to the uppermost part of a body of water. There are two types of surface wave that cause the ground surface to move in a rolling up-and-down motion like a cork on a water wave (a **Rayleigh wave**) and a horizontal side-to-side motion (a **Love wave**). All we really need to know about surface waves for the purposes of this lab is that surface waves travel more slowly than body waves and that surface waves cause the damage that is directly related to the earthquake because they distort the ground surface.

Effects of Earthquakes on Structures

Much of the structural damage caused by earthquakes is related to the rolling and side-to-side shearing motion of surface waves, which distort the foundations of buildings. Seismic waves can cause a structure to vibrate at a particular frequency, called a **resonant frequency**, at which the structure is particularly vulnerable to damage. An example of resonance is the party trick in which someone moistens a finger and slowly rubs it around the rim of a glass until the glass starts to vibrate, emitting a sound (also known as *making the glass sing*). This is the basis for a musical instrument invented by Benjamin Franklin called a *glass harmonica* for which Mozart and other composers wrote pieces. Another example of resonance is when a pure tone is sung by an opera singer or emitted by an electronic speaker long enough for a glass to vibrate at its resonant frequency with enough intensity that it shatters.

A double-decked section of the Nimitz Freeway (Interstate 880) called the Cypress Structure collapsed during the Loma Prieta earthquake in 1989, killing many people who were driving along its lower deck (**Fig. 16.2**). While part of the structural failure was due to foundation conditions that we consider in Activity 16.1 (**Fig. 16.3**), seismologists and structural engineers think that the seismic waves caused resonance in the Cypress Structure that contributed to its collapse.

Earthquakes can have very different effects on buildings in different parts of the same city. To understand why, geologists compare earthquake damage to geologic maps that show the type of rock, sediment, or soil that occurs

Figure 16.2 Collapsed Freeway after the Loma Prieta earthquake. This double-deck section of I-880 in Oakland, California, called the Cypress Structure, collapsed during the Loma Prieta earthquake of 1989. Foundation failure and seismic shaking at the resonant frequency of the structure are interpreted to have contributed to the collapse.

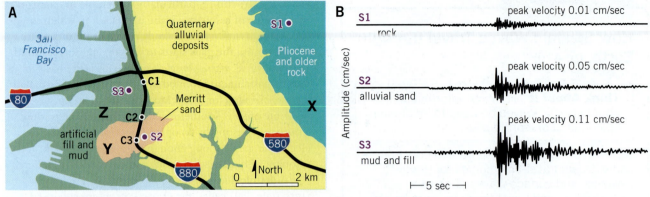

Figure 16.3 Geologic map and seismographs used in Activity 16.1. A. Geologic map of an area of Oakland, California, centered on latitude 37.822°N, longitude 122.281°W. This area sustained damage from the Loma Prieta earthquake. **B.** The seismograms show the north–south horizontal component of a strong (magnitude 4.1) aftershock of the Loma Prieta earthquake as recorded at sites S1, S2, and S3 on the map. (Based on work by Tom Holzer, Susan Hough, and others.)

at the ground surface on which the building foundations are built. They perform a variety of laboratory analyses on these geological materials to measure their strength and ability to transmit seismic waves and the effect of seismic vibrations on their characteristics. Earthquake engineers can simulate earthquakes in the laboratory to observe their effects on models of construction sites as well as buildings, bridges, hillsides, tunnels, and other structures. All of this information is used to design safer structures, revise building codes in earthquake-prone regions, and construct earthquake hazard maps to assist in public planning.

Sensing Earthquakes

A **seismometer** is an instrument that detects seismic waves and converts those waves to an electrical signal that can be recorded, usually in digital form. A **seismograph** is a system of devices including a seismometer, a recording device, and a very accurate clock whose time signal is recorded along with the seismic data. Many seismographs use time data transmitted by GPS satellites to calibrate their clock systems. Most seismographs used for routine earthquake detection are part of a seismic network that includes several seismographs placed in different locations. Networked seismographs record seismic data digitally and transmit their data in real time via phone lines, a cellular network, or a satellite uplink to a network facility where seismologists can examine the results. The place where a seismograph has been permanently sited is sometimes called a **seismograph station** although the seismometer is usually placed in a quiet place underground where humans do not visit unless maintenance is needed. A **seismogram** is a plot showing the record of ground motion caused by the passage of seismic waves (**Fig. 16.3B**). Time is represented along the horizontal axis of a seismogram, and the vertical axis indicates displacement and direction of displacement (i.e., whether up or down, north or south, east or west). The curve on a seismogram is called a **seismic trace.**

ACTIVITY 16.3

Locate the Epicenter of an Earthquake, (p. 421)

Think About It How can information about seismic waves be used to locate the epicenter of an earthquake?

Objective Graph seismic data to construct and evaluate travel-time curves for P-waves, S-waves, and surface waves, then use seismograms and travel-time curves to locate the epicenter of an earthquake.

Before You Begin Read the following section: Interpreting Seismograms.

Plan Ahead You will need a pencil, ruler, calculator, and a simple compass for drawing circles.

Interpreting Seismograms

Figure 16.4 is a seismogram from an earthquake in New Guinea recorded at a seismic station in Australia 1800 km away from the epicenter. The first impulse recorded on the seismogram from this earthquake is the first P-wave, which arrived at the seismograph at 14.2 minutes after 7:00. That is, the **P-wave arrival time** was 7:14.2. The second large impulse of seismic waves is attributed to the slower S-waves, which arrived at the seismograph at 7:17.4. The final large impulse of seismic energy was from the surface waves that traveled along Earth's surface, which did not begin to arrive until 7:18.3. The origin time of the earthquake at its focus beneath New Guinea was 7:10.4. Therefore the time it took the first P-wave to travel from the earthquake source to the Australian seismic station

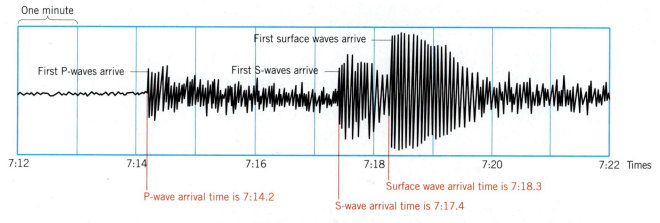

One minute

First surface waves arrive

First P-waves arrive

First S-waves arrive

7:12   7:14   7:16   7:18   7:20   7:22 Times

P-wave arrival time is 7:14.2

S-wave arrival time is 7:17.4

Surface wave arrival time is 7:18.3

Figure 16.4 Seismogram of an earthquake in New Guinea. This is a sketch of a seismogram from an earthquake in New Guinea as recorded in Australia. It shows vibrations associated with the P-wave, S-wave, and surface waves.

(that is, the **travel time**) was 3.8 minutes (7:14.2 minus 7:10.4). The travel time for the first S-wave was 7.0 min, and it took 7.9 min for the first surface waves to arrive (**Fig. 16.5**).

Similar travel-time observations for many earthquake records from many seismograph stations around the world have allowed seismologists to develop average travel-time curves (**Fig. 16.5**). The lowest curve is the average travel-time response for P-waves, which take the shortest amount of time to travel from a distant earthquake to a given seismic station. The curve provides us with information about how the velocity of P-waves change as they pass through deeper parts of Earth before traveling back to the surface. The highest curve is actually a straight line or nearly so because the velocity of surface waves does not change much as they pass along Earth's surface. This travel-time graph allows us to correlate the difference in arrival time between P, S, and surface waves with the distance from a seismic station to the epicenter.

ACTIVITY 16.5

New Madrid Seismic Zone, (p. 425)

Objective Interpret seismograms and focal mechanism studies to infer fault motion in the New Madrid Seismic Zone within the North American Plate.

Before You Begin Read the following section: Focal Mechanism Analysis.

Plan Ahead You will need a pencil, ruler, and calculator.

Focal Mechanism Analysis

The relative motions of blocks of rock on either side of a fault zone can be determined in several ways through analysis of earthquake records. We will work with a simplified example of a **focal mechanism solution**, which is a physical explanation of the cause of a recorded vibration that might turn out to be an earthquake or perhaps a very large explosion.

Imagine that an earthquake is about to occur at a particular focus point along a fault (**Fig. 16.6A**). We will track what happens to a number of points located on a small sphere around the earthquake focus, called the *focal sphere*. As the earthquake occurs, a P-wave emerges from the focus, traveling outward in all directions at the speed of sound in rock—several thousand m/s. As the wave front from the P-wave reaches the focal sphere, we take note of the direction in which each of the points on the focal sphere moves as that wave reaches them. This is the *first motion* caused by the passing P-wave.

An earthquake generated by slip along a fault exhibits a different pattern of first motions than is generated by an explosion. An explosion causes the first motions

ACTIVITY 16.4

San Andreas Fault Analysis at Wallace Creek, (p. 423)

Think About It How do geologists use landscape features and focal mechanism studies to analyze fault motions?

Objective Analyze and evaluate active faults using remote sensing and geologic maps.

Before You Begin Read the following section: Focal Mechanism Analysis.

Plan Ahead You will need a pencil, ruler, and calculator.

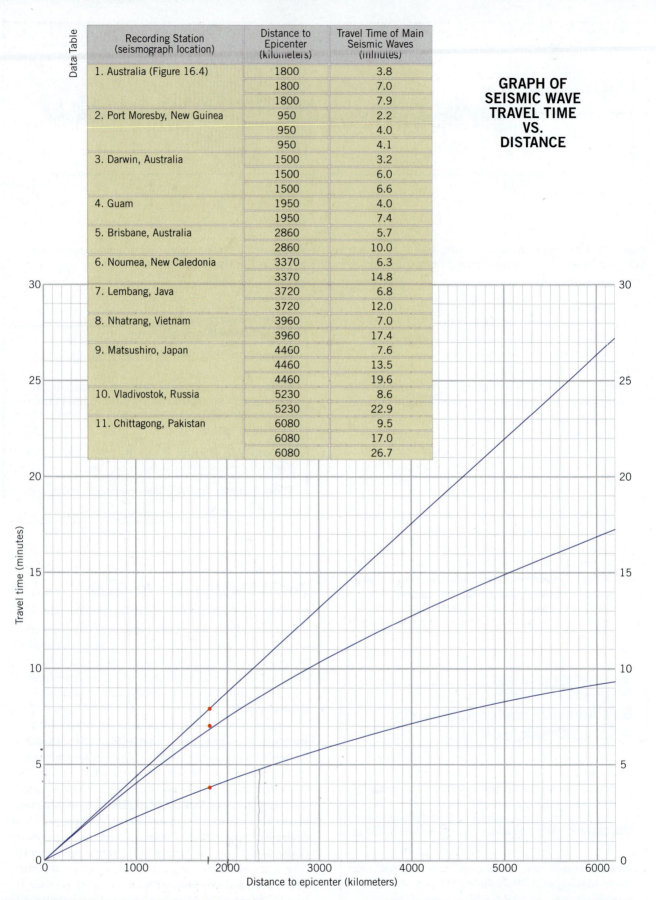

Recording Station (seismograph location)	Distance to Epicenter (kilometers)	Travel Time of Main Seismic Waves (minutes)
1. Australia (Figure 16.4)	1800	3.8
	1800	7.0
	1800	7.9
2. Port Moresby, New Guinea	950	2.2
	950	4.0
	950	4.1
3. Darwin, Australia	1500	3.2
	1500	6.0
	1500	6.6
4. Guam	1950	4.0
	1950	7.4
5. Brisbane, Australia	2860	5.7
	2860	10.0
6. Noumea, New Caledonia	3370	6.3
	3370	14.8
7. Lembang, Java	3720	6.8
	3720	12.0
8. Nhatrang, Vietnam	3960	7.0
	3960	17.4
9. Matsushiro, Japan	4460	7.6
	4460	13.5
	4460	19.6
10. Vladivostok, Russia	5230	8.6
	5230	22.9
11. Chittagong, Pakistan	6080	9.5
	6080	17.0
	6080	26.7

GRAPH OF SEISMIC WAVE TRAVEL TIME VS. DISTANCE

Figure 16.5 Seismic data and travel time–distance graph. The table shows data for an earthquake that occurred in New Guinea and was recorded at 11 seismic stations. The red dots show where the first arrivals at the seismic station in Australia plot on the travel-time graph. The black curves represent average travel times for many earthquakes recorded at many seismic stations worldwide.

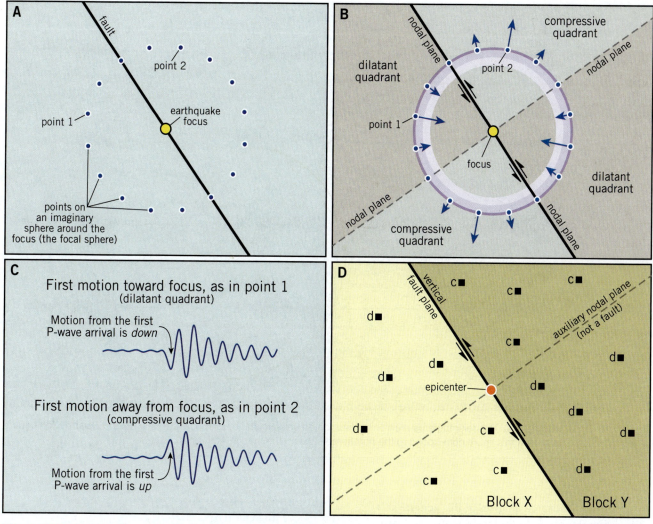

Figure 16.6 Visualization of the focal mechanism of an earthquake. A. Points (blue circles) along an imaginary small sphere around an earthquake focus. **B.** Slip on the fault results in an initial P-wave (purple circle) that causes some points on the focal sphere to move away from the focus and others to move toward the focus (blue arrows). The different directions of P-wave first motions allow us to define two compressive quadrants and two dilatant quadrants. **C.** Interpretation of seismic traces to indicate whether the P-wave first motion was toward or away from the focus. **D.** Map of a hypothetical region showing a fault along which an earthquake has occurred, the epicenter, and the P-wave first motions (c = compressional, d = dilational) observed for the earthquake at seismic stations adjacent to the fault (black squares). The relative motion of Block X and Block Y along the fault is indicated by the paired half-arrows.

of all the points on the focal sphere to be directed away from the focus. An earthquake generated by frictional slip along a fault causes a different pattern of motion associated with the passage of the first P-wave. P-wave first motions caused by faulting define four distinct quadrants around the earthquake focus. Two quadrants have points whose initial motion is *away from* the focus and the other two quadrants have points that move *toward* the focus (**Fig. 16.6B**). We will refer to a quadrant in which the first motion is *away from* the focus as a **compressive** or **compressional quadrant**, and a quadrant in which first motion is *toward* the focus as a **dilatant quadrant**—also called a **rarefactional quadrant** or a **dilatational quadrant**.

Seismographs located in a dilatant quadrant of an earthquake will record the first P-wave impulse as a

motion toward the earthquake focus—seen as a downward trace on the seismogram associated with the vertical component of the record (**Fig. 16.6C**). Seismographs located in a compressive quadrant of an earthquake will record the first P-wave impulse as an upward trace on the seismogram. By noting the first-motion behavior at several seismographs located around the epicenter, it is possible to map the boundaries or **nodal planes** that separate these quadrants (**Figs. 16.6C** and **16.6D**). One of the nodal planes corresponds to the **fault plane** along which the earthquake was generated, and the other is simply a plane called the **auxiliary plane** that is perpendicular to the fault through the earthquake focus. You will have a chance in Activity 16.5 to use the seismic traces in **Fig. 16.7** to define the four quadrants and determine the direction of fault slip associated with an earthquake in the New Madrid Seismic Zone.

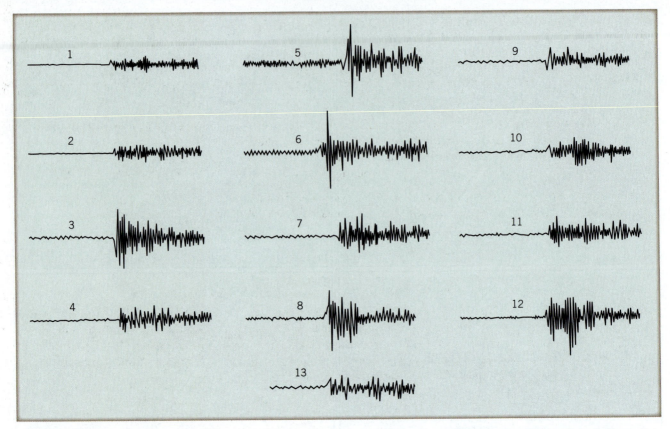

Figure 16.7 Seismograms. The seismograms are from 13 numbered seismic stations around the New Madrid Seismic Zone after an earthquake. Numbers in this figure correspond to the numbered sites on the map in Activity 16.5.

More sophisticated analyses of this general sort allow seismologists and other geoscientists to learn a substantial amount about a seismic event. The automated process of determining focal mechanism solutions from data recorded by networks of seismographs allows us to quickly determine whether a seismic event is an earthquake or an explosion. Verification of international nuclear test ban treaties depends substantially on the existence of global and regional networks of seismic stations.

If the event is an earthquake, rapid analysis of records from networked seismographs enables notification of public officials of potential seismic disasters while the earthquake is occurring. The rapid availability of focal mechanism solutions can trigger alerts that save lives, particularly by helping us recognize earthquakes that are likely to generate tsunamis. Earthquake early warning systems that utilize data from networked GPS and seismograph stations can provide seconds or even minutes of prior warning to major urban areas in the event of a great earthquake along the San Andreas Fault or Cascadia Subduction Zone.

MasteringGeology™

Looking for additional review and test prep materials? Visit the Study Area in MasteringGeology to enhance your understanding of this chapter's content by accessing a variety of resources, including Pre-Lab Videos, Self-Study Quizzes, Geoscience Animations, Mobile Field Trips, *Project Condor* Quadcopter videos, *In the News* articles, glossary flashcards, web links, and an optional Pearson eText.

Name: _____ **Course/Section:** _____ **Date:** _____

A Obtain a small plastic or paper cup. Fill it ¼ full of water. Then without moving the cup, *slowly* sprinkle quartz sand grains into the cup, allowing the grains to sink to the bottom of the cup. Keep adding sand a little at a time just until the sand gets to the top of the water. Place two or three coins upright in the sand so they resemble vertical walls. The coins are meant to represent buildings constructed on a substrate of uncompacted sand. This is model 1.

1. *Simulate an earthquake* by tapping the cup on a tabletop while you rotate the cup counterclockwise. Describe what happened to the vertical coins in the uncompacted sediment of model 1 when you simulated an earthquake.

2. Now make model 2. Remove the coins from model 1, and press down on the sediment in the cup so that it is well compacted. You might need to drain some of the water from the cup because you want the flat upper surface of the sand to be above the water level in the cup. Then place the coins into this compacted sediment just as you placed them in model 1 earlier. *Simulate an earthquake* as you did for model 1. What happened to the vertical coins in the compacted sediment of model 2 when you simulated an earthquake?

3. Based on your experimental models 1 and 2, which kind of Earth material is more hazardous to build on in earthquake-prone regions: compacted sediment or uncompacted sediment? (Justify your answer by citing evidence from your experimental models.)

4. Consider the moist, compacted sediment in model 2. Do you think this material would become *more* hazardous or *less* hazardous to build on if it became totally saturated with water during a rainy season? To find out, design and conduct another experimental model of your own. Call it model 3, describe what you did, and explain what you learned.

B **REFLECT & DISCUSS** Write a statement about whether wet *compacted* sand is stronger, weaker, or about the same strength as wet *uncompacted* sand, and what the reason for this is.

417

C The San Francisco Bay area of California is located in a tectonically active region, so it can be subjected to strong earthquakes. **Fig. 16.3A** is a map showing the kinds of geologic materials upon which buildings have been constructed in a portion of Oakland at the west end of the Oakland Bay Bridge. These materials include hard compact Pliocene and older rock, Quaternary alluvial sand and gravel in coastal terrace deposits, a sandy alluvial formation called the Merritt Sand, and an uncompacted artificial fill that is mostly mud. The artificial fill was used in the late 19th and early 20th centuries to create a seaport in an area that was originally dominated by tidal flats and creeks.

Imagine that you have been hired by an insurance company to assess the geologic risks in buying newly constructed apartment buildings located at X, Y, and Z in **Fig. 16.3A**. Your job is to rank the risk of property damage during strong earthquakes at the three sites, ranging from *low* (little or no damage expected) to *high* (damage can be expected). The only thing that you have as a basis for reasoning is **Fig. 16.3A** and knowledge of your experiments with models in part **A** of this activity.

1. Which site has the highest risk of damage during a future earthquake? Why?

2. Which site has the lowest risk of damage during a future earthquake? Why?

D On October 17, 1989, a strong earthquake occurred at Loma Prieta, California, and shook the entire San Francisco Bay area. Seismographs temporarily placed at locations S1, S2, and S3 (**Fig. 16.3A**) after the main shock recorded the shaking of a significant aftershock, and the resulting seismograms that show north–south horizontal motion are given in **Fig. 16.3B**. More intense shaking is indicated by the larger amplitude of the seismic trace, which diverges more from the midline of the seismogram. The records in **Fig. 16.3B** show that the intensity of shaking was less at location S1 than at locations S2 or S3.

1. The Loma Prieta earthquake caused little significant damage at location X, but there was moderate damage to buildings at location Y and severe damage at location Z. For example, the double-deck portion of I-880 collapsed between points C1 and C2, causing many deaths (**Figs. 16.2** and **16.3A**) and was damaged but did not collapse between C2 and C3. Explain how this damage report compares to your risk assessment in part C.

2. The Loma Prieta earthquake shook the entire San Francisco Bay region. Yet **Fig. 16.3** provides evidence that the earthquake had very different effects on structures located quite near to each other. Explain how the properties of geologic material on which buildings are constructed (for example, strong versus weak or compacted versus uncompacted) influence how much the buildings are shaken in an earthquake.

E **REFLECT & DISCUSS**   Imagine that you are an elected member of the city council in one of the cities around San Francisco Bay. Name two actions that you could propose to decrease the damaging effects of future earthquakes in your community, such as the damage that occurred at locations Y and Z in the Loma Prieta earthquake.

Name: _____ **Course/Section:** _____ **Date:** _____

Notice the seismic data provided with the graph in **Fig. 16.5**. There are data for 11 seismic stations that recorded the same earthquake in New Guinea near latitude 3°N and longitude 140°E. The distance from the epicenter (surface distance between the recording station and the epicenter) and travel time of the main seismic waves are provided for each recording station. Notice that there are three lines of data from most of the recording stations. They show the travel times for the three main kinds of seismic waves (P-waves, S-waves, and surface waves). However, instruments at some locations recorded only one or two kinds of direct waves (P-waves, or P- and S-waves). Location 1 on the data table in **Fig. 16.5** is the Australian recording station where the seismogram in **Fig. 16.4** was obtained.

A On the graph in **Fig. 16.5**, plot points from the data table in pencil to show the travel time of each main seismic wave in relation to its distance from the epicenter (when recorded on the seismogram at the recording station). For example, the data for location 1 have already been plotted as red points on the graph. Recording station 1 was located 1800 km from the earthquake epicenter and the main waves had travel times of 3.8 minutes, 7.0 minutes, and 7.9 minutes. Plot points in pencil for data from all of the remaining recording stations in the data table and then examine the graph.

Notice that your points do not produce a *random pattern*. They fall in *discrete paths* close to the three narrow black lines or curves already drawn on the graph. These black lines or curves were drawn by plotting many thousands of points from hundreds of earthquakes exactly as you just plotted your points. Explain why you think that your points and all of the points from other earthquakes occur along three discrete lines or curves.

B Study the three discrete, narrow black curves in **Fig.16.5**. Label the curve that represents travel times of the P-waves. Label the curve that connects the points representing travel times of the S-waves. Label the line or curve that connects the points representing travel times of the surface waves. Why is the S-wave curve steeper than the P-wave curve?

C Why do the surface wave data points that you plotted on **Fig. 16.5** form a straight line whereas data points for P-waves and S-waves form curves? (*Hint:* P- and S-waves are body waves that travel through Earth's interior, whereas surface waves travel along Earth's surface.)

D Notice that the origin on your graph (where travel time is zero and distance is zero) represents the location of the earthquake epicenter and the start of the seismic waves. The time interval between the first arrival of P-waves and the first arrival of S-waves at the same recording station is called the *S-minus-P time interval*. How does the S-minus-P time interval change with distance from the epicenter?

E Imagine that an earthquake occurred this morning. The first P-waves of the earthquake were registered at a recording station in Houston at 6:12.6 AM, and the first S-waves arrived at the same Houston station at 6:17.1 AM. Use the travel-time graph (**Fig. 16.5**) to answer each of the following questions.

1. What is the S-minus-P time interval of the earthquake?

2. How far from the earthquake's epicenter is the Houston recording station located?

F **REFLECT & DISCUSS** You have determined the distance between Houston and the earthquake epicenter. What additional data would you require to determine the location of the earthquake's epicenter on a map?

Locate the Epicenter of an Earthquake

Name: _Journie D. Ferguson_   **Course/Section:** _____   **Date:** _10/22/2020_

A single earthquake produced the seismograms in **Fig. A16.3.1** at three different locations (Alaska, North Carolina, and Hawaii). Times have been standardized to Charlotte, North Carolina, to simplify comparison. See if you can use these seismograms and a seismic-wave travel-time curve (**Fig. 16.5**) to locate the epicenter of the earthquake that produced the seismograms.

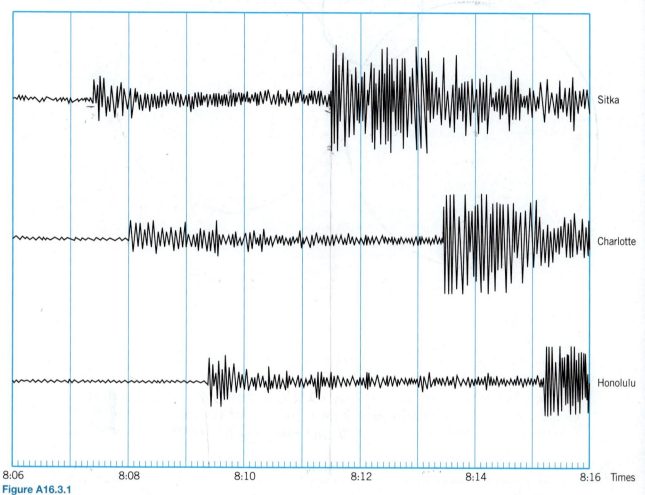

Figure A16.3.1

A Using the seismograms in **Fig. A16.3.1**, estimate to the nearest tenth of a minute the times that P-waves and S-waves first arrived at each seismic station (Sitka, Charlotte, Honolulu). You might want to review **Fig. 16.4** to help pick out the P- and S-waves. Then subtract P from S to get the S-minus-P time interval.

	First P arrival	First S arrival	S-minus-P
Sitka, AK	807.4	8:11.4	4 mins
Charlotte, NC	8:08	8:13.4	5.4 mins
Honolulu, HI	8:09.3	8:15.2	5.9 mins

B Using the S-minus-P time intervals determined in part **A** and the travel-time curves in **Fig. 16.5**, determine the distance from the epicenter (in kilometers) for each seismic station.

Sitka, AK: **2500** km   Charlotte, NC: **3900** km   Honolulu, HI: **4000** km

C Next find the earthquake's epicenter on the map in **Fig. A16.3.2** using the distances just obtained by completing the following steps

Scale bar = 6.8

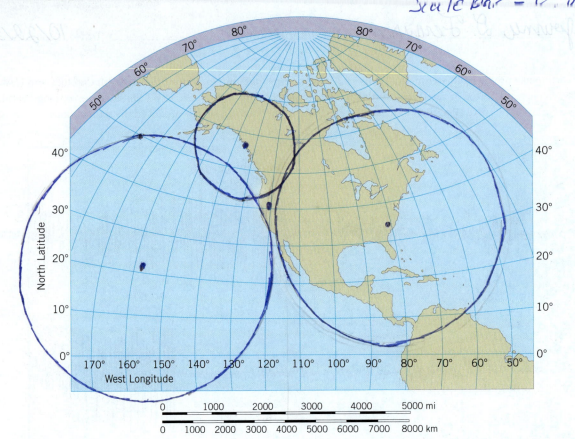

Figure A16.3.2

1. Use the following geographic coordinates to locate and mark the three recording stations on the world map. Plot these points as exactly as you can.

   Sitka, AK: 57°N latitude, 135°W longitude *6.8cm · 2500 = 12000/8000 = 1.5 cm*
   Charlotte, NC: 35°N latitude, 81°W longitude *6.8cm · 3900 = 26520/8000 = 3.3 cm*
   Honolulu, HI: 21°N latitude, 158°W longitude *6.8cm · 4000 = 27200/8000 = 3.4 cm*

2. Use a drafting compass to draw a circle around each recording station. Make the radius of each circle equal to the *distance from epicenter* determined for the station in part **B**. (Use the scale on the map to set this radius on your drafting compass.) The circles you draw should intersect approximately at one point on the map. This point is the epicenter. (If the three circles do not quite intersect at a single point, then find a point that is equidistant from the three edges of the circles, and use this as the epicenter.) Record the location of the earthquake epicenter:

   N Latitude ____*41*____ W Longitude ____*121*____

3. What is the name of a major fault that occurs near this epicenter? You can consult the Quaternary Fault and Fold Database of the United States (http://earthquake.usgs.gov/hazards/qfaults/), which has an interactive map that can help you find active faults.

   The San Andreas

D **REFLECT & DISCUSS** Your three circles may not have intersected exactly at a single point. How could you improve your results?

My drawing compass is a very flimsy and inaccurate, so often circles would be drawn bigger or smaller than intended. My measurements could also be off by just a bit.

Name: _____ **Course/Section:** _____ **Date:** _____

Figure A16.4.1 is a digital hillshade map of the area around Wallace Creek in central California along the San Andreas Fault, centered around 35.272°N, 119.827°W. This image and the lidar data used to make it are available via OpenTopography.org.

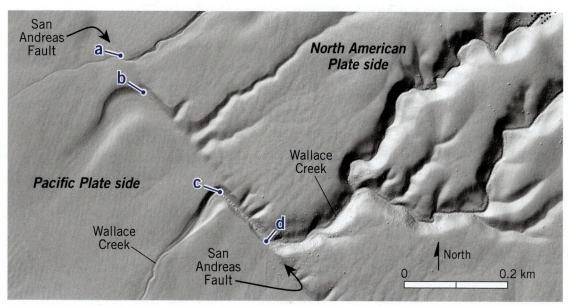

Figure A16.4.1

A The most recent large earthquake along this part of the San Andreas Fault occurred in 1857 when about 9.5 m of slip occurred there during a magnitude ~7.9 earthquake.

1. Use a pencil to trace the fault between the two arrows on **Fig. A16.4.1** as carefully as you can.

2. Describe the visual evidence in the hillshade image that you used to decide where to draw the trace of the fault.

3. Recall that geologists use a pair of arrows along the trace of a strike-slip fault to indicate the direction of relative motion across the fault. Interpret the sense of displacement across the San Andreas Fault (i.e., whether its motion is right lateral or left lateral), and use a pencil to draw paired arrows along the fault trace.

B Wallace Creek is an ephemeral stream that is dry most of the time but flows during wet seasons. Flow is from the North American side toward the Pacific side. Points *a*, *b*, and *c* on the Pacific Plate side mark points where Wallace Creek has flowed across the fault at various times in the past from point *d* on the North American Plate side. Geologists Kerry Sieh and Richard Jahns interpreted the approximate ages of initial development of the Wallace Creek channels as follows: channel *a* ~13,000 yr; channel *b* ~10,000 yr; and channel *c* ~3700 yr.

1. What is the length in mm on **Fig. A16.4.1** that corresponds to 200 m of distance in the Wallace Creek area?
   _____ mm

2. Recalling that 200 m = 200,000 mm, divide 200,000 by the number you measured in part **B1** and state the fractional scale of the hillshade map. 1:_____

3. Measure the following distances on **Fig. A16.4.1** in mm, convert them to distance in the Wallace Creek area in mm using the fractional scale you calculated in part **B2**.

   a to *b* is _____ mm on the map or _____ mm in the field.

   b to *c* is _____ mm on the map or _____ mm in the field.

   c to *d* is _____ mm on the map or _____ mm in the field.

   a to *d* is _____ mm on the map or _____ mm in the field.

4. Recalling that 1 m = 1000 mm, complete the following table (**Fig. A16.4.2**).

	Distance (m)	Displacement time (yr)	Displacement rate (m/yr)
a to *b*		13,000 − 10,000 = _____	
b to *c*		10,000 − 3,700 = _____	
c to *d*		3,700	
a to *d*		13,000	

Figure A16.4.2

C The total displacement rate of the Pacific Plate relative to the North American Plate at Wallace Creek is ~0.051 m/yr. according to the MORVEL model of Charles DeMets and coauthors. Calculate the percentage of the motion of the Pacific Plate relative to North American Plate that was due to slip along the San Andreas Fault at Wallace Creek over the last 13,000 years by dividing the displacement rate you calculated in part **B4** by the total displacement rate. _____ %

D **REFLECT & DISCUSS** Many people think that the San Andreas Fault is the only major active fault in the western United States and is the only plate-boundary fault between the North American and Pacific Plate. Your work in part **C** indicates that only part of the motion between these two plates occurs along the San Andreas Fault. What might you do to investigate where the rest of the motion between these two plates occurs?

Name: _____   **Course/Section:** _____   **Date:** _____

The New Madrid Seismic Zone (NMSZ) has produced some of the strongest earthquakes in the United States. In 1811 and 1812, three earthquakes with magnitudes estimated to be as great as ~7.5 occurred along the NMSZ along with one magnitude-7 aftershock. The future potential for strong earthquakes and its proximity to major cities such as Memphis, St. Louis, and Nashville make this fault system a focus of continuing study.

The great earthquakes along the NMSZ are inferred to have originated in Paleozoic and Precambrian rock that is buried by approximately 1 km of younger sediments and sedimentary rock within the *Mississippi Embayment*. Reworking of the ground surface and deposition of sediment by the Mississippi River tends to obscure or bury landscape features caused by faulting, making it difficult or impossible to see NMSZ faults at the ground surface. Geophysical surveys and precise location of smaller earthquakes are helping us to make better fault maps and improve our understanding of the NMSZ. The approximate trace of a primary fault in the NMSZ is indicated with the thick dashed curves in **Fig. A16.5.1**, although the system also includes other faults of different types and orientations.

A **Fig. 16.7** displays 13 seismograms from an earthquake along the NMSZ and are numbered to correspond to the 13 seismic stations whose locations are plotted in **Fig. A16.5.1**. Analyze the seismograms in **Fig. 16.7** to determine if their P-wave first motions indicate compression or dilation (refer to **Fig.16.6** as needed). Plot this information on **Fig. A16.5.1** by writing a *C* beside the stations where compression occurred and a *D* beside the stations where dilation occurred.

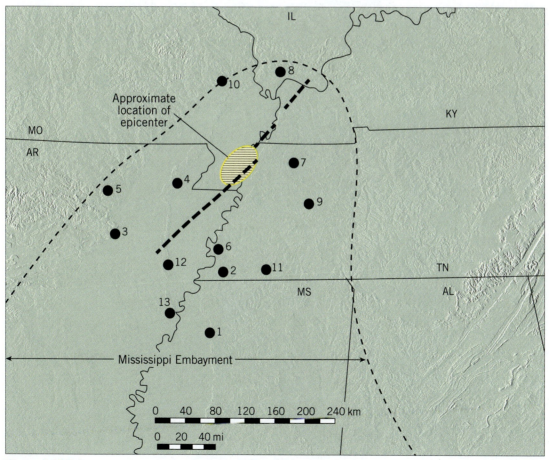

Figure A16.5.1

D When you have finished plotting the letters in part **A**, draw half-arrows on the map to indicate the relative motions of the blocks of rock on either side of the main fault. Referring to **Figs. 16.6B** and **16.6D**, notice that the half arrow that shows relative displacement on a given side of a fault is parallel to the fault and points toward the compressive quadrant (away from the dilatant quadrant) on that side of the fault.

C **REFLECT & DISCUSS** Does the main fault have a right-lateral motion or a left-lateral motion? How do you know?